Introduction to Sensors for Ranging and Imaging

Introduction to Sensors for Ranging and Imaging

Dr. Graham Brooker

SciTech Publishing, Inc
Raleigh, NC
www.scitechpub.com

Printed in the U.S.A.
10 9 8 7 6 5

ISBN13: 9781891121746

SciTech President: Dudley R. Kay
Production Director: Susan Manning
Production Coordinator: Robert Lawless
Cover Design: Kathy Gagne

This book is available at special quantity discounts to use as premiums and sales promotions,
or for use in corporate training programs. For more information and quotes, please contact:

Director of Special Sales
SciTech Publishing, Inc.
911 Paverstone Dr. – Ste. B
Raleigh, NC 27613

Phone: (919)847-2434
E-mail: sales@scitechpub.com
http://www.scitechpub.com

Library of Congress Cataloging-in-Publication Data

Brooker, Graham.
    Introduction to Sensors for ranging and imaging / Graham Brooker.
        p.   cm.
    Includes bibliographical references and index.
    ISBN 978-1-901121-74-6 (hardcover : alk. papter)   1. Detectors–Scientific applications.
2. Radar.   3. Imaging systems.   I. Title.
    TA165.B727 2009
    621.36'7–dc22
                                                                        2008030149

Table of Contents

Acknowledgements

I would like to thank my wife, Mandi, for the time that I stole from her to write this book and for the good grace with which she bore my lack of attention. The affair is now over!

This is not my book alone, it is partly the distillation of eight years of teaching, feedback from students and support from colleagues. But mostly it reflects the help and support I have received from my family. Corrections to the style and grammar were made by Mandi and my mom-in-law, Mary Owens, and the artistic talents of my elder daughter, Jennie, breathed life into the drawings. My younger daughter, Sarah, and my Mom were always interested and always asked questions.

I am just sorry we didn't finish it in time for my Dad to read it.

1

Introduction to Sensing

1.1 INTRODUCTION

Sensors in the natural world include those which equip us with our five senses—sight, hearing, smell, taste, and touch. These convert the various and diverse inputs to electrochemical signals that can be used to inform or control the living organism.

In a similar way, in man-made devices, sensors are also used to measure various stimuli. However, because of the broad range of potential inputs and outputs, the accepted definition of a sensor is refined. In this definition, all devices that convert input energy into output energy are referred to as *transducers*, and sensors form a small subset of the group as defined below: "A sensor is a transducer that receives an input signal or stimulus and responds with an electrical signal bearing a known relationship to the input" (Fraden 2003).

Systems of sensors and transducers are constructed for a variety of applications, including surveillance, imaging, mapping, and target tracking. In some cases, the sensors provide their own source of *illumination* and are referred to as active sensors. Passive sensors, on the other hand, do not provide illumination and depend on variations of natural conditions for detection.

1.1.1 Active Sensors

Active sensors require the application of external power for their operation. This *excitation signal* is modified by the sensor to produce an output signal in the form of a beam of energy or a field. They are restricted to frequencies that can be generated and radiated fairly easily and therefore exclude part of the far infrared (IR) band (above 3×10^{12} Hz), parts of the ultraviolet (UV) band, and the gamma ray region. However, inroads are being made into these regions with the development of terahertz sources and detectors based on artificial photonic crystals.

Semiactive sensors use a man-made excitation signal generated by (or radiated from) a source that is not coincident with the sensor and is not generated specifically for that purpose. Such systems include multistatic radar that uses mobile phone or commercial broadcasts as the source. This sensor modality is both energy efficient and covert and is proving to be particularly effective in detecting stealth aircraft.

1.1.2 Passive Sensors

Passive sensors directly generate an electric signal in response to a stimulus. That is, the input stimulus is converted by the sensor into output energy without the need for an additional source

Table 1.1 Advantages of active and passive sensors

Passive	Active
Low power requirement (long life)	Measurement requirements matched to transmitter characteristics
Cannot be detected (covert)	Range measurement by temporal correlation
Simple principles (sometimes)	Radiation pattern constrains observation
Good reliability due to simplicity	High range and angle resolutions possible
Field of view constrains observation	Long-range operation possible
High angular resolution possible	
Large variety	

Table 1.2 Disadvantages of active and passive sensors

Passive	Active
Targets of interest must radiate or modify the field (electrical, gravity)	Large power requirement
	Easy to detect (not covert)
Prone to feature ambiguity and errors of scale	Complex reception and transmission processes
Typically short range, though not always	Reliability determined by two elements, the receiver and the transmitter
Availability not guaranteed (no light, contrast, etc.)	

of power to illuminate the environment. The salient characteristics of most passive sensors are that they do not emit radiation (or an excitation signal) as part of the measurement process but rely on a locally generated or natural source of radiation (light from the sun), an available energy field (gravity), or even a chemical gradient.

Passive sensors can exploit electromagnetic (EM) radiation of any frequency in which some natural phenomenon radiates and for which a detection mechanism exists. This can extend from extra-low frequency (ELF) (below 3×10^3 Hz) up to gamma rays (above 3×10^{19} Hz). In the acoustic domain, they can exploit acoustic energy (vibration) from infrasound frequencies less than 1 Hz from earthquakes or explosions up to the ultrasound. Tables 1.1 and 1.2 list the relative advantages and disadvantages of active and passive sensors.

1.2 A BRIEF HISTORY OF SENSING

1.2.1 Sonar

Just before midnight on April 14, 1912, the *RMS Titanic* struck an iceberg in the North Atlantic during her maiden voyage to New York City. She sank 2 hours and 40 minutes later at 02:20 on Monday, April 15, with great loss of life. This tragedy precipitated a number of major changes in maritime practices and ship design. These included the establishment of the International Ice Patrol, new regulations related to the number of life boats (the *Titanic* did not have enough for everyone) and a requirement for 24-hour radio watch-keeping. In addition, this stimulated the development of the first successful acoustic "echo location" devices.

The use of acoustics for underwater applications has a long history. As far back as 1490, Leonardo da Vinci described a tube that could be placed over the ear and partially immersed in water to detect the sound of passing vessels, and by the 19th century, underwater bells were used in conjunction with lighthouses to warn of hazards.

A number of attempts had been made to use the principle of echo location for depth sounding prior to the sinking of the *Titanic*. In 1855, M. Maury, Superintendent of the National Observatory, attempted unsuccessfully to determine the depth of the ocean by detonating a charge of gunpowder on the surface and listening for an echo. Using the same technique, a German physicist, Alexander Behm, measured the depth of Lake Plön in 1912 and obtained a patent for his device the following year. A more sophisticated and successful application was the British Admiralty depth-sounding machine. In this case, a steel hammer striking a steel plate in the bottom of a ship generated a short sound pulse, and a microphone detected any echoes.

The *Titanic* disaster led to Lewis Richardson's proposal that high-frequency beams could be used for depth measurement as well as obstacle detection at sea, and in 1913, a Canadian, Reginald Fessenden, built a "fathometer" which used an oscillator to generate a short pulse of sound at a few kilohertz that could detect an iceberg at a distance of 2 miles. Transducers based on this technology were fitted to British submarines during World War I.

In France, Paul Langevin and his Russian colleague Constantin Chilowsky developed what they termed a "hydrophone" as a method for detecting icebergs. But it was the threat of submarine warfare during World War I that drove this research. In 1915 Langevin began working on active detection methods using an electrostatic transmitter and a carbon-button microphone receiver and by 1916 he was able to detect echoes from the bottom and from a sheet of metal at a distance of 200 m.

Meanwhile, the sinking of the cruise liner *Lusitania* by a German submarine in 1915 prompted Robert Boyle, then head of physics at the University of Alberta, to write to his mentor, Ernest Rutherford, complaining that there had been no attempt to mobilize Empire scientific effort to combat the U-boat scourge. A few months later the Admiralty formed the Board of Inventions and Research (BIR), whose mandate was to find "devices and methods that could shorten the war." By April 1916, Boyle was on his way to England to join the scientific group.

Prior to Boyle's arrival, Rutherford, who had been appointed to the BIR, and his team had identified four potential means of submarine detection: utilizing sound, heat, electromagnetic disturbances, and visual characteristics. The team had concentrated on passive detection using sound, and Rutherford himself had focused on using quartz, a piezoelectric material for this purpose (Collins 2007).

In June 1917, an Allied submarine conference took place in Washington, D.C., where Britain, France, and the United States agreed to work cooperatively. Scientists from Britain and France met and exchanged information and Boyle was asked to duplicate Langevin's work. The results were at first disappointing, and he was sent to France in 1917, where he spent several months working in Langevin's labs to improve his understanding of the French technology. Ultimately, by merging the British quartz piezoelectric transducer technology with French vacuum tube amplifiers, both teams were able to make substantial progress. According to one source, by 1918, Langevin was able to detect submarines at a range of 1500 m (Graaf 1981), while another source suggests that Boyle was successful in detecting submarines out to 500 yards (Collins 2007). To maintain the secrecy of their successes, no mention was made of either sound or quartz and the word Asdics was coined to replace "supersonics," which had been used to describe the research. In 1939 the Admiralty made up a story that Asdic stood for Anti-Submarine Detection Investigation Committee, a committee that has never existed.

Though this technology was too late to be of any use in World War I, the push for its development paid great dividends, and by the start of World War II more than 165 British destroyers, sloops, and trawlers were fitted with Asdic equipment. These devices initially operated in the 14–22 kHz frequency range, being a compromise between transducer size, which increases with decreasing frequency, and attenuation, which increases as the frequency is increased. In a typical installation, the transducer was installed in a streamlined pod mounted below the ship and flooded with seawater so that the sound could propagate into the ocean, unhindered. It was also designed to generate as little noise as possible from water flow so that the device could be operated while the ship was traveling.

The original Asdic "Searchlight" produced a conical pattern (known as a pencil beam) with a beamwidth of 16° that was directed downwards at an angle of 10° and could be manually scanned in azimuth. It was capable of detecting a submarine out to a range of about 2500 yards under good conditions, but this range was reduced in rough or high-salinity seas or if a temperature inversion existed. As the war progressed, two fan beam transducers ($65° \times 4°$), code named "Q-attachment" and "Sword," were developed to improve area coverage and to detect deep-diving submarines, respectively. They operated at higher frequencies and had a shorter operational range. British Asdic technology continued to be exported to the United States throughout the war, and even as late as mid-1944, fifty of the new advanced "Sword" units were fitted to ships in the Pacific fleet.

The acronym SONAR (Sound Navigation and Ranging), coined as an equivalent to radar, was introduced by the U.S. Navy to describe their systems, and the name became standard with the formation of the North American Treaty Organization (NATO), but it was not until the 1960s that the name was accepted by the Royal Navy (Mason 1962).

In the last 60 years, a large number of active and passive sonar systems have been developed for various military and commercial applications. Depth sounders and fish finders are common on all but the smallest craft. Multibeam sonar generates a fan beam at right angles to the direction of travel and uses time-of-flight techniques to generate depth maps over a wide swathe (typically two to four times the water depth). Side-scan sonar, typically mounted on a towfish, can produce photographic quality strip-map images of the sea floor by recording the intensity of the echo signal as a function of range as the boat moves. The theoretical performance of a side-scan sonar is analyzed in detail later in the book.

Based on research initiated in the early 1970s using a 16×16 hydrophone array (Harmuth 1979), the advent of sufficiently fast computers has made it possible to produce extremely high resolution two- and three-dimensional (2D and 3D, respectively) sonic images with larger arrays. This imaging principle is more akin to the operation of the eye than that of conventional sonar, as it involves the simultaneous collection of reflected acoustic energy by a large number of closely spaced sensors to form an image. An example of this, discussed later in the book, is a system developed by Thomson Marconi Sonar using a randomly filled array of 100×100 hydrophones operating above 1 MHz to generate 3D images with voxol[1] resolutions down to 1 mm × 1 mm × 1 mm.

1.2.2 Radar

The use of electromagnetic signals for target detection had to await practical methods to both generate and receive such signals. Following on the work of Michael Faraday and James Clerk

---

1 A voxol is the volume equivalent of an image picture element (pixel).

Maxwell, between 1885 and 1889, Heinrich Hertz, a professor of physics at Karlsruhe University, generated "Hertzian waves" in his laboratory. The waves were produced by an induction coil coupled to a resonator, and a similar resonator detected them by sparking across a small gap. He showed, through a series of experiments, that these waves traveled at the speed of light and could be reflected, refracted, and polarized as were light waves. The initial experiments were conducted at a wavelength of about 10 m, as determined by the size of the resonator/antenna, while the later experiments, using cylindrical metal reflectors to focus the beam, were conducted at the much shorter wavelength of 66 cm, where the optical properties of the radiation were easier to observe indoors.

By using a more sensitive detector, the filings coherer perfected by Edouard Branly in 1890 (Larsen 1971), more powerful transmitters, and larger antennas, Guglielmo Marconi was able to increase the detection range significantly. His outdoor experiments culminated in the first transatlantic broadcast on December 12, 1901, from Poldhu in Cornwall, England, to Signal Hill in St. John's, Newfoundland, a distance of 3500 km.

More interesting from a sensor perspective were the experiments undertaken by Jagadish Chandra Bose. Though aware of the greater penetrating power of low-frequency signals, he realized that, to study the light-like properties of this recently discovered radiation, it was necessary to work at a higher frequency. His apparatus operated at 60 GHz, and by 1895 he was demonstrating its transmission capabilities to astounded audiences in Kolkata, India. In the following year he went on a lecture tour of the United Kingdom and soon after presented his work to the Royal Society in London. Bose used waveguides, horn antennas, dielectric lenses, and various polarizers at frequencies as high as 60 GHz. He improved on the coherer for signal detection and also worked on semiconductors for the same purpose.

The first applications of the technology to "echo location" followed soon after. In 1904 Christian Huelsmeyer demonstrated ship detection using this principle. His transmitter used a spark gap and a parabolic antenna to direct the beam, while a simple dipole and coherer coupled to a bell formed the receiver. The patented system, called a telemobilescope, could detect the presence, but not the range, of other ships out to 3 km. Later amendments to the patent included mounting the device on a tower so that the elevation angle of arrival could be used to estimate the range.

Nicola Tesla was the next to enter the picture, and in 1917 he proposed to use the standing waves set up between a transmitter and a reflector (such as a vessel at sea) to determine the object's position and speed. He also proposed to perform the same function using pulsed reflected surface waves.

In the following years, a number of researchers investigated aspects of interference in which a moving target situated between a transmitter and a receiver could introduce multipath fluctuations in the received signal level. During experiments with a transmitter tuned to operate at a wavelength of 5 m, Albert Taylor and Leo Young, from the Naval Aircraft Research Laboratory (later the Naval Research Lab [NRL]), identified this phenomenon for a wooden steamer on the Potomac River in 1922, and Lawrence Hyland, also from the NRL, detected an airplane in June 1930.

Similar experiments were conducted in the United Kingdom, and on February 26, 1935, Robert Watson-Watt and Arnold Wilkins demonstrated the detection of an aircraft to a representative of the Air Ministry Committee. Wilkins had set up his receiver the previous day, and using broadcasts from the British Broadcasting Corporation (BBC) shortwave (49 m) transmitter at Daventry, he was able to detect returns from a Handley Page Hayford bomber at ranges of up to 8 miles (13 km). This demonstration was considered by Edward Appleton, then Professor of Physics at London University, to be "pure theatre," as similar experiments had been conducted in the United States, France, and Russia. However, the theatre aspect notwithstanding, these methods were in any event unsatisfactory, as they were only able to detect the presence of an aircraft, not its location or velocity.

To determine the range to a target requires the transmission of a pulse and a method to measure the time until an echo is detected. Robert Page, from the NRL, was instrumental in developing this technology in the United States, and on April 28, 1936, his pulsed radar successfully detected a small airplane flying up and down the Potomac out to a range of 2.5 miles (4 km). Within 2 months, the range had been extended to 25 miles.

A decade prior to this, physicists in the United Kingdom and the United States had been investigating both pulsed and continuous wave methods to measure the height of reflecting layers in the atmosphere. Towards the end of 1924, Appleton and his colleague M. Barnett began a series of experiments to prove the existence of what is now called the ionosphere. With the cooperation of the BBC, they beamed radio waves directly upward and were able to detect an echo from a reflecting layer in the upper atmosphere. By modulating the frequency of the transmitted signal they were also able to determine the time taken for the signal to travel up to this "target" and to return, from which the height (about 100 km) could be calculated. The ionosphere was probably the first object detected by radiolocation, and their research was the precursor to the development of radar, which proved to be of great importance during World War II.

It was only 6 months after the Appleton experiments that Merle Tuve, a graduate student from Johns Hopkins University, and Gregory Breit, a physicist with the Carnegie Institution's Department of Terrestrial Magnetism, measured the height of the ionosphere by transmitting short-duration radio pulses, at a wavelength of 71.3 m, upward and measuring the round-trip time. This pulsed technique is more practical, as the transmitter is turned off during the period that the echo is expected, and hence the tiny return signal is much easier to detect.

These initial experiments transmitted for about half of the time, allowing a similar period for reception, which was not ideal. The introduction of the multivibrator, which had been invented back in 1918 by H. Abraham and E. Bloch in France, allowed researchers to develop apparatuses that transmitted for shorter intervals and listened for longer periods so that smaller targets could be detected and range ambiguity resolved. Of almost equal importance was the introduction of the cathode ray tube in place of the mechanical oscillograph, which made the display and subsequent interpretation of fast-changing pulsed signals easier. This device, which used gas discharge as a source of electrons, had been invented by Ferdinand Braun at the University of Strasbourg in Germany in 1897 and was retained in commercial oscilloscopes for another 25 years.

Other developments that were instrumental in making radar more useful were the development of the superheterodyne process in 1912, vacuum tube oscillators such as the Klystron in 1939, robust crystal detectors, and the transmit-receive (TR) switch, among others. The multi-cavity magnetron, which is discussed later in this section, played a pivotal role in the success of the Allied radar systems deployed during World War II (Dummer 1977).

The early radar systems developed in the United Kingdom and the United States operated at fairly long wavelengths. However, in 1927, French engineers Camille Gutton and Emile Pierret performed experiments at wavelengths as short as 16 cm. These experiments were resumed by

Gutton's son, Henri, in 1934 and a patent was awarded for a "New system of location of obstacles and its applications" for a device for detecting obstacles (icebergs, ships, and planes) using ultra-short wavelengths produced by a primitive magnetron. The radar was tested aboard the cargo ship *Oregon*, with transmitters operating at 80 cm and 16 cm, and coasts were detected at a range of 10–12 nautical miles. The shorter wavelength proved to be more effective and a radar based on this technology was fitted to the *SS Normandie* in 1935 (Guerlac 1987).

In 1933 a microwave department was initiated at Telefunken, and in 1936 the first "Würzburg" radar was produced. It operated at a wavelength of 53 cm and had a 3 m parabolic dish with a rotating dipole that produced a squinted beam that could pinpoint an aircraft to an angular accuracy of about 2° and a range accuracy of 100 m. The original radars had an operational range of 10 km, but that was improved to 28 km within the year. Giant Würzburg radars were later produced that used a 7.4 m parabolic dish which increased the operational range still further, to 80 km.

At about the same time, two German scientists, Hans Hollmann and Hans-Karl von Willisen, and businessman Guenther Erbsloeh formed a company called GEMA to develop radar technology. A pulsed radar operating with a wavelength of 60–80 cm and with a range accuracy of 50 m was first tested in May 1935. This radar demonstrated a detection range of 10 km for ships and 28 km for an aircraft flying at 500 m. Production versions of the land-based variation of this radar, code-named "Freya," and seaborne equivalent, "Seetakt," were first delivered to the German armed forces in 1938 and saw service throughout World War II. These sets had a maximum range of 75 miles (120 km) but could not measure an aircraft's altitude, rendering both its range- and height-finding capabilities inferior to Britain's Chain Home stations. In its favor, however, was both full mobility and 360° coverage (Buderi 1998). The Seetakt was an excellent radar for sea search, and it was more than 2 years before Royal Navy ships carried its equivalent.

The Freya and Würzburg radars were generally paired, with the Freya operating as a designator for the Würzburg, which then provided accurate range and height information to direct night fighter interception of incoming bombers. These radars were very effective, and by the end of the war some 50,000 bombers had been shot down. However, as the war continued, Allied jamming became more effective and the age of electronic warfare had begun.

Across the English Channel, Watson-Watt had gained the full support of Air Vice-Marshal Hugh Dowling, who understood the advantages to be gained for fighter aircraft by long-range detection of enemy aircraft. By the end of July 1935, a tracking range of 60 km had been achieved, and in September the Air Defense Research Committee had approved the construction of Chain Home.

Chain Home transmitters consisted of pairs of dipoles suspended between 100 m towers that radiated more like floodlights than the narrow beams other radar designers strove for. The receivers used pairs of crossed dipoles mounted on 75 m high wooden towers. One antenna detected the east-west (E-W) component of the signal and the other the north-south (N-S) component. This allowed the operator to determine the direction of arrival from the relative amplitudes of the two signals. Operation was at a wavelength of 7.5–15 m with a pulse rate of 25 Hz, the latter allowing adjacent stations to be synchronized using the mains frequency.

By the start of the war in September 1939, a chain of 20 Chain Home stations had been commissioned, each able to detect aircraft out to a range of 80 miles (150 km) in good weather. The elevation coverage of the stations extended from 1.5° to 16° and therefore were unable to detect low-flying aircraft.

In reality, the Chain Home system was so archaic, even in 1940, that it confused German electronic intelligence gathering exercises which were expecting more sophisticated technology and a higher operating frequency. That notwithstanding, it proved to be crucial to winning the Battle of Britain (Brown 1999).

By early 1940 it was obvious that Britain's production capability could not match that of Germany and the countries in her possession. This was particularly true in the field of electronics, which was already playing a crucial role in the war. Sir Henry Tizard proposed that Britain should trade wartime secrets to the United States in exchange for research and production capability. Items to be traded included radar technology and particularly the top secret magnetron that had been developed by John Randall and Harry Boot at Birmingham University in February 1940.

There is some controversy regarding this invention because its basic principles had been understood for the previous 20 years and patents for multicavity resonant magnetrons had already been granted to both German and American scientists. In fact, even the Japanese, the Swiss, and the Russians had worked on the devices, with two Russian scientists, N. Alekseev and D. Malairov, having published a complete description of the magnetron in the open literature (Brown 1999; Hollmann 2007). There is no doubt, however, that the Allies exploited the small size, high operational frequency, and high pulsed power output of the device to their advantage.

One of the first radars to use the magnetron was the H_2S bomb aiming radar. Frederick Lindermann (Lord Cherwell), Winston Churchill's scientific adviser, was determined to find a method that would allow pinpoint bombing accuracy at long range. Existing triangulation and beam-riding techniques were line-of-sight and hence limited to a maximum range of 250 miles, which was insufficient to reach deep into the heart of Germany. In early 1942, Bernard Lovell was made head of a program to develop a bombing radar. The device, code-named H_2S, derived either from the idea that "it stinks" or from the fact that it would allow bombers to home in on their targets: "home sweet home." The radar comprised a cosecant-squared fan beam antenna that rotated within a dome on the belly of the bomber. A pulsed magnetron operating at a wavelength of 10 cm was the transmitter and a plan position indicator (PPI) was used to display the ground returns. From the initial trial on April 23, the prototype radar performed well. Unfortunately, the Halifax bomber carrying it crashed on June 7, killing most of Lovell's technical team and destroying the radar. That disaster notwithstanding, Churchill insisted that the radar be mass produced and installed in Pathfinder aircraft as soon as possible. On January 30, 1943, just 13 months after the start of the program, Pathfinders flew their first mission to Hamburg. The crews returned elated, as they had been able to identify all the key landmarks on the way to the target from ranges in excess of 20 miles before dropping marker flares on Hamburg (Buderi 1998).

On February 2, during a raid on Cologne, a bomber was shot down near Rotterdam and the top secret magnetron was soon in the hands of German engineers. Countermeasures eventually appeared, but by that time the Allies were on their way to developing higher frequency systems.

One of the outcomes of the Tizard mission to the United States was the MIT Radiation Laboratory. Known affectionately as the "Rad Lab," it was initiated to develop and exploit microwave technology in general and radar technology in particular. Among his other projects, Luis Alvarez, one of the research scientists, had started working on an advanced bombing radar which, it was hoped, would have a resolution comparable to that of the Norden optical bombsight. He wanted to see individual bridges and factories, not just the blobs that were visible on H_2S. The "Eagle," or AN/APQ-7, consisted of a double array of reversed dipoles (to eliminate grating lobes) with a length of 20 feet (6 m). It operated at a wavelength of 3 cm, and instead of being mechanically scanned, used electronic phase shifters to scan a 60° sector in front of the aircraft. The first prototype tests occurred on June 16, 1943, and were followed by a very successful production model manufactured by Western Electric that had an azimuth beamwidth of 0.4° and could pick up cities out to a range of 250 km (Brown 1999).

Another very successful radar to come out of the Rad Lab was the gun-laying or fire control radar (FCR), SCR-584. It was developed by Ivan Getting and Lee Davenport, starting with a rooftop prototype first tested on May 31, 1941. Its development was championed by Brigadier General Colton, of the Signal Corps, who had a contract for its manufacture placed with Bell Labs. As with the German Würzburg radar, it used a rotating dipole at the focal point of a 2 m parabolic dish to measure the angular offset of the aircraft with great accuracy. This technique is known as conical scanning, or "conscan," and is described in detail later in this book. Unlike its German counterpart, the antenna of the SCR-584 was fitted with a servomechanism that allowed for automatic tracking of the target once it had been detected. It operated at a wavelength of 10 cm and transmitted 300 kW pulses with a duration of 0.8 μs that enabled it to detect bomber-size targets from a range of 65 km and to track them automatically from 30 km. It was first used in combat early in 1944 on the Anzio beachhead in Italy, where its introduction was timely, since the Germans by that time had learned how to jam its predecessor, the SCR-268.

By the end of the war, radar technology was mature and hundreds of different radar systems had been designed and deployed by both Allied and Axis forces. Just how substantial this effort had been can be judged by the fact that the United States had spent $2819 million on radar development during that time.

Many other countries had also developed radar systems in haste and in secret during World War II. These included South Africa (Austin 1992), Australia, Canada, Japan, and the Soviet Union, among others (Brown 1999).

With the reduction in postwar budgets and the redeployment of many of the research scientists back to their own universities, radar progress slowed, with the remainder of the 1940s being used for consolidation. Two techniques that did receive attention were monopulse tracking radar, which had been invented at the NRL in 1934, and Moving Target Indicator (MTI) radar, both of which are evaluated later in this book.

Monopulse radar reached maturity in the 1950s with the introduction of the AN/FPS-16, which had an angular tracking accuracy of about 0.1 mrad. It was used for military applications and to guide the first U.S. space launches. Pulsed Doppler radar also made its debut as the seeker in an air-to-air missile, and the concept for Synthetic Aperture Radar (SAR) was formulated by Carl Wiley of the Goodyear Aircraft Corporation.

Wiley observed the relationship between the along-track coordinate of an object being linearly traversed by a radar beam and the instantaneous Doppler shift of the echo from the object. He concluded that a frequency analysis of the signal could be used to synthesize a narrower effective beamwidth than that of the physical beam. This Doppler beam-sharpening (DBS) concept was further developed by a group at the University of Illinois in 1953 using data from a coherent X-band pulsed airborne radar. The major problem with SAR, which was recognized early, is the data processing overhead.

A number of important theoretical concepts were developed that helped take radar design from a qualitative to a quantitative science, including the statistical theory of detection of targets in noise, and matched filtering, which allows a designer to configure a radar receiver to maximize the signal-to-noise ratio (SNR). The ambiguity diagram, which quantifies the relationship between the range and Doppler performance of a radar processor, was also introduced.

Exploitation of the Doppler shift by radar systems, which commenced in the 1950s, is essential for the operation of some continuous wave (CW) radars, MTI, and pulsed Doppler radar, which must detect moving targets in the presence of large amounts of clutter. It is the basis for SAR and inverse SAR, which allows radars to produce high-resolution images, as well as weather radars and police speed traps.

The height of the cold war in the 1960s saw the development of the first electronically steered phased array radars and introduction of the precursors to the Airborne Warning and Control System (AWACS) radars, which came of age in the following decade with the advent of powerful digital signal processors (Brookner 1985).

By the 1980s, mass production methods and improvements in both solid state radar and signal processing hardware led to a proliferation of phased array systems. These included air defense systems, such as the Patriot and Aegis systems, and massive radars for ballistic missile detection, the best known of which are the Pave Paws radars. Sophisticated remote sensing radars and radiometers started appearing on satellites; these were capable of measuring precipitation, temperature, wave height and hence wind velocity, among other useful environmental characteristics.

In the last 20 years radar technology has improved considerably. This is primarily due to advances in the speed and sophistication of digital signal processing. SAR processors are now sufficiently small to fit into small unmanned airborne vehicles (UAVs); millimeter wave radars with complex multibeam processing are now beginning to appear in cars for collision warning and autonomous cruise control. Networks of unattended radars perform air traffic control functions and monitor the weather, and sophisticated radars have even mapped the surface of our two nearest neighbors, Venus and Mars, with unparalleled resolution. Most of these concepts, including target detection, matched filtering SAR, Doppler processing, MTI, and phased arrays are covered in later sections of this book.

1.2.3 Lidar

After demonstrating the first successful microwave amplification by stimulated emission of radiation (maser) using ammonia molecules, Charles Townes and Arthur Schawlow at Bell Labs began an investigation into the feasibility of developing an optical maser in 1957. Within a year, a paper with their theoretical calculations was published and a patent application was filed for the idea.

At about the same time, Gordon Gould, a graduate student at Columbia University, was working on the energy levels of excited thallium. He and Townes met and discussed aspects of radiation emission, after which, in November 1957, he made notes about his ideas for a "laser" which included the idea of an open resonator. In the following year, Aleksandr M. Prokhorov published his proposal for using an open resonator and Townes and Schawlow also settled on the concept, apparently unaware of the published work of Prokhorov and the contents of Gould's logbook.

In 1959, in a conference paper, Gould introduced the acronym LASER, for light amplification by stimulated emission of radiation, to the public, even though he had coined the term 2 years earlier. His logbooks also include possible applications for the device, including radar, spectrometry, interferometry, and even nuclear fusion. He filed a patent application in April 1959, but the U.S. Patent Office denied his application and awarded the patent to Bell Labs. This event set into motion a legal battle that continued for 28 years, and it was only in 1987 that he was issued patents for the optically pumped and gas discharge lasers.

The first working laser was made by Theodore Maiman on May 16, 1960, at the Hughes Research Laboratories, beating other researchers including Townes at Columbia University, Schawlow at Bell Labs, and Gould at TRG (Technical Research Group). Maiman used a flash lamp to pump a synthetic ruby crystal to produce red light at a wavelength of 694 nm. The partially mirrored ends of the ruby rod, parallel to within one-third of a wavelength, formed the open resonator (Fabry-Perot) structure. The laser was only capable of pulsed operation due to its three energy-level pumping scheme and for reasons of heat dissipation (Hecht 1994).

A short while after the initial announcement of the first successful laser, other researchers jumped on the bandwagon trying out many different substances. Yttrium-aluminum-garnet (YAG) proved to be particularly effective, and even scotch whisky was shown to be capable of lasing. Ali Javan, William Bennet, and Donald Herriot made the first gas laser using a mixture of helium and neon (HeNe) in 1961, which operated in the IR spectral band at a wavelength of 1150 nm. Early the following year, a visible version of the HeNe laser operating at 623.8 nm was produced. This represented the first continuous visible laser and was to become the most common laser type for many years (Ready 1978). In the same year, Don Nelson and Willard Boyle constructed the first continuous lasing ruby by replacing the flash lamp with an arc lamp.

The first injection laser diode used gallium arsenide (GaAs) and emitted radiation at 850 nm in the near IR. It was developed by Robert Hall in September 1962, based on the theoretical work by Nikolai Basov's group at the Lebedev Physics Institute in Moscow. Later that same year, on October 10, Nick Holonyak, Jr. demonstrated the first semiconductor laser in the visible range. It used the alloy gallium arsenide phosphide (GaAsP) and was an efficient and compact source of visible coherent light which would ultimately lead to the first commercially available light-emitting diodes (LEDs). These early semiconductor and gas lasers were little more than scientific curiosities, as they could only be operated when cooled with liquid nitrogen to 77 K.

Before the end of 1962, GE was offering both GaAs and GaAsP laser diodes for sale at $1600 and $3200, respectively, about 10 times the price that Texas Instruments was charging for their IR LEDs. Now, some 40 years later, high-performance red InAlGaP DVD lasers cost less that $1 in packaged form (Dupuis 2003). However, to reach that milestone took some time. It took until 1970 before Izuo Hayashi and Morton Panish of Bell Labs and Zhores Alferov in the Soviet Union independently developed the first heterojunction lasers that operated at room temperature. But even then it was still some time before they were reliable.

Semiconductor lasers are now ubiquitous, with applications ranging from fiber-optic communications to read and write heads for CDs and DVDs, while other solid state and gas lasers fill niche applications. Higher power lasers most often still use an electrical discharge to pump a gas medium, as was the case with the first HeNe devices. CO_2 lasers operate at wavelengths between 9 and 11 μm and can produce powers from a few watts up to megawatts. Excimer lasers (KrF, ArF, XeF) produce outputs in the UV spectral band and are typically used for medical applications or for communicating with submerged submarines.

From an active sensing perspective, lasers form the core of high-resolution range measurement devices, commonly known as laser range finders, LADAR (LAser Detection and Ranging), or LIDAR (LIght Detection and Ranging). The first laser range finder was demonstrated within a year of the demonstration of the first laser in 1960, and the military was quick to embrace the technology with neodymium (Nd):YAG being the material of choice because of its good thermal conductivity and, hence, high power capability.

The operational principles of a laser range finder are straightforward. A short laser pulse (10–30 ns) is emitted through some collimating optics and directed towards a reflective target. A sensitive photodiode as the focal point of similar optics operates as a receiver which detects the reflected light and converts it to a current pulse. The time taken from the transmission of the pulse to the reception of the echo is used to determine the range.

A typical high-power military Nd:YAG unit, such as the M-70B, produces 20 ns pulses with an energy of 150 mJ at a repetition rate of either 1 or 10 pulses per second. The maximum range is 10 km with a resolution of ±5 m. Unfortunately, a combination of the relatively high peak power and an operational wavelength of 1064.5 nm pose a serious danger of eye damage (Ready 1978). However, lower power eye-safe range finders are now produced in large numbers for recreational, industrial, and military use. These devices generally use laser diodes operating at around 900 nm in the near IR and produce 10 ns pulses with energy of less than 1 mJ. To achieve good range performance, these systems rely on exceptionally sensitive detectors based on PIN or Avalanche photodiodes, with the latter approaching single-photon detection capability.

Two other common active applications include laser interferometry and laser-induced fluorescence spectroscopy (LIFS). In the former, the coherent monochromatic nature of the laser light generates clear interference patterns which can be used to measure extremely short distances—typically smaller than one-quarter of the wavelength of the light. LIFS relies on the high power density of the laser pulse to excite fluorescence in the illuminated object. The spectrum of this fluorescence is then used to determine some of its chemical properties. Spectroscopy is commonly used in medical diagnostics and to measure pollution in the air or water. Later chapters of this book will consider these laser applications in more detail.

1.3 PASSIVE IR SENSING

Sir William Herschel, an astronomer, discovered IR radiation in 1800. He knew that sunlight could be broken down into a spectrum and also that it provided heat, so he arranged a bank of thermometers with blackened bulbs to measure the heat across the spectrum. To his surprise, he noticed an increase in the temperature from violet through the spectrum which peaked beyond the visible on the red side. He called this invisible radiation, which we now know as IR, "calorific rays" (Dummer 1977).

His son, Sir John Herschel, managed to record this radiation by creating an evaporograph image by using a carbon-black suspension in alcohol. He called this a thermogram, and it laid the foundations for the development of IR photography and more recently the sophisticated thermal imaging equipment used in industrial, medical, and military applications.

In 1821, Thomas Seebeck discovered that a circuit made from two dissimilar metals, with junctions at different temperatures, would deflect a compass needle. He soon realized that the deflection was caused by an electric current induced by the temperature difference. The Seebeck voltage does not depend on the distribution of temperature along the metals between the junctions, but only on the difference between those temperatures. This is the physic al principle for the thermocouple, which was invented by Leopoldo Nobili in 1829 (Dummer 1977).

Macedonio Melloni, an Italian physicist, soon used this technology to produce a device called a thermopile, using a number of hot and cold junctions in series. He was able to detect a person at a range of 30 feet by focusing their thermal energy onto the hot junctions of the pile. Detectors comprising a horn collector and the thermopile were soon being manufactured by scientific instrument makers for sale to laboratories and universities worldwide. A typical example, from the 1881 catalogue of James W. Queen & Co. of Philadelphia, shows their top-of-the-line model with 49 pairs of junctions at a cost of $40.00 (Greenslade 2007).

At about this time, Samuel Langley, an American astronomer, needed a sensitive device for his research to measure the distribution of heat in the sun's spectrum. As none of the thermopile instruments available to him were sufficiently sensitive, in 1878 he invented the bolometer, a device which converts changes in temperature into changes in resistance. His device, when connected as one arm of a Wheatstone bridge, could measure the temperature of celestial bodies to an accuracy of $10^{-5}\,^{\circ}\mathrm{C}$ and was capable of detecting the thermal radiation from a cow at a range of about 300 m.

It was not until the beginning of the 20th century that normal photographic processes were extended into the IR with the discovery of kryptocyanine. Although IR emulsions were not available commercially until the 1930s, Professor Robert Wood of Johns Hopkins University was able to publish IR photographs of landscapes in 1910 (Williams and Williams 2006).

During World War I, Case was the first person to experiment with photoconducting detectors. These used thallium sulfide and produced an output due to the direct interaction of IR photons and the material. They were therefore much more sensitive and also faster than other thermal detectors, such as the bolometer, that relied on an increase in the temperature of the sensing element. Experiments conducted during World War I and pursued through the mid-1930s, employed a parabolic mirror to concentrate IR radiation from the plane's hot engine onto a sensor. On a clear day it could pick up planes more than 50 km away, but fog, smoke, or clouds rendered it useless.

Between the wars, the U.S. Signal Corps Laboratories devoted a great deal of effort to target detection using IR. William Blair, the director of the laboratories, never really believed that it would be the answer because of its inability to penetrate fog and rain. However, he did hire Dr. S. Anderson, the country's leading expert on the subject, to push the technique, but even he could not overcome the attenuation problem (Brown 1999).

It is interesting to note that John Baird, who is considered by many to be the inventor of television, was first to demonstrate thermal imaging using a television tube. Details of the tube, which he called the Noctovisor, were published in the February 1927 issue of *Nature*. By 1929 he had built a self-contained outdoor version which was demonstrated to the public (Williams and Williams 2006).

During World War II, photoconductive detectors were further refined and used in a number of military applications such as target location, tracking, and weapon guidance. However, compared to the radar effort, very little investment was made in this field. It was only in the late 1950s and early 1960s that Texas Instruments, Honeywell, and Hughes Aircraft developed single-element, cryogenically cooled detectors that scanned scenes and produced line images. Because of their expense, these were used only by the military (Sparrius 1981).

Application areas expanded to include surveillance and intrusion during the Vietnam War, and shortly thereafter the first space-based applications for natural resource, pollution monitoring, and astronomy were developed. Towards the end of the decade, this technology started to appear in a few commercial applications, and though the first IR cameras were bulky and temperamental, they still found markets in research and development (R&D), preventive maintenance, and surveillance.

In the 1970s the pyroelectric vidicon tube was developed by Philips and EEV and became the core of a new product for firefighting, an application which was first used by the Royal Navy on board ships. The fact that it used an uncooled sensor made it more robust and less expensive than cooled scanned-detector-based systems, but its resolution and sensitivity were not as good. Portable mechanically scanned imagers were available for more demanding applications and their performance and reliability continued to improve with the development of small detector arrays to increase frame rate, improved photo detectors, and superior optics. However, they remained complicated and expensive to make because of the mechanical scanning and cryogenic cooling required.

In 1978 the Raytheon R&D group, then part of Texas Instruments, patented ferroelectric IR detectors using barium strontium titanate (BST) which were demonstrated to the military the following year. In the late 1980s, the federal government awarded contracts to Honeywell and Raytheon to develop high-density focal plane arrays (FPAs) for practical military applications. Raytheon went on to commercialize their uncooled BST technology and Honeywell developed vanadium oxide (VOx) microbolometer technology, which was finally granted a patent in 1994. Both these companies went on to develop the technology into equipment such as rifle sights and drivers' viewers in time for the 1991 Gulf War.

As volumes increased and prices dropped, many other companies licensed the technology and started to produce low-cost IR imagers for nonmilitary applications. These include surveillance, hot-spot detection on switchgear and power lines, firefighting, and a myriad of medical uses.

1.4 SENSOR SYSTEMS

The previous section considered the development of individual active and passive sensor types, all of which can be used in isolation to perform a particular measurement function. For example, a golfer may use his range finder to measure the exact distance to the next green. But more often than not, sensors do not operate in isolation, but are part of a larger system which may include other transducers, signal processing networks, and often some form of actuation or data recording capability. The autonomous agricultural vehicle shown in Figure 1.1 is a good example of a typical system (Hague et al. 2000).

Agriculture offers an extremely challenging environment for autonomy. The area of operation is large, cluttered, and may be uneven; wheel slippage may be significant, particularly if the ground is wet; and environmental conditions (dust, rain, fog, etc.) will almost certainly affect sensor observations.

Two categories of sensors must be considered in this application. In the one group are those that monitor the internal states of the system, for example, those that report articulation angles. Such sensors are usually restricted to the measurement of position, velocity, and acceleration. The second group are those that make external observations. These include navigation sensors

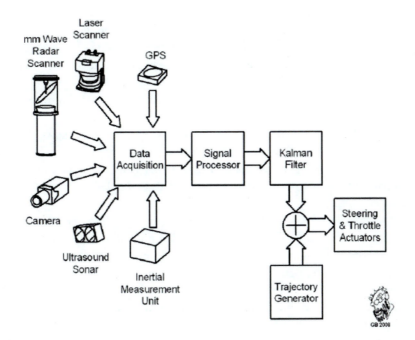

Figure 1.1 Sensors and actuators in an autonomous agricultural vehicle.

and those that make observations relevant to the vehicle's function. The latter can include crop health and the locations of individual plants for a vehicle that follows plant rows.

Typically the sensors that are used for navigation include odometers (including Doppler radar), inertial sensors, and compasses. An odometer generally uses the rotation of one or more wheels to measure vehicle motion. Unfortunately, such low-cost devices are prone to errors from a number of sources, including slippage and changes in the wheel's effective diameter, and because the errors are cumulative, position accuracy degrades very quickly. These sensors are therefore rather impractical for rough terrain agricultural applications and are often augmented by non-contact velocity measurements based on Doppler radar.

Inertial sensors are also often used to augment, or as an alternative to, odometer-based dead reckoning (Nebot et al. 1997). These are assemblies of accelerometers and rate gyros that are typically integrated into a single module from which the linear and angular rates and can be obtained. Unfortunately, these sensors are prone to drift and need other sensors to constrain the errors (Durrant-Whyte 1996). Because dead reckoning sensors integrate motion information to give position, any small biases in the sensor output accumulates. They do have a high bandwidth (typically > 100 Hz) and are reasonably accurate in the short term.

Range and bearing measurements made to artificial landmarks (such as poles or corner reflectors) mounted in strategic positions are a common method of correcting for drifts in the dead reckoning sensors. Line-scan laser systems, such as the SICK LMS 200, produce such data at regular intervals. For superior all-weather performance and longer range, the laser system can be replaced by a line-scanning millimeter wave radar (Brooker et al. 2007). These sensors operate at a much lower update rate (typically < 5 Hz) than the dead reckoning sensors but do not suffer from drift.

An alternative to landmark sensors is, of course, the global positioning system (GPS). However, unless the differential correction is implemented, the accuracy is not as good as that provided by radar or scanning laser systems. The update rate of typical GPS units is also usually only a few measurements per second.

Local feature detection can rely on machine vision (visible and IR) and ultrasound sonar. The primary advantages of the former are that the angular resolution is excellent and the update rate is high. That notwithstanding, vehicle motion and vibration can result in image blurring if an electronic shutter is not used, and variations in light levels and shadowing can make image interpretation difficult. Because the contrast between green plants and the ground is higher in the IR it is sometimes advantageous to convert commercial charge-coupled device (CCD)-based video cameras to operate at these longer wavelengths.

Ultrasound systems have been used to characterize plant types (Harper and McKerrow 1999) and can easily determine their presence, but they too are not without their shortcomings. Their performance outdoors and from a moving platform is compromised by air movement, while loud engine or machine noise can degrade their sensitivity.

In the absence of one single good localization sensor, it is necessary to use a combination of sensors to achieve acceptable accuracy. The Kalman filter provides an optimum method of fusing the information from the high-bandwidth dead reckoning sensors and the slower, drift-free response of landmark-based systems to obtain the best possible estimate of the vehicle position. This position, in conjunction with inputs from the local feature sensors, is used to control the throttle and steering to guide the vehicle along any specified trajectory.

1.5 FREQUENCY BAND ALLOCATIONS FOR THE ELECTROMAGNETIC SPECTRUM

As the previous sections have shown, sensors can be made to operate over a very broad band of frequencies for both electromagnetic (EM) and acoustic applications. In theory, although the EM spectrum extends below 1 Hz, in practice, radiation, detection, and data rate limitations make the use of such low frequencies impractical.

At the one extreme, an existing extra-low-frequency (ELF) electromagnetic application is communication with submerged submarines at about 80 Hz (a wavelength of 3750 km) (Rowe 1974). In contrast to this, positron emission tomography (PET) scanners generate and detect gamma rays with an energy of 511 keV, equivalent to a frequency of 1.23×10^{20} Hz or a wavelength of only 2.43×10^{-12} m. As a matter of convenience, rather than referring to the actual wavelength, or frequency, the EM spectrum is broken up into the following overlapping bands:

- Gamma rays
- X-rays
- UV
- Visible
- IR
- Microwave
- Radio

Many of these general bands are further subdivided to convey more information, as with near and far IR, or from a historical motive, such as with the radar band designations, as shown in Figure 1.2 to Figure 1.5.

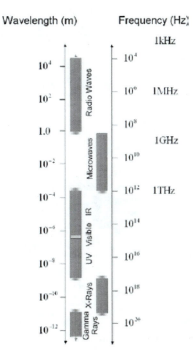

Figure 1.2 The electromagnetic spectrum.

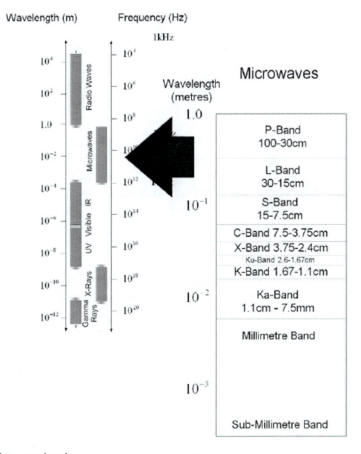

Figure 1.3 The microwave band.

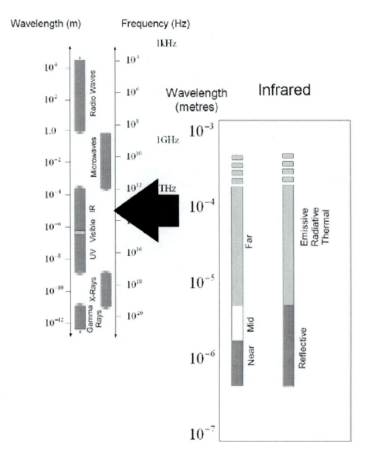

Figure 1.4 The infrared band.

The subdivision of the frequency allocations in the microwave band can be confusing, as there are various different standards in use. In this book, the U.S. microwave and radar nomenclature is generally used (see Figure 1.6). Until recently, the region above 40 GHz was underutilized, and the whole band from 30 to 300 GHz was referred to as the millimeter wave region (wavelength between 10 mm and 1 mm). The submillimeter wave region then encompassed everything with a wavelength between 1 mm and the start of the far IR band at 50 μm.

As new technologies evolve, and these bands are opened up, smaller subdivisions are being used. These include V-band for the region around 77 GHz, where automotive radar applications have been licensed, and W-band for the region around 94 GHz, where most military and experimental radar work occurs. The recently named terahertz band is generally considered to include all frequencies from 1 THz to 6 THz (300 μm to 50 μm), but is sometimes considered to extend right up to 100 THz to encompass both the far and the long-wave IR bands.

It is interesting to speculate on the relevance of the letters used for the various frequency designations. The original H_2S radar operated at a wavelength of 10 cm, and by 1943 the United States was developing bombing radar that operated at 3 cm, which was mooted to replace the

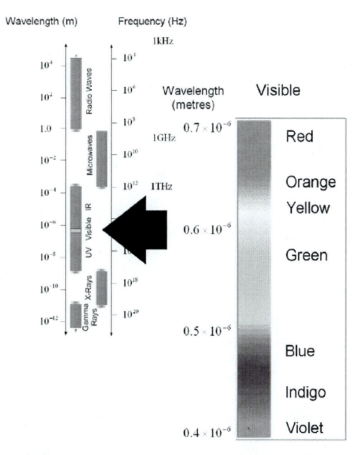

Figure 1.5 The visible band.

existing H_2S radars in the Pathfinder force. This new radar was codenamed H_2X. It is for these reasons that the band from 7.5 to 15 cm was designated the S-band and that from 2.4 to 3.75 cm was designated the X-band (Lovell 1991).

Useful EM radiation spans 16 orders of magnitude, from the ELF band below 3×10^3 Hz up to gamma rays above 3×10^{19} Hz. Some of the applications and their frequencies of operation are shown in Table 1.3.

1.6 FREQUENCY BAND ALLOCATIONS FOR THE ACOUSTIC SPECTRUM

Useful acoustic signals span eight orders of magnitude, as summarized in Table 1.4. This extends from the lower infrasound region at below 1 Hz for earthquake and structural vibration measurements, up to extreme ultrasound at about 1 GHz for microscopy.

Most active sensors operate in the audio and ultrasonic ranges, with long-range sonar, as used for submarine detection, operating at around 5 kHz and extending up to about 500 kHz for short-range imaging. Because of the high signal attenuation in the air, many industrial ranging

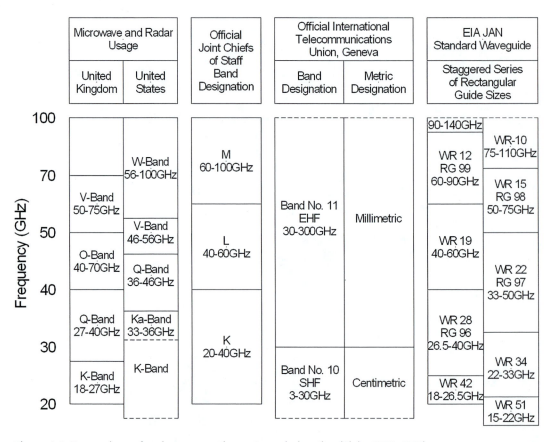

Figure 1.6 Nomenclature for electromagnetic spectrum designation (Blake 1995–1996).

Table 1.3 Frequency bands and typical applications for electromagnetic systems

Band	Frequency	Wavelength	Applications
VHF	30–300 MHz	1–10 m	Over the horizon radar, communications
UHF	300–1000 MHz	30–100 cm	Ground penetrating, communications
L	1–2 GHz	15–30 cm	Ground surveillance, astronomy
S	2–4 GHz	7.5–15 cm	Ground surveillance
C	4–8 GHz	3.75–7.5 cm	Space-based SAR
X	8–12.5 GHz	2.4–3.75 cm	Fire control, proximity, airborne/space SAR
Ku	12.5–18 GHz	16.7–24 mm	Collision avoidance, speed traps
K	18–26.5 GHz	11.3–16.7 mm	Fire control radar, collision avoidance
Ka	26.5–40 GHz	7.5–11.3 mm	Fire control radar, surveillance
Millimeter	30–300 GHz	1–10 mm	Astronomy, collision avoidance, missile seekers
Submillimeter		50 μm–1 mm	Astronomy, detection of explosives
Far IR		14–50 μm	Properties of molecules
Long-wave IR		8–14 μm	Laser radar, forward-looking IR
Near IR		1–3 μm	Personnel detection
Very near IR		0.76–1 μm	Imaging, laser ranging (industrial)
Visible		380–760 nm	Imaging, astronomy
UV		100–380 nm	Missile plume detection, gas fire detection

Table 1.4 Frequency bands and typical applications for acoustic systems

Designation	Frequency range	Applications
Infrasound	Below 50 Hz	Weather monitoring, earthquakes, nuclear explosion detection
Audio	50 Hz–20 kHz	Audio communications, acoustic modem, range measurement, hydroacoustic positioning, seismic prospecting
Ultrasound	20–200 kHz	Range measurement, imaging
	200–500 kHz	Short-range underwater sonar for imaging
	≈1 GHz	Microscopy

applications also operate within the audio band. However, this can be distracting in environments where people are present, so for short-range applications, frequencies above 20 kHz are preferred. At the upper extreme, acoustic microscopy operates through water at ranges measured in microns, and therefore, to obtain the required resolution, frequencies up to 1 GHz must be used.

1.7 References

Austin, B. (1992). Radar in World War II: the South African contribution. *Engineering Science and Education Journal* 1(3):121–130.

Blake, B., ed. (1995–1996). *Jane's Radar and Electronic Warfare Systems*, 7th ed. Alexandria, VA: ITP.

Brooker, G., Widzyk-Capehart, E., Hennessey, R., Bishop, M., and Lobsey, C. (2007). Seeing through dust and water vapor: millimeter wave radar sensors for mining application. *Journal of Field Robotics* 24(7):527–557.

Brookner, E. (1985). Phased-array radars. *Scientific American* 252(2):76–84.

Brown, L. (1999). *A Radar History of World War II*. Bristol: Institute of Physics Publishing.

Buderi, R. (1998). *The Invention that Changed the World: The Story of Radar from War to Peace*. London: Abacus.

Collins, A. (2007). Early sonar developed by UofA engineering professor. Viewed July 2007. Available at http://www.engineering.ualberta.ca/news.cfm?story=61396.

Dummer, G. (1977). *Electronics Inventions, 1745–1976*. Oxford: Pergamon Press.

Dupuis, R. (2003). The diode laser—the first thirty days forty years ago. Viewed July 2007. Available at http://www.ieee.org/organizations/pubs/newsletters/leos/feb03/diode.html.

Durrant-Whyte, H. (1996). An autonomous guided vehicle for cargo handling applications. *International Journal Robotics Research* 15(5):407–440.

Fraden, J. (2003). *Handbook of Modern Sensors: Physics, Designs and Applications*. Berlin: Springer Verlag.

Graaf, K. (1981). Physical acoustics, in *A History of Ultrasonics*. New York: Academic Press; chap. 1.

Greenslade, T. (2007). Instruments for natural philosophy. Viewed December 2007. Available at http://physics.kenyon.edu/EarlyApparatus/index.html.

Guerlac, H. (1987). *The History of Modern Physics: Radar in World War II*. New York: Henry E. Guerlac.

Hague, T., Marchant, J., and Tillett, N. (2000). Ground-based sensing systems for autonomous agricultural vehicles. *Computers and Electronics in Agriculture* 25(1–2):11–28.

Harmuth, H. (1979). *Acoustic Imaging with Electronic Circuits*. New York: Academic Press.

Harper, N. and McKerrow, P. (1999). Recognizing plants with ultrasonic sensing for mobile robot navigation. Third European Workshop on Advanced Mobile Robots (Eurobot '99).

Hecht, J. (1994). *Understanding Lasers: An Entry Level Guide*, 2nd ed. New York: IEEE Press.

Hollmann, M. (2007). Radar world. Viewed July 2007. Available at http://www.radarworld.org.

Larsen, E. (1971). *A History of Invention*. New York: J. M. Dent & Sons.

Lovell, B. (1991). *Echoes of War: The Story of H₂S Radar*. London: Adam Hilger.

Mason, W. (1962). Uses of ultrasonics in radio, radar and sonar systems. *Proceedings of the IRE* 50(5): 1374–1384.

Nebot, E., Sukkarieh, S., and Durrant-Whyte, H. (1997). Inertial navigation aided with GPS information. Proceedings, Fourth Annual Conference on Mechatronics and Machine Vision in Practice, Toowoomba, Qld., Australia; pp. 169–174.

Ready, J. (1978). *Industrial Applications of Lasers*. New York: Academic Press.

Rowe, H. (1974). Extremely low frequency (ELF) communication to submarines. *IEEE Transactions on Communications* 22(4):371–384.

Sparrius, A. (1981). Electro-optical imaging target trackers. *Transactions of the South African Institute of Electrical Engineers* 72(11):278–284.

Williams, R. and Williams, G. (2006). Medical and scientific photography: infrared photography. Viewed July 2007. Available at http://msp.rmit.edu.au/Article_03/index.html.

2

Signal Processing and Modulation

2.1 THE NATURE OF ELECTRONIC SIGNALS

Electronic signals, whether they are constrained within wires, or distributed more ephemerally in a field, convey information encoded in their levels and fluctuations. The measurement and manipulation of these levels forms the basis of signal processing. These signals can be classified into broad classes dependent on the rate and nature of the variations that take place (Carr 1997).

2.1.1 Static and Quasi-Static Signals

Static signals are, by definition, unchanging over a long period of time. Such signals are essentially direct current (DC) levels and in isolation convey very little information. Quasi-static signals are those that change very slowly with time; examples of such signals include the long-term drift on a sensor and the decreasing voltage on a slowly discharging battery.

2.1.2 Periodic and Repetitive Signals

Periodic signals are those that repeat themselves on a regular basis, though the timescale for repetition can be from femtoseconds to days, or even years. These include sine, square, and sawtooth waves, among others, with their defining nature being that each waveform is identical. Repetitive signals are periodic in nature, but the exact shape may change slightly with time. Electrocardiograph (ECG) signals and the outputs of tide gauges are examples of this type. Ultimately, very few signals are truly periodic, so there is some uncertainty regarding the transition region between these two classes.

2.1.3 Transient and Quasi-Transient Signals

Transient signals are, by definition, one time only signals, while quasi-transient signals are those which are periodic but with a duration that is very short compared to the period of the waveform. Once again, this classification is rather arbitrary, with the differences between quasi-transient and periodic signal types being rather vague. Typically a quasi-transient signal is one whose duty cycle is less than 0.1%, so a pulsed radar signal is a good example of this.

2.2 NOISE

Noise can be defined as unwanted signals, usually of a random nature, that interfere with the detection or analysis of a signal carrying information. The nature of the noise depends on the sensor type and is usually classified as an external or an internal source.

A wide range of external noise sources exists. These include natural sources like wind-generated rumble across the transducer in an ultrasonic ranging system and electromagnetic noise generated by sunspot activity that degrades the performance of a long-range radar system. Man-made sources of both acoustic and electromagnetic noise are ubiquitous, ranging from ship engine noise that degrades sonar performance to fluorescent lights and car ignition systems that generate wide bandwidth electromagnetic interference.

Internal sources of noise are generated within the sensor electronics and include thermal noise, shot noise, and avalanche noise, among others. These sources of noise are reasonably easy to quantify in a statistical sense and form the basis for the analysis of target detection and false alarm calculations.

2.2.1 Thermal Noise

Thermal noise, sometimes also known as Johnson noise, is electrical noise generated by random fluctuations of the voltage or current due to the thermal agitation of electrons within a conductor. More formally, if $v(t)$ is the thermal noise voltage across the terminals of a resistor, R, then if this voltage, v_j, is measured at regular intervals over a long period the mean value, $\bar{v}$, is

$$\bar{v} = \frac{1}{m} \sum_{j=1}^{m} v_j. \tag{2.1}$$

Measurements show that in the limit as $m \to \infty$, the mean value approaches zero. This result can be justified by considering the random motion of large numbers of electrons which produce fluctuations in the potential. These must average out to zero in the long term, otherwise they would result in the flow of a current.

The time average squared signal $\bar{v}^2$ is determined in a similar way:

$$\bar{v}^2 = \frac{1}{m} \sum_{j=1}^{m} v_j^2. \tag{2.2}$$

In the limit as $m \to \infty$, this mean squared value, $\bar{v}^2$ (V^2), can be shown to approach

$$\bar{v}^2 = 4kTR\Delta f, \tag{2.3}$$

where k is Boltzmann's constant (1.38×10^{-23} J/K), T (Kelvin) is the absolute temperature, R (ohms) is the resistance value, and Δf (Hz) is the bandwidth (Young 1990).

If samples of the noise voltage are taken over a long period and the results plotted as a histogram with bin widths dV, a distribution of the form shown in Figure 2.1 is produced.

The probability $p\{V\}dV$ that any future measurements will fall in the range $V \to V + dV$ is given by this plot, which is known as the probability density function (PDF). This function

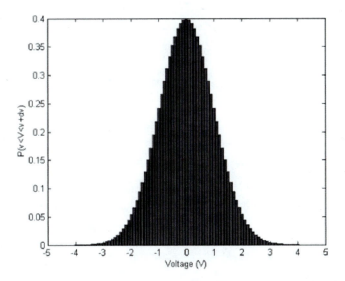

Figure 2.1 Histogram showing the probability density function (PDF) for thermal noise with unity variance.

approximates the normal or Gaussian distribution (Walpole and Myers 1978) which can be described by

$$p\{V\}dV = \frac{1}{\sigma\sqrt{2\pi}}e^{-V^2/2\sigma^2}. \tag{2.4}$$

The time average squared value, $\bar{v}^2$, equates to the variance, σ^2, because the distribution is Gaussian. Its value is a measure of how wide the distribution is, hence it is a useful indicator of the amount of noise present. However, in practice it is more common to specify the noise level in terms of the root mean square (RMS) quantity, where v_{rms} is

$$v_{rms} = \sqrt{\bar{v}^2}. \tag{2.5}$$

2.2.1.1 Noise Power Spectrum for Thermal Noise In theory, the noise power spectrum for thermal noise is completely flat. This is known as "white noise," as an analogy to white light, which comprises a uniform mix of all the colors. Strictly speaking, however, it is not possible to produce a power spectrum that is truly white over an infinite frequency range, as the total power integrated over this bandwidth would be infinite. In reality, all noise generating processes are subject to some band limiting mechanism, which produces a finite noise bandwidth. In addition, the measurement process is also band limited, which limits the measured value for the total noise power still further.

It is often convenient to remove the bandwidth dependence on the power spectrum; this is referred to as the power spectral density (PSD) or voltage variance per hertz, V^2/Hz. Electronic noise levels are often quoted as volts per root hertz ($V/\sqrt{Hz}$), as this serves as a reminder that the RMS voltage increases with the square root of the noise bandwidth.

Example: Determine the noise power spectral density of a 100 kΩ resistor at a temperature of 25°C.

$$\bar{v}^2 = 4kTR \quad (\text{V}^2/\text{Hz})$$
$$= 4 \times 1.38 \times 10^{-23} \times (273 + 25) \times 100 \times 10^3$$
$$= 1.65 \times 10^{-15} \quad \text{V}^2/\text{Hz}.$$

By taking the square root, it is possible to obtain the PSD in the more common form

$$v_{\text{rms}} = 40.56 \, \text{nV}/\sqrt{\text{Hz}}$$

A typical oscilloscope has a 20 MHz bandwidth and a very high input impedance compared to the resistor value, so it would measure the RMS voltage across the resistor to be

$$v_{rms} = 40.56 \times 10^{-9} \times \sqrt{20 \times 10^6}$$
$$= 181 \, \mu V.$$

2.2.2 Shot Noise

Shot or Schottky noise is typically generated by the current flowing across a barrier such as a PN junction. The noise is generated by the migration of individual charge elements across the barrier at random intervals, so even though, on average, the current flow may be constant, fluctuations around the average take the form of a Poisson distribution (Walpole et al. 1978),

$$p(n, \gamma) = \frac{e^{-\gamma} \gamma^n}{n!}, \tag{2.6}$$

where e is the base of the natural log (2.71828), n is the actual number of occurrences, and γ is the expected number of occurrences during a given time period.

Figure 2.2 shows a number of examples of the distribution for different values of γ. Note that the occurrences must be discrete integers, so the lines joining these points are for illustration only.

The Poisson distribution has a number of interesting characteristics, one of which is that the mean and the variance are both equal to γ. In addition, the figure shows that as the expected number of occurrences in a given time interval increases, the distribution becomes more normal. For $\gamma > 1000$, a Gaussian distribution with both mean and variance equal to γ is an excellent approximation of the Poisson distribution.

One of the results of this relationship is that as the mean current, I_{dc} (A), through a PN junction increases, the RMS noise current, i_{rms} (A), increases proportionally,

$$i_{rms} = \sqrt{2qI_{dc}\Delta f}, \tag{2.7}$$

where q is the electron charge (1.6×10^{-19} C) and Δf (Hz) is the bandwidth. As with the thermal noise case, the magnitude of shot noise is also proportional to the measurement bandwidth (Young 1990). If this current flows through a load resistor, R_{load} (Ω), then the RMS noise voltage, v_{rms} (V) will be

$$v_{\text{rms}} = i_{\text{rms}} R_{\text{load}}. \tag{2.8}$$

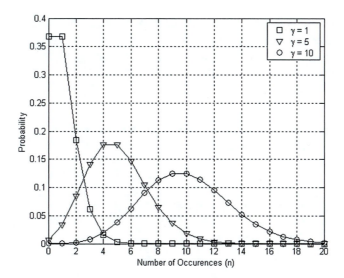

Figure 2.2 Poisson distributions for differing occurrence expectations.

2.2.2.1 Noise Power Spectrum for Shot Noise   The power spectral density for shot noise can be determined by examining the noise generation process. As each charge element flows across a PN junction, it generates a current pulse. Because the duration of each pulse is extremely short, its effective bandwidth is very wide. Therefore it can be concluded that shot noise is "white." A good analogy of the process is the sound of rain on a corrugated tin roof.

Example: Determine the noise power spectral density, i_{rms} (A/$\sqrt{Hz}$) of the shot noise for a forward-biased diode carrying a current of 0.5 A.

$$i_{rms} = \sqrt{2qI_{dc}}$$
$$= \sqrt{2 \times 1.6 \times 10^{-19} \times 0.5}$$
$$= 0.4\,nA/\sqrt{Hz}$$

If this current flows through a load resistor with R_{load} = 1 kΩ, the RMS voltage spectral density will be v_{rms} = 400 nV/$\sqrt{Hz}$.

The RMS voltage measured by an oscilloscope with a 20 MHz bandwidth will be

$$v_{rms} = 400 \times 10^{-9} \times \sqrt{20 \times 10^{6}} = 1.79\,mV.$$

2.2.3 1/f Noise

Known as "one-over-*f* noise," "flicker noise," or sometimes as "pink noise," it has the characteristic that its power spectrum $p(f) = 1/f^{a}$, where typically $0.5 < \alpha < 2$. This form of noise is extremely common in nature and is observed in all semiconductor devices (Young 1990).

Although $1/f$ noise is dominant at low frequencies, typically below 10 Hz, thermal and shot noise become dominant as the frequency increases. In oscillators, however, the noise is mixed up to frequencies close to the carrier, where it becomes phase noise.

2.2.4 Avalanche Noise

This form of noise occurs in gas discharge tubes and in reverse-biased PN junctions at breakdown. It is commonly used to electronically generate white noise.

2.3 SIGNALS

Most acoustic and electromagnetic sensors exploit the properties of periodic signals. In the time domain, such signals are constructed from sinusoidally varying voltages or currents constrained within wires:

$$v_c(t) = A_c \cos \omega_c t = A_c \cos 2\pi f_c t, \tag{2.9}$$

where
  $v_c(t)$ = signal,
  A_c = signal amplitude (V),
  ω_c = frequency (rad/s),
  f_c = frequency (Hz),
  t = time (s).

Sinusoidal electrical signals can be generated by the appropriate frequency selective feedback (shown in Figure 2.3) or by feedback across an inductive-capacitive (LC) tank circuit.

2.4 SIGNALS AND NOISE IN THE FREQUENCY DOMAIN

In the frequency domain, a continuous sinusoidal signal of infinite duration can be represented in terms of its position on the frequency continuum and its amplitude. However, most practical signals are not of infinite duration and so there is some uncertainty in the measured frequency, and this is represented in the frequency domain by a finite spectral width.

From a mathematical perspective, this is equivalent to windowing the continuous sinusoidal signal using a rectangular pulse of duration τ (s). Because windowing, or multiplication, in the time domain becomes convolution in the frequency domain, the continuous signal spectrum must be convolved by the spectrum of a rectangular pulse to obtain the spectrum of the windowed signal.

The spectrum of a rectangular pulse is the Sync function (Carlson 1998),

$$F(\omega) = \tau \frac{\sin(\omega\tau/2)}{\omega\tau/2}, \tag{2.10}$$

and the spectrum of a continuous sinusoidal signal is an impulse $\delta(\omega)$, so the resultant convolution is just the Sync function.

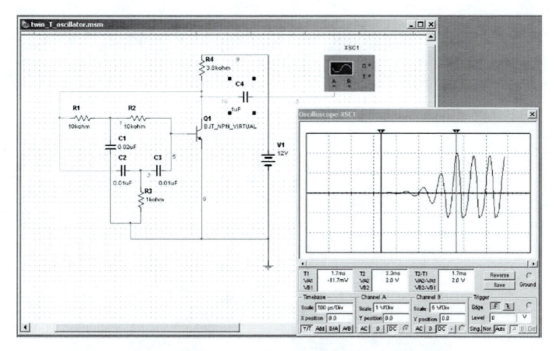

Figure 2.3 Sinusoidal voltage signal generated by an oscillator (Electronics Workbench 2003).

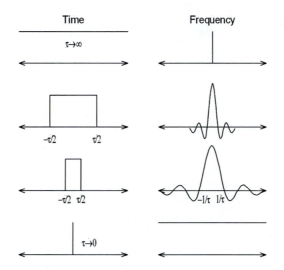

Figure 2.4 Mapping the relationship between the duration of a pulse and its spectrum.

It can be seen from equation (2.10) that as the duration of the signal decreases, $\tau \to 0$, its spectral width increases until, in the limit, when the signal can be represented by an impulse $\delta(t)$, the spectral width is infinite. This relationship is shown graphically in Figure 2.4.

More complex signals can usually be made up of a number of sinusoidal components of varying amplitudes. These can be calculated using the Fourier series, so it is often easier to identify the

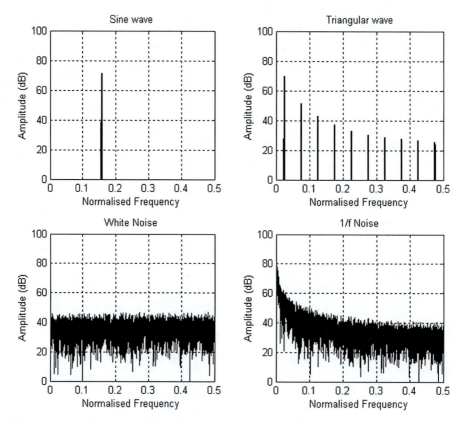

Figure 2.5 Spectra of various types of signal and noise.

spectrum of a time domain signal by processing it through a Fourier transform and then examining the amplitudes of the resultant components. Some examples of this process are shown in Figure 2.5.

2.4.1 The Fourier Series

The spectra in Figure 2.5 were obtained using Fourier techniques. All continuous periodic signals can be represented by a fundamental frequency sine wave and a collection of sine and/or cosine harmonics of that signal.

The Fourier series for any waveform can be expressed as (Edminster 1972; Carr 1997)

$$v(t) = \frac{a_0}{2} + \int_{n=1}^{\infty} a_n \cos(n\omega t) + \int_{n=1}^{\infty} b_n \sin(n\omega t), \tag{2.11}$$

where a_n, b_n are the amplitudes of the harmonics, which can be zero, and n is an integer.

The amplitude terms for each spectral component can be calculated by integrating the product of the time domain signal with a sample phasor of the correct frequency:

$$a_n = \frac{2}{T} \int_0^t v(t) \cos(n\omega t) \, dt$$

$$b_n = \frac{2}{T} \int_0^t v(t) \sin(n\omega t) \, dt \tag{2.12}$$

If there is a component of the signal at, or near, the selected frequency, this phase sensitive integration process will produce a nonzero amplitude. This process only examines the signal at discrete integer frequencies determined by the value of n and at DC, where the term $a_0/2$ is the average value of $v(t)$ over a complete cycle.

As can be seen from the harmonics for the triangular wave signal shown in Figure 2.5, even though the series is infinite, the coefficients decrease in amplitude and eventually become so small that their contribution is considered to be negligible. For example, the electrocardiogram trace shown in Figure 2.6, with a fundamental frequency of about 1.2 Hz, can be reproduced with 70 to 80 harmonics, which equates to a bandwidth of about 100 Hz (Carr 1997).

Depending on the application, a square wave may require up to 1000 harmonics to reproduce the sharp transitions that define the switching points (Carr 1997). The harmonic analysis in Figure 2.7 shows that the amplitudes of the coefficients are reduced by progressively smaller amounts, with the result that their individual contributions remain important.

The effect of truncating the series is demonstrated in Figure 2.8, where the waveform is reconstructed by summing the appropriately scaled sinusoidal components. For example, in the case where five components are used, the signal is reconstructed as follows:

$$v(t) = \sum_{n=1}^{5} a_n \cos(n\omega t) + b_n \sin(n\omega t) \tag{2.13}$$

Figure 2.8 shows the effectiveness of the reconstruction for 5, 50, and 500 coefficients. It can be seen that the transitions are still not perfect, even after the largest number.

The following MATLAB code is used to generate Figure 2.8, but it can easily be modified to analyze a single cycle of any other waveform.

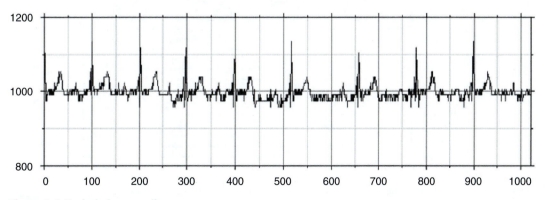

Figure 2.6 Typical electrocardiogram trace.

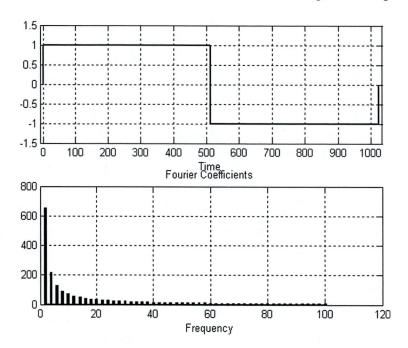

Figure 2.7 Amplitudes of Fourier coefficients to produce a square wave.

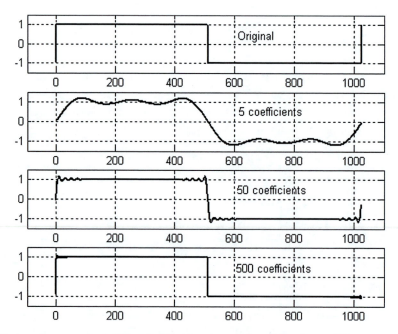

Figure 2.8 Effect on the reconstructed signal of limiting the number of Fourier coefficients.

```
% generate a square wave and look at the reconstruction from the
Fourier coefficients
% square_recon.m
%
% variables
t = (0:1023);                      % time (samples)
w0 = (2*pi)/1024;                  % fundamental frequency

% generate the square wave
sig=[-1,ones(1,511),-ones(1,511),1];
subplot(411), plot(t,sig,'k'), axis([-100,1100,-1.5,1.5])
grid

% determine the Fourier coefficients;
f = 2*fft(sig)/1024;
a = real(f);
b = imag(f);

% reconstruct the square wave
out = a(2)*cos(w0*t)+b(2)*sin(w0*t);
for n=2:500
    out = out + a(n+1)*cos(n*w0*t)+b(n+1)*sin(n*w0*t);
    if n==5
        subplot(412), plot(t,-out,'k'), axis([-100,1100,-1.5,1.5])
        grid
    elseif n==50
        subplot(413), plot(t,-out,'k'), axis([-100,1100,-1.5,1.5])
        grid
    elseif n==500
        subplot(414), plot(t,-out,'k'), axis([-100,1100,-1.5,1.5])
        grid
    end
end
```

2.5 SAMPLED SIGNALS

To process signals within a computer (digital signal processing [DSP]) requires that they be sampled periodically and then converted to a digital representation using an analog-to-digital converter (ADC). The result is that a continuous signal is reduced to one that is only defined at discrete points, as indicated by circles in Figure 2.9. To ensure accurate representation the signal

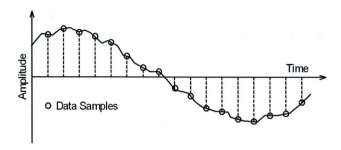

Figure 2.9 Digitizing a signal.

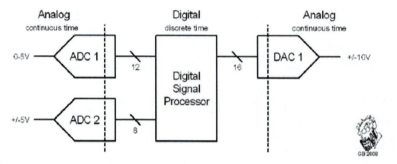

Figure 2.10 Typical configuration for a digital signal processing application.

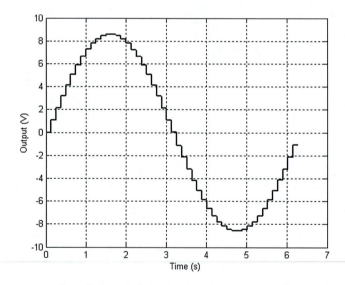

Figure 2.11 Analog reconstruction of a sampled signal using a zero-order hold.

must be sampled at the Nyquist rate, which is defined as at least double the highest significant frequency component of the signal. In addition, the number of discrete levels to which the signal is quantized must also be sufficient to represent variations in the amplitude to the required accuracy. Most ADCs quantize to 12 or 16 bits which represent $2^{12} = 4096$ or $2^{16} = 65536$ discrete levels, respectively.

The signal can then be manipulated within a computer in various ways. This process often produces intermediate results that are much larger than the input values, requiring that a larger word size be used if fixed point values are used or, as is more common, a floating point representation is used. Once the signal has been processed, the final result is often passed through a digital-to-analog converter (DAC) to convert it from the internal binary representation back to a voltage. It is therefore possible to represent a complete DSP operation, as shown in Figure 2.10, with ADCs, a processor, and DACs.

The DAC only outputs new values at discrete times, and so a continuous signal needs to be reconstructed. This generally involves holding the signal constant (zero-order hold) during the period between samples, as shown in Figure 2.11. This signal is then cleaned up by passing it through a low-pass filter to remove high-frequency components generated by the sampling process.

2.5.1 Generating Signals in MATLAB

MATLAB includes a number of built-in functions that make it easy to generate both periodic and aperiodic signals. A small sample of these is shown in the following section.

```
% Generate a square wave
% Fundamental freq w0 rad/s
% Duty cycle rho percent
A = 1;
w0 = 10*pi;
rho = 50;
t = 0:0.001:1;
sq = A*square(w0*t, rho);
plot(t,sq)
axis([0,1,-1.1,1.1]);
```

```
% Generate a triangular wave
% Amplitude A
% Fundamental frequency w0 rad/s
% Width W
A = 1;
w0 = 10*pi;
W = 0.5;
t = 0:0.001:1;
tri = A*sawtooth(w0*t, W);
plot(t,tri)
grid
axis([0,1,-1.1,1.1]);
```

```
% Generate a sine wave
% Amplitude A
% Fundamental frequency w0 rad/s
% Phase shift phi rad
A = 1;
w0 = 10*pi;
phi = pi/4;
t = 0:0.001:1;
sine = A*sin(w0*t + phi);
plot(t,sine)
grid
axis([0,1,-1.1,1.1]);
```

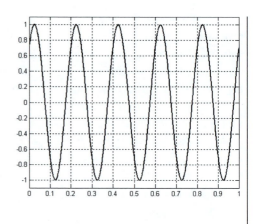

An exponentially damped sine wave is easily generated by taking the product of an exponential function and a sine wave:

$$x(t) = A\sin(\omega_0 t + \varphi)e^{-at}. \tag{2.13}$$

```
% Generate an exponentially
% decaying sine wave
% Amplitude A
% Fundamental frequency w0 rad/s
% Phase shift phi rad
% Exponent a
A = 1;
w0 = 10*pi;
phi = pi/4;
a = 6;
t = 0:0.001:1;
expsine = A*sin(w0*t + phi).*exp(-a*t);
plot(t,expsine)
grid
axis([0,1,-1.1,1.1]);
```

Other useful MATLAB functions include the following: COS, CHIRP, DIRAC, GAUSPULS, PULSTRAN, RECTPULS, SINC, and TRIPULS.

2.5.2 Aliasing

If the analog signal is not sampled at at least twice the frequency of the highest frequency component, then these high-frequency signals are folded or *aliased* down to a lower frequency, as illustrated in the time domain representation shown in Figure 2.12.

In the frequency domain, a generic analog signal may be represented in terms of its amplitude and total bandwidth, as shown in Figure 2.13a. A sampled version of the same signal can be represented by a repeated sequence spaced at the sample frequency, generally denoted f_s, and shown in Figure 2.13b and Figure 2.13c. If the sample rate is not sufficiently high, then the sequences will overlap and high-frequency components will appear at a lower frequency (albeit with reduced amplitude).

In reality, the finite roll-off of the filter response requires that a guard band be maintained between the spectra. This is achieved by selecting an anti-aliasing filter with a cutoff frequency that is less than 0.4 times the sample frequency. Using this ratio as a rule of thumb, a typical third-order low-pass filter will attenuate these unwanted signals by between 40 and 60 dB (1/100 to 1/1000) in voltage.

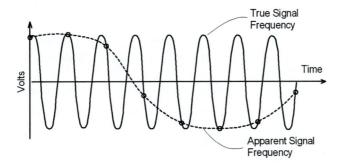

Figure 2.12 Interpretation of aliasing effects in the time domain.

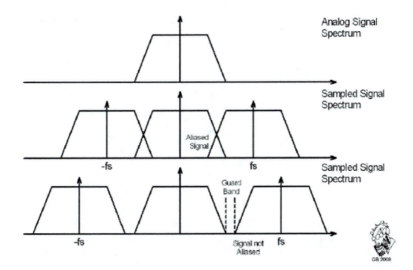

Figure 2.13 Interpretation of aliasing effects in the frequency domain.

2.6 FILTERING

A filter is a frequency selective network that passes certain frequencies of an input signal and attenuates others. It can be constructed of real components such as capacitors and inductors and operate on real voltages, or it can be implemented as an algorithm that runs within a DSP system. The three common types of filter are

- High pass,
- Low pass,
- Band pass.

A high-pass filter blocks signals below its cutoff frequency and passes those above the cutoff frequency unattenuated, while a low-pass filter passes signals below its cutoff frequency and attenuates those above. Band-pass filters pass a range of frequencies while attenuating those both above and below that range. Finally, a fourth, less common configuration is a band-stop or notch filter that attenuates signals at a specific frequency or over a narrow range of frequencies and passes all other frequencies (Williams and Taylor 1988).

Filters can be characterized by their impulse responses in the time domain, but are usually characterized by their frequency domain amplitude response $|H(\omega)|$ or gain, which is the ratio of the output to input voltage over the frequency range of interest. Here we will only use sampled data filters that can be synthesized in MATLAB. An example of the transfer function of such a filter is shown in Figure 2.14.

Note that the upper and lower limits of the pass band represent the half-power points (0.707 of the peak voltage gain). If the gain is plotted in decibels, it would be calculated using $20\log_{10}(\text{gain})$ to convert to power.

MATLAB implements filters as the "Direct Form II Transposed" of the standard difference equation

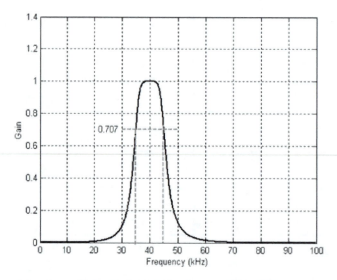

Figure 2.14 Butterworth band-pass filter transfer function generated by MATLAB.

$$a(1)\,y(n) = b(1)\,x(n) + b(2)\,x(n-1) + \ldots + b(n_b + 1)\,x(n - n_b)$$
$$-a(2)\,y(n-1) - \ldots - a(n_a + 1)\,y(n - n_a),$$

where the coefficients $A = [a(1), a(2), \ldots, a(n_a + 1)]$ and $B = [b(1), b(2), \ldots, b(n_b + 1)]$ are generated when the filter is synthesized, as shown in the following MATLAB code.

```
% Band-pass filter
% band-pass1.m
% variables
fs = 200e+03;              % sample frequency (Hz)
ts = 1/fs;                 % sample period (s)
fmat = 40.0e+03;           % center frequency (Hz)
bmat = 10.0e+03;           % bandwidth (Hz)

wl=2*ts*(fmat-bmat/2);   % lower band
wh=2*ts*(fmat+bmat/2);   % upper band
wn=[wl,wh];

% coefficients for 6th order Butterworth band-pass filter
[B,A]=butter(3,wn);

% determine the transfer function
[h,w]=freqz(B,A,1024);
freq=(0:1023)/(2000*ts*1024);

% plot the frequency response
plot(freq,abs(h));
grid
xlabel('Frequency (kHz)');
ylabel('Gain');
```

2.6.1 Filter Categories

The major filter categories are as follows:

- Butterworth (maximally flat),
- Chebyshev (equiripple)
- Bessel (linear phase),
- elliptical.

Note from Figure 2.15 that the cutoff frequency (200 Hz in this case) specified in MATLAB is equal to the 3 dB point for the Butterworth filter and to the pass-band ripple for the Chebyshev and elliptic filters.

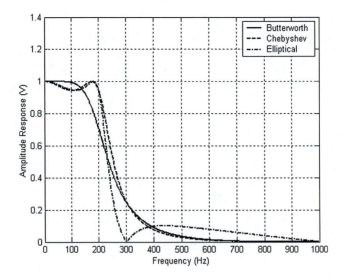

Figure 2.15 Comparison between low-pass filter transfer functions.

2.6.1.1 Butterworth This approximation to an ideal low-pass filter is based on the assumption that a flat response at zero frequency is most important. The transfer function is an all-pole type with roots that fall on the unit circle. It exhibits fairly good amplitude and transient characteristics, but the roll-off is quite slow for the filter order.

2.6.1.2 Chebyshev The transfer function is also all-pole, but with roots that fall on an ellipse. This results in a series of equal-amplitude ripples in the pass band and a sharper roll-off than the Butterworth filter. It exhibits good selectivity, but poor transient behavior.

2.6.1.3 Bessel These filters are optimized to obtain a linear phase response which results in a step response with no overshoot or ringing and an impulse response with no oscillatory behavior. However, it suffers from poor frequency selectivity compared to the other response types.

2.6.1.4 Elliptical These filters have zeros as well as poles which create equiripple behavior in the pass band similar to Chebyshev filters. Zeros in the stop band reduce the transition region so that extremely sharp roll-off characteristics can be achieved; for this reason they are commonly used in anti-aliasing filters.

2.6.2 Filter Roll-off

The rate at which the signal is attenuated as a function of frequency is proportional to the order of the filter. Figure 2.16 shows the roll-off for Butterworth filters. The theoretical slope is $6n$ dB/octave.[1]

---

1 An octave is a doubling in frequency.

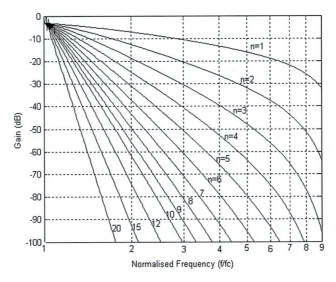

Figure 2.16 Roll-off for Butterworth low-pass filters.

2.6.3 The Ear as a Filter Bank

Filters need not be implemented using electrical components or computer algorithms, and many are physically realized mechanically. A good example of the latter is a car suspension, which performs a low-pass function. Biological systems have also evolved to be excellent filters. In the ear, for example, sound waves are transmitted into the cochlea, which tapers in size like a cone. Through the middle of this cone stretches the basilar membrane, which gets wider as the cochlea gets narrower and conveys the vibratory movement as a traveling wave (much like a snapping rope). The amplitude of this wave reaches a peak at a location which is dependent on frequency, as shown in the two examples in Figure 2.17.

High-frequency peaks occur toward the base of the basilar membrane, up to 20 kHz (where the membrane is stiffest and narrowest), while the low-frequency peaks occur toward the apex, down to 20 Hz. Hair cells rest on the basilar membrane and convert these vibrations to electro-chemical signals that are transmitted to the brain for interpretation.

A cochlear implant consists of a string of electrodes, each excited by a narrow band of frequencies, which stimulate the hair cell nerves directly. This allows some hearing to be restored in cases where either the cilia or basilar membrane has been damaged.

2.7 ANALOG MODULATION AND DEMODULATION

A continuous unmodulated signal cannot be used to transfer information and therefore cannot be used to measure range, as there is no way of determining when the signal was transmitted. For most sensor applications, this transmitted signal is "marked" in some way, usually by altering its amplitude or frequency. The round-trip time from the moment the "mark" is transmitted to when it is received can then be used to determine the range to the target if the speed of propagation is known.

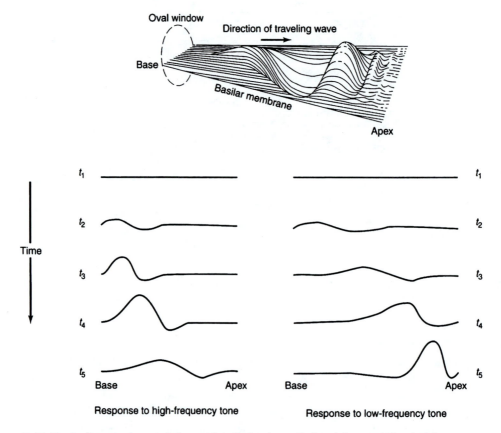

Figure 2.17 The basilar membrane of the cochlea depicted uncoiled and flattened showing the resonance for traveling waves of different frequencies.

The high-frequency signal is often referred to as the carrier, because is carries the information conveyed by the modulation. The low-frequency signal that is modulated onto the carrier is often referred to as the baseband signal (Young 1990).

2.7.1 Amplitude Modulation

Amplitude modulation (AM) is a modulation technique in which the amplitude of the carrier is varied in accordance with some characteristic of the baseband modulating signal. It is the most common form of modulation because of the ease with which the baseband signal can be recovered from the transmitted signal.

The AM process can be achieved reasonably easily by using an amplifier or an attenuator with a gain (or attenuation) that is proportional to the control applied voltage. For an unmodulated carrier, $\cos 2\pi f_c t$, described earlier, and for a baseband signal $v_b(t)$, the AM signal $v_{am}(t)$ is described by (Taub and Schilling 1971)

$$v_{am}(t) = A_c[1 + v_b(t)]\cos 2\pi f_c t. \tag{2.14}$$

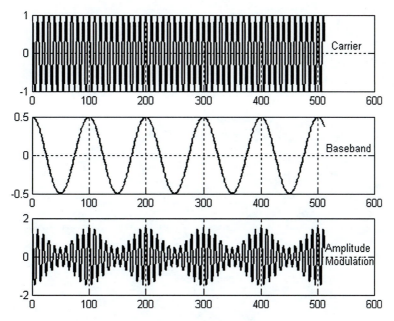

Figure 2.18 Time domain representation of amplitude modulation.

Where the modulating signal is a sinusoid with a frequency f_a and amplitude A_{am}, then equation (2.14) becomes

$$v_{am}(t) = A_c[1 + A_{am}\cos 2\pi f_a t]\cos 2\pi f_c t. \tag{2.15}$$

This process is illustrated in Figure 2.18 for $A_{am} < 1$. The extent to which the carrier has been amplitude modulated is expressed in terms of a percentage modulation which is calculated by multiplying A_{am} by 100. If a modulation depth of more than 100% is applied ($A_{am} > 1$), a phase reversal occurs at the crossover and the resultant signal becomes far more difficult to demodulate.

To determine the characteristics of the signal in the frequency domain, it can be rewritten in the following form (using a trig identity):

$$v_{am}(t) = A_c \cos 2\pi f_c t + \frac{A_c A_{am}}{2}[\cos 2\pi (f_c - f_a)t + \cos 2\pi (f_c + f_a)t]. \tag{2.16}$$

It can be seen that the signal is, in fact, made up of three independent frequency components:

- The original carrier at a frequency of f_c,
- A frequency at the difference between the carrier and the baseband, $f_c - f_a$, and
- A frequency at the sum of the carrier and the baseband, $f_c + f_a$.

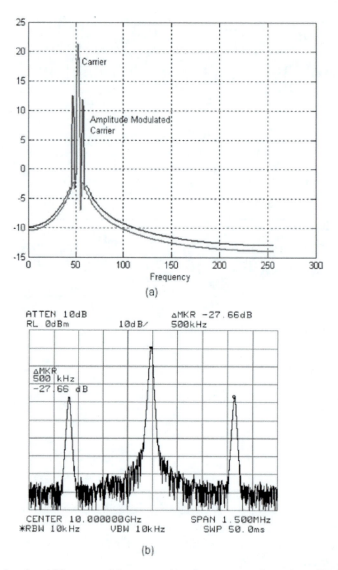

Figure 2.19 (a) Simulated and (b) measured frequency domain representations of amplitude modulation.

The spectrum of an arbitrary AM signal generated and displayed using MATLAB is shown in Figure 2.19a, and the 10 GHz output of a signal generator modulated at 500 kHz and measured using a spectrum analyzer is shown in Figure 2.19b.

As stated earlier, demodulation, also know as detection, of an AM signal is a straightforward process. The "tank" circuit in the crystal radio shown in Figure 2.20 is tuned to resonate at the carrier frequency and thus operates to select only a single radio program which passes into the demodulation stage.

Demodulation is then achieved with the use of a simple diode rectifier (crystal) and low-pass filter (capacitor). Finally, a pair of high-impedance headphones converts the resultant envelope

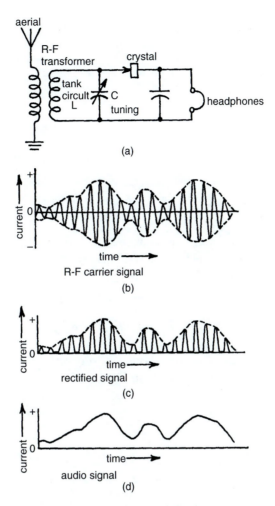

Figure 2.20 Demodulation of an AM signal by a crystal radio (Jacobowitz 1965).

current to an audio signal that accurately reproduces the original modulation. A simple "crystal set" with a good long-wire antenna can extract sufficient energy from the carrier to drive the headphones without amplification. These radios can therefore operate without batteries.

2.8 FREQUENCY MODULATION

Frequency modulation (FM) is a modulation technique in which the frequency of the carrier is varied in accordance with some characteristic of the baseband modulating signal. This process is achieved using a device called a voltage controlled oscillator (VCO). The governing equation for this process is (Taub et al. 1971)

$$v_{fm}(t) = A_c \cos\left[\omega_c t + k \int_{-\infty}^{t} v_b(t)\, dt \right].$$

(2.17)

The reason that the modulating signal is integrated is because variations in the modulating term equate to variations in the carrier phase. To confirm that this is correct, the instantaneous angular frequency can be obtained by differentiating the instantaneous phase term

$$\omega = \frac{d}{dt}\left[\omega_c t + k \int_{-\infty}^{t} v_b(t)\,dt\right] = \omega_c + k v_b(t). \tag{2.18}$$

The deviation of the instantaneous frequency from the carrier frequency, $f_c = \omega_c/2\pi$, is

$$\delta f = f - f_c = \frac{k}{2\pi} v_b(t). \tag{2.19}$$

This shows that the deviation of the instantaneous frequency is directly proportional to the amplitude of the modulating signal. Hence the combination of an integrator and phase modulator produces frequency modulation.

Rewriting equation (2.17) for sinusoidal frequency modulation gives

$$v_{\mathrm{fm}}(t) = A_c \cos[\omega_c t + \beta \sin \omega_a t], \tag{2.20}$$

where β, which is the maximum phase deviation, is usually referred to as the modulation index. This modulation process is shown graphically in Figure 2.21 for a very large modulation index.

The instantaneous frequency in this case is again obtained by taking the derivative of the instantaneous phase term

$$\begin{aligned} f &= \frac{\omega_c}{2\pi} + \frac{\beta\omega_a}{2\pi}\cos\omega_a t \\ &= f_c + \beta f_a \cos\omega_a t. \end{aligned} \tag{2.21}$$

Therefore the maximum frequency deviation, defined as Δf, occurs when $\cos\omega_a t = 1$, and it is

$$\Delta f = \beta f_a. \tag{2.22}$$

It should be realized that even though the instantaneous frequency lies within the range $f_c \pm \Delta f$, the spectral components of the signal do not lie within this range. Some manipulation of the formula for $v_{\mathrm{fm}}(t)$ shows that the spectrum comprises a carrier with amplitude $J_0(\beta)$ (Bessel Function of the first kind) with sidebands spaced symmetrically on either side of the carrier at offsets of ω_a, $2\omega_a$, $3\omega_a$, ..., as shown in Figure 2.22. As with the AM case, Figure 2.22a is MATLAB simulation and Figure 2.22b is the measured spectrum for a 10 GHz frequency modulated carrier. Whereas in the AM case the total bandwidth of the signal is 1 MHz, in the FM case it is 9.2 MHz, even though the same information is being conveyed.

Theoretically the bandwidth is infinite, however, for any β, most of the power is confined within a finite bandwidth. Carson's rule states that the bandwidth containing 98% of the signal energy is typically twice the sum of the maximum frequency deviation plus twice the modulating frequency.

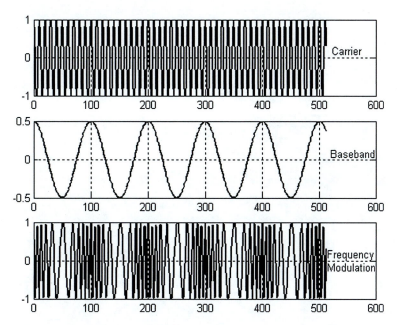

Figure 2.21 Time domain representation of frequency modulation.

There are a number of different methods to demodulate an FM signal, but the easiest is to convert the modulation to AM using the so-called slope detection method, and then to envelope detect it. Conversion to AM can be achieved by passing the signal through a frequency-sensitive circuit such as a low-pass or band-pass filter. The circuitry to perform this function is known as a frequency discriminator.

In the block diagram shown in Figure 2.23, the FM signal is split and passed through two band-pass filters with center frequencies just below and just above the carrier frequency. The two signals are then detected and filtered to remove the residual carrier before the difference is determined.

The system transfer function in the frequency domain can be obtained by subtracting the band-pass characteristics of the two filters, as shown in Figure 2.24. It can be seen that the voltage output at a frequency of 50 Hz will be zero, with the output going positive as the frequency decreases to 45 Hz, and going negative as the frequency increases to 55 Hz. If the frequency excursion exceeds the linear region between 45 and 55 Hz, distortion will occur because the output voltage will no longer bear a linear relationship with the frequency deviation.

In the time domain, the FM signal after detection and filtering produces two symmetrical demodulated signals with DC offsets, as shown in Figure 2.25. The difference between these reproduces the original 1 Hz baseband sinusoidal signal with minimal distortion.

Alternative techniques used in most modern radio receivers to demodulate the FM signal are based on phase-lock loops (PLLs) or quadrature detection (Straw 1990). In the PLL implementation shown in Figure 2.26, the FM signal is one of the inputs to a phase detector. When the loop is tracking, the VCO provides a signal with the same frequency and phase as this FM input signal, but this will only occur when the output, $v_o(t)$, is at the correct voltage. It is obvious, therefore, that $v_o(t)$ will be directly proportional to the input frequency.

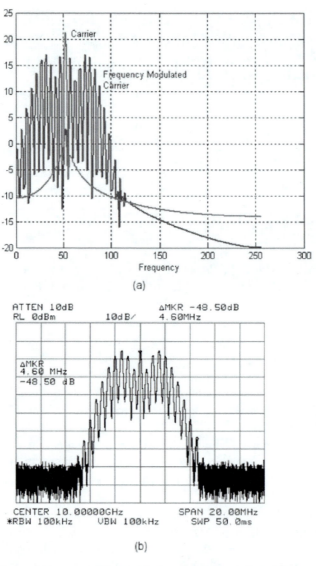

Figure 2.22 (a) Simulated and (b) measured frequency domain representation of frequency modulation.

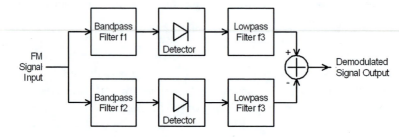

Figure 2.23 A discriminator converts the FM signal into an amplitude variation and an envelope detects the resulting AM signal.

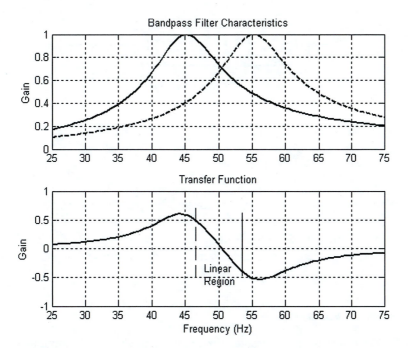

Figure 2.24 The difference signal from a pair of offset band-pass filters produces a symmetrical transfer function to convert variations in frequency to variations in amplitude.

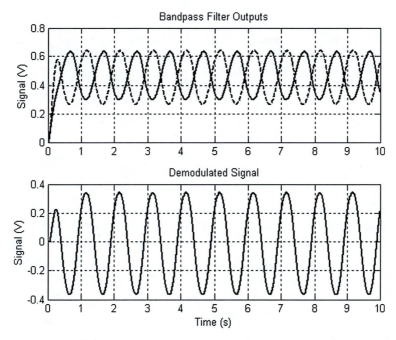

Figure 2.25 Stages of FM demodulation showing (a) the detected and filtered outputs from the band-pass filters and (b) the final demodulated output obtained by taking the difference between these signals.

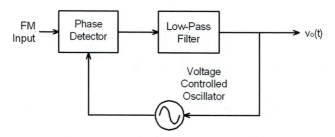

Figure 2.26 FM demodulation using the phase-lock loop.

Quadrature detectors use a reactance to produce two signals with a 90° phase difference. The phase-shifted signal is then applied to an LC tuned circuit resonant at the carrier frequency. Frequency changes result in additional leading or lagging phase shifts that can be detected by comparing zero crossings (known as a coincidence detector) or by analog phase detection (Straw 1990; Young 1990).

2.9 LINEAR FREQUENCY MODULATION

In most active sensors that operate using the frequency modulated continuous wave (FMCW) principle, the frequency, ω_b (rad/s), is not modulated sinusoidally, but in a linear manner with time. Writing the equation for a linear increase in frequency

$$\omega_b = A_b t, \qquad (2.23)$$

and substituting equation (2.23) into the standard equation for FM results in the following:

$$
\begin{aligned}
v_{fm}(t) &= A_c \cos\left[\omega_c t + A_b \int_{-\infty}^{t} t\,dt \right] \\
&= A_c \cos\left[\omega_c t + \frac{A_b}{2} t^2 \right]
\end{aligned}
\qquad (2.24)
$$

It can be seen that the phase modulation component now follows a quadratic with time. However, it is not possible to continue to increase the frequency indefinitely, so the modulation period is limited.

The MATLAB simulation for an arbitrary frequency sweep is shown in Figure 2.27a and the linear modulation of a 10 GHz signal measured by a spectrum analyzer is shown in Figure 2.27b.

In FMCW systems, a portion of the transmitted signal is mixed with (multiplied by) the returned echo. The transmit signal frequency will have shifted from that of the received echo because of the round-trip time to the target, τ (s),

$$v_{fm}(t - \tau) = A_c \cos\left[\omega_c(t - \tau) + \frac{A_b}{2}(t - \tau)^2 \right]. \qquad (2.25)$$

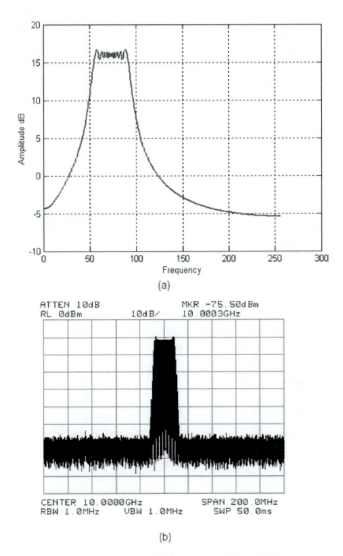

Figure 2.27 Simulated and measured frequency domain representation of linear FM.

The beat frequency is determined by taking the product of the transmitted signal and the echo,

$$v_{fm}(t-\tau)v_{fm}(t) = A_c^2\cos\left[\omega_c t + \frac{A_b}{2}t^2\right]\cos\left[\omega_c(t-\tau) + \frac{A_b}{2}(t-\tau)^2\right]. \tag{2.26}$$

This result can be simplified using the trigonometric identity that relates the product of two cosines, $\cos A\cos B = 0.5[\cos(A + B) + \cos(A - B)]$,

$$v_{out}(t) = \frac{1}{2}\left[\cos\left\{(2\omega_c - A_b\tau)t + A_b t^2 + \left(\frac{A_b}{2}\tau^2 - \omega_c\tau\right)\right\} + \cos\left\{A_b\tau t + \left(\omega_c\tau - \frac{A_b}{2}\tau^2\right)\right\}\right]. \tag{2.27}$$

The first cosine term describes a linearly increasing FM signal (chirp) at about twice the carrier frequency, with a phase shift that is proportional to the delay time, τ. This term is generally filtered out.

The second cosine term describes a number of phase terms that can be ignored and a beat signal at a fixed frequency

$$f_{beat} = \frac{A_b}{2\pi}\, \tau.$$

(2.28)

It can be seen that the signal frequency is directly proportional to the round-trip time, τ, and hence is directly proportional to the range to the target. This technique has been perfected for use in short-range radar systems and is discussed in detail in Chapter 6. The spectrum of the complete output signal, including the chirp term and the constant frequency term, is shown in Figure 2.28.

2.10 PULSE CODED MODULATION TECHNIQUES

Many sensors marry digital and analog techniques by modulating the carrier in a discrete rather than continuous manner. These include various forms of pulse modulation, including pulse amplitude, pulse width, and pulse position modulation, as well as the closely related frequency-shift and phase-shift methods in common use today, among others.

2.10.1 Pulse Amplitude Modulation

Pulse amplitude modulation (PAM) is a technique in which the amplitude of individual, regularly spaced pulses in a pulse train is varied in accordance with some characteristic of the modulating

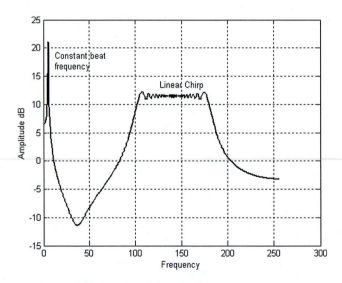

Figure 2.28 Frequency domain representation of the FMCW sensor output.

signal. For time-of-flight sensors, the amplitude is generally constant, as the primary objective is to measure the range to a target. Because the ability of a pulsed sensor to resolve two closely spaced targets is determined by the pulse width, τ (s), selecting the correct value is important. The pulse width also determines the bandwidth requirements for the receiver, as there is a close relationship between the width of a single pulse and its spectral content, as seen in Figure 2.29.

Note that the width of the peak to the first zero crossings is $1/2\tau$ and that frequency components extend out symmetrically in both directions. This relationship can easily be explored using MATLAB, where it is a simple matter to generate a fixed amplitude carrier signal with various pulse widths and use the fast Fourier transform (FFT) procedure to obtain a spectrum. In the results shown in Figure 2.30, the magnitude of the spectrum is in decibels.

Note that the width of the main lobe in the frequency domain increases as the pulse width decreases. If, instead of taking the spectral width as the distance between zero crossings, but rather as the distance between the half-power (3 dB) points, which is more usual in the definition of a pulse width, then the relationship between the pulse width, τ (s), and the spectral width, β (Hz), is

$$\beta \approx 1/\tau. \tag{2.29}$$

In a time-of-flight sensor, this relationship determines the bandwidth required by a receiver to receive pulses of a specific duration. A filter that conforms to this relationship is known as a "matched" filter. It is formally defined in Chapter 6.

In a repeated sequence of pulses, the fine structure of the spectrum is determined by the total length of the observed sequence, as shown in Figure 2.31.

Figure 2.32 shows the measured spectra of a number of pulsed 10 GHz signals with progressively increasing pulse widths of 50 ns, 100 ns, and 500 ns compared to a continuous signal. In Figure 2.32a and Figure 2.32b, the minima on either side of the main lobe are marked with the cursor pair and the frequency difference is measured. The results can be read at the ΔMKR label above each spectrum. As expected, the results are consistent with those shown in Figure 2.29.

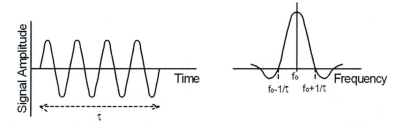

Figure 2.29 Relationship between pulse width and frequency for a single pulse [adapted from (Mahafza 2000)].

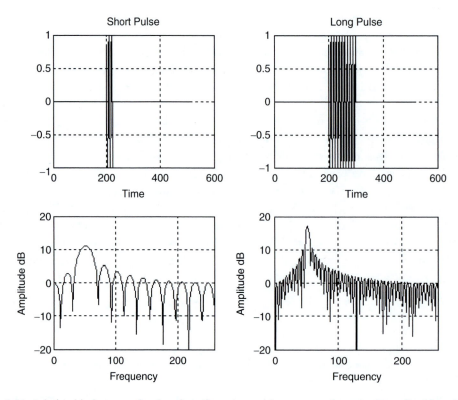

Figure 2.30 Relationship between the duration of a pulse and its spectrum for pulsed amplitude modulation.

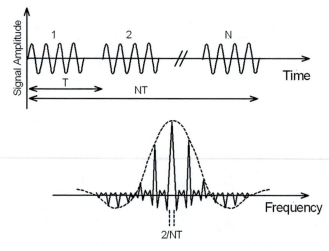

Figure 2.31 Relationship between pulse width and frequency for a sequence of pulses [adapted from (Mahafza 2000)].

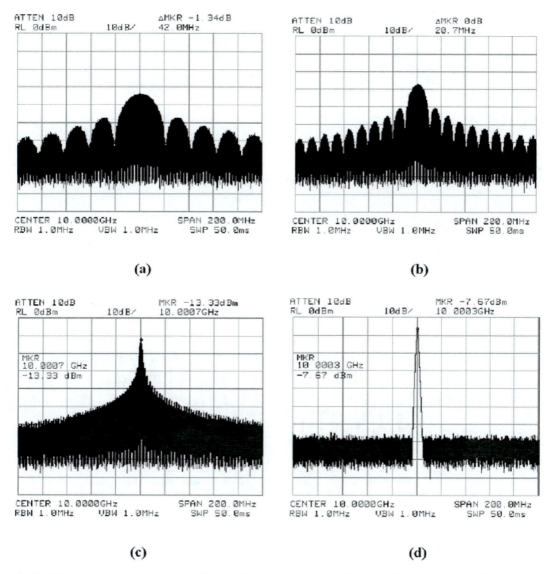

Figure 2.32 Spectra of pulsed signals with durations of (a) 50 ns, (b) 100 ns, (c) 500 ns, and (d) continuous.

It can be seen that if the two points 3 dB down on the peak were taken instead of the minima, then the relationship defined in equation (2.29) would be satisfied.

2.10.2 Frequency-Shift Keying

Frequency-shift keying (FSK) is the digital equivalent of frequency modulation in which only two different frequencies, f_1 and f_2, are utilized (Taub et al. 1971):

$$v_{\text{fsk}}(t) = A_{\text{c}} \cos(\omega_{\text{c}} \pm \Omega)t, \tag{2.30}$$

where $f_1 = (\omega_{\text{c}} + \omega_{\text{m}})/2\pi$ and $f_2 = (\omega_{\text{c}} - \omega_{\text{m}})/2\pi$, and ω_{m} is a constant angular frequency offset from the carrier.

From a communications perspective, a single bit of information can be represented by a single cycle of the carrier if the maximum data transfer rate is important, or if the data rate is not critical, then multiple cycles can be used as illustrated in Figure 2.33.

Two common methods of generating the signal are shown in Figure 2.34. In Figure 2.34a, a VCO frequency is directly modulated using a digital signal stream. In this case, because the VCO outputs a signal continuously with a finite transition time between the two frequencies, there is a finite period during which the frequency changes. In Figure 2.34b, a digitally controlled switch

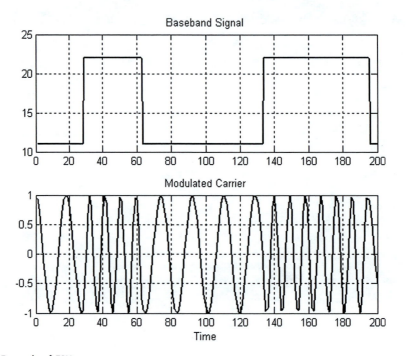

Figure 2.33 Example of FSK.

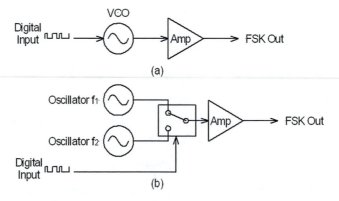

Figure 2.34 Two common methods of generating FSK.

toggles between the outputs of two oscillators, which results in discrete breaks between the transitions. The latter will generate additional spectral components introduced by the switching transients, but the accuracy with which the two frequencies are generated will be better than that produced by the VCO method.

Demodulation can be achieved by detecting the outputs of a pair of filters centered at the two modulation frequencies, f_1 and f_2, as shown in Figure 2.35. An alternative method is to use a PLL similar to that shown in Figure 2.26, which will generate at its output the exact digital sequence (Carlson 1975).

It is possible, in some cases, to synchronize the transition so that it always occurs at the start of a carrier frequency cycle so that there are no phase discontinuities. This is known as synchronous FSK. However, in most cases there is no synchronization and large phase discontinuities can occur which widen the spectrum.

If the spectrum of the FSK stream is examined, it can be seen, where the number of cycles per bit is large, that two distinct peaks occur, as illustrated in Figure 2.36. However, as the number of cycles per bit decreases, the spectrum shifts to a flatter, broad, almost uniform peak, shown in Figure 2.37. In the latter case, because the spectral content is more uniformly spread, the peak power density of the signal is reduced proportionally.

Frequency-shift keying is a very simple modulation technique and is still extremely popular. It was originally used by teleprinters which operated at about 45 bps, and was introduced in 1962 for a Bell modem which operated at up to 300 bps. Early personal computers (PCs) used a Kansas City interface which used FSK to store programs and data on audio cassettes at up to 1200 bps. It is now used in touch-tone phones and a myriad of other communications systems, operating at speeds in excess of 1 Mbps (Dorf 2006).

2.10.3 Phase-Shift Keying

Phase-shift keying (PSK) is a common modulation scheme in which a digital sequence is represented by changes in phase of the carrier, relative to some reference. The most common form is binary phase-shift keying (BPSK), or binary coding as it is sometimes called. In this scheme, the carrier phase is switched between 0 and π according to a digital sequence, $\varphi_m(t)$,

$$v_{\text{bpsk}}(t) = A_c \cos\left[\omega_c t + \varphi_m(t)\right]. \tag{2.31}$$

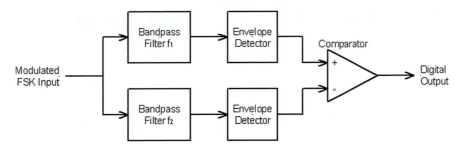

Figure 2.35 Band-pass filter method of demodulating FSK signals.

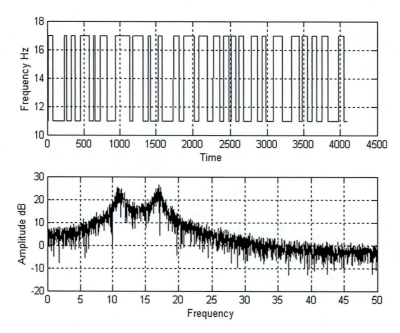

Figure 2.36 FSK spectrum, 5 cycles per bit.

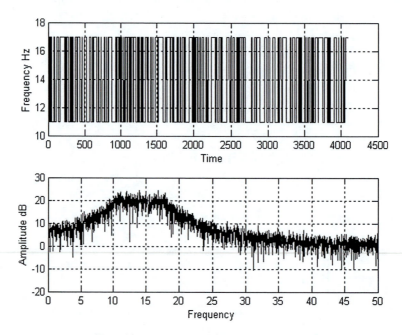

Figure 2.37 FSK spectrum, 1 cycle per bit.

This modulation technique can be implemented quite easily using a balanced mixer, as shown in Figure 2.38, or with a dedicated BPSK modulator.

When the modulation signal f_m is high (+1), diodes A and B are forward biased and the carrier f_c is coupled directly to the output transformer. However, when f_m is low (−1), then diodes C and D are forward biased and f_c is coupled to the opposite terminals of the output transformer, which results in a reversal of the phase (Young 1990). An example of this process in action is shown in Figure 2.39.

Demodulation is achieved by multiplying the modulated signal by a coherent carrier (a carrier that is identical in frequency and phase to the carrier that originally generated the BPSK signal). This produces the original BPSK signal plus a signal at twice the carrier, which can be filtered out.

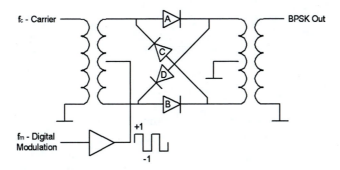

Figure 2.38 Implementing BPSK using a balanced mixer.

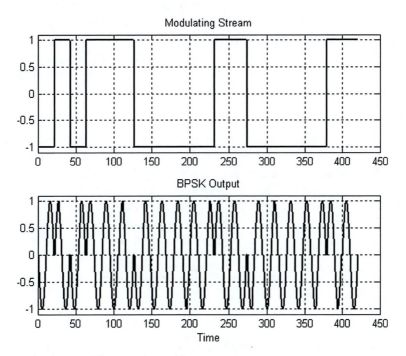

Figure 2.39 Example of BPSK with one cycle per bit.

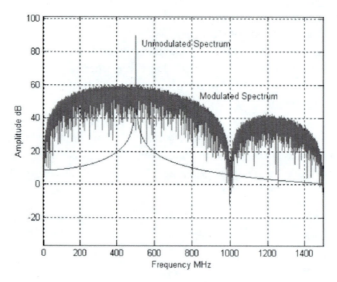

Figure 2.40 PSK spectrum for a 500 MHz carrier and 1 cycle per bit modulation.

As with the FSK process, the spectral width of the carrier is widened by the BPSK modulation process, as can be seen from the spectra shown in Figure 2.40. In this case a carrier at 500 MHz is shown along with the output spectrum after modulation at one cycle per bit. In this case the peak power spectral density is reduced by 30 dB compared to the unmodulated carrier, even though the total power transmitted, and therefore the range performance, remains unaltered.

When used as a sensor, the range resolution is determined by the transmitted bandwidth, which is, in turn, determined by the duration of each bit. Therefore, for the best range resolution, the modulation rate should be at one cycle per bit, as in the example.

This spread spectrum modulation scheme offers a number of advantages in regard to performance and covertness when compared to other techniques, and for that reason it is discussed in more detail in Chapter 11.

2.10.4 Stepped Frequency Modulation

As discussed in more detail in Chapter 11, the maximum detection range that can be achieved by an active sensor is determined by the average power that it radiates. Therefore, it is often advantageous to transmit long-duration pulses even though they offer reduced range resolution (which is inversely proportional to the pulse width). The stepped frequency modulation technique is a way around this dilemma. In this case a sequence of pulses, each at a slightly different frequency, is transmitted and then later received and processed to obtain a range resolution determined not by the pulse duration, but by the bandwidth of the total transmitted sequence (Currie and Brown 1987).

If a signal is transmitted at a carrier frequency ω_c (rad/s) and it reflects off a target at a range R (m), then the round-trip delay time is

$$\tau = 2R/c. \tag{2.30}$$

The phase difference between the transmitted signal reference and the echo can be determined by mixing a portion of the transmitted signal, as with the FMCW case discussed earlier, with the echo. One component of the mixer output is at twice the carrier frequency, which can be filtered out, but the other component is a phase shift which is proportional to the product of the carrier frequency and the round-trip time:

$$v_{\text{out}}(t) = \cos \omega_c t \cos \omega_c (t - \tau)$$
$$v_{\text{out}}(t) = \frac{1}{2}[\cos \omega_c (2t - \tau) + \cos \omega_c \tau]. \tag{2.31}$$

Substituting equation (2.30) into equation (2.31) and filtering the high-frequency components leaves the phase term only:

$$v_{out}(t) = \frac{1}{2}\cos \frac{2\omega_c R}{c}. \tag{2.32}$$

The actual phase shift, φ (rad), can be obtained by replacing ω_c with $2\pi f_c$:

$$\varphi = 4\pi f_c R/c. \tag{2.33}$$

It can be seen that the phase shift for a target at a fixed range, R, will change with each new transmitted frequency, f_c (Hz). In addition, if the frequency step is selected appropriately, this phase change is sinusoidal in nature and can be considered to be a synthetic Doppler signal. The frequency of this Doppler can therefore be used to determine the range of the target to a fraction of the pulse width. In addition, if a number of closely spaced reflectors are present, they will each have their own unique Doppler frequency and so can each be resolved as a separate target.

2.11 CONVOLUTION

The convolution process is a particularly useful one for modeling the time domain performance of systems, including time-of-flight sensors.

2.11.1 Linear Time Invariant Systems

It is convenient to describe the relationship between the input, $x(t)$, and output, $y(t)$, signals of linear time invariant (LTI) systems in terms of their impulse response, $h(t)$. In fact, an LTI system is completely characterized by its impulse response, $x(t) = \delta(t)$, applied at time $t = 0$, or in the discrete time domain at $n = 0$.

Applying an impulse to the input of an unknown LTI system is therefore a good method for determining its characteristics. This is easily achieved in the discrete time case where the input is set equal to an impulse $\delta(n)$. However, in the continuous time case, it is not possible to produce a true impulse with zero width and infinite amplitude, and an approximation must be used.

To determine why this impulse response idea is so useful, consider that an arbitrary input signal can be expressed as the weighted superposition of a sequence of time-shifted impulses, $x(t) = x(\tau)\delta(t - \tau)$. The system output is then just the weighted superposition of the time-shifted impulse responses. This can be written as an equation (Haykin and Van Veen 2003):

$$y(t) = \int\limits_{\tau=-\infty}^{t} h(t-\tau)x(\tau)d\tau = \int\limits_{0}^{\infty} h(\tau)x(t-\tau)d\tau. \tag{2.34}$$

This weighted superposition is termed the *convolution integral* for continuous time systems and the *convolution sum* for discrete time systems. Therefore, given a system defined by its impulse response, $h(t)$, the output, $y(t)$, is found by convolving the input, $x(t)$, with this impulse response, $h(t)$.

If the Laplace transform is taken of this convolution process, it can be shown that

$$Y(s) = H(s)X(s), \tag{2.35}$$

that is, convolution in the time domain is equivalent to multiplication in the Laplace domain. In addition, if s is replaced by $j\omega$, to consider the relationship only along the imaginary axis, then the equation becomes (Haykin et al. 2003)

$$Y(\omega) = H(\omega)X(\omega), \tag{2.36}$$

which describes the frequency response of the system.

It is the magnitude of this transfer function, $|H(\omega)|$, that was considered when the characteristics of various filters were discussed earlier in this chapter. In this case, the mapping between the time domain and the frequency domain is through the Fourier transform.

In the example shown in Figure 2.41, the impulse response of a third-order Butterworth bandpass filter (normalized passband 0.2 to 0.3) is generated. As this characterizes the filter completely, its Fourier transform should reproduce the frequency response accurately, and it does!

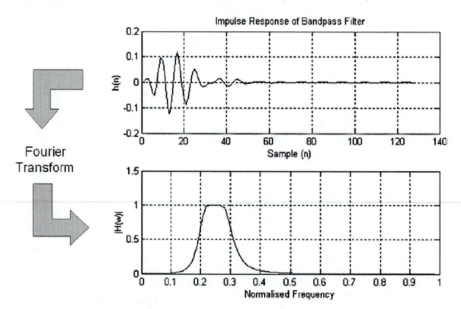

Figure 2.41 MATLAB example showing the relationship between the impulse response of a band-pass filter and its Fourier transform.

The MATLAB code to produce this figure is shown below.

```
% relationship between filters in the time and frequency domain
% filter_rel.m

% generate a Butterworth band-pass filter
wn=[0.2,0.3];
[b,a]=butter(3,wn)

% generate the impulse response of the filter
x=zeros(1,128);
x(1)=1;
y=filter(b,a,x);

subplot(211), plot(y),grid, xlabel('Sample (n)'), ylabel('h(n)')
title('Impulse Response of Band-pass Filter')

% take the Fourier transform of the impulse response
f=fft(y);
freq=(0:63)*1/64;

subplot(212), plot(freq, abs(f(1:64))), grid, xlabel('Normalized
Frequency'),ylabel('|H(w)|')
%title('Frequency Response')
```

2.11.2 The Convolution Sum

Most modern systems are digital, which requires that the input and output data be sampled. In this case the convolution sum replaces the convolution integral and is defined by the following equation where $y(n)$ is the output, $x(n)$ is the input, and $h(n)$ is the system impulse response (Haykin et al. 2003):

$$y(n) = \sum_{k=-\infty}^{\infty} x(k)h(n-k). \tag{2.37}$$

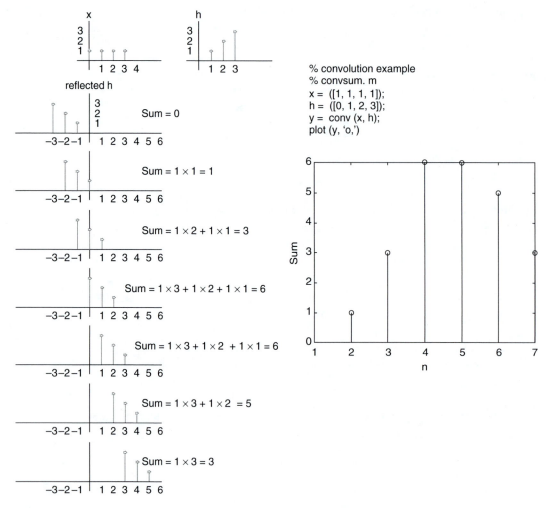

Figure 2.42 Comparison between a manual implementation of the convolution sum and that performed using MATLAB.

However, to be tractable, $x(n)$ and $h(n)$ are nonzero only over a finite interval. The convolution process can be performed manually if the shape of the two functions is known, or if the points that describe the two functions have been tabulated. In the example shown in Figure 2.42, $N_x = 4$ and $N_h = 3$, which makes the total duration of the convolved sequence $N_x + N_h - 1 = 6$.

To perform this function manually, the order of $h(n)$ is first reversed before being shifted across $x(n)$ one sample at a time. The output, $y(n)$, is equal to the sum of the products of each of the aligned terms.

2.11.3 Worked Example: Pulsed Radar Echo Amplitude

Because a pulsed time-of-flight sensor can be considered to be a form of network, the convolution process works quite well in determining the amplitude and structure of the echo from a distributed target. In this application the impulse response of the round-trip propagation to the point target is a signal that is delayed in time by τ (s) and attenuated in amplitude by the factor a:

$$h(t) = a\delta(T - \tau).$$

Consider an aircraft flying toward the sensor, as shown in Figure 2.43, and assume that only the few parts of the aircraft listed in Table 2.1 generate significant echoes. Determine the magnitude of the echo signal if a pulse with a length of 3 m is transmitted.

- The transmit signal is treated as a sequence of impulses separated by the sample interval of 10 cm covering a total of 3 m in range.
- The target is treated as a sequence of four impulses with separations as listed in Table 2.1.
- The reflected signal received back at the radar is well described by the convolution of the transmitted pulse and the array of point reflectors.

Radar

Figure 2.43 Radar illuminating an aircraft target showing some of the scattering points.

Table 2.1 Reflection contributions from various parts of the aircraft

Contribution	Distance from datum (m)	Magnitude
Nose	2	1
Ordnance	6	1
Wing	8	1
Tail	14	1

```
% convolution demo
% convdemo.m
%
% generate the 3m transmit pulse with a resolution of 100mm
a=([zeros(1,70),ones(1,30),zeros(1,70)]);

% generate the point target reflectors with a resolution of 100mm
b=zeros(1,170);
b(20)=1;                % Nose
b(60)=1;                % Ordnance
b(80)=1;                % Wing
b(140)=1;               % Tail

% take the convolution to determine the return from all of the
reflectors
c=conv(a,b);

% plot the results
x=(1:170)/10;
subplot(311) ,plot(x,a,'k',x,a,'ko'), grid;
title('Transmit Pulse'),ylabel('Amplitude');
subplot(312), plot(x,b,'k',x,b,'ko'),grid;
title('Target Reflectors'),ylabel('Amplitude');
xx=(1:339)/10;
subplot(313),plot(xx,c,'k',xx,c,'k+');
grid
title('Echo Pulse')
xlabel('Range(m)')
ylabel('Amplitude')
```

The results in Figure 2.44 show the relative sizes of the transmit pulse and the positions and sources of the four point echoes from a datum point 2 m in front of the nose. The results are shown in Figure 2.44b, in which it can be seen that the convolution process will model the pulse overlap where the reflectors are less than one pulse width apart.

This simple example does not consider that the transmitted pulse is just the envelope for the sinusoidal carrier, and so the convolution should include both amplitude and phase effects, neither does it pay regard to the effects of the round-trip distance, which doubles the effective spacing between the reflectors, as is explained in Chapter 6.

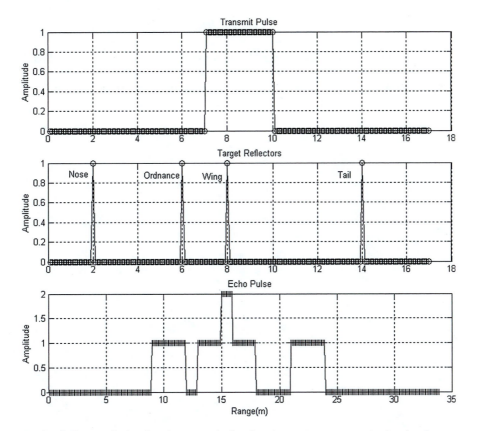

Figure 2.44 Simulation result showing the transmit signal and target structure and echo signal.

2.12 References

Carlson, A. (1975). *Communication Systems: An Introduction to Signals and Noise in Electrical Communication*, 2nd ed. New York: McGraw-Hill.

Carlson, G. (1998). *Signal and Linear System Analysis*, 2nd ed. New York: John Wiley & Sons.

Carr, J. (1997). *Electronic Circuit Guidebook*, Vol. 1, *Sensors*. San Diego, CA: Prompt Scientific.

Currie, N. and Brown, C. (1987). *Principles and Applications of Millimeter-Wave Radar*. Boston: Artech House.

Dorf, R. (2006). *The Electrical Engineering Handbook*. Boca Raton, FL: CRC Press.

Edminster, J. (1972). *Shaum's Outline Series: Theory and Problems of Electric Circuits*. New York: McGraw-Hill.

Electronics Workbench (2003). *Multisim 7 Simulation and Capture*. Toronto, Ontario, Canada: Interactive Image Technologies Ltd.

Haykin, S. and Van Veen, B. (2003). *Signals and Systems*, 2nd ed. New York: John Wiley & Sons.

Jacobowitz, H. (1965). *Electronics Made Simple*. New York: Doubleday.

Mahafza, B. (2000). *Radar Systems Analysis and Design Using MATLAB*. Boca Raton, FL: CRC.

Straw, R., ed. (1990). *The ARRL Handbook for Radio Amateurs*. Newington, CT: ARRL.

Taub, H. and Schilling, D. (1971). *Principles of Communication Systems*. Tokyo: McGraw-Hill Kogakusha, Ltd.

Walpole, R. and Myers, R. (1978). *Probability and Statistics for Engineers and Scientists*, 2nd ed. New York: Macmillan.

Williams, A. and Taylor, F. (1988). *Electronic Filter Design Handbook*. New York: McGraw-Hill.

Young, P. (1990). *Electronic Communication Techniques*, 3rd ed. Newark, NJ: Merrill.

3

Infrared Radiometers and Image Intensifiers

3.1 INTRODUCTION

Radiometers are instruments for detecting or measuring radiant energy, and although the term can be applied to sensors operating over any band in the electromagnetic spectrum, it is most often applied to devices used to measure infrared (IR) radiation. The radiation observed by such a sensor is either emitted by the object being observed or reflected from it. In this chapter, these relationships will be quantified in terms of measurable characteristics of the object.

The concept of blackbody radiation is introduced in terms of its temperature-dependent spectral characteristics as described by the Planck function (after Max Planck). The total radiated (or absorbed) power density by a surface can be determined by integration and, if the bandwidth is sufficiently wide, this is proportional to the temperature in Kelvin (K) of the body raised to the fourth power as described by Boltzmann's law (after Ludwig Boltzmann). This relationship is exploited by many radiometric sensors to measure temperature.

For example, a noncontact thermometer consists of a single radiation-sensitive element, called a bolometer, and some form of lens or antenna to constrain its field of view. The radiated energy from the scene heats the element, which results in a measurable change in resistance that can be used to determine the temperature.

Early thermal imaging systems scanned a single temperature-sensitive element over the focal plane in a raster fashion to build up an image, while more modern systems use a two-dimensional (2D) array of elements to produce thermal images similar to those of digital cameras. Because the energy of IR photons is much lower than their visible counterparts, each element only receives a miniscule amount of power, so these arrays must be made from particularly sensitive materials and are often also cryogenically cooled.

The general principles used by image intensifiers or "night vision" cameras are completely different from those used by thermal imagers. In such systems an image is projected onto a photocathode which releases electrons that are then accelerated by a strong electric field. When each high-energy electron strikes a phosphor at the output of the device, it generates many photons and so the image is intensified. These devices can be made sensitive to the IR by the selection of appropriate photocathodes.

3.2 THERMAL EMISSION

In any object, every atom and molecule vibrates. The average kinetic energy of the vibrat-ing particles is represented by the absolute temperature (K). According to the laws of electro-dynamics, a moving electric charge is associated with a variable electric field that in turn produces a changing magnetic field. In essence, this interaction produces an electromagnetic wave that radiates from the body at the speed of light.

3.2.1 Blackbody Radiation

Kirchoff's law (after Gustav Kirchoff) states that if radiation is incident on an object, the sum of the absorbed, reflected, and transmitted radiation must equal the total incident radiation. The fraction absorbed is a function of the wavelength, and a body that absorbs all the radiation at all wavelengths has an absorptivity $\alpha = 1$ and is defined as a black body. Kirchoff went on to determine that any body that is capable of absorbing all radiation is equally capable of the emission of a similar proportion and that its emissivity, ε, will therefore be equal to its absorp-tivity. A practical example of the principle is a box that is lightproof except for a small hole. Any radiation that enters the hole is scattered and absorbed by repeated reflections from the walls so that almost no energy escapes. If such a box is heated, it becomes what is known as a "cavity radiator," and it radiates energy with characteristics determined only by the temperature.

Classical theory states that as the temperature is increased, more modes of vibration should be possible and so any black body should radiate huge amounts of energy in the blue and ultra-violet (UV) spectral bands. This is contrary to measured data, and is known as "the ultraviolet catastrophe."

3.2.2 The Planck Function

Planck suggested that the emission should somehow be constrained so that the hot body could not continuously emit radiation, but should only emit in quanta of a definite magnitude, the size of which would increase with increasing frequency:

$$E = hf = hc/\lambda, \tag{3.1}$$

where

E = energy (J),
h = Planck's constant (6.625×10^{-34} Js),
f = frequency (Hz),
c = speed of light (m/s),
λ = wavelength (m).

Using this quantum theory, Planck went on to produce a formula for the spectral irradi-ance (power density per unit wavelength) that fitted the measured relationship between radiated wavelength and temperature even though he did not at first have any theoretical justification for it.

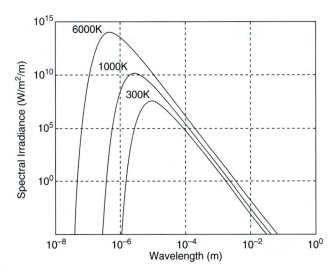

Figure 3.1 Blackbody spectra at different temperatures.

The spectral irradiance, $B_\lambda(T)$ is formally defined as the energy (J) emitted per second per unit wavelength from one square meter of a perfect black body at a temperature T (K) and is described by

$$B_\lambda(T) = \frac{2\pi hc^2/\lambda^5}{e^{hc/\lambda kT} - 1},$$ (3.2)

where

h = Planck's constant (6.625×10^{-34} Js),
k = Boltzmann's constant (1.3804×10^{-23} J/K),
λ = wavelength (m),
c = speed of light (3×10^8 m/s),
T = temperature (K).

This equation, shown graphically in Figure 3.1, generates a spectrum with a peak that increases in both magnitude and frequency as the temperature increases.

3.2.3 Properties of the Planck Function

Because the radiation process is statistical in nature and a large number of particles are involved, an extremely wide range of wavelengths is radiated. However, the broad spectral peak, λ_{max}, around which most power is radiated is fixed and can be determined by equating to zero the first derivative with respect to the wavelength of Planck's function. This relationship, illustrated in Figure 3.2, is known as Wien's displacement law (after Wilhelm Wien) and states in simple terms that high temperature sources emit most of their power at short wavelengths:

$$\lambda_{max} = 2898/T,$$ (3.3)

where λ_{max} is the wavelength (μm) at which most power is radiated and T is the temperature (K).

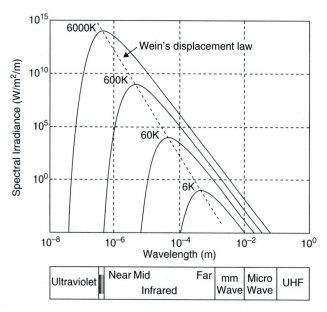

Figure 3.2 The peak of the blackbody spectrum is a function of the temperature, a relationship known as Wein's displacement law.

At long wavelengths, where the photon energy is much smaller than the thermal energy contribution ($hc/\lambda \ll kT$), the power density per unit wavelength becomes proportional to temperature. This relationship is known as the Rayleigh-Jean law:

$$B_\lambda(T) \approx 2\pi kcT/\lambda^4. \tag{3.4}$$

This formula is obtained by using a Taylor expansion on the denominator of the Planck function and setting the higher order terms to zero.

```
function pltlam = planck(temp, lambda)
%
% Calculate the Planck function for the given temperature and
wavelength.
%
% Arguments:
%
%         temp              (input)      temperature of blackbody (Kelvin)
%
%         lambda            (input)      wavelength at which to evaluate,
                                         can be a vector
%                                        (meters)
%
%
```

```
%            pltlam            (output)     value of Planck function
%                                                    (Joules per sec per
%                                                     square meter
%                                                     per unit wavelength)
  % error checks
  if (temp <= 0)
    error('temperature must be greater than zero');
  end
  if (min(lambda) <= 0)
    error('wavelength must be greater than zero');
  end

  % some physical constants:
  %
  c = 3.00e8;              % speed of light (metres/sec)
  h = 6.625e-34;           % Planck's constant (Joules-sec)
  k = 1.3804e-23;          % Boltzmann constant (Joules/Kelvin)

  % check to see if we can use the Rayleigh-Jeans approximation
  tiny = 0.001;
  arg = (h*c)./(lambda*k*temp);
  if (arg < tiny)
    % yes, can use the Rayleigh-Jeans approximation
    pltlam = (2*pi*k*temp*c)./(lambda.^4);

  else
    % not valid, must use the full formula
    num = 2*pi*h*c*c ./ (lambda.^5);
    den = exp((h*c)./(lambda.*k.*temp)) - 1;

    if (den == 0)
      pltlam = 0;
    else
      pltlam = num./den;
    end

  end
```

The total power density, Φ^0 (W/m^2), within a particular bandwidth is determined by integrating B_λ over that bandwidth. This is normally solved numerically or by approximation, as there is no closed form:

$$\Phi^0 = \int_{\lambda_1}^{\lambda_2} \frac{2\pi hc^2/\lambda^5}{e^{hc/\lambda kT}-1} d\lambda. \tag{3.5}$$

If the bandwidth includes much more than 50% of the total radiated power, then the Stefan-Boltzmann law can be used to approximate the value. This law states that the radiated power density is proportional to the temperature raised to the fourth power:

$$\Phi^0 \approx \sigma T^4, \tag{3.6}$$

where σ is the Stefan-Boltzmann constant (5.67×10^{-8} W/m²/K⁴).

In the long-wavelength case where the Rayleigh-Jean law is valid, the integral is easy to solve in closed form to give

$$\Phi^0 \approx \int_{\lambda_1}^{\lambda_2} \frac{2\pi kcT}{\lambda^4} d\lambda = \frac{2\pi kcT}{3} \left(\frac{1}{\lambda_1^3} - \frac{1}{\lambda_2^3} \right). \tag{3.7}$$

3.2.4 Confirmation of Stefan-Boltzmann and Rayleigh-Jean Laws

To confirm these relationships, the total radiated power density (W/m²) is determined by integrating numerically over the band from 10^{-8} to 10^{-1} m (10 nm to 10 cm) at a number of different temperatures. The same curves are also integrated over the microwave and millimeter wave band from 10^{-3} to 10^{-1} m (1 mm to 10 cm).

From the shapes of the two graphs shown in Figure 3.3, it can be seen that the total power density (in W/m²) is indeed a function of T^4 over the full band, and a function of T over the millimeter and microwave band.

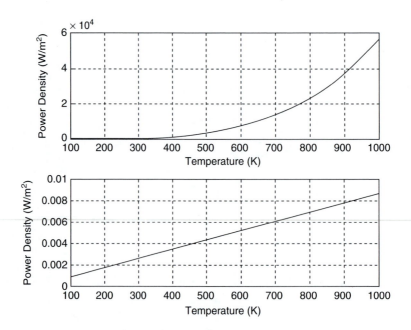

Figure 3.3 Total radiated power density with temperature (upper) over the full bandwidth and (lower) in the microwave region.

```
function binteg = int_planck(temp, lam1, lam2)
%
% Integrate the Planck function for the given temperature over a range
of wavelengths
% from lam1 to lam2 (meters) to obtain the total power density
%
% Arguments:
%
%        temp     (input)      temperature of blackbody (Kelvin)
%        lam1     (input)      start wavelength (meters)
%        lam2     (input)      end wavelength (meters)
%
%        binteg   (output)     value of integrated Planck function
%                                              (W/sqm)
%
% error checks
if temp<0
    output('temp < 0');
    return;
end;
if lam1<=0
    output('wavelength <= 0');
    return;
end;
if lam2<=0
    output('wavelength <= 0');
    return;
end;

% generate 10000 points in logspace between lam1 and lam2
  lam = logspace(log10(lam1),log10(lam2),10000);

  pltlam =[];
% call the planck function to determine the magnitudes of the emission
  pltlam=planck(temp,lam);

% integrate over the span
  dy=diff(lam);              % find the y-interval which is not constant
  n=length(dy);              % because it was defined using logspace
  pltlam=pltlam(1:n);
  da=dy.*pltlam;             % take the product of the magnitude with the
                             %   y-interval
  binteg = sum(da);          % sum the small areas to obtain the total
```

3.3 EMISSIVITY AND REFLECTIVITY

Different materials absorb radiation in different ways. Metals reflect most of the energy away, so they do not absorb much of the incident radiation, while a black material (like soot) absorbs most of the incident energy and gets warm. To understand how this occurs, it is easiest to consider the interaction of the incident electromagnetic radiation with the charges (electrons) in the material.

Metal has electrons that are free to move through the entire solid: that is why they can conduct electricity. These free electrons also oscillate together in response to the electric field of an incoming light wave. By oscillating, they radiate electromagnetic energy, just like the oscillating current in an antenna. The radiation from these oscillating electrons produces the "reflected" light. In this situation, little of the incoming radiant energy is absorbed, it is just reradiated.

Soot will conduct electricity, but not as well as a metal will. There are unattached electrons that can move about in the solid, but they keep bumping into atoms in the lattice (they have a short mean free path). When they bump, they cause a vibration and so give up energy as heat. These free electrons are effective intermediaries in transferring energy from the radiation into heat, with the result that very little is reradiated. Soot will therefore have a lower reflectivity than metal.

Emissivity (ε): A measure of the ability of a body to radiate heat, which is given by the ratio of the power radiated by the body per unit area to the power radiated per unit area of a black body at the same temperature. Most materials exhibit emissivities ranging from about 0.1 up to about 0.95, as can be seen in Table 3.1. The emissivity of a highly polished surface falls below 0.1, while an oxidized or painted surface will be much higher. Oil-based paint, regardless of the color in the visible spectrum, has an emissivity greater than 0.9. Human skin, because of its texture, exhibits an emissivity close to 1.

Reflectivity (ρ): The ratio of power reflected per unit area to the power incident per unit area. For an opaque object, all the incident radiation must either be reflected or absorbed, therefore $\rho + \alpha = 1$, and from Kirchoff, $\alpha = \varepsilon$. Therefore the relationship between the reflectivity and the emissivity is $\rho = 1 - \varepsilon$.

Table 3.1 IR emissivity for various materials

Material	Emissivity (ε)
Skin	0.98
Wet soil	0.95
Paint (average)	0.94
Heavy vegetation	0.93
Dry soil	0.92
Dry grass	0.91
Sand	0.90
Dry snow	0.88
Asphalt	0.83
Oxidized steel	0.79
Concrete	0.76
Polished steel	0.07

The Stefan-Boltzmann law is generally written to include the surface area of the object, A, and its emissivity, ε, to produce a value for the total radiated power (or flux),

$$\Phi = A\varepsilon\sigma T^4, \tag{3.8}$$

where
  Φ = total power radiated (W),
  A = surface area (m^2),
  ε = emissivity,
  σ = Stefan-Boltzmann constant (5.67×10^{-8} W/m^2/K^4),
  T = temperature (K).

This formula determines the power that would radiate from a body with a temperature T (K) towards an infinitely cold space (0 K). It is, for most practical purposes, irrelevant, as most objects radiate into environments where the temperature is much higher than this, and the net power radiated by the body is reduced.

To determine the equilibrium point, consider a sensor with a temperature T_s and an emissivity ε_s facing an object with a temperature T_o and an emissivity ε_o, as shown in Figure 3.4. The net power that flows into the sensor (which can be measured) is Φ (W).

The flux radiated from the target toward the sensor is Φ_{TO}, but some of this flux will be reflected, $\Phi_{TR} = -\Phi_{TO}(1 - \varepsilon_s)$, leaving $\Phi_T = \Phi_{TO} + \Phi_{TR} = \Phi_{TO} - \Phi_{TO}(1 - \varepsilon_s) = \Phi_{TO}\varepsilon_s$. Rewriting Φ_{TO} in terms of the emissivity, the temperature and area of the target become

$$\Phi_T = A_T\sigma\varepsilon_T T_T^4\varepsilon_s. \tag{3.9}$$

By similar reasoning, the net flux radiating from the sensor toward the target will be $\Phi_S = \Phi_{SO}\varepsilon_T$. Rewriting Φ_{SO} in terms of the emissivity, the temperature and area of the sensor become

$$\Phi_S = A_S\sigma\varepsilon_S T_S^4\varepsilon_T. \tag{3.10}$$

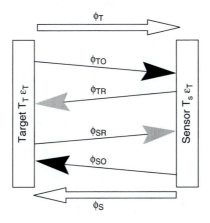

Figure 3.4 Flux balance between sensor and target.

Because the two fluxes, described by equation (3.9) and equation (3.10), propagate in opposite directions, they are combined into a net flux that flows between the two surfaces:

$$\Phi = \Phi_T + \Phi_S = A\varepsilon_T\varepsilon_S\sigma\left(T_T^4 - T_S^4\right). \tag{3.11}$$

3.3.1 Worked Example: Blackbody Radiation from Human Body

Calculate the following:

- The total power radiated by a human being radiating into cold space.
- The excess power radiated by a human being in a room at 20°C.
- The power radiated by a human being over a 2 GHz band at 100 GHz.

To determine the total power radiated by a human being over all wavelengths we can use the Stefan-Boltzmann approximation: $\Phi = A\varepsilon\sigma T^4$.

The body surface area (BSA) (m²) of a human can be determined empirically from their height and mass using Gehan's formula (Gehan and Georges 1970). This was derived using 401 subjects ranging in age from less than 5 years to more than 20 years old and determined using coating, surface integration, and triangulation:

$$BSA = 0.02350 h^{0.42246} m^{0.51456}, \tag{3.12}$$

where h is the height (cm) and m is the mass (kg) of the human subject.

Substituting into this formula using a height of 180 cm and a mass of 80 kg produces a BSA of 2 m². The emissivity is 0.98, as tabulated above, and the temperature of a healthy human being is 37°C (310 K). The total radiated power is then calculated using the Stefan-Boltzmann equation:

$$\Phi = 2 \times 0.98 \times 5.67 \times 10^{-8} \times 310^4 = 1026\,W.$$

This power loss could not be sustained and would result in the subject quickly suffering from hypothermia and then freezing if it were not for the compensating absorption of radiation from surrounding surfaces.

If the human subject is then placed in a room in which all the surfaces are painted and have an emissivity of 0.94 and are at a temperature of 20°C, then the net power radiated is calculated using equation (3.11):

$$\Phi = \Phi_T + \Phi_S = A\varepsilon_T\varepsilon_S\sigma\left(T_T^4 - T_S^4\right)$$
$$= 2 \times 0.98 \times 0.95 \times 5.67 \times 10^{-8} \times \left(310^4 - 293^4\right)$$
$$= 195\ W.$$

To determine the power radiated in the millimeter wave band, the integrated form of the Rayleigh-Jean approximation, equation (3.7), can be used:

$$\Phi \approx \frac{2\pi A\varepsilon p k c T}{3}\left(\frac{1}{\lambda_1^3} - \frac{1}{\lambda_2^3}\right).$$

For $\lambda_1 = 3.031 \times 0^{-3}$ m and $\lambda_2 = 2.97 \times 10^{-3}$, the total radiated power equates to 11.7×10^{-6} W. This result can be verified by using the MATLAB function listed earlier to perform the integration graphically and then scaling by the area and the emissivity of the body.

3.4 DETECTING THERMAL RADIATION

The long history behind the measurement of both IR and visible light has spawned a profusion of units to describe the origination and transfer of this form of electromagnetic energy, some of which are listed in Table 3.2. Radiometric quantities are based on the physical characteristics of the process, while photometric quantities include their effectiveness to stimulate the eye.

From a sensor perspective, there are three different basic interactions that can be exploited to detect thermal radiation:

- External photoeffect,
- Internal photoeffect, and
- Heating.

Most IR detectors operate using quantum mechanical interaction between incident photons and electrons and the detector material. Photoconductive detectors absorb photons to elevate an electron from the valence band to the conduction band of the material, changing the conductivity of the detector. For this to happen, a photon must be sufficiently energetic to excite an electron. Photovoltaic detectors absorb photons to create an electron-hole pair across a PN junction, which can produce a small current. Such devices can be manufactured as part of an array that includes a capacitor that stores a charge proportional to the incident radiation, in a manner similar to that of a charge-coupled device (CCD).

Quantum mechanical detectors with band-gap energies small enough to respond to the longer wave IR radiation (8–12 μm region) must be cooled to cryogenic temperatures between 77 K and 25 K to eliminate thermally generated carriers. To maintain these temperatures, the detector must be enclosed in a vacuum housing called a Dewar with a window transparent at the required IR wavelengths.

Microbolometers absorb thermal energy over all wavelengths and do not require cryogenic cooling, even when operating in the far-IR range. They are often used for low-cost commercial applications. Their main disadvantage is a very poor detectivity.

Table 3.2 Quantities used to describe radiation transfer (Zissis 1993)

Radiometric (physical)	Photometric (psychophysical)
Radiant flux, power Φ_e watt (W)	Luminous flux, φ_v lumens (lm)
Radiant intensity, I_e (W/sr)	Luminous intensity, I_v (lm/sr) or candela
Radiant excitance (emittance), M_e (W/m²)	Luminous excitance, M_v (lm/m²)
Radiance, L_e (W/m²/sr)	Luminance, L_v (lm/m²/sr)
Irradiance, E_e (W/m²)	Illuminance, E_v (lm/m²) or lux

Golay cells and capacitor microphones are pneumatic detectors. In Golay cells, the sealed xenon gas expands when it is warmed by incident IR radiation. The resultant variation of pressure shifts a mirror located between a light source and a photocell, varying the amount of light entering the photocell and thus changing the output of the photocell. In capacitor microphones, the varying expansion of the gas affects the capacitor film, which in turn produces variations in the electrostatic capacity. These are extremely sensitive and broadband devices.

3.4.1 External Photoeffect

If light with a photon energy hf (J) exceeding the work function W falls on a metal surface, some of the incident photons will transfer their energy to electrons, which will then be ejected from the metal. Since hf is greater than W, the excess energy, given by $hf - W$, transferred to the electrons will be observed as their kinetic energy after they have escaped. The relation between electron kinetic energy, E, and the frequency (i.e., $E = hf - W$) is known as the Einstein relation (after Albert Einstein), and its experimental verification helped to establish the validity of quantum theory. The energy of the electrons depends on the frequency of the light, while the intensity of the light determines the rate of photoelectric emission. The amount of energy that must be supplied to free electrons to enable them to escape from the metal is known as the work function; some of these are listed in Table 3.3.

3.4.2 Internal Photoeffect

The photon has sufficient energy to create a free electron, free hole, or both in the material (usually a semiconductor).

3.4.2.1 Photoconductive Detectors

- Monitor voltage change across R.
- Monitor current change through sample.

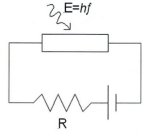

In a semiconductor the valence band of energy levels is almost completely full while the conduction band is almost empty. The conductivity of the material derives from the few holes present in the valence band and the few electrons in the conduction band. Electrons can be excited from the valence to the conduction band by

Table 3.3 Work functions of some common metals

Metal	Work function (eV)
Barium	2.5
Cesium	1.9
Copper	4.5
Potassium	2.2
Silver	4.6
Sodium	2.3
Tungsten	4.5

$1\ eV = 1.61 \times 10^{-19}\ J$

light photons with an energy h that is larger than the energy gap E_g between the bands. The process is an internal photoelectric effect. The value of E_g varies from semiconductor to semiconductor.

The cutoff wavelength, λ_c (m), beyond which no emission will occur is

$$\lambda_c = hc/E_g. \tag{3.13}$$

Values of the long-wave cutoff are listed in Table 3.4, along with their energy gap and the maximum temperature at which the process can operate effectively before internal thermal effects swamp it.

Visible radiation produces electron transitions with almost unity quantum efficiency ($\eta \approx 1$) in silicon. Each transition yields a hole-electron pair (i.e., two carriers) that contributes to electrical conductivity. For example, if 1 mW of light strikes a sample of pure silicon in the form of a thin plate 1 cm^2 in area and 0.03 cm thick (which is thick enough to absorb all incident light), the resistance of the plate will be decreased by a factor of about 1000. In practice, photoconductive effects are not usually as large as this, but this example indicates that appreciable changes in conductivity can occur even with low illumination.

3.4.2.2 Photovoltaic Detectors
- Sensors measure open circuit voltage or short circuit current.
- The sensitive region of the detector is at the junction.
- Homojunctions use PN semiconductors.
- Heterojunctions use different materials with similar lattice spacing.
- Metal-semiconductor interfaces include Schottky barrier detectors and gallium arsenide (GaAs) photodiodes.

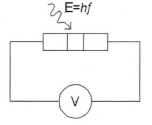

The photovoltaic effect consists of the generation of an electromotive force as a consequence of the absorption of radiation; that is, a current will flow across the junction of two dissimilar materials when light falls upon it. The primary effect is photo-ionization (i.e., the production of equal numbers of positive and negative charges). One or both charges can then migrate to a region in

Table 3.4 Band-gap and long-wave cutoff of some materials

Material	Band-gap energy, E_g (eV)	Long-wave cutoff, λ_c (μm)	Operating temperature (K)
Si	1.09	1.1	300
Ge	0.81	1.4	300
PbS	0.49	2.5	77
InSb	0.22	5.5	77
HgCdTe	0.25	22	77
Ge:Hg	0.087	14	
Si:Ge	0.065	17	

1 eV = 1.6×10^{-19} J

which charge separation can occur. This charge separation happens normally at a potential barrier between two layers of the solid material.

3.5 HEATING

Photon detectors measure the rate at which quanta are absorbed. They are sensitive to the frequency of the photons, as can be seen in the figure showing D^*. Thermal detectors usually measure the rate at which energy is absorbed by its heating effect and are insensitive to frequency over a wide range. However, they are much less efficient than photon detectors.

These sensors can operate using a number of different principles and come in three main types:

- Bolometers that change their resistance with temperature,
- Pyroelectric sensors that change their electrical polarization with temperature, and
- The Seebeck effect that describes the generation of a temperature-dependent potential difference across the junction of two dissimilar metals.

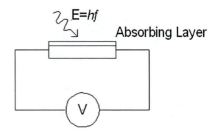

3.5.1 Bolometers

Bolometers are thermal detectors made from strips of blackened material in which the resistance has a sufficiently large temperature coefficient for the temperature change due to absorbed radiation to result in a measurable change in that resistance. These resistors can be made of metal or a semiconductor. Metal types are generally used for temperature measurement, while thermistor types are preferred for IR measurements because of their higher responsivity.

The essential components of a bolometer are the sensing element and a package fitted with an IR window, as illustrated in Figure 3.5. The sensing element is typically very thin to minimize heat capacity and to maximize sensitivity. In some cases it is mounted on a heat sink to provide high-speed response, but in others it is a miniature thermistor bead suspended from tiny wires. Most commercial detectors consist of two identical devices that are connected into a bridge circuit in which one element is irradiated and the other shielded. This configuration can compensate for drift due to changes in the ambient temperature and is therefore more accurate than the single element variety (Meijer and van Herwaarden 1994). Uncooled bolometers typically have specific

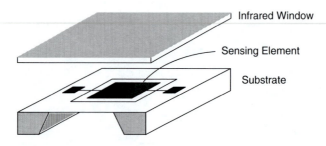

Figure 3.5 Construction of a modern bolometer.

detectivities, D^*, between 1.1×10^8 and 1.3×10^8 cmHz$^{1/2}$/W, while cryogenically cooled types are more sensitive than this.

3.5.2 Pyroelectric Sensors

Pyroelectric materials are crystalline substances that have an internal electrical dipole moment and change their electrical polarization with temperature. In highly insulating materials, even slow changes in the sample's temperature which produce changes in the internal dipole moment generate changes in the surface charge. A small capacitor manufactured by depositing a pair of electrodes onto opposite faces of a sliver of pyroelectric material creates a sensor in which the changing charge on the capacitor can be detected with changes in the intensity of the incident thermal radiation. Because these sensors only detect changes in temperature, they must be used with chopped radiation. This can be introduced using a mechanical chopper, or by using a mask and the motion of the hot object to perform this function.

The relationship between a temperature change ΔT and the varying charge ΔQ or the varying voltage ΔV across the device is described by

$$\Delta Q = P_Q A \Delta T, \tag{3.14a}$$

$$\Delta V = P_V b \Delta T. \tag{3.14b}$$

For a sensor element with a capacitance C_e which is defined as

$$C_e = \frac{\Delta Q}{\Delta V} = \varepsilon_r \varepsilon_0 \frac{A}{b}, \tag{3.15}$$

the relationship between the equation (3.14a) and equation (3.14b) is

$$\Delta V = P_Q \frac{A}{C_e} \Delta T = P_Q \frac{\varepsilon_r \varepsilon_0}{b} \Delta T, \tag{3.16}$$

where
  P_Q and P_V = pyroelectric coefficients of the material,
    A = area over which the energy is absorbed (m^2),
    b = thickness of the material (m),
    ε_r = relative dielectric constant of the material,
    ε_0 = permittivity of free space (8.85×10^{-12} F/m).

The pyroelectric coefficient is determined by the material from which the sensor is made. It describes either the charge (coulombs) or the voltage that is generated by a temperature change of 1 K (Meijer et al. 1994; Fraden 2003).

A simple way of looking at the phenomenon is to consider the generation of voltage as a secondary effect of thermal expansion. Since all pyroelectric materials are also piezoelectric, heating the lattice causes the front side of the sensing element to expand, which in turn induces a piezoelectric charge across the sensor electrodes.

Three of the more common materials used in these detectors are strontium barium niobate (SBN), triglycine sulfate (TGS), and the polymer film polyvinylidene fluoride (PVDF). TGS is the most sensitive, but it has a low Curie temperature (49°C) at which the dipole moment disappears, so other less sensitive materials are often used in its stead. Because PVDF film is very low cost it is often used in household burglar alarms.

3.5.3 Thermopiles

A thermocouple consists of a combination of two different materials bonded together which will generate a potential difference, ΔV, proportional to the temperature difference, ΔT, between the hot and cold terminals. This characteristic is known as the Seebeck effect (after Thomas Seebeck) and can be mathematically expressed as

$$\Delta V = \alpha_s \Delta T, \tag{3.17}$$

where α_s is the Seebeck coefficient, and its value depends on the difference between the thermoelectric coefficients of the two materials used.

A modern thermopile sensor is constructed in the form of a "pill" in which a thin membrane supports the hot junctions of the circuit, while the cold junctions are attached to the cold annular base, as illustrated in Figure 3.6. Typically vacuum deposition techniques are used to apply up to a few hundred junctions in a star form.

Materials should have a high thermoelectric coefficient, low thermal conductivity, and low resistivity. Unfortunately, materials like gold, silver, and copper that have low resistivity also have poor thermoelectric coefficients, while those with high thermoelectric coefficients, like bismuth (Bi) and antimony (Sb), have high resistivity, as can be seen from Table 3.5. For metal thermopiles, the latter are used, and with careful doping Bi-Sb junctions can achieve thermoelectric coefficients of up to 230 μV/K. Efficient thermopile sensors can also be made using semiconductors rather than metal junctions, as the thermoelectric coefficient of crystalline silicon is very

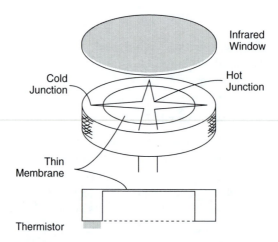

Figure 3.6 Construction of a modern thermopile sensor.

high and is often used to form extremely sensitive sensors (Fraden 2003). Table 3.6 gives some of the specifications for a TI-110 Bi-Sb thermopile.

3.6 PERFORMANCE CRITERIA FOR DETECTORS

The key detector performance parameters are responsivity (R), signal-to-noise characteristics or noise equivalent power (NEP), and specific detectivity (D^*).

3.6.1 Responsivity

The responsivity of a sensor can be defined as the ratio of RMS signal current (or voltage) to the RMS input power. For an IR detector, if the probability that a photon with energy $E = hf$ will produce an electron is η (quantum efficiency), then the average rate of production of electrons $\langle r \rangle$ for an incident beam of optical power Φ is

$$\langle r \rangle = \eta \Phi / hf . \tag{3.18}$$

Table 3.5 Thermoelectric coefficients and volume resistivities of selected elements

Element	Thermoelectric coefficient, α (μV/K)	Resistivity, ρ ($\mu\Omega$ m)
p-Si	100–1000	10–500
Antimony	32	18.5
Copper	0	0.0172
Silver	−0.2	0.016
Aluminum	−3.2	0.028
Bismuth	−72.8	1.1
n-Si	−100 to −1000	10–500

Table 3.6 Specifications of two commercial bismuth-antimony thermopiles

Parameter	Model 1.0DR	Model 2.5DR
Element size (mm)	1.0	2.5
Number of junctions	72	72
Resistance (kΩ)	25	20
Noise voltage (nV/Hz$^{1/2}$)	21	8
Responsivity (V/W)	100	20
NEP (nW)	0.2	0.9
D^* (cmHz$^{1/2}$/W)	3.5×10^8	3.5×10^8
Time constant (ms)	35–55	55–80
Field of view (deg)	80	80

The signal current, i_{sig} (A), that is generated will be the product of the rate of production and the electron charge, e,

$$i_{sig} = \langle r \rangle e = \eta e \Phi / hf. \tag{3.19}$$

It can be seen from the equation that the output current is proportional to the input power, making this a "square-law" detector. This characteristic is exploited when the terminals of the photodiode drive a short circuit, as occurs in a current-to-voltage converter. The diode is said to be operating in "current mode" and the relationship between the incident optical power, Φ, and the current flow, i_{sig}, tracks along the current axis ($V = 0$) in the V-I curves shown in Figure 3.7. The term i_o refers to the "dark current," which is a current that flows in the absence of light and is attributed to thermal generation of hole-electron pairs.

The responsivity, R (A/W), is the ratio of the photocurrent (into a short circuit) generated for each watt of incident power. This can be derived from equation (3.19) by taking the ratio of the output current to the incident power:

$$R = \frac{i_{sig}}{\Phi} = \frac{\eta e}{hf} = \frac{\eta e \lambda}{hc}. \tag{3.20}$$

In reality, responsivity has an added dependence because the quantum efficiency is also a function of wavelength, as shown in Figure 3.8. At some frequencies, where the material does not absorb strongly, photons may pass through the material or penetrate too deeply to produce photoionization close enough to the junction to be detected. In a silicon photodiode, the responsivity increases with increasing wavelength to reach a peak at just over 900 nm before falling off sharply.

To compare different sensors more easily, the specific responsivity, R^0 (A/W/m^2) can be quoted, in which the performance is normalized with respect to the area of the detector:

$$R^0 = i_{sig} / \Phi(\lambda) A_{det}, \tag{3.21}$$

where i_{sig} is the signal current (A), $\Phi(\lambda)$ is the incident power (W), and A_{det} is the detector area (m^2).

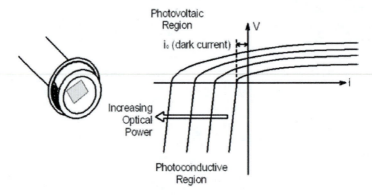

Figure 3.7 Packaged photodiode and its V-I characteristics.

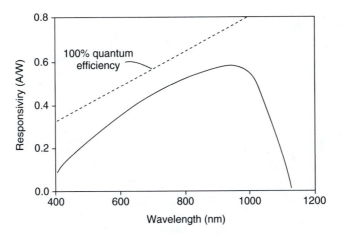

Figure 3.8 Responsivity of a typical silicon photodiode as a function of wavelength.

3.6.2 Noise Equivalent Power

The NEP (W) is defined as the optical power input to the detector that will create a signal-to-noise ratio (SNR) of 1. In this case, the noise will be a combination of thermal noise and shot noise generated by the random fluctuations in photocurrent as carriers are created and recombined. Dividing the RMS noise current, i_{noise}, of the detector by the responsivity gives the NEP:

$$NEP = i_{noise}/R. \qquad (3.22)$$

3.6.3 Detectivity and Specific Detectivity

The detectivity, D (per W) is the reciprocal of the NEP:

$$D = 1/NEP. \qquad (3.23)$$

Because different detectors have different areas and signal bandwidths, β, a term D^* (dee star) is often used. It refers to the detectivity over a bandwidth of 1 Hz with a detector area of 1 cm^2 (cmHz$^{1/2}$/W). A high detectivity value indicates a low-noise photodiode or detection system:

$$D^* = \frac{\sqrt{A_{det}\beta}}{NEP}. \qquad (3.24)$$

Figure 3.9 shows the typical spectral responses for a number of detectors along with the recommended operating temperatures. The operational principles of these cryogenically cooled detectors is no different from other thermal sensors except that they can operate out to much longer wavelengths, and their sensitivities are generally much higher. Table 3.7 lists the characteristics of a mercury cadmium telluride (HgCdTe) IR detector as a function of its temperature.

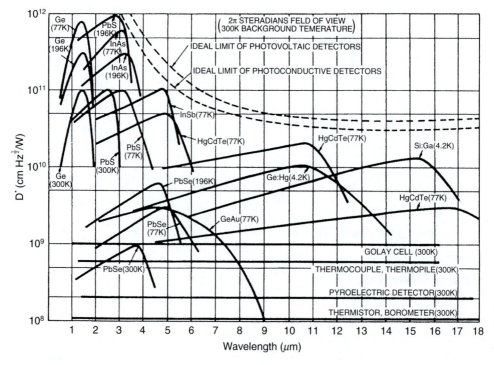

Figure 3.9 Operating ranges for some IR detectors (Fraden 2003).

Table 3.7 Specifications of HgCdTe far-IR detectors with an area of 1 mm × 1 mm

Temperature	Spectral peak (λ_p)	Cutoff (λ_c)	Dark resistance	Rise time	Maximum current	D^* @ λ_p
243 K	3.6 μm	3.7 μm	1 kΩ	10 μs	3 mA	10^9
77 K	15 μm	16 μm	20 kΩ	1 μs	30 mA	3×10^9

3.7 NOISE PROCESSES AND EFFECTS

There are three common sources of noise that affect IR systems:

- Johnson noise is generated by the random thermal motion of current carriers (electrons and holes). It is broadband, and therefore the RMS noise current, $i_{johnson}$, is expressed as noise per unit bandwidth (A/Hz$^{1/2}$/W):

$$i_{johnson} = \sqrt{\frac{4kT}{R_{shunt}}}, \tag{3.25}$$

where k is Boltzmann's constant, T is the temperature (K), and R_{shunt} is the shunt resistance of the detector.

- Shot noise is generated from fluctuations in the rates of thermal generation and recombination of carriers. The RMS current, i_{shot} (A/Hz$^{1/2}$), is

$$i_{shot} = \sqrt{2e\left(i_{photo} + i_{dark}\right)}, \tag{3.26}$$

where i_{photo} and i_{dark} are the mean photo- and dark currents, respectively, and e is the charge on an electron. Note that the shot noise increases with increasing photocurrent.

- Surface and contact effects "$1/f$ noise." This is not well understood and is empirically determined for individual detector families. It is only dominant at frequencies below 100 Hz.

Thermal detectors also exhibit temperature noise due to random temperature changes that arise from fluctuations in the rate of heat transfer from the detector to the surroundings. They also suffer from spatial noise called "fixed pattern noise," which arises from element-to-element differences.

Avalanche noise occurs in reverse-biased PN junctions at breakdown, and its magnitude is considerably larger than that of shot noise. Devices that operate using avalanche breakdown include zener diodes and avalanche photodiodes (APDs). Reverse biasing the base-collector junction of a transistor into avalanche breakdown is a common method of generating white noise.

There is a fundamental limit to a detector's performance due to "radiation" noise caused by statistical fluctuations in the incident background radiation. At room temperature, a signal out of a quantum detector would be swamped by thermally generated carriers. This can only be overcome by mounting the detector on the cold finger of a Dewar assembly; 8–12 μm systems require cooling to 77 K and 3–5 μm systems require cooling to 193 K.

3.8 APPLICATIONS

3.8.1 Passive UV Sensor (External Photoeffect)

These sensors are used to detect missile plumes or gas flames because both produce spectra that extend into the UV spectrum below 200 nm. After passing through the atmosphere, sunlight loses a large portion of its UV spectrum below 250 nm, so this is a good region in which to operate.

The UVtron detector is a transparent gas-filled tube with a high voltage applied across its electrodes. Upon being exposed to UV light, the high-energy photons strike the cathode and release free electrons by the external photoeffect. The electrons gain kinetic energy as they are accelerated toward the anode by the potential difference between the two plates.

If an electron strikes a gas molecule with sufficient energy, the molecule will be ionized and release more UV radiation. This results in an avalanche multiplication effect and the tube becomes conductive. A relaxation oscillator circuit is used to produce a pulse train in the presence of UV radiation. Because of the good sensitivity of the photocathode and the high gain of the avalanche effect, these detectors are more sensitive than smoke detectors for outdoor applications (Fraden 2003).

The spectral response of the cathode material shown in Figure 3.10 is selected because it is sensitive to UV light in the region below 250 nm to be away from the residual UV present in sunlight. The lowest suitable work function (J) for the cathode can be calculated from the photon energy:

$$E = hf = hc/\lambda. \tag{3.27}$$

A photon of light with a wavelength of 250 nm will have an energy of

$$E = 6.625 \times 10^{-34} \times 3 \times 10^{8}/250 \times 10^{-9} = 7.95 \times 10^{-19} \text{J} (4.97 \text{ e}V)$$

It can be seen that none of the materials listed in Table 3.4 would be suitable, and that silver ($E = 4.6$ eV) be the most appropriate, with a sensitivity extending up to a wavelength of 270 nm.

3.8.2 Radiation Thermometer (Internal Photoeffect: Thermopile)

A noncontact temperature sensor uses a sensing element strongly responsive to IR radiation, such as a thermopile. A good example is the Melexis MLX90601 module, shown in Figure 3.11, which has been developed to simplify the process. It consists of an MLX90247 thermopile sensor as an IR sensing element and a close-coupled thermistor to measure the cold junction temperature, with an MLX90313 processor to calculate the target temperature.

A wide-bandwidth silicon window, transparent to thermal IR energy with $\lambda > 5.5$ μm ensures that the Stefan-Boltzmann law is valid. Therefore the net flux (power) that flows between the sensor and the target is given by $\varphi = A\varepsilon_T\varepsilon_S\sigma(T_T^4 - T_S^4)$. Because the temperature increase in the hot junction is directly proportional to this incident power, the output voltage is

$$V_t = \alpha\left(T_T^4 - T_S^4\right), \tag{3.28}$$

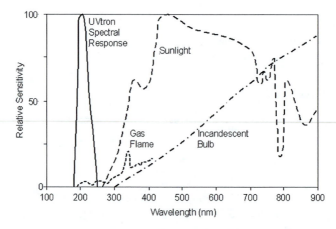

Figure 3.10 The UVtron spectral response compared to solar, incandescent, and gas flame spectra showing that the sensor operates in the solar blind region [adapted from (Fraden 2003)].

where T_T is the target temperature (K), T_S is the sensor, or cold junction, temperature (K), and α is a device constant.

In reality, the power radiated by the target is the sum of the power emitted and the power reflected, so these thermometers are only accurate when the target has a high emissivity ($\varepsilon > 0.9$) in the thermal IR band. It is interesting to note that because window glass is opaque to IR, its temperature can be measured reasonably accurately, whereas a polished metal surface has a low emissivity and so the measured temperature will be incorrect. If the temperature of a metal surface must be measured, then a small portion of the surface should be painted matt black (Melexis 2007).

3.8.3 Passive IR Sensor (Internal Photoeffect: Pyroelectric)

Passive infrared (PIR) sensors are responsive to far-IR radiation within the spectral range from 4 to 20 μm, where most of the thermal power radiated by human beings is concentrated. An interesting pyroelectric material that is commonly used is the polymer film PVDF, which was discussed earlier in this chapter. Though it is not as sensitive as most solid-state crystals, it has the advantages of being both flexible and low cost.

A typical intruder detector, as shown in Figure 3.12, consists of a Fresnel lens that focuses the IR image onto a PVDF film with multiple interdigital electrodes. As an object moves across the sensor's field of view, as illustrated in Figure 3.13, the image traces across the film, moving from one pair of electrodes to the next and generating an alternating current with an amplitude of about 1 pA. A field-effect transistor (FET) follower with an input impedance of about 50 GΩ converts this to a 50 mV signal that can be amplified and detected.

The Fresnel lens and the PVDF film are often curved with equal radii to ensure that the surface of the film is always in focus. As the lens has a larger aperture, it captures more IR radiation than the sensor alone and also focuses it to a small point. A Fresnel lens can extend the detection range of the sensor to about 30 m.

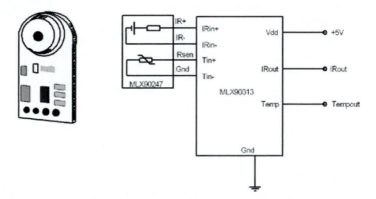

Figure 3.11 Melexis ML9061 noncontact thermometer module made from a single thermopile sensor with a close-coupled cold junction temperature sensor and a linearization module.

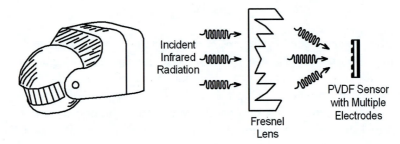

Figure 3.12 Drawing of a generic PIR sensor and a schematic showing how the radiation is focused onto the detector.

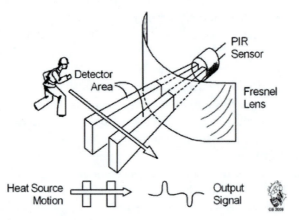

Figure 3.13 Operational principle of a PIR sensor.

3.8.4 Crookes' Radiometer

William Crookes invented the radiometer in 1875 while investigating the behavior of hot bodies in a vacuum. His hope that the radiometer would be used to measure heat radiation was never fulfilled, and the way it works was not explained for many years, but it remains a fascinating scientific toy.

A radiometer consists of a glass bulb from which most of the air has been removed. The rotor, mounted on a vertical support within the bulb, bears four light horizontal arms to which are attached vertical metal vanes. Each vane has one side polished and the other side blackened.

When radiant energy (at any frequency) strikes the polished surfaces, with high reflectivity, most of it is reflected away, but when it strikes the blackened surface, with low reflectivity, most of it is absorbed, raising the temperature of those surfaces. The air near a blackened surface is heated slightly, which increases the mean velocity of the gas molecules, which in turn exerts more pressure on the blackened surface, causing the rotor to turn.

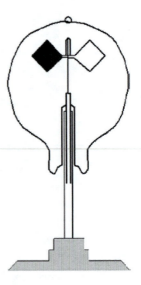

3.9 INTRODUCTION TO THERMAL IMAGING SYSTEMS

Electro-optical thermal imaging systems (TISs) of various types and applications have been developed. Forward-looking infrared (FLIR) systems are typically high-resolution devices with a small instantaneous field of view. IR search and tracking (IRST) systems locate and track targets using low target-to-background contrasts. They are typically enhanced with sensors that incorporate a large instantaneous field of view and an ability to scan over a 360° hemisphere in azimuth and 50° to 90° in elevation. A third class of systems is the IR spectrometer, which breaks the spectrum into many narrow 10 to 20 nm bands that are individually processed to aid the identification of geologic, ocean, soil, vegetation, and atmospheric features.

All of these systems use the temperature gradient of the object to produce TV-like images at night as well as during the day. They are often called night vision devices, but should not be confused with image intensifiers that use the same name. Their usefulness is due mainly to the following characteristics (Hovanessian 1988; Klein 1997):

- It is a completely passive technique that requires no external source of illumination. This allows for day and night operation and is covert.
- It is an ideal method to detect hot or cold spots, or areas of different emissivity within a scene.
- The longer wavelength of thermal radiation allows for the penetration of mist, smoke, or dust better than visible radiation and therefore allows for the imaging of visually obscured objects.

Every object not at absolute zero (0 K) emits electromagnetic radiation with a wavelength that is a function of the temperature of the object. The short wavelengths between 0.45 and 1 μm (the human eye sees between 0.46 and 0.65 μm) used by visible cameras are emitted by very hot sources such as the sun or incandescent light bulbs, whereas the longer IR wavelengths between 3 and 14 μm are emitted by cooler objects at around 300 K (27°C). This relationship, calculated using Planck's function, is shown graphically in Figure 3.14.

3.9.1 Scattering and Absorption

Radiation is either scattered or absorbed as it propagates through the atmosphere. Scattering is a phenomenon that causes a change in the direction of the radiation, and is caused by the absorption and reradiation of the energy by suspended particles. For particles substantially smaller than the wavelength of the radiation, scattering is proportional to λ^{-4}, while for larger particles, the scattering is independent of the wavelength.

Particles less than about 2 μm in diameter hardly scatter IR radiation at all, and therefore the radiation can penetrate smoke and light mist far better than visible radiation, as illustrated in Figure 3.29 at the end of this chapter. However, larger particles such as rain, fog, and dust scatter visible and IR radiation by about the same amount.

In the IR band, absorption by gas molecules is usually the dominant attenuation process. For example, Figure 3.15 shows that there are three good IR transmission windows at 1–3 μm,

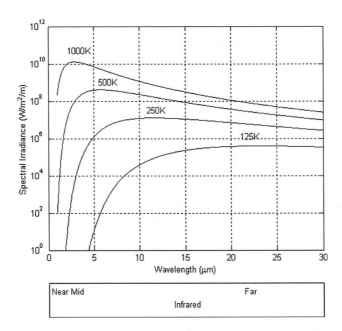

Figure 3.14 Planck's function in the IR.

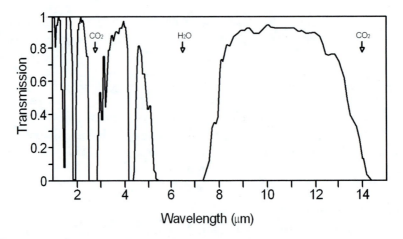

Figure 3.15 Transmission of IR radiation through the atmosphere.

3–5 μm, and 8–12 μm, with major water and carbon dioxide absorption bands between them. These windows dictate the choice of wavelengths used in IR sensor design. The absorption bands also show why carbon dioxide and water vapor are such potent greenhouse gases.

Much of the IR band is of interest for scientific, industrial, and military applications. Thematic imagers typically operate in the visible and near-IR band below 3 μm, while the longer wavelengths are used for industrial and medical temperature measurement. The bands used by missile seekers are influenced by target radiation characteristics, atmospheric attenuation, and detector spectral response. Five bands are usually used, spanning a range from 1.5 to 12.5 μm. Modern seekers generally use a combination of two bands to improve target detection probability and to minimize the effects of countermeasures (flares). To improve performance still further, one of the channels is often in the millimeter-wave band.

3.9.2 Scanning Mechanisms and Arrays

Early thermal imagers called serial scan sensors used either a single or a small array of detector elements preceded by horizontal and vertical mirrors or prism-based scanners to produce an image. In the configuration shown in Figure 3.16, two scanning mirrors are used to build up a rasterized image of the scene, pixel by pixel. The primary advantage of this technique is that it allows the small detector area to be cooled easily and effectively. Because the detector requires a finite time to detect the IR radiation and to stabilize, a trade-off had to be made between the frame update rate and the image quality.

In more modern parallel scan sensors, a vertical column of between 112 and 180 detectors (first-generation sensors) or 240 to 960 detectors (second-generation sensors) is scanned horizontally using a single mirror. In these scanners, the column is often displaced by one pixel on alternate scans to produce an interlaced display. As the number of individual detectors in the array increases, the frame update rate can be improved or the integration time can be increased, with a resultant improvement in image quality.

In series-parallel scan mechanisms, a linear detector array scans the image in the direction of the row of detectors, as with the serial scanner. In this way all of the detectors scan over the same image point, although at slightly different times. Synchronized integration ensures that the signals from each point are added coherently, with the result that the signal level increases linearly with the number of elements. In contrast, the noise power, which is added incoherently, increases with the square root of the number of detectors. This improves the overall SNR and hence the image quality. What is more, the outputs from the various detector elements need not be well matched, and in fact, even if a few elements are dead, the sensor will still perform adequately (Sparrius 1981).

More recently, unscanned 2D staring arrays (focal plane arrays), similar to a TV's CCD that sample the scene image at the focal plane of the sensor optics, have become available. These

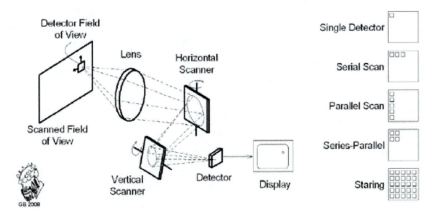

Figure 3.16 Scanning mechanism for an electro-optical sensor.

arrays are constructed of between 480 and 300,000 (often 320×240 pixels) detectors, along with amplifiers, multiplexers, and other supporting electronics on a single chip called a sensor chip assembly (SCA). These staring arrays can use either high-performance cooled detectors, which are generally photovoltaic in nature and based on mercury cadmium telluride (HgCdTe) or platinum silicide (PtSi), or they can use an uncooled microbolometer. Systems based on vanadium oxide (VOx) thermistors offer reduced sensitivity at much lower cost (Boeing 2000).

3.9.3 Microbolometer Arrays

The individual sensor elements in a microbolometer array use the change in electrical resistance of a VOx thermistor deposited onto the tiny "platelets" which are fabricated by silicon micromachining in a silicon foundry. Incoming target radiation heats the VOx, causing a change in electrical resistance which is read by measuring the resulting change in bias current. More than 80,000 sensors can be fabricated together into a 2D array. The structure can be dimensioned to operate at 30 Hz. That is, the thermal conductance of the isolating legs can be adjusted to match the time constant for 30 Hz operation.

Each detector element consists of a two-layer structure. Interconnecting readout circuitry is applied to the silicon process wafer and then the microbolometer structure is built on top of the readout circuitry. First, a pattern of islands is deposited on the readout circuitry. The islands are made of a material that can be selectively etched away later to form a bridge structure. Three layers—silicon nitride, VOx, and silicon nitride—are deposited over the sacrificial islands. The sacrificial islands are then etched away, leaving the thermally isolated bridge structure of VOx, as shown in Figure 3.17 (Infrared Solutions 2007).

Most of today's IR camera manufacturers use the 320×240 microbolometer array. However, an alternative for cost-sensitive commercial applications is the smaller 160×120 array, which can be produced at a much lower cost because far more arrays can be produced on a single wafer and the yield is higher. In addition, one of the most expensive components of an IR camera is the lens, and its cost is proportional to the array size.

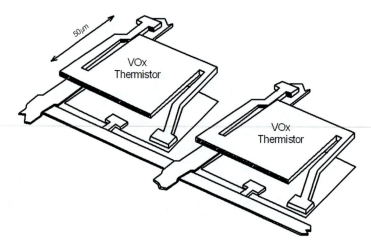

Figure 3.17 A drawing of two elements of a microbolometer array.

The only advantage of the larger array is a larger field of view (FOV). With the same f number and focal length lens and the same detector size, a camera with 320×240 or 160×120 will have identical spatial resolution. The difference is that the target size for a fixed distance between the camera and target will be twice as large in both dimensions for the camera with the larger array.

3.9.4 Key Optical Parameters

The following five parameters are key to describing the optical performance of an imaging system (Crawford 1998).

The Aperture Diameter For a circular aperture, such as the objective lens of an IR camera, a point source will appear as a disk surrounded by a number of faint rings. This pattern is often referred to as a point-spread function, in which the central disk, called the Airy disk, has an angular radius, Φ_{lim} (rad), which is determined by the aperture diameter, D_o (m):

$$\Phi_{lim} = 1.22\lambda/D_o . \tag{3.29}$$

According to the Rayleigh criterion for resolution, two point sources cannot be resolved if their separation is less than the radius of the Airy disk, as illustrated in Figure 3.18. Therefore, equation (3.29) can also be used to describe this angular resolution.

f number The f number is the ratio of the focal length and the limiting aperture. It determines the light-gathering capability, or how much energy is focused on the detector. The scene energy reaching the focal plane is proportional to $(1/f)^2$.

Focal length The focal length, f_1, is the distance from the optical center (or pole) to the principal focus of a lens or curved mirror. From the definition of the f number, $f_1 = fD_o$.

Transmission Ordinary glass does not transmit much beyond 2.5 μm in the near-IR range, so lenses for thermal imaging applications must be made from less common materials such as those listed in Table 3.8. The semiconductors germanium and silicon are commonly used to make lenses and other optical components. Silicon is typically used in the 3–5 μm region and germanium in the 8–13 μm region. Being semiconductors, these materials generate charge carriers and become opaque at high temperatures. Two chalcogenides have also been developed as lenses. These are zinc selenide and zinc sulfide. The latter is brittle and does not transmit well beyond

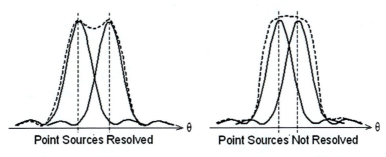

Figure 3.18 Rayleigh criterion for resolution illustrated using the point-spread function for two aperture-limited point sources.

Table 3.8 Lens material characteristics for IR operation

Material	Transmission band (μm)	Refractive index
Germanium	2–20	4
Silicon	1.5–6	3.4
Sapphire	0.4–5	1.63
Zinc selenide	0.5–20	2.4
Zinc sulfide	0.6–15	2.2

15 μm, but it can operate at elevated temperatures. In contrast, zinc selenide is both low loss and wide band, but it is much softer and more expensive, so it is only used in high-performance applications. A number of refractory materials have been developed for high-temperature lenses and radomes in military applications, and, apart from sapphire, are not used in the more cost-sensitive commercial applications (Burnay et al. 1988).

Light is lost in an optical system through absorption and reflection. As it passes from one refractive medium to another, some portion of the energy is reflected at the medium boundary and some absorbed within the medium. At normal incidence, the fraction of light reflected, ρ, is determined by the refractive index, n, of the lens material:

$$\rho = \frac{(1-n)^2}{(1+n)^2}.$$

(3.30)

The fraction transmitted into the lens, $\tau = 1 - \rho$, is

$$\tau = \frac{4n}{(1+n)^2}.$$

(3.31)

A small amount of light is absorbed as it passes through the lens, but because lenses are generally very thin, this can often be neglected. As the light exits the lens, the proportion calculated in equation (3.30) is again reflected, with the remainder exiting the lens. The total transmittance through the optics, τ_o, is the product of the transmittances of all of the elements in the optical chain.

From Table 3.8, it can be seen that the refractive indices of the common IR lens materials are rather high, so losses due to reflection are significant. To increase the transmittance, multilayer coatings are deposited on the lens to grade the refractive index change from air to that of the material. For example, uncoated germanium has a transmittance of about 47%, but by using multilayer antireflection coatings, this can be increased to 97%.

Mirrors do not contribute to absorption losses, but their reflectances are not 100%, particularly at the shorter wavelengths, as shown in Figure 3.19. In this case, the total transmission through the optical chain is the product of the reflectance of all the mirrors in the system.

3.10 PERFORMANCE MEASURES FOR IR IMAGERS

3.10.1 Detector Field of View

The definitions for field of view (FOV) and instantaneous field of view (IFOV) are explicit for a single-element scanned electro-optical sensor. In that case, the FOV is the scanned field of view, while the IFOV is the detector field of view. However, as the number of elements in the detector array increases, the effective IFOV increases until, for a staring array, the effective IFOV and FOV are identical. In this book, this distinction is not made, and for convenience, the IFOV remains the field of view for a single detector element and the FOV is the total field of view of the whole array as defined in Figure 3.20. It should be remembered that in optics, the lengths are generally measured in centimeters and areas in square centimeters.

From the construction in Figure 3.20, it can be seen that the lens aperture, D_o, determines the FOV (rad), while the size of the individual detector elements, d_o, determines the IFOV. These quantities are determined from these apertures and the focal length, f_l, by

$$FOV = 2 \tan^{-1} \frac{D_o}{2f_l}. \tag{3.32}$$

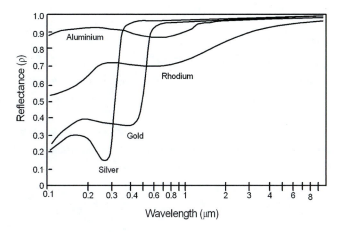

Figure 3.19 Spectral reflectance of some mirror coatings.

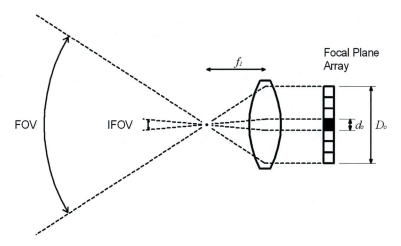

Figure 3.20 Definitions of FOV and IFOV.

Using the small angle approximation for $\tan^{-1}$ for IFOV (rad) allows the equation to be simplified to

$$IFOV = 2\tan^{-1}\frac{d_o}{2f_l} \approx \frac{d_o}{f_l}. \tag{3.33}$$

These equations are often written in terms of the limiting aperture of the sensor, A_s, and the area of the individual detector element, A_{det}, rather than their lengths, in which case, the FOV is taken to be the sensor field of view, Ω_s (steradian):

$$\Omega_s = A_s/f_1^2, \tag{3.34}$$

and the IFOV is taken to be the solid angle, α_d (steradian) subtended by the element:

$$\alpha_d = A_{\text{det}}/f_1^2. \tag{3.35}$$

3.10.2 Spatial Frequency

Spatial frequency, as illustrated in Figure 3.21, is defined as the number of intensity cycles or line pairs that exist in an image per degree (or milliradian) of subtended angle. A line pair consists of one dark and one light line. It is one of the critical measures that define the performance of an imager.

Because most IR imaging devices are still operated by people viewing the output image on a screen, it is important to take into account the characteristics of human visual perception. Experi-

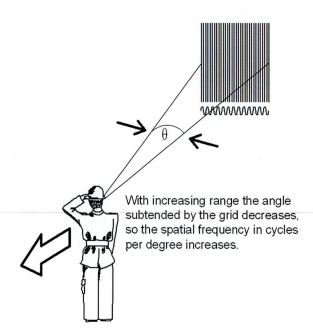

With increasing range the angle subtended by the grid decreases, so the spatial frequency in cycles per degree increases.

Figure 3.21 Definition of spatial frequency.

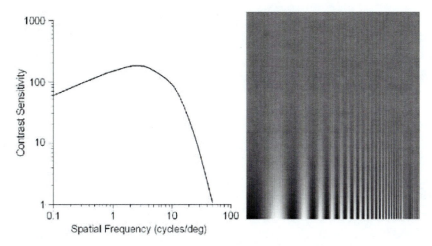

Figure 3.22 Human visual response to spatial frequency and contrast.

ments have shown that there is a nonlinear relationship between contrast sensitivity and spatial frequency, as shown in Figure 3.22. This, and other effects, makes quantitative analysis of target detection quite difficult to achieve.

The MATLAB code shown below generates the spatial frequency image shown in Figure 3.22.

```
% spatial_freq.m
% Generate a spatial frequency and contrast image.
%
% Arguments:
%
%       hsize            horizontal size of image
%
%       vsize            vertical size of image
%
%       spread           fpread in contrast cycles
%
% some physical constants:

hsize = 500;
vsize = 400;
spread= 8;
```

```
% generate the increasing spatial frequency along the x axis
freq = linspace(0.01^(1/spread),1, hsize).^spread * 0.3;
% freq is 1/2/pi * dw / dx
w = 2*pi*cumsum( freq );
hline = sin(w);

% generate the contrast along the y axis
vline = linspace(0,1,vsize).^3;

% take the product to form an image
im = vline'*hline;
imagesc(im);
colormap gray;
```

3.10.3 Signal-to-noise Ratio for a Point Target[1]

Equation (3.24) can be rewritten to give the NEP (W) in terms of the specific detectivity, D^*, the area of each detector, A_{det}, and the signal bandwidth, β:

$$NEP = \frac{\sqrt{A_{det}\beta}}{D^*}. \tag{3.36}$$

The signal level from an isolated point target with radiation intensity $\mathcal{J}_t$ (W/sr) results in a power density H_t (W/cm^2) at the sensor of

$$H_t = \mathcal{J}_t R_{cm}^2, \tag{3.37}$$

where R_{cm} is the range to the target (cm). The total power received by the sensor, S_{rec} (W), is the product of the power density and the aperture area:

$$S_{rec} = A_s \frac{\mathcal{J}_t}{R_{cm}^2}. \tag{3.38}$$

The SNR is the ratio of the received power, S_{rec}, to the NEP:

$$SNR = \frac{S_{rec}}{NEP} = \frac{A_s D^* \mathcal{J}_t}{\sqrt{A_{det}}\sqrt{\beta}R_{cm}^2} \tag{3.39}$$

In general, the bandwidth is determined from the illumination time of each detector element, which can be derived from the total number of elements and the frame rate of the sensor.

For the sensor FOV of Ω_s (steradians), the total number of pixels in the scene will be this value divided by the IFOV, α_d, of each detector:

$$n_r = \Omega_s / \alpha_d. \tag{3.40}$$

---

1 The analysis in this section is based on work by (Hovanessian 1988).

For a frame rate, T_f (sec), the illumination time, t_d (s) for each element is

$$t_d = T_f/n_r = T_f \alpha_d / \Omega_s. \tag{3.41}$$

In most modern sensors, more than one detector element is used, so for n detectors, the illumination time will increase proportionally to

$$t_d = T_f \alpha_d n / \Omega_s. \tag{3.42}$$

The approximate bandwidth, β (Hz), of the signal is equal to the reciprocal of this illumination time:

$$\beta \approx 1/t_d = \Omega_s / T_f \alpha_d n. \tag{3.43}$$

Substituting back into equation (3.39), the equation for the SNR becomes

$$SNR = \frac{A_s D^* \mathfrak{J}_t}{\sqrt{A_{\det}} R_{cm}^2} \left(\frac{T_f \alpha_d n}{\Omega_s} \right)^{1/2}. \tag{3.43}$$

The area of the individual detector elements is often unknown, but the IFOV and the f number are often quoted, so it is more convenient to rewrite equation (3.43) using these two parameters. From equation (3.35), $A_{\det} = f_1^2 \alpha_d$, and from the definition of the f number, $f_1 = f D_o$, therefore

$$A_{\det} = f^2 D_o^2 \alpha_d. \tag{3.44}$$

In addition, there are a number of losses that will significantly reduce the SNR. These include optics, electronics, detector, and scanning losses and can be incorporated into the equation for SNR as an efficiency, η. Substituting equation (3.44) into equation (3.43) and including the efficiency gives

$$SNR = \frac{A_s D^* \mathfrak{J}_t}{D_o f R_{cm}^2} \eta \left(\frac{T_f n}{\Omega_s} \right)^{1/2}. \tag{3.45}$$

3.10.4 Worked Example: IRST System SNR

Infrared search and tracking systems are used to detect and track aircraft targets. Calculate the SNR at a range of 100 km for the following sensor and target characteristics:

Sensor field of view (Ω_s):	$20° \times 10°$
Detector IFOV	1.36 mr $\times$ 1.36 mr
Frame rate (T_f):	2 s
Number of detectors (n):	32
Aperture diameter (D_o):	15 cm
f number:	1
Detectivity (D^*):	3×10^{10} cmHz$^{1/2}$/W
Total efficiency (η):	0.1

Assume that the aircraft is radiating a power of 1 kW.

From the aperture diameter, D_o, determine the area, $A_s = \pi D_o^2/4 = 176.7 \text{ cm}^2$.

$$SNR = \frac{176.7 \times 3 \times 10^{10} \times 10^3}{15 \times 1 \times \left(100 \times 10^5\right)^2} \times 0.1 \times \left(\frac{2 \times 32}{20 \times 10 / 57.3^2}\right)^{1/2}$$
$$= 3.11 \times 0.1 \times 32.4$$
$$= 10.08 \quad (10 \text{ dB})$$

If this sequential scan sensor is replaced with a staring array with the same D^*, then the improved SNR can be determined from the ratio of the total number of detector elements in the staring array to the number in the sequential scan sensor.

The total number of elements in the staring array will be

$$n_r = \frac{\Omega_s}{\alpha_d} = \frac{20 \times 10 / 57.3^2}{\left(1.36 \times 10^{-3}\right)^2}$$
$$\approx 32768 \quad (256 \times 128).$$

Therefore the increase in the SNR will be $(n_r/n)^{1/2} = 32$.

$$SNR = 10.08 \times 32 = 322.6 \quad (25 \text{ dB})$$

Note that these results do not include atmospheric losses, and these would be significant over a range of 100 km.

3.10.5 Signal-to-noise Ratio for a Target in Ground Clutter

In many applications, the target will not be isolated in space, as in the previous example, but rather surrounded by other objects at a similar temperature. For example, most IR surveillance sensors are required to identify people or vehicles on the ground. In these cases, the radiation emitance of the background and the target temperature gradient must be taken into account.

Again using (Hovanessian 1998), for a target with a spectral emitance $W(\lambda,T)$ (W/cm^2 sr μm), the total power received, S (W), in each detector element can be calculated by integrating over the wavelength range from λ_1 to λ_2,

$$S = A_s \alpha_d \int_{\lambda_1}^{\lambda_2} W(\lambda,T)\,\mathrm{d}\lambda, \tag{3.46}$$

where A_s is the sensor aperture (cm^2) and α_d is the IFOV (steradian). The SNR is calculated as before:

$$SNR = \frac{A_s \alpha_d}{D_o f} \eta \left(\frac{T_f n}{\Omega_s}\right) \int_{\lambda_1}^{\lambda_2} W(\lambda,T) D^*(\lambda)\,\mathrm{d}\lambda. \tag{3.47}$$

In this equation the specific detectivity, $D^*(\lambda)$, becomes part of the integral to take into account its wavelength sensitivity.

To obtain the change in SNR with changes in temperature it is necessary to determine the relationship between the spectral emittance and the temperature,

$$\Delta W(\lambda, T) = \frac{\partial W(\lambda, T)}{\partial T} \Delta T,$$ (3.48)

from which can be obtained the changes in signal power, ΔS (W),

$$\Delta S = A_s \alpha_d \int_{\lambda_1}^{\lambda_2} \frac{\partial W(\lambda, T)}{\partial T} \Delta T,$$ (3.49)

and finally the change in SNR can be obtained:

$$\Delta SNR = \frac{A_s \alpha_d}{D_o f} \eta \tau_o \tau_a \left(\frac{T_f n}{\Omega_s} \right) \int_{\lambda_1}^{\lambda_2} \frac{\partial W(\lambda, T)}{\partial T} D^*(\lambda) \Delta T \, d\lambda.$$ (3.50)

Note that it is common practice to separate the optical, τ_o, and the atmospheric, τ_a, transmission losses from the other efficiency factors, as has been done in equation (3.50).

3.10.6 Noise Equivalent Temperature Difference

One of the important performance measures for an IR imager is its ability to detect small changes in temperature. The smallest temperature difference that a system can detect, and therefore display as a different color/shade, is called the thermal resolution. This is sometimes described by the noise equivalent temperature difference (NETD), which is the temperature change that alters the collected flux by an amount equal to the NEP. This can be determined from equation (3.50) by setting $\Delta SNR = 1$ and solving for ΔT:

$$NETD = \frac{D_o f}{A_s \alpha_d \tau_a} \left(\frac{\Omega_s}{T_f n} \right)^{1/2} \frac{1}{M^*},$$ (3.51)

where M^* is the figure of merit for the sensor. This figure of merit includes not only the specific detectivity, D^*, of the sensor, but also the spectral dependence of the emitted radiation, $\partial W(\lambda, T)/\partial T$, and the efficiency factors η and τ_o:

$$M^* = \int_{\lambda_1}^{\lambda_2} \eta(\lambda) \tau_o(\lambda) \frac{\partial W(\lambda, T)}{\partial T} D^*(\lambda) \, d\lambda.$$ (3.52)

Tables for the thermal emittance integral at specific temperatures and over specific wavelength bands are often provided in the literature (Klein 1997). Alternatively, the actual response curves can be provided, as shown in Figure 3.23. The latter are intuitively more meaningful and can be more accurate, as they incorporate the wavelength sensitivity of the detector (as shown) and also the various losses.

3.10.7 Example

Calculate M^* over the spectrum from 8 to 14 μm for a HgCdTe detector operating at 77 K which has a peak D^* of 2×10^{10} with the normalized response shown in Figure 3.23. Assume that

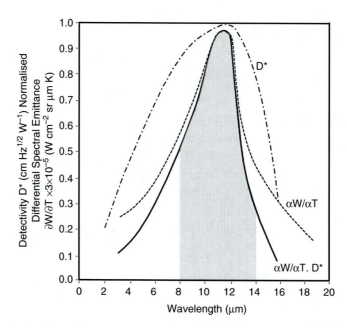

Figure 3.23 Spectral emittance and detectivity as a function of wavelength [adapted from (Hovanessian 1988)].

Table 3.9 Differential spectral emittance scaled by the normalized D^* at a number of wavelengths

Wavelength (μm)	$\dfrac{\partial W}{\partial T} D^*$
8	0.5
9	0.64
10	0.8
11	0.96
12	0.88
13	0.45
14	0.37

$\eta = 1$ and $\tau_{\mathrm{o}} = 1$. The data in Table 3.9 are calculated from the figure and subsequently used to determine the area under the curve.

Using trapezoidal integration between 8 and 14 μm, M^* can easily be calculated to be

$$M^* = 1 \times 1 \times \left(\frac{0.5 + 0.37}{2} + 0.64 + 0.8 + 0.96 + 0.88 + 0.35 \right) \times 3 \times 10^{-5} \times 2 \times 10^{10}$$
$$= 2.5 \times 10^6.$$

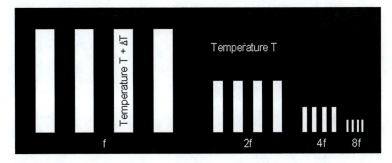

Figure 3.24 Four-bar targets of increasing spatial frequency used to determine MRTD.

3.10.8 The Minimum Resolvable Temperature Difference

It is not possible to achieve high thermal and spatial resolutions simultaneously. As neither, on its own, is a good measure of the overall performance of a thermal imaging system, a different metric is used. A single quantity called the minimum resolvable temperature difference (MRTD) measures both performance factors simultaneously. Because it is determined experimentally, it takes into account all of the factors that matter, even the performance of the operator.

The measurement is performed by slowly heating a series of targets, shown in Figure 3.24, at a fixed range from the sensor. The MRTD is defined as the minimum temperature difference in the scene needed by an operator to resolve a standard four-bar pattern that represents the fundamental spatial frequency of the sensor (Klein 1997).

The fundamental spatial frequency is determined from the number of line pairs, each comprising one dark and one light line resolved by the sensor at a given range. To measure the MRTD as a function of spatial frequency, the sensor being tested views the largest target in the figure, where the bars and spaces are both blackbody radiators at different temperatures. Because the bar height is seven times its width, the darks and lights form a square so the operator has no visual cue regarding the orientation of the bar.

Initially the system gain is adjusted so that thermal noise is clearly visible on the screen. The temperature difference starts at zero and then the difference, ΔT, is increased until the bars are confidently visible by a trained operator. This ΔT is the MRTD at the fundamental spatial frequency of the sensor. The process is repeated for the four-bar targets at the higher spatial frequencies to produce a curve such as the one shown in Figure 3.25.

Note that the MRTD axis follows a logarithmic scale, while the measured performance is nearly linear on those axes. This curve shows that human resolution of higher spatial frequency information requires increasingly large temperature contrasts between features and the target background.

3.11 TARGET DETECTION AND RECOGNITION

The Johnson criterion relates the number of line pairs across a target critical dimension to the probability that the operator can detect, orientate, recognize, or identify the target. These criteria are listed in Table 3.10 as a function of the operator task and confidence. They are used along with the MRTD curve to determine the maximum range at which detection (or recognition) will occur.

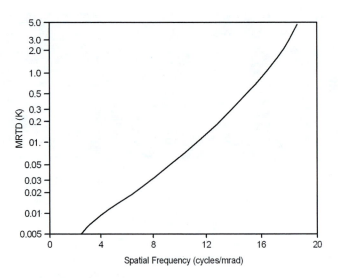

Figure 3.25 Measured MRTD versus spatial frequency for a FLIR.

Table 3.10 Required resolution for detection, orientation, recognition, and identification (Johnson 1958)

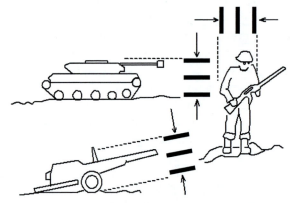

Target	Resolution per minimum dimension in line pairs			
Broadside view	*Detection*	*Orientation*	*Recognition*	*Identification*
Truck	0.9	1.25	4.5	8.0
M-48 tank	0.75	1.2	3.5	7.0
Stalin tank	0.75	1.2	3.3	6.0
Centurion tank	0.75	1.2	3.5	6.0
Half-track	1.0	1.5	4.0	5.0
Jeep	1.2	1.5	4.5	5.5
Command car	1.2	1.5	4.3	5.5
Soldier (standing)	1.5	1.8	3.8	8.0
105 Howitzer	1.0	1.5	4.3	6.0
Average	1 ± 0.25	1.4 ± 0.35	4.0 ± 0.8	6.4 ± 1.5

3.11.1 Example of FLIR Detection

Calculate the detection and recognition ranges for a tank target (3 m × 3 m) with a temperature contrast of 2.5 K to the background. Use the measured MRTD curve shown in Figure 3.26.

The MRTD curve is reproduced in Figure 3.26, from which it can be seen that for an MRTD of 2.5 K the spatial frequency that can be resolved is 17.6 cycles/mrad. From the Johnson criteria, detection requires 1 line pair (cycle) and recognition about 4 line pairs (cycles). As a 3 m target subtends an angle of 1 mrad at a range of 3 km, the range at which the imager can put one line pair across this target is the detection range:

$$R_{\text{det}} = 3 \text{ km} \times 17.6 = 52.8 \text{ km}.$$

For recognition, the imager needs to put 4 line pairs across the target, therefore the detection range decreases proportionally:

$$R_{\text{rec}} = 3 \text{ km} \times 17.6/4 = 13.2 \text{ km}.$$

Atmospheric loss is significant and must be included if an accurate measure of the performance is to be made. Even for clear air, with a visibility of greater than 4 km, transmittance losses of between 0.72 and 0.97 dB/km are experienced, depending on relative humidity and air temperature. A value of 0.9 dB/km will be used in this example. In fog, with a visibility of 1 km, transmittance losses of between 3 and 4 dB/km can be expected. For the example, the higher extreme of 4 dB/km will be used.

The easiest way to solve for these additional losses is graphically. Because there is a linear relationship between detection range and spatial frequency resolution, it is possible to replot the MRTD curve with R_{det} replacing spatial frequency. A scaling factor equal to 3 must be used so that the detection range of 52.8 km corresponds to the spatial frequency resolution of

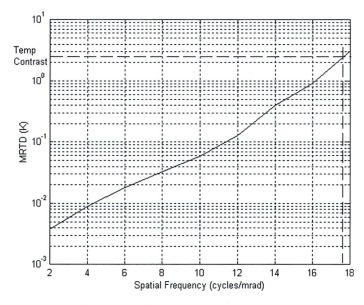

Figure 3.26 Measured MRT curve for a thermal imaging system.

17.6 cycles/mrad, as shown in Figure 3.27. The transmittance losses through the air are then plotted on this graph as a function of range, starting at 0 km, with the actual temperature difference of 2.5 K and with slopes of −0.9 dB/km and −4 dB/km, respectively, for the clear air and fog cases.

From the intersection of the loss lines with the MRTD graph it can be seen that the detection range in clear air reduces to 22 km and in fog it reduces to 7 km. To confirm that this method is valid, consider the case where there is no loss. In this case, the loss line is horizontal and it intersects the MRTD curve at 52.8 km as expected.

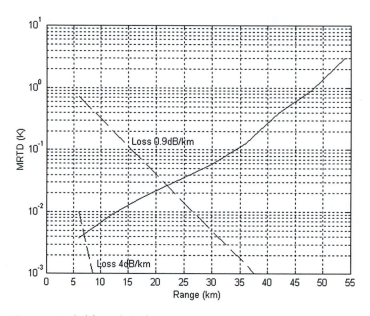

Figure 3.27 An MRTD curve scaled for a detection range of 52.8 km.

```
% mrtd_calc.m
%
% Calculate the detection range for a FLIR from the MRTD curves.
%
% Arguments:
%
%          xtar              size of the target (m)
%
%          mrtd              MRTD curve given for the FLIR (K)
%
%          delta_t           temperature contrast between target and
                                       background
```

```
% input data
xtar = 3.0;                              % target 3m across
mrtd =
([0.00384,0.00909,0.01812,0.03253,0.05919,0.12918,0.40287,0.92253,3.11792]);
sfreq =(2:2:18);                         % spatial frequency
delta_t = 2.5;                           % contrast temperature (K)


% determine the resolvable spatial frequency at a temperature contrast
of delta_t
sfdet = interp1(mrtd,sfreq,delta_t);


% plot the MRTD curve
semilogy(sfreq,mrtd,'k');
grid
xlabel('Spatial Frequency (cycles/mrad)')
ylabel('MRTD (K)')
pause


% for detection we need a spatial frequency of only 1 cycle at the
detection range
rdet = sfdet*xtar/1.0;                    % detection
range = sfreq*rdet/sfdet;                 % scale the range axis to suit
dB1 = -0.9*range;
loss1 = delta_t*10.0.^(dB1/10);           %clear air transmittance loss
dB2 = -4*range
loss2 = delta_t*10.0.^(dB2/10);           % fog 1km visibility transmittance
                                              loss


% plot the scaled MRTD curve and the transmittance loss
semilogy(range,mrtd,'k',range,loss1,'k-',range,loss2,'k-')
grid
axis([0,55,0.001,10])                     % detection
xlabel('Range (km)');
ylabel('MRTD (K)');
pause


% for recognition we need a spatial frequency of 4 cycles at the
detection range
rdet = sfdet*xtar/4.0;                    % recognition
range = sfreq*rdet/sfdet;                 % scale the range axis to suit
dB1 = -0.9*range;
loss1 = delta_t*10.0.^(dB1/10);           %clear air transmittance loss
dB2 = -4*range
loss2 = delta_t*10.0.^(dB2/10);           % fog 1km visibility transmittance
                                              loss
```

```
% plot the scaled MRTD curve and the transmittance loss
semilogy(range,mrtd,'k',range,loss1,'k-',range,loss2,'k-')
grid
axis([0,15,0.001,10])                    % recognition
xlabel('Range (km)');
ylabel('MRTD (K)');
```

In the recognition case, the new graph (Figure 3.28) is generated with a scaling factor of ¾ to achieve correspondence between the recognition range of 13.2 km and the spatial frequency resolution of 17.6 cycles/mrad. As before, the loss lines start at a temperature difference of 2.5 K and follow the −0.9 dB/km or the −4 dB/km slopes. The intersection points are now 10 km for recognition in clear air and 5 km for recognition through fog.

3.12 THERMAL IMAGING APPLICATIONS

In addition to military applications, thermal imaging has a wide variety of industrial uses, including the detection of gas and oil leaks, plant inspection, and power line monitoring. Skin temperature observations on human subjects by medical specialists can give insight into many physiological problems, particularly those concerned with thermoregulation and metabolism. Airborne prospecting, using multispectral and hyperspectral imagers that use characteristic IR spectra to identify specific minerals, is also common, while space-based thematic imagers monitor a number of IR and visible frequency bands (see Table 3.11) to observe specific characteristics of the biosphere.

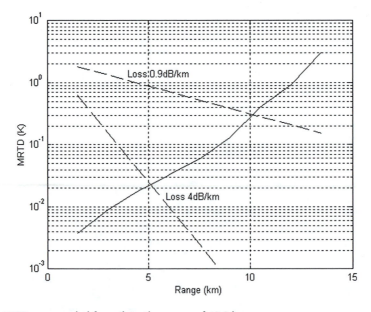

Figure 3.28 An MRTD curve scaled for a detection range of 13.2 km.

Table 3.11 Wavelengths used for specific applications by the Landsat thematic mapper

Band	Range (μm)	Characteristic
1	0.45–0.52	Coastal water mapping
2	0.52–0.60	Healthy vegetation
3	0.63–0.69	Chlorophyll absorption
4	0.76–0.9	Water delineation, vegetation vigor
5	1.55–1.75	Snow or cloud differentiation
6	10.4–12.5	Plant heat stress measurement
7	2.08–2.35	Hydrothermal mapping

Figure 3.29 Comparison between a visible image and a long-wave IR image showing its ability to penetrate through smoke (Buss 2003).

Another recent application of the technology includes thermoelastic stress analysis. These techniques are based on the use of thermal methods to determine the stress levels in dynamically loaded structures and can be used to analyze elastic, plastic, and fracture phenomena (Burnay et al. 1988). Acoustic thermography is an extension of this principle in which the object under test is excited by a low-power acoustic/ultrasound source. Flaws in the structure dissipate the acoustic energy as heat, which can be detected using an IR imager.

Figure 3.29 shows a thermal image made using an uncooled microbolometer-based long-wave IR camera. Note that the contrast of natural terrain is quite low because the temperatures of the various objects and their emissivities are similar. It is only objects such as vehicles and people, which are hotter than the surroundings, that stand out. Because of the long wavelength of this radiation compared to the size of the smoke particles, it is able to penetrate the smoke and expose the two people who were obscured in the visible image.

At this point it is interesting to consider the difference between operating a radiometer in the IR band and in the microwave band. In Figure 3.29, the image brightness is proportional to the amount of energy received by each detector element. Because this energy is integrated over a wide bandwidth around the peak of the radiated spectrum, the Stefan-Boltzmann law can be applied, and therefore the brightness is proportional to T^4. However, in the case of a microwave radiometer (analyzed in Chapter 4), the bandwidth is narrow and away from the peak, making the Rayleigh-Jean approximation valid. In this case, the brightness is proportional to the effective T and so differences in emissivity and reflectivity play a major role in coloring the image.

3.13 IMAGE INTENSIFIERS

Image intensifiers are passive sensors that amplify ambient UV, visible, or near-IR radiation from a scene and then redisplay it in the visible spectrum. The scene is imaged onto a photocathode which has an energy-band structure such that on absorption of light, electrons are emitted from the surface. The process of amplification involves either adding kinetic energy to each electron or multiplying the number of electrons using an avalanche effect, so that many more photons are emitted when the electrons strike a phosphor screen than were used in the generation process.

3.13.1 First-Generation Tubes

In first-generation devices, illustrated in Figure 3.30, the photo electrons are focused and accelerated by an electric field. Because the accelerated electrons possess increased kinetic energy, each electron gives rise to many photons (in the visible part of the spectrum) when it strikes the phosphor. This results in an amplification or gain (Proxitronic 2007).

To maintain crisp images, the distance traveled by each electron is limited in proximal-focus tubes. This limits the spacing between the electrodes, and hence the allowed acceleration voltage, to avoid arcing, is not very high. Single-stage devices offer gains of between 50 and 100, which give satisfactory performance down to moonlight illumination levels. For operation under starlight, the illumination level is much lower, as shown in Table 3.12, and higher gains must be used. Gains greater than 50,000 can be achieved by cascading multiple stages, as shown in Figure 3.31, but there is a limit to the number of stages that can be cascaded to maintain acceptable image quality. Depending on the required gain, first-generation tubes feature especially high

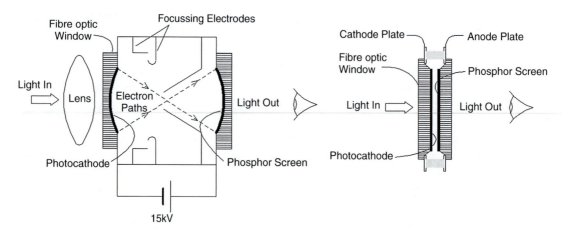

Figure 3.30 First-generation (a) electrostatic-focus inverting tube and (b) proximal-focus noninverting tube.

Table 3.12 Natural illumination levels

Condition	Illumination (lux)	Condition	Illumination (lux)
Direct sunlight	10^5	Deep twilight	1
Full daylight	10^4	Full moon	10^{-1}
Overcast day	10^3	Quarter moon	10^{-2}
Very dark day	10^2	Clear starlight	10^{-3}
Twilight	10	Overcast starlight	10^{-4}

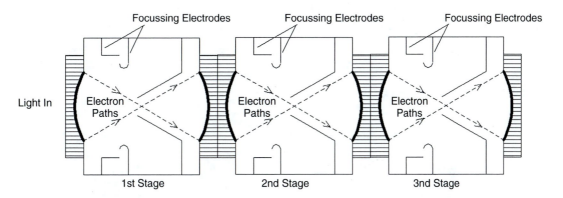

Figure 3.31 Cascading tubes to increase gain.

image resolution, a wide dynamic range (the ability to reproduce the ratio between the bright and dark parts of the image), and low noise.

3.13.2 Second-Generation Tubes

Second-generation tubes use a microchannel plate (MCP) between the photocathode and the phosphor, as illustrated in Figure 3.33. Early MCPs consisted of about a million hollow glass tubes fused together into a disc. These tubes are about 10 μm in diameter and 1 mm long, with the inside walls of each coated with a secondary emitting material to act as electron multipliers (Lampton 1981). A single-stage tube with an MCP can produce gains of up to 5×10^4. Modern MCPs as illustrated in Figure 3.32, contain between 2 and 6 million holes, but this is still a limiting factor, and their resolutions are poorer than their first-generation counterparts.

In the configurations shown in Figure 3.33, the potentials applied to the MCP are intermediate to those of the anode and the cathode. The proximally focused intensifier is similar to a two-electrode intensifier in as far as the gaps between the photocathode and the MCP and the MCP and screen are sufficiently short to minimize electron dispersion. In the focused intensifier case, the electrons emitted by the photocathode are focused by an electric field onto the MCP.

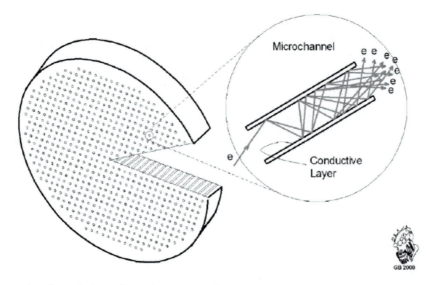

Figure 3.32 Microchannel plate schematic.

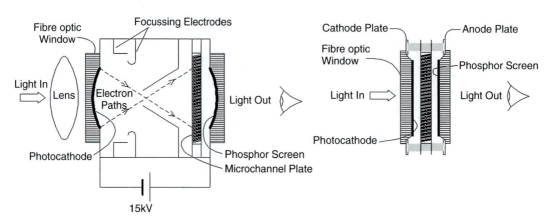

Figure 3.33 Second-generation tubes using microchannel plates.

3.13.3 Limitations of Microchannel Plates

Four major physical constraints limit MCP performance (Proxitronic 2007):

- The average output signal cannot exceed the maximum current that can be sustained within the walls of the microchannel. When the electron flux is too large, the electric charge removed from the glass is not replaced immediately and the gain is reduced. Bright parts of an image can saturate and lose contrast.
- If one of the molecules of gas that remains in the tube becomes ionized, it becomes accelerated toward the input of the tube, where it could strike the wall and initiate a new cascade of electrons. This cascade can mask the signal. Curved microchannel paths minimize this effect.

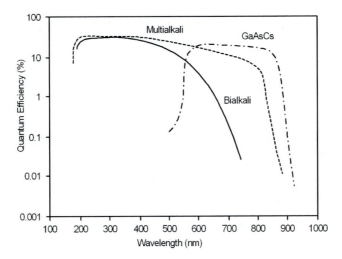

Figure 3.34 Third-generation photocathode responses.

- The density of the electron cloud in the tube is limited to about $10^7/mm$ before mutual electrostatic repulsion returns additional secondary electrons to the surface of the channel before the field can accelerate them.
- The surface area of the channels is less than the surface area of the entire plate. Geometric constraints limit this to a maximum of 91%, but because the walls must be a finite thickness, this is generally lower (typically about 55%). This means that about half of the input flux hits the area between the channels. Channels can be made funnel shaped to minimize this effect.

3.13.4 Third-Generation Tubes

Third-generation tubes are similar in construction to second-generation devices but have a more sensitive GaAs or multialkali photocathode which extends into the near-IR band (450–950 nm), as shown in Figure 3.34.

3.13.5 Spectral Characteristics of the Scene

Although the spectral content of sky illumination, for sunlight and moonlight, peaks in the visible region, the spectral content of clear starlight is weighted toward the near-IR, as shown in Figure 3.35. Hence, for very-low-light applications, third-generation photocathodes are made sensitive in this region.

Another advantage of operating in the near-IR is that vegetation exhibits higher reflectivity above 800 nm and hence target background contrasts can be larger (Craig 1994). Figure 3.36 shows images produced by third-generation tubes which illustrate the good contrast with foliage (left) and good dynamic range (right) that can be achieved.

3.13.6 Time Gating Microchannel Plates

Because MCPs offer huge amplifications, it is possible to operate these devices over very short illumination periods. A useful technique developed to exploit this is illumination of the scene

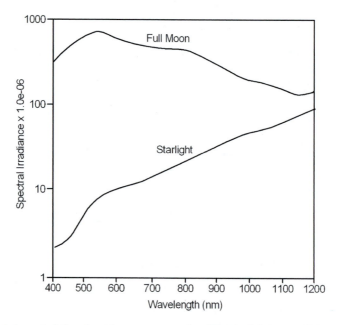

Figure 3.35 Spectral characteristics of outdoor scenery under different lighting conditions.

Figure 3.36 Image intensifier images showing the good resolution and high gain available from modern third-generation night vision systems.

with a short burst of laser light (typically <10 ns) and then time gating the MCP for the same period, delayed by a fixed period. This will capture the reflections from a 1.5 m slice of the scene at a range determined by the selected delay. This technique can be used for underwater imaging through turbid conditions, as the gating process minimizes the backscatter from suspended particulates except at the range of interest and therefore results in cleaner images. Even faster gating speeds are possible, and images of the internal structure of the human body are being made by gating MCPs in 200 ps or less.

3.14 References

Boeing. (2000). FLIR. Viewed July 2000. Available at http://www.boeing.com/defence-space/infoelect/flir.

Burnay, S., Williams, T., and Jones, C., eds. (1988). *Applications of Thermal Imaging*. London: Adam Hilger.

Buss, J. (2003). *EO/IR Sensors*. Arlington, VA: Office of Naval Research.

Craig, D. (1994). *RSAF Platform Engineers Course: Electro-Optics Module*. Edinburgh, SA, Australia: Defence Science and Technology Organization.

Crawford, F. (1998). Electro-optical sensors overview. *IEEE AES Systems Magazine*, October, pp. 17–24.

Fraden, J. (2003). *Handbook of Modern Sensors: Physics, Designs and Applications*. Berlin: Springer Verlag.

Gehan, G. and Georges, S. (1970). Estimation of human body surface area from height and weight. *Cancer Chemotherapy Report* 54:225–235.

Hovanessian, S. (1988). *Introduction to Sensor Systems*. Boston: Artech House.

Infrared Solutions. (2007). Micro-bolometer technology. Viewed August 2007. Available at http://www.infraredsolutions.com/html/technology/microbolometerF.shtml.

Johnson, J. (1958). Analysis of image-forming systems. Image Intensifier Symposium, paper AD-220160, Fort Belvoir, VA.

Klein, L. (1997). *Millimeter-Wave and Infrared Multisensor Design and Signal Processing*. Norwood, MA: Artech House.

Lampton, M. (1981). The microchannel image intensifier. *Scientific American* 245(November):46–55.

Meijer, G. and van Herwaarden, A., eds. (1994). *Thermal Sensors*. Bristol, UK: Institute of Physics Publishing.

Melexis. (2007). MLX90601 family IR thermometer modules. Viewed July 2007. Available at http://www.melexis.com/prodfiles/0004774_MLX90601_10.pdf.

Proxitronic. (2007). Introduction to image intensifier tubes. Viewed August 2007. Available at http://www.proxitronic.de.

Sparrius, A. (1981). Electro-optical imaging target trackers. *Transactions of the South African Institute of Electrical Engineers* 72(11):278–284.

Zissis, G. (1993). *The Infrared & Electro-optical Systems Handbook*, Vol. 1, *Sources of Radiation*. Bellingham, WA: SPIE; chap. 4.

4

Millimeter Wave Radiometers

The concept of blackbody radiation was introduced in Chapter 3. It was shown that the total power emitted by an object is a function of the temperature, and the emissivity of the material is proportional to T^4 as described by the Stefan-Boltzmann law. It was also shown that if the power is measured in a region far from the emission peak, the source brightness, B_f (W/m²/Hz/sr), is directly proportional to the temperature, T (K), according to the relationship described by the Rayleigh-Jean law (Currie and Brown 1987):

$$B_f = 2kT/\lambda^2, \tag{4.1}$$

where k is Boltzmann's constant (1.3804×10^{-23} J/K), T is the source temperature (K), and λ is the wavelength (m).

This approximation is accurate to within 1% for frequencies below 100 GHz, and to within 3% for frequencies below 300 GHz. It can therefore be applied to both microwave and millimeter wave systems (Ulaby 1987).

4.1 ANTENNA POWER TEMPERATURE CORRESPONDENCE

Consider a lossless antenna with area A_r and beamwidth Ω_r at a distance R from a source of radiation with an area A_t. Assuming R is sufficiently great that the power density, S_t, is uniform over the solid angle, Ω_t, as shown in Figure 4.1, then the power, P (W), intercepted by the antenna is the product of the target power density, S_t (W/m²), and the receiver antenna aperture, A_r (m²):

$$P = S_t A_r. \tag{4.2}$$

The power can also be expressed in terms of the radiation intensity, F_t (W/sr):

$$P = F_t A_r/R^2. \tag{4.3}$$

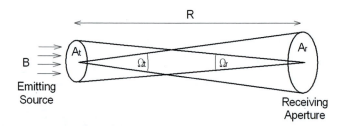

Figure 4.1 Representation of the spatial parameters that govern the equations of radiometry.

If the brightness, B (W/m^2/sr), is defined as

$$B = F_t / A_t, \tag{4.4}$$

and the solid angle, Ω_t (sr), subtended by the source of the radiation is given by

$$\Omega_t = A_t / R^2, \tag{4.5}$$

then the power can be written in terms of the target brightness:

$$P = B A_r \Omega_t. \tag{4.6}$$

For a differential solid angle weighted by the normalized antenna gain, equation (4.6) becomes

$$\partial P = A_r B(\theta, \varphi) F_n(\theta, \varphi) \partial \Omega, \tag{4.7}$$

where $B(\theta,\varphi)$ (W/m^2/sr) is the source brightness as a function of solid angle and $F_n(\theta,\varphi)$ is the normalized radiation pattern of the antenna as a function of solid angle. The power incident on the antenna, in terms of the brightness of the source and the gain pattern, can be obtained by integrating over 4π steradians and across the frequency band f_1 to f_2:

$$P = \frac{A_r}{2} \int_{f_1}^{f_2} \iint_{4\pi} B(\theta, \varphi) F_n(\theta, \varphi) \partial \Omega \partial f. \tag{4.8}$$

Note that this received power is reduced by one-half because the direct polarization from the source is random and it is assumed to be received by a linearly polarized antenna.

If this antenna is placed within a black body, and if the detected power is limited to a small bandwidth such that the brightness is constant with frequency, then the Rayleigh-Jeans approximation of equation (4.1) can be substituted for $B(\theta,\varphi)$ to obtain the received power:

$$P_{bb} = \frac{kT(f_2 - f_1) A_r}{\lambda^2} \iint_{4\pi} F_n(\theta, \varphi) \partial \Omega. \tag{4.9}$$

From antenna theory the pattern solid angle, Ω_p (sr), is equal to the integral in equation (4.9), and can therefore be written in terms of the wavelength and the antenna aperture:

$$\Omega_p = \lambda^2 / A_r. \tag{4.10}$$

Substituting into equation (4.9) results in an equation for the received power which is a function of the temperature and the bandwidth:

$$P_{bb} = kT(f_2 - f_1). \tag{4.11}$$

This is known as the fundamental equation of radiometry, and the following points regarding it need to be noted:

- The detected power is independent of the antenna gain because the source of radiation is extended and uniform and not a point source.
- The equation is independent of the distance from the radiating target.
- The temperature of the antenna structure has no effect on the output power.
- Temperature and power are interchangeable so all the gain calculations can be applied directly to the measured "temperature."
- The power detected is directly proportional to the bandwidth.

4.1.1 Example of Power Received from a Black Body

Determine the power received by an antenna operating at 150 GHz with a bandwidth of 2 GHz observing a black body with a temperature 310 K:

$$
\begin{aligned}
P_{bb} &= kT(f_2 - f_1) \\
&= 1.38 \times 10^{-23} \times 310 \times 2 \times 10^9 \\
&= 8.56 \times 10^{-12} \, \text{W} \, (-80.7 \text{ dBm}).
\end{aligned}
$$

4.2 BRIGHTNESS TEMPERATURE

All real bodies are to some extent "grey," as they radiate less than a black body. In addition, the brightness temperature, $T_b(\theta,\varphi)$, for a grey body can also be angle dependent because of variations in emissivity. The brightness temperature for the grey body is thus defined as the effective temperature of the body as if it were a black body. It can be obtained from the physical temperature,

$$T_b(\theta,\varphi) = \varepsilon(\theta,\varphi)T, \tag{4.12}$$

where $\varepsilon(\theta,\varphi)$ is the emissivity and T (K) is the physical temperature of the radiating object. In the example above, if the target has an emissivity, ε, of 0.8, then the brightness temperature, T_b, is $0.8 \times 310 = 248$ K, and the received power is reduced accordingly to −81.64 dBm.

4.3 APPARENT TEMPERATURE

In radiometry, the apparent antenna temperature, T_{AP}, replaces the received power, P, as the measure of signal strength, where T_{AP} is defined as the temperature of a matched resistor with noise power density output $P = kT_{AP}$ at the antenna port. The apparent antenna temperature, T_{AP}, is calculated from the brightness temperature, including atmospheric and antenna losses, as illustrated in Figure 4.2.

The radiation received in the main lobe of the antenna is made up of a number of components:

- The brightness temperature, T_B, from terrain emissions.
- The scatter temperature, T_{SC}, which is the radiation reflected from terrain in the main lobe but not generated by it. Radiation from both the atmosphere, T_{DN} (the downward or downwelling temperature), and galactic radiation may be reflected. At frequencies greater than 10 GHz, only the downward radiation from the atmosphere need be considered.
- The upward or upwelling temperature, T_{UP}, from that portion of the atmosphere that is below the radiometer. In this example, the radiometer is shown to operate in space, but radiometers also operate from aircraft.

These contributors to the total radiation are then attenuated by the portion of the atmosphere through which the radiation passes before it reaches the antenna:

$$T_{AP}(\theta,\varphi) = T_{UP}(\theta,\varphi) + \frac{1}{L_A}[T_B(\theta,\varphi) + T_{SC}(\theta,\varphi)], \qquad (4.13)$$

where

T_{AP} = apparent temperature (K),
T_{UP} = upwelling temperature from the atmosphere (K),
L_A = atmospheric loss factor,
T_B = brightness of the observation area (K),
T_{SC} = brightness of the radiation scattered from the observation area (K).

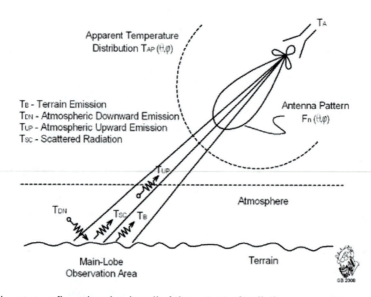

Figure 4.2 Radiometer configuration showing all of the sources of radiation.

4.4 ATMOSPHERIC EFFECTS

4.4.1 Attenuation

Atmospheric attenuation is a function of the air density, and for horizontal or oblique paths through the atmosphere, it is usually calculated by integration. Figure 4.3 shows the attenuation through the atmosphere at different grazing angles. For approximate results, interpolation can be used to determine the value at intermediate angles.

The upward and downward brightness temperatures of the atmosphere vary with frequency and will obviously be higher where the attenuation is higher, as the atmosphere is more opaque. At 94 GHz, the attenuation through the entire atmosphere, L_A (dB), is determined primarily by the water vapor concentration, ρ_o (g/m^3):

$$L_A = 0.17 + 0.06\rho_o. \tag{4.14}$$

For aircraft-based radiometers, the equations governing the attenuation are more complex and are considered in some detail later in this chapter.

4.4.2 Downwelling Radiation

In most surface imaging radiometers, window frequencies are selected in the 35 GHz, 94 GHz, or 160 GHz bands. However, to probe the atmospheric water content, the region around 240 GHz can be used, as the brightness temperature is proportional to the water vapor concentration (see Figure 4.4). For operation at 94 GHz, typical values of the downwelling temperature as a function of atmospheric conditions are shown in Table 4.1.

4.4.3 Upwelling Radiation

For space-borne radiometers, the upwelling radiation is that of the entire atmosphere and is equal to the downwelling radiation. For aircraft, only part of the atmosphere contributes to the upwell-

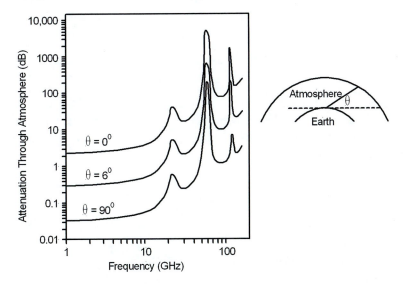

Figure 4.3 Attenuation through the atmosphere.

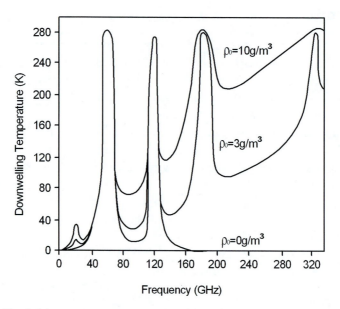

Figure 4.4 Downwelling brightness temperature as a function of frequency with water vapor concentration as a parameter.

Table 4.1 Downwelling temperature under different weather conditions

Conditions	Downwelling temperature (K)
Clear sky	10–60
Thick fog	120
Overcast	150
Fog	180
Thick clouds	180
Moderate rain	240

ing radiation, and as the atmosphere can be stratified and complex, it is easiest to treat it as an attenuator and calculate the upwelling radiation in those terms. To maintain thermal equilibrium, any medium that absorbs radiation (attenuates) must also radiate, therefore the effective temperature is

$$T_e = \left(1 - \frac{1}{L_A}\right)T,$$ (4.15)

where
   T_e = effective temperature of attenuator (atmosphere),
   L_A = attenuator loss factor = $10^{\alpha/10}$,
    T = physical temperature of the attenuator (K).

4.5 TERRAIN BRIGHTNESS

Various forms of terrain have dissimilar brightness temperatures, therefore, from an imaging perspective, it is the temperature difference between objects and not the absolute temperature that determines the image quality.

- Metallic objects: These are lossless and opaque and so are perfectly reflecting. As a result, their brightness will equal the downwelling radiation, if they are smooth.
- Water: The brightness of water is dependent on polarization, angle of view, and, to a lesser extent, temperature, purity, and surface conditions. Because it is also reflective, its brightness is also dependent on the downwelling temperature. At 94 GHz, the reported brightness for water (vertical polarization) varies between 150 K and 300 K.
- Soil: As with water, soil is dependent on polarization and angle of view. It is also dependent on moisture content and surface roughness. At 94 GHz, the reported brightness for soil (vertical polarization) varies between 160 K and 280 K.
- Vegetation: The brightness of vegetation depends on its type and moisture content. At 94 GHz it is reported to vary between 230 K and 300 K.
- Built-up areas: These will be complex. However, at 94 GHz, asphalt is reported to be 260 K to 300 K.

Although there is a significant overlap between brightness temperatures in these cases, due to the fact that the data were taken under a variety of weather conditions, there will, in general, be a significant contrast between different materials under identical weather conditions.

4.6 WORKED EXAMPLE: SPACE-BASED RADIOMETER

A space-based radiometer operating at 94 GHz with a bandwidth of 2 GHz looks directly downward to the ground at a temperature of 27°C which has an average emissivity, ε (over the footprint), of 0.9. What is the received power?

As discussed in Chapter 3, the sum of the reflectivity and the emissivity is unity if there is no absorption. The reflectivity is therefore $\rho = 1 - \varepsilon = 0.1$.

From Figure 4.3, the total attenuation directly downward through the atmosphere at 94 GHz is about 1.2 dB. The loss is $L_A = 10^{dB/10} = 1.31$.

Assume that the air has a water content of 3 g/m³. From Figure 4.4, the downwelling brightness temperature at 94 GHz is 30 K. Because the radiometer is space-based, it can be assumed that the upwelling and downwelling temperatures are equal therefore:

$$T_{AP} = T_{UP}(\theta, \varphi) + \frac{1}{L_A}[\varepsilon T_B + \rho T_{SC}], \tag{4.16}$$

$$\begin{aligned} T_{AP} &= 30 + \frac{1}{1.31}[300 \times 0.9 + 30 \times 0.1] \\ &= 30 + 208.4 \\ &= 238.4 \end{aligned}$$

Table 4.2 Temperature contrast of metallic objects and other materials under different weather conditions

Material	Atmospheric conditions		
	Clear	*Fog*	*Rain*
Vegetation	220 K	200 K	40 K
Water	120 K	100 K	30 K
Concrete	190 K	170 K	40 K

For a bandwidth of 2 GHz, the received power, P (dBm), can be calculated using equation (4.11). Note that the factor of 30 added to the total converts from dBW to dBm:

$$P = 30 + 10\log^{10} kT\beta$$
$$= -81.8 \text{ dBm}.$$

4.6.1 Temperature Contrast

Typical temperature contrasts of metallic objects to other materials depend on the difference between the reflected downwelling temperature and the temperature of the surrounding low-reflectivity terrain. These are summarized in Table 4.2. Note that the contrast is largest when the air is clear and at its lowest when the air is opaque with rain.

4.7 ANTENNA CONSIDERATIONS

4.7.1 Beamwidth

The 3 dB beamwidth of the antenna, θ_{3dB} (deg), can be approximated by

$$\theta_{3dB} \approx 70\lambda/D, \tag{4.17}$$

where D (m) is the diameter of the antenna and λ (m) is the wavelength.

4.7.2 Efficiency

In the previous discussion it was assumed that the antenna is lossless, however, in reality an antenna absorbs a certain amount of the power incident on it, and hence it also radiates. Therefore the apparent temperature, T_{AO} (K), at the antenna output port is a function of the radiation efficiency of the antenna, η_1 (typically 0.6), as well as the scene temperature, T_A (K), and the physical temperature, T_P (K), of the antenna:

$$T_{AO} = \eta_1 T_A + (1 - \eta_1) T_P. \tag{4.18}$$

Note that η_1 is considered to be equivalent to the surface reflectivity, ρ, of the antenna.

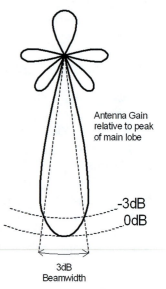

Antenna Gain relative to peak of main lobe

-3dB
0dB

3dB Beamwidth

4.7.3 Fill Ratio

The size of the antenna footprint does not affect the terrain's brightness temperature. However, the footprint area is important when an object with a different emissivity or physical temperature from that of the terrain is present in the footprint. If such an object is completely enclosed by the antenna, the observed brightness temperature can be calculated to be

$$T_B = T_{BG}(1 - F) + T_{BT}F, \tag{4.19}$$

$$F = A_T/A, \tag{4.20}$$

where

F = fill-in ratio,
T_{BG} = ground brightness temperature (K),
T_{BT} = target brightness temperature (K),
A_T = target area (m^2),
A = antenna footprint (m^2).

4.8 RECEIVER CONSIDERATIONS

4.8.1 Mixer Implementations for Microwave Receivers

Until recently, low-noise amplifiers at frequencies above 50 GHz were both expensive and had poor noise figures. In many cases, radiometers used (and still use) mixers fed directly from the antenna port, as illustrated in Figure 4.5. In the schematic shown, the mixer generates a difference and a sum frequency, $f_{RF} - f_{LO}$ and $f_{RF} + f_{LO}$, and the difference frequency becomes the intermediate frequency (IF) output.

Two different frequencies satisfy the requirement for $f_{IF} = f_{RF} - f_{LO}$. If $f_{RF} = f_{LO} + f_{IF}$, then the output of the mixer will be $f_{LO} + f_{IF} - f_{LO} = f_{IF}$. However, if $f_{RF} = f_{LO} - f_{IF}$, then the output of the mixer will be $f_{LO} - f_{IF} - f_{LO} = -f_{IF}$. This latter response is called the image response of the mixer and is indistinguishable from the direct response in this configuration. In the example shown, f_{RF} = 93 + 1 = 94 GHz for the direct response and f_{RF} = 93 − 1 = 92 GHz for the image response.

If the radiometer receiver is implemented as shown in the diagram, it will receive signals over the band from 91.5 to 92.5 GHz and from 93.5 to 94.5 GHz, both of which will be down-converted to the IF band from 0.5 to 1.5 GHz.

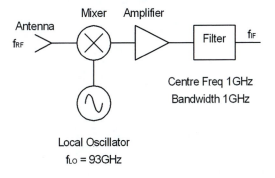

Figure 4.5 The down-conversion process.

4.8.1.1 Mixer Specifications The primary disadvantage of the configuration shown is that the conversion from RF to IF incurs a loss in output power and hence a reduction of signal-to-noise ratio (SNR). The mixer conversion loss, L_c (dB), is defined as follows:

$$L_c = 10 \log_{10} (\text{available RF input power/IF output power}). \tag{4.21}$$

Practical mixers usually have a conversion loss between 4 and 8 dB, which usually increases with increasing frequency, as shown in Table 4.3. It is also a function of local oscillator (LO) drive (or pump) power, which is generally between 7 and 13 dBm.

Mixer noise characteristics are important, so when specifying or reading mixer specifications a distinction must be made as to whether the input is a single-sideband (SSB) or a double-sideband (DSB) signal. It was shown above that the mixer produces an IF output from two input frequencies and will therefore collect noise power from both frequencies. When used with a DSB input, the mixer will have desired signals at both radio frequencies, while an SSB input provides the desired signal at only one of those frequencies. The DSB noise figure will therefore be 3 dB lower than the SSB noise figure.

4.8.2 Noise Figure

In calculating the noise figure (NF), it is useful to represent the down-converter block by two separate inputs (signal and image) to the mixer, as shown in Figure 4.6. It can also be shown that

Table 4.3 Typical TRG mixer specifications with increasing frequency

Model number	960	960A	960B	960U	960V	960E	960W	960F
Frequency range (GHz)	18–26.5	26.5–40	33–50	40–60	50–75	60–90	75–110	90–140
Waveguide	WR-42	WR-28	WR-22	WR-19	WR-15	WR-12	WR-10	WR-8
DSB noise figure (dB)[1]	3.5	4.0	4.0	4.5	4.5	5.0	5.0	5.5
Conversion loss (dB)[2]	5.0	5.5	5.5	6.0	6.0	6.5	6.5	7.0

[1] DSB noise figure assumes a +7 dBm LO, IF of 10–1000 MHz, and a 1.5 dB IF amplifier noise figure.
[2] Conversion loss SSB (dB) assumes a +7 dBm LO. Starved or high LO drive versions are available (e.g., 0 dBm < LO < +16 dBm).

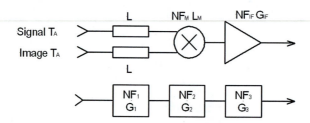

Figure 4.6 Schematic for a down-converter block showing both the signal and image inputs and its representation as a change of cascaded gain blocks.

the NF for a cascaded receiver chain made up of a number of stages each with gain and individual NFs is

$$NF = NF_1 + \frac{NF_2 - 1}{G_1} + \frac{NF_3 - 1}{G_1 G_2}. \tag{4.22}$$

For the double-sided mixer implementation shown in Figure 4.6, substitute $L = NF_1$, $L_M/2 = NF_2$, $1/L = G_1$, $2/L_M = G_2$, and $NF_{IF} = NF_3$ to obtain the total NF:

$$NF_{DSB} = L + L\left(\frac{L_M}{2} - 1\right) + (NF_{IF} - 1)\frac{L.L_M}{2} = \frac{L.L_M NF_{IF}}{2}. \tag{4.23}$$

Similarly for SSB operation,

$$NF_{SSB} = LL_m NF_{IF}. \tag{4.24}$$

4.9 THE SYSTEM NOISE TEMPERATURE

Even without any external input, the radiometer will produce an output because the receiver is not at absolute zero. This output can be defined in terms of an equivalent noise temperature, T_{sys}, of a matched resistor at the antenna port (Currie et al. 1987). The available noise power, P_N (W), from such a resistor is expressed as

$$P_N = kT_{sys}\beta G, \tag{4.25}$$

where k is Boltzmann's constant (1.38×10^{-23} J/K), β (Hz) is the system bandwidth, and G is the system power gain.

The receiver introduces additional noise into the system that is incorporated into the equation for T_{sys} (K):

$$T_{sys} = T_o(NF - 1), \tag{4.26}$$

where NF is the noise figure for the receiver and is defined as the ratio of the input SNR to the output SNR with the input terminated at $T_o = 290$ K. Equation (4.23) and equation (4.24) can be written in terms of their effective temperature, T_{sys} (K):

$$T_{sys} = T_o(NF_{DSB} - 1) \text{ for DSB operation,} \tag{4.27a}$$

$$T_{sys} = T_o(NF_{SSB} - 1) \text{ for SSB operation.} \tag{4.27b}$$

4.10 RADIOMETER TEMPERATURE SENSITIVITY

The ability of a radiometer to detect changes in the input temperature, ΔT, is determined from the analysis of the detector output when the input is band-limited white noise (Bhartia and Bahl

1984). The processes involved in converting the thermal input at the antenna to a voltage output are described in Figure 4.7. Assuming a rectangular filter, the double-sided spectrum at IF has a bandwidth, β_{IF}, and a height $P_{IF} = 0.5kT_{sys}G_{IF}$, as shown in Figure 4.8.

A square-law detector produces an output signal proportional to the square of the input envelope. It can be shown that the postdetection probability density function includes a DC component, P_{DC}, and a double-sided triangular noise component, P_{AC}, as shown in Figure 4.9. The magnitude of the DC power component is given by

$$P_{DC} = (kTG/2)^2 \beta_{IF}^2, \tag{4.28}$$

and the AC power density has height $P_{AC} = (kTG/2)^2\beta_{IF}$ and width β_{IF}.

The signal is then passed through a final low-pass filter with a bandwidth β_{LF} which does not alter P_{DC} but reduces the AC component to an almost rectangular density function (because $\beta_{LF} \ll \beta_{IF}$) of height $P_{AC} = (0.5kTG)^2\beta_{IF}$ and width β_{LF}, as illustrated in Figure 4.9.

The AC power output for a double-sided spectrum after the low-frequency filter is the area of these spectral components and can be written as

$$P_{AC} = 2(kTG/2)^2 \beta_{IF}\beta_{LF}. \tag{4.29}$$

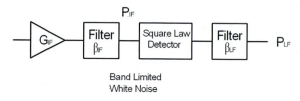

Figure 4.7 Processing noise through a radiometer.

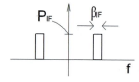

Figure 4.8 Band-limited white noise spectrum output.

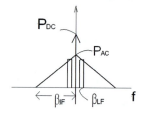

Figure 4.9 Postdetection output.

The ratio of the AC power component to the DC power component is the ratio of equation (4.28) and equation (4.29):

$$\frac{P_{AC}}{P_{DC}} = \frac{2\left(\frac{kTG}{2}\right)^2 \beta_{IF}\beta_{LF}}{\left(\frac{kTG}{2}\right)^2 \beta_{IF}^2} = \frac{2\beta_{LF}}{\beta_{IF}}. \tag{4.30}$$

In terms of voltages this can be rewritten as

$$\frac{V_{AC}}{V_{DC}} = \sqrt{\frac{2\beta_{LF}}{\beta_{IF}}}. \tag{4.31}$$

Since the temperature change, ΔT, is measured by V_{AC} while the sum of the antenna and system temperatures, $T_A + T_{sys}$, determines V_{DC}, then the two ratios will be the same and

$$\frac{\Delta T}{T_A + T_{sys}} = \sqrt{\frac{2\beta_{LF}}{\beta_{IF}}}, \tag{4.32}$$

therefore

$$\Delta T = \left(T_A + T_{sys}\right) = \sqrt{\frac{2\beta_{LF}}{\beta_{IF}}}. \tag{4.33}$$

If the low-pass filter is implemented as an ideal integrator with a time constant, τ, then $\beta_{LF} = 1/(2\tau)$ and the temperature change, ΔT, can be rewritten as

$$\Delta T = \frac{T_A + T_{sys}}{\sqrt{\beta_{IF}\tau}}. \tag{4.34}$$

Note that β_{IF} is not the 3 dB bandwidth, but the reception bandwidth, for a two-pole RC filter, $\beta_{IF} = 1.96\beta_{3dB}$.

Because ΔT is derived from the AC component of the band-limited thermal noise output by the radiometer, it represents the extent of the natural variations in the signal level. If a change in the temperature, T_A, is to be detected, it must exceed this natural variation.

If the change in temperature results in a change in the output equal to ΔT, this is equivalent to a SNR equal to 1. However, to obtain a reasonable probability of detection, the required SNR will be much higher, typically 13 dB. The acceptable temperature difference will therefore have to be scaled appropriately. This is discussed in Chapter 9.

4.11 RADIOMETER IMPLEMENTATION

A number of different radiometer configurations exist. The simplest of these is the total power radiometer, which consists of a high-gain receiver followed by a square-law detector, as shown in Figure 4.10a. The Dicke radiometer shown in Figure 4.10b is slightly more sophisticated, as it includes a synchronous detector coupled to a reference load which can compensate for gain variations. Other radiometer configurations use noise injection to implement the Dicke configuration over a wide range of temperatures at the expense of sensitivity or minimum detectable temperature difference (Klein 1997).

The increasing requirements for small, multiple element radiometers and improvements in the performance of millimeter wave integrated circuit (MMIC) technology have promoted the development of direct-detection radiometers. These consist of a single integrated module containing an antenna followed by a high-gain (40–60 dB) amplifier chain, a band-pass filter, and a detector (Appleby et al. 2003).

4.11.1 Total Power Radiometer

A square-law detector cannot distinguish between an increase in the signal power (an increase in T_A) and an increase in the predetection gain, G. If the gain varies by ΔG around the average gain G, then the minimum detectable temperature change ΔT_{min} is

$$\Delta T_{min} = \left(T_A + T_{sys}\right)\left[\frac{1}{\beta_{IF}\tau} + \left(\frac{\Delta G}{G}\right)^2\right]^{1/2}. \tag{4.35}$$

To minimize the effect of this variation in gain, an internal reference load with a known temperature, T_c, is switched into the input on a regular basis and can be used as a calibration reference.

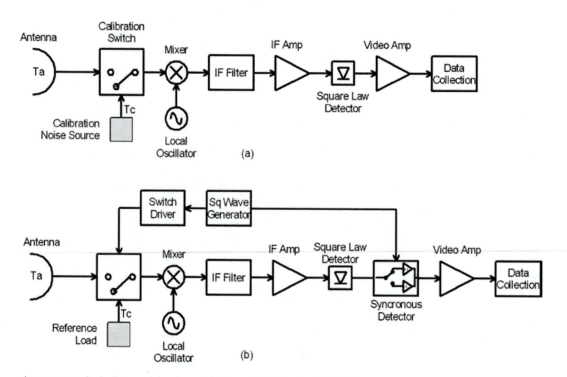

Figure 4.10 Block diagram of radiometer types: (a) total power, (b) Dicke.

4.11.2 Dicke Radiometer

Dicke radiometers are generally used when signal integration times are long and gain fluctuations are of concern, as in applications where atmospheric temperature and water vapor profiles are being measured (Klein 1997). In Figure 4.10b it can be seen that the receiver input is switched at a constant rate (much faster than the integration time) between the antenna port and a reference load maintained at a constant temperature. The output of the square-law detector is then synchronously detected, such that the final output is proportional to the difference between the temperature of the antenna and the reference load.

The derivation of this formula is beyond the scope of this course, but it should be noted that it is based on the premise that only half of the switching time, τ, is used to view the antenna, while the other half is used to view the Dicke reference:

$$\Delta T_{\min} = \left(T_A + 2T_{sys} + T_C\right)\left[\frac{1}{\beta_{IF}\tau} + \left(\frac{\Delta G}{G} \cdot \frac{T_A - T_C}{T_A + 2T_{sys} + T_C}\right)^2\right]^{1/2}. \tag{4.36}$$

4.11.3 Performance Comparison between Radiometer Types

If the performance of these two radiometer implementations is compared for the following realistic scenario: $T_A - T_C = 10$ K, $T_C = 300$ K, $T_{sys} = 1000$ K, $\beta_{IF} = 1$ GHz, and $\tau = 0.1$ s, then the following results are obtained (Currie et al. 1987):

Total power $\Delta T_{\min} = 0.185$ K for $\Delta G/G = 0.01\%$,
              $\Delta T_{\min} = 1.87$ K for $\Delta G/G = 0.1\%$.
Dicke $\Delta T_{\min} = 0.241$ K for $0.01\% < \Delta G/G < 0.1\%$.

Hence if $\Delta G/G > 0.0172\%$, then the Dicke configuration is superior.

4.12 INTERMEDIATE FREQUENCY AND VIDEO GAIN REQUIREMENTS

The required IF gain is determined from the signal level required by the square-law detector, and the video gain is then determined from the output of the square-law detector and the voltage requirements for the display or signal data logger (Figure 4.11). The power at the antenna of a typical uncooled radiometer is −75 dBm, so to make it compatible with the square-law detector, a gain of 65 dB is required. In that case, the detector output would be 10 mV, so for an operating voltage of 1 V, additional video voltage amplification of 100 times (40 dB) is required.

4.13 WORKED EXAMPLE: ANTITANK SUBMUNITION SENSOR DESIGN

Millimeter wave antitank missiles, mortar shells, and other submunitions often resort to radiometric tracking or detection for the final phases of the engagement. This is illustrated in Figure 4.12.

In this example, each skeet is released from a height of 25 m with an upward velocity of 50 m/s and a horizontal velocity of 10 m/s. It can be assumed that the cone angle remains constant at 10° and that it spins at a constant rate of 2 rps. The skeet is fitted with a radiometric seeker with a 50 mm aperture, which operates at 95 GHz with a receiver bandwidth of 2 GHz.

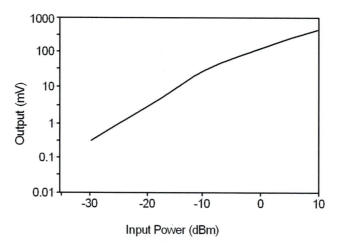

Figure 4.11 Typical microwave square-law detector characteristics.

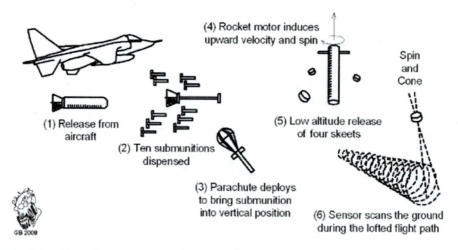

Figure 4.12 Deployment of submunitions containing millimeter wave radiometers.

A MATLAB procedure is used to generate the position of the skeet relative to the target during the 10 s after release. This is then used to determine the beam footprint that is generated on the ground to produce Figure 4.13.

The next stage of the process is to determine the radiometric temperature difference between the target and the surrounding ground as follows. The average sky temperature at 95 GHz is assumed to be 60 K over a 140° cone around the vertical, 150 K over a 10° segment, and 300 K up to 10° above the surface around the target, as shown in Figure 4.14 (Currie et al. 1987).

Assume that the tank has been driving, its temperature is 35°C (308 K), and its emissivity, ε, is 0.1. The average target temperature over the full hemisphere will be the sum of reflected temperatures, scaled by their various areas and the tank reflectivity, and the emitted temperature, scaled by the emissivity.

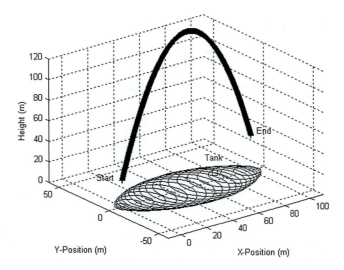

Figure 4.13 Skeet position and search footprint on the ground.

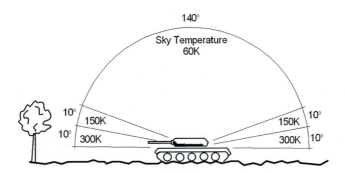

Figure 4.14 Passive target detection temperature scenario.

The area, A (sr), of each of the 10° sections can be found by integration:

$$A_1 = \int_0^\alpha 2\pi r^2 \cos\theta d\theta,$$
$$A_1 = A_2 \approx 1.09 \text{ sr.} \tag{4.37}$$

Therefore the area of the remaining 140° section is

$$A_3 = 2\pi - (A_1 + A_2) = 4.10 \text{ sr.}$$

The reflected (scattered) temperature is

$$T_{SC} = \rho\left[\frac{A_1}{2\pi}T_1 + \frac{A_2}{2\pi}T_2 + \frac{A_3}{2\pi}T_3\right] = 0.9\left[\frac{1.09}{2\pi}300 + \frac{1.09}{2\pi}150 + \frac{4.10}{2\pi}60\right] = 105.5\,K.$$

The radiated (brightness) temperature of the tank is

$$T_{BT} = \varepsilon T_T = 0.1 \times 308 = 30.8\,\text{K}.$$

The apparent temperature is the sum of the reflected and radiated brightness temperatures modified by the loss through a portion of the atmosphere, plus the upwelling temperature. For a path length of 100 m and a clear air attenuation of 0.2 dB/km, the loss is very small and $L_A \approx 1$. The upwelling temperature, which is related to the attenuation, is also very small ($T_{up} \approx 1$ K), so can also be ignored, and the temperature can be calculated as

$$T_{AP} = T_{UP}(\theta, \varphi) + \frac{1}{L_A}[T_{BT} + T_{SC}] \approx 30.8 + 105.5 = 136.3\,K.$$

Assuming that the surrounding terrain is at a temperature of 20°C (293 K) and the emissivity of the ground is 0.92 (typical for grass and soil), then the apparent temperature of the ground is calculated in the same way as it is for the target and is

$$T_{APG} = \varepsilon_G T_G + \rho T_{GSC} = 0.92 \times 293 + 0.08 \times 117 = 279\,K.$$

Note that the scattered contribution is very small, as the reflectivity of the ground is low, and also that the tank appears to be much colder than the surrounding terrain because it scatters the cold sky temperature. The actual brightness temperature seen by the radiometer is determined by these two temperatures weighted by the beam-fill factor at each range.

An antenna looking straight down will illuminate a circular footprint on the ground. The diameter of the footprint will be a function of the antenna beamwidth (generally to the half-power or 3 dB contour) and the distance to the ground. A reasonable approximation for the beamwidth, θ_{3dB} (rad), is $\theta_{3dB} = 1.22\lambda/D$, where λ (m) is the wavelength and D (m) is the antenna diameter. The footprint area on the ground is determined by the beamwidth and the range, R (m), to the ground:

$$A_B = (\pi/4)d^2 = \pi/4\,(R\theta_{3dB})^2,$$

for $D = 50$ mm and $\lambda = 3.16$ mm, $A_B = 4.67 \times 10^{-3}R^2$.

The cross-sectional area of a tank as seen from above, $A_T = 20$ m^2, means that the scene temperature measured by the antenna will be the sum of the background and tank brightness temperatures scaled by their relative areas:

$$T_A = T_{APG}\frac{A_B - A_T}{A_B} + T_{APT}\frac{A_T}{A_B}.$$

Figure 4.15 shows scene temperature as the range to the ground varies, assuming that the tank is in the center of the beam. To keep the model as simple as possible, it is assumed that both the beam footprint on the ground and the tank are circular.

As the antenna scans across the terrain, it will measure the background temperature of 279 K. However, when it encounters the tank, the measured temperature will dip to the apparent temperature shown in the figure. The actual temperature difference will depend on the range to the target and the distance of the tank from the center of the beam. Figure 4.16 shows that the antenna sweeps across the target a number of times during its flight (assuming it does not detonate its warhead), and that each time this occurs the measured temperature dips significantly from the 279 K background toward the lower temperature of the tank.

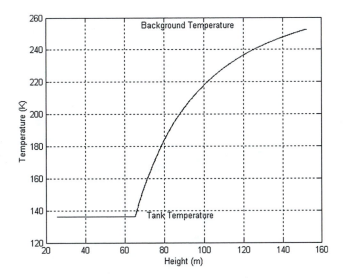

Figure 4.15 Temperature variation due to beam-fill effects.

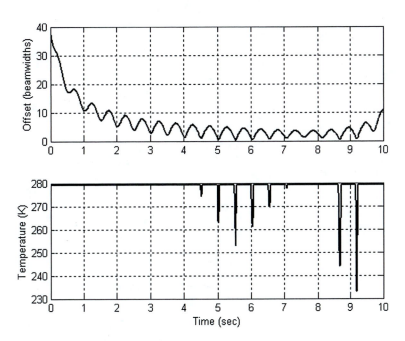

Figure 4.16 Antenna temperature as a function of time, showing instances when the beam scans through the cold target.

```
% Calculate the temperature measured by a 94GHz radiometer
% mounted on a skeet
% skeet.m

% Arguments

atank = 20;                      % Tank cross-sectional area sq m
tbt = 136.3;                     % Tank temperature K
tbg = 279;                       % Background temperature K
xtank = 80;                      % Tank position
ytank = 10;
ztank = 0;

% radiometer characteristics
dant = 0.05;                     % antenna diameter m
freq = 95.0e+09;                 % frequency Hz

% initial state of skeet at t=0
xinit = 0;                       % m
yinit = 0;
zinit = 25;

xvinit = 10.0                    % m/s
yvinit = 0;
zvinit = 50.0;

xainit = 0;                      % m/s.s
yainit = 0;
zainit = -9.8;

cone = 10.0;                     % cone angle deg
rot = 2.0;                       % cone rotation rate rps

% constants
dt = 0.01;                       % step size s
degrad = pi/180;                 % degrees to radians
c = 3.0e+08;                     % speed of light m/s
lam = c/freq;                    % wavelength of radiation m
theta3db = 1.22*lam/dant;        % antenna 3dB beamwidth rad

% time up to 10s
t = (0:dt:10);
len = length(t);
```

```
% generate the skeet position with time
xa=ones(size(t))*xainit;
ya=ones(size(t))*yainit;
za=ones(size(t))*zainit;

xv=xvinit+cumsum(xa)*dt;
yv=yvinit+cumsum(ya)*dt;
zv=zvinit+cumsum(za)*dt;

x=xinit+cumsum(xv)*dt;
y=yinit+cumsum(yv)*dt;
z=zinit+cumsum(zv)*dt;

plot3(x,y,z,'k+');
hold

% generate the antenna pointing vector due to the cone angle
theta = t*rot*2*pi;
offset = z.*tan(cone*degrad);
xoff = x+offset.*cos(theta);
yoff = y+offset.*sin(theta);
zoff = zeros(size(xoff));

plot3(xoff,yoff,zoff,'k');

% Plot the tanks position
plot3(xtank,ytank,ztank,'ks');
grid
axis([-10,110,-60,60,0,120])
xlabel('X-Position (m)');
ylabel('Y-Position (m)');
zlabel('Height (m)')
%title('Skeet Position and Scanned Footprint on the Ground')
hold;
pause;

% Calculate the effective scene temperature
abeam = (z.*theta3db).^2.*pi./4;    % area of the beam footprint on
the ground

% if the area of the beam on the ground is less than the area of the
tank
% then beamfill occurs and there is no longer a background contribution
% assumes that the beam is centered on the tank
```

```
abeam  = max(abeam,atank);
tback  = tbg.*(abeam-atank)./abeam;
ttank  = tbt.*atank./abeam;
ta  = tback+ttank;


% Angular Distance in beamwidths of the tank from the center of the
beam
rbeam=sqrt((xoff-xtank).^2+(yoff-ytank).^2)./(z.*theta3db);


% if part of the tank is within the beam, then the temperature
% depends on how close to the center of the beam it is
k  = find(rbeam<1);
sig(k)  = ta(k).*(1-rbeam(k))  + tbg.*rbeam(k);


% if none of the tank is in the beam, then the temperature
% is equal to the background
n=find(rbeam>1);
sig(n)  = tbg;


subplot(211),plot(t,rbeam,'k'),grid;
ylabel('Offset  (beamwidths)')
subplot(212),plot(t,sig,'k'),grid;
xlabel('Time  (sec)')
ylabel('Temperature  (K)')
```

4.13.1 Radiometer Implementation

To keep the cost of the skeet as low as possible, a total power radiometer is used with an uncooled front end. The allowable integration time is made equal to the dwell time of the antenna on a target.

The circumference of the circle scanned by the beam with a cone angle of 10° (0.17 rad) is $circ = 2\pi(R\theta_{cone})$. The size of the antenna footprint on the ground is

$$D_{foot} = R(1.22\lambda/D).$$

As the skeet scans at 2 rps, the dwell time is

$$\tau = \frac{D_{foot}}{circ}\frac{1}{2} = \frac{1.22\lambda}{4\pi D_{cone}} = \frac{1.22 \times 3.21}{4\pi \times 50 \times 0.17} = 37\,ms.$$

4.13.2 Receiver Noise Temperature

For a system that operates using one sideband with the mixer placed just behind the feed horn at the focal point of the antenna so that the waveguide losses are minimized, then the following are reasonable assumptions:

$L = 0.2$ dB $= 1.05$ (feed loss from the antenna to the mixer),
$L_m = 6$ dB $= 3.98$ (mixer conversion loss),
$NF_{IF} = 1.5$ dB $= 1.41$ (low noise amplifier noise figure),
$NF_{SSB} = LL_MNF_{IF} = 1.05 \times 3.98 \times 1.41 = 5.88$
$T_{sys} = T_o(NF_{SSB} - 1) = 290(5.88 - 1) = 1415$ K.

4.13.3 Minimum Detectable Temperature Difference

The formula to determine the minimum temperature difference is given by

$$\Delta T_{min} = (T_A + T_{sys})\left[\frac{1}{\beta_{IF}\tau} + \left(\frac{\Delta G}{G}\right)^2\right]^{1/2}.$$

Assuming that the system gain is completely stable, so $\Delta G = 0$, the equation reduces to

$$\Delta T_{min} = \frac{T_A + T_{sys}}{\sqrt{\beta_{IF}\tau}} = \frac{275 + 1415}{\sqrt{2 \times 10^9 \times 37 \times 10^{-3}}} = 0.2 \, K,$$

where
T_A = background temperature (279 K),
T_{sys} = receiver system temperature (1415 K),
β_{IF} = receiver bandwidth (2 GHz),
τ = integration time (37 ms).

Thus a 0.2 K temperature drop should just be detectable. However, for a good probability of detection, a SNR of at least 13 dB is required, and so a temperature drop of at least $0.2 \times 10^{13/10} = 4$ K is required.

Going back to the formula for the beam-fill effects, a temperature difference of 4 K will occur at a range of 420 m, which exceeds the height reached by the skeet, so the tank will always be detectable.

The received power from the scene temperature will be

$$P = 30 + 10\log_{10} kT\beta = 30 + 10\log_{10}\left(1.38 \times 10^{-23} \times 279 \times 2 \times 10^9\right) = -81 \text{ dBm}.$$

The actual output power will be higher than that, as it includes the noise generated within the receiver as well:

$$P = 30 + 10\log_{10} k\left(T_A + T_{sys}\right)\beta = -73 \text{ dBm}.$$

From the graph for the square-law detector, shown in Figure 4.11, a signal level of -10 dBm is needed, and therefore an IF gain, G_{IF}, of $-10 + 73 = 63$ dB. For the detector power shown in Figure 4.11, this power level corresponds to an output of 10 mV, so if a final DC level of 1 V is needed, then a voltage gain of 100 is required.

4.14 RADIOMETRIC IMAGING

Imaging radiometers for both aircraft and satellite based systems commonly consist of a radiometer (or cluster of radiometers operating at different frequencies) that illuminates an oscillating scan mirror as illustrated in Figure 4.17a. The extent of the oscillation determines the swathe width, while the forward motion of the sensor generates the push-broom scan pattern. To ensure continuous coverage of the swathe, the scan rate and the size of the footprint on the ground must be matched to the forward velocity.

With recent advances in phased array technology, it is becoming practical to generate radiometric images using aperture synthesis techniques. The idea is to replace the large, high-resolution antenna with a thinned array of smaller antennas as illustrated in Figure 4.17b. The measurement process involves cross-correlating all of the received signals in pairs depending on their separation. The image can be then obtained by applying the inverse Fourier transform to these measurements. The main advantages of this technology are no moving parts and the ability to utilize the structure of the platform as a carrier (Peichl et al. 2003).

As with the examples from Chapter 6, a combined GPS and INS is usually mounted on the aircraft to register its position and attitude along with the scan angle of the mirror so that an accurate representation of the terrain can be reconstructed.

4.14.1 Image Processing

A radiometric image swathe was made of an airstrip from an altitude of 760 m. The image is 500 m wide by 1800 m long. Figure 4.18 shows a composite aerial photograph above the uncorrected radiometric image (Suss et al. 1989).

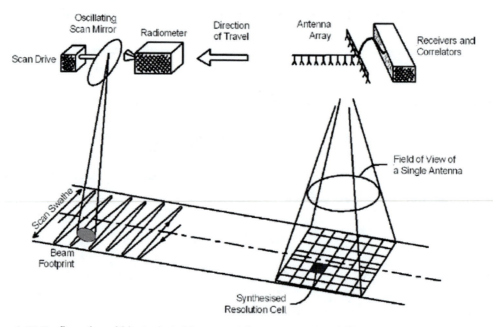

Figure 4.17 Configuration of (a) a typical airborne push-broom scanner and (b) an aperture synthesis imager.

Because no registration data were logged, high-frequency roll and low-frequency drift are superimposed on the radiometric image. Postprocessing can be used to remove the low-frequency components because the true shape of the target is known. This is achieved by fitting a spline (shown in Figure 4.19) through one distorted edge of the runway and using that to displace the image column appropriately. The final corrected image is shown in Figure 4.20.

Figure 4.18 Photograph and a raw radiometric image prior to motion compensation (Suss et al. 1989).

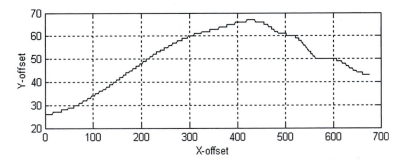

Figure 4.19 Spline fit to the edge of the runway used to correct low-frequency image drift.

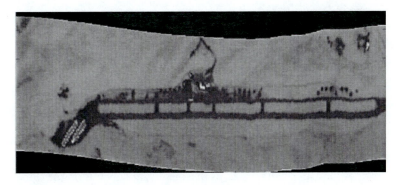

Figure 4.20 Radiometric image corrected for aircraft drift.

```
% radiometric image processing
% rad_proc.m

% read and display the image
 i_raw=imread('runway','bmp');
 imagesc(i_raw);
 [y,x,z]=size(i_raw);
 xs=(1:x);

 % enter a number of points along what is known to be
 % a straight line (terminate by hitting Enter)
 [X,Y]=ginput;

 % fit a spline to the points
 ys = spline(X,Y,xs);
 ys = ys-mean(ys);
 dat = ceil(50-ys);

 % plot the spline
 plot(xs,dat)
 xlabel('X-offset')
 ylabel('Y-offset')
 grid
 pause

 % displace the columns of the image up or down according
 % to the spline value of that column
 i_trans = uint8([]);
 for n = 1:x
      i_trans(dat(n):dat(n)-1+y,n,1)=i_raw(:,n,1);
 end

 % display the corrected image
 imagesc(i_trans)
 colormap(gray)
```

Enhancement of the contrast over the temperature range of interest can be achieved by expanding a small percentage of the output voltage to correspond to the full range of colors or shades in the display. An alternative is to perform a nonlinear mapping, which in Figure 4.21 is based on the cumulative distribution of brightness temperatures across the image. It can be seen that this transform suppresses small targets, as the weighting depends on the slope of the cumulative distribution and is therefore proportional to the number of pixels at that temperature.

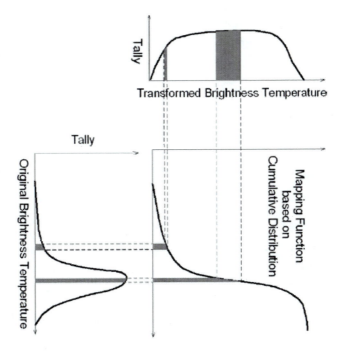

Figure 4.21 Transforming the brightness temperature histogram to enhance the region of interest.

Figure 4.22 Processed radiometric images showing contrast enhancement.

An example of the application of this processing technique is shown in the transformed image in Figure 4.22. It can be seen that the contrast in the region surrounding the runway is enhanced because a large proportion of the image is at that temperature. In contrast, the two rows of aircraft parked on the apron are all but eliminated because that feature contains only a few pixels that are at a significantly different temperature to the remainder of the image.

4.15 APPLICATIONS

Microwave and millimeter wave radiometers are used in a large variety of imaging and nonimaging applications from the ground, the air, and from space. The following sections outline a number of these applications.

4.15.1 Airborne Scanned Millimeter Wave Radiometer

An experimental scanned millimeter wave radiometer developed by the German Aerospace Research Establishment (DLR) is typical of the genre (Suss et al. 1989). It has the following characteristics:

- Function   Experimental
- Height   >80 m
- Aircraft speed   50 m/s
- Scan width   ±14.5°
- Ground resolution (nadir)   <1.5 × 1.5 m
- Frequency   140 GHz
- Sensitivity   <0.4K
- Type   Total power with cooled front-end

The image reproduced in Figure 4.23 shows a 1730 m long swathe with a width of 60 m produced at an altitude of 125 m. Note from the image that flat metal roofs of hangars appear white (as they reflect the cold sky directly above). The aircraft are slightly warmer than the hangars, as they reflect a broader range of sky temperatures because they are curved. Concrete and grass-covered areas have the highest brightness temperatures, as their emissivity is higher than that of metallic objects.

Forward-looking radiometers mounted on aircraft can be used to produce images of the ground through mist and fog or rain. An example of such as system is the 1040 element focal plane camera developed by TRW, shown in Figure 4.24. It generates 10° × 15° field of view images at video frame rates (Moffa et al. 2000).

Figure 4.23 Radiometric images of an airport showing a 1730 m swathe and two magnified sections of interesting regions in the image (Suss et al. 1989).

Figure 4.24 TRW millimeter wave radiometric camera developed for low-visibility landing (National Aeronautics and Space Administration 1997).

4.15.2 Scanning Multichannel Microwave Radiometer

Satellite-based radiometers are some of the most useful sensors in providing information for atmospheric and terrestrial research. The scanning multichannel microwave radiometer (SMMR) was a typical multichannel radiometer that operated for nearly a decade measuring sea surface temperatures (National Snow and Ice Data Center 2007).

- Manufacturer   Jet Propulsion Laboratory
- Satellite   NIMBUS 7
- Operational   October 1978 to August 1987
- Function   Sea surface temperature, wind stress, and sea ice cover
- Slant range   1380 km
- Height   960 km
- Beam nadir   42°
- Beam incidence   50.3°
- Satellite speed   6.5 km/s
- Scan width   ±25° (780 km)
- Ground resolution 17×25 km at 0.81 cm to 100×150 km at 4.54 cm
- Band   0.81, 1.36, 1.66, 2.8, 4.54 cm
- Sensitivity   Not available
- Type   6× Dicke radiometer
- Calibration   Ambient RF termination and a deep space horn
- Polarization   Alternating for four low-frequency channels
  Dual for 0.81 and 1.36 cm channels

Figure 4.25 shows the schematic diagram of the SMMR that is representative of a number of space-based satellites used for remote sensing of the earth surface. Figure 4.26 shows a map of the sea temperature measured by a space-based radiometer.

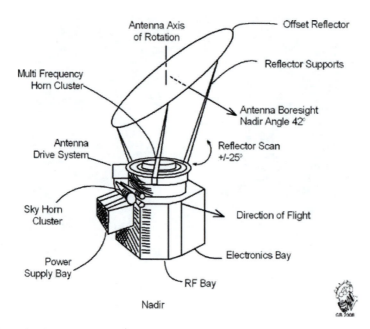

Figure 4.25 Schematic of the scanning multichannel microwave radiometer (National Snow and Ice Data Center 2007).

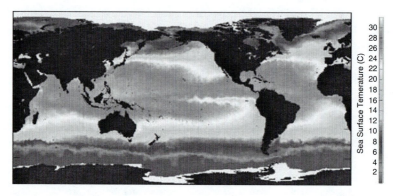

Figure 4.26 Nighttime sea temperature measured using a space-based radiometer.

4.15.3 Ground-Based Millimeter Wave Radiometers

4.15.3.1 Low-Visibility Imaging   Because millimeter wave energy is hardly attenuated as it passes through mist and fog, for the reasons discussed in Chapter 7, it can be used for imaging through low-visibility conditions. An example of this capability is the image pair in Figure 4.27, which shows a view of the Malvern Hills at a range of 2 to 2.5 km made with a 35 GHz radiometer. It can be seen that the mist in the visible image completely masks the hills, whereas at 35 GHz they are clearly visible.

High-speed image enhancement and superresolution techniques have been developed in the United Kingdom, the United States, and Russia which are capable of producing "photographic"

Figure 4.27 Visible and radiometric images of the Malvern Hills made at 35 GHz (DERA 2001).

(a)   (b)

Figure 4.28 Comparison of (a) a photograph and (b) a high-resolution 95 GHz radiometric image enhanced using superresolution techniques.

quality images such as the one shown in Figure 4.28. The diffraction-limited resolution of a radiometric imager was discussed earlier in this chapter, and this requires that large apertures be used if good resolution images are required. For example, a radiometer with a 1 m × 1 m aperture and a 40° field of view has a figure of merit (number of resolvable pixels across the image) about one-quarter as good as that of a typical thermal imager. A factor of two sharpening can be achieved using linear restoration techniques, but to achieve factors of four or more requires nonlinear methods (Lettington et al. 2003).

The noncoherent imaging process can be thought of as the convolution of the true scene temperature by the antenna beam pattern (commonly referred to as the point-spread function). This function limits the detected spatial frequency, and above this frequency, no information is recorded about the scene. Linear deconvolution can be used to restore an image, but the result is that the missing high-frequency components result in spatial ringing at contrast boundaries. Nonlinear processes are therefore required to restore these high-frequency components to suppress the ringing and further enhance sharp features of the image. A number of different methods have been derived that use a priori knowledge about the image to apply constraints to the process of solving a set of simultaneous equations which perform this function (Lettington et al. 2003).

4.15.3.2 Concealed Weapon Detection   Since the World Trade Center attack, many institutions including Millitech, Farran/Smiths, ThruVision and DERA/QinetiQ have been developing high-speed scanned radiometers that can be installed in the entrances to airports, stations, banks, sports arenas, and other areas where security is important. Radiometric images such as those shown in Figure 4.29 can see weapons concealed beneath clothing.

In Figure 4.29c, it would seem that the figure casts a shadow. However, because the image is in fact a negative, the higher the temperature, the darker the color, so what is visible is in fact the reflected image from the smooth ground. The knife, hidden under the newspaper, is reflecting the cold sky and is therefore at the lowest temperature in the image.

One of the main advantages of this technology is that it is also able to produce images of nonmetallic low-density materials, and because it is totally passive (unlike X-ray techniques), it is not harmful.

To speed up the image generation process, scanners using clusters of receivers with high-speed mechanical scanning have been developed. A typical example, developed by QinetiQ produces

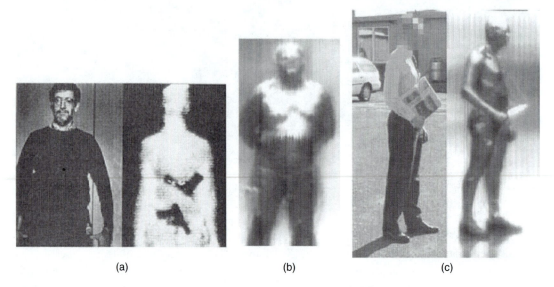

(a)   (b)   (c)

Figure 4.29 Advances in radiometric images of human beings from (a) the first Millivision results to (b,c) recent images made by DERA/QinetiQ.

wide field of view images (40° × 20°) based on only 32 direct detection receivers, is shown in Figure 4.30 (Appleby et al. 2003). The ultimate goal is to eliminate the requirement for mechanical scanning by completing a staring array, using phased array technology, or applying some more exotic imaging technology to produce a true radiometric camera.

A number of products based on this and similar technologies are beginning to appear in airports and other public places. One example, shown in Figure 4.31, is the Tadar imager that uses a 16- or 32-channel array and scanned optics to produce radiometric images at a 10 Hz frame rate (Smiths Detection 2007).

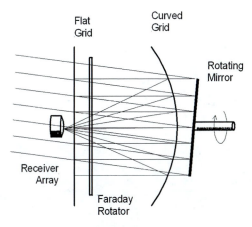

Figure 4.30 Mechanically scanned imager based on Schmidt optics.

Figure 4.31 Farran/Smiths Tadar passive imager operates with a 10 Hz frame rate using a 16- or 32-channel array and scanned optics (Smiths Detection 2007).

4.15.3.3 Surveillance and Law Enforcement   In addition to its ability to penetrate mist, fog, and clothing, millimeter wave radiation can also penetrate thin layers of wood, plasterboard (dry wall), and plastics. Objects hidden behind walls made of these materials can often be viewed. To illustrate this capability, the images in Figure 4.32 show a sport utility vehicle (SUV) with the garage door open and the garage door closed. In this case, the door was constructed of two layers of plywood with a foam core. Another example of this penetration capability is Figure 4.33, which shows a photograph and a 94 GHz radiometric image of the same scene in which the partially hidden figures become visible through the wall. In conjunction with active Doppler and ultrawideband radar, such imaging capabilities are extremely useful to the military in urban warfare situations, where it is often of vital importance that the number and disposition of people within a building be determined.

4.15.3.4 Medical Imaging   Because millimeter wave radiation can penetrate the top millimeter or so of skin, medical radiometers are useful in identifying skin cancers and the like. The early stages of basal cell carcinoma (BCC) are often undetectable in the visible and infrared bands, but longer wavelength millimeter wave techniques offer a promising method of improving early detection.

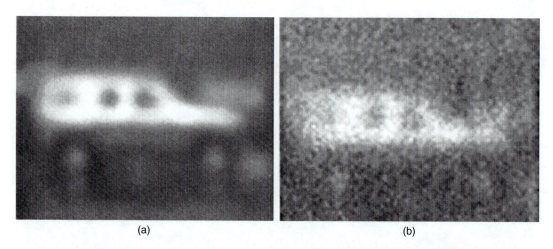

(a)   (b)

Figure 4.32 Radiometric images of an SUV showing the penetration capabilities of millimeter wave radiation where (a) is the direct image and (b) is the image through a closed garage door (Smith et al. 1996).

Figure 4.33 Photograph and 94 GHz radiometric image of the same scene showing the penetration capability of the longer wavelength (Olsen et al. 1997).

Other applications of the technology are being investigated using the prototype radiometer shown in Figure 4.34. This sensor, built at the University of St. Andrews, St. Andrews, Scotland, for medical research (Robertson 2004), works in the close-focus regime and has a diffraction-limited spot diameter of only 5 mm (the pixels are 2.24 mm × 2.24 mm).

One added benefit of millimeter wave radiation is its ability to penetrate surgical dressings and bandages. This penetration capability, in conjunction with its extremely good spatial resolution, is being evaluated by the St Andrews unit in hospital trials. These include looking through multilayer bandages and dressings to investigate leg ulcers and skin grafts, as well as imaging through plaster casts.

4.15.4 Radio Astronomy

The millimeter wave radiometer is a basic tool of radio astronomy, and radiometers have been used to detect many species of molecules in interstellar clouds. The absorption and emission of molecular lines is primarily governed by their rotational motions, and the resonance lines are more abundant and intense in the millimeter wave region than the centimeter region.

Minimum detectable antenna temperatures on the order of tenths of a degree or less and the inherently weak signals from resonances that might be light-years away, coupled with

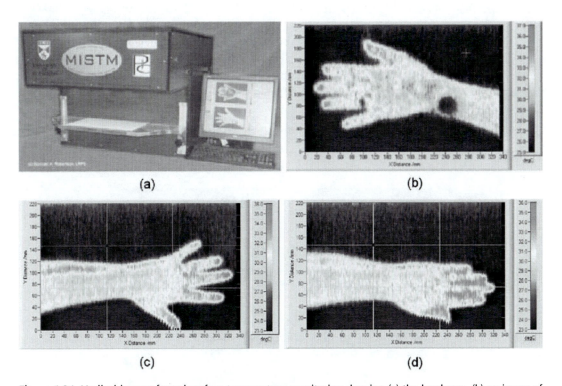

Figure 4.34 Medical imager for subsurface temperature monitoring showing (a) the hardware, (b) an image of a human hand with watch, (c) a naked hand, and (d) a hand swathed in bandages. (Courtesy University of St. Andrews).

earth atmospheric noise, require systems with extremely low noise temperature and high sensitivity.

4.15.4.1 Single-Dish Telescopes In the 1980s, only single-dish antennas were available for use at millimeter wavelengths. They were the 13.7 m dish at the University of Massachusetts (Amherst, MA), usable to 300 GHz, an 11 m dish at Kitt Peak (Tucson, AZ), usable to 140 GHz, and a 20 m dish on the Onsala Peninsula in Sweden, usable to 150 GHz. Most of the antennas built for radio astronomy work are not suitable for millimeter wave work because the surface of the dish is not sufficiently smooth at such high frequencies. However, more recently a number of telescopes have come on line, of which a good example is the James Clerk Maxwell Telescope (JCMT) at the Mauna Kea Observatory in Hawaii. It operates in four bands in the millimeter wave and submillimeter wave region: 210–280 GHz, 300–380 GHz, 460–520 GHz, and 800–900 GHz. The primary task of such single telescopes is to survey large regions of the sky looking for objects suitable for scrutiny by the large millimeter wave arrays.

4.15.4.2 Telescope Arrays A number of arrays have been developed or are under development for high-resolution millimeter wave and terahertz astronomy. The Combined Array for Research in Millimeter-Wave Astronomy (CARMA) is the merger of two American university-based arrays: the Owens Valley Radio Observatory (OVRO) and the Berkeley-Illinois-Maryland Association (BIMA) array. The BIMA array contribution to CARMA is 10 telescopes, each of which is 6.1 m in diameter, and the OVRO array contribution is six 10.4 m telescopes. These telescopes have been relocated to a new high-altitude site above 1200 m on Cedar Flat in the Inyo Mountains of California, as shown in Figure 4.35. The maximum baseline of this array is limited to 2 km and will provide a resolution of 0.1 arc second at 230 GHz.

Figure 4.35 The CARMA millimeter wave telescope array (CARMA 2007).

The Atacama Large Millimeter/Submillimeter Array (European Organization for Astronomical Research in the Southern Hemisphere 2007) is under construction on the Chajnantor plain at an altitude of 5000 m in the Chilean Andes. The array will ultimately consist of up to eighty 12 m telescopes operating at wavelengths between 0.3 and 9.6 mm. This site was chosen because it is above most of the earth's atmosphere and is very dry. The area will provide for baselines from 150 m up to 18 km. A resolution of 0.005 arc second will be possible at 1 THz, a factor of 10 better than that achieved by the Hubble space telescope in the optical band.

4.15.4.3 Applications   A combination of the resolving power, dust penetration capability, and the spectrum covered by these arrays will permit the examination of interstellar clouds and investigation of the formation of stars and protoplanets in regions of space too dusty for optical and infrared observations. In addition, many organic molecules produce spectra in these bands. For example, prior to incorporation into CARMA, the BIMA array was able to map the abundance of the different molecules in specific regions around stars and in the remnants of supernovas. Because of the abundance of quite complex organic molecules in space, there is speculation that life evolved there and not on earth as was once thought.

The spectra of linear molecules are characterized by a series of almost harmonically related frequencies, f_s (Hz), given by

$$f_s = nh/4\pi^2 I, \tag{4.38}$$

where n is an integer, h is Planck's constant, and I is the molecular moment of inertia. Table 4.4 lists the resonant frequencies of some of the molecules detected in interstellar clouds by the BIMA array. Millimeter wave radio astronomy has been used to measure the brightness temperature of the sun, moon, and the other planets. Planets appear brighter at millimeter wavelengths than at lower wavelengths and hence provide information about their surfaces and atmospheres.

The part of the sun viewed by a radiometer depends on the wavelength, since the absorption of electromagnetic energy by the solar constituents is frequency dependent. As the depth of penetration also depends on the frequency, millimeter wave observations provide information not easily obtained at other frequencies.

Table 4.4 Some molecules detected in interstellar clouds using the BIMA array

Molecule	Frequency (GHz)	Molecule	Frequency (GHz)
SiO	130.246	OCS	108.463
CN	113.492	HNCO	87.925
$C_{12}O_{16}$	115.271	CH_3OH	85.521
$C_{13}O_{16}$	110.201	CH_3CN	110.331–110.383
$C_{12}O_{18}$	109.782	CH_3C_2H	85.457
CS	146.969	X-ogen	89.190
$HC_{12}N_{14}$	88.630–88.634	HNC	90.665

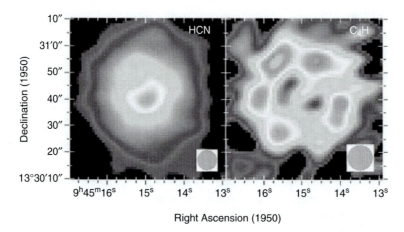

Right Ascension (1950)

Figure 4.36 Organic molecule distribution in the area around the star IRC+10216 mapped by the BIMA array (BIMA 1993).

4.16 References

Appleby, R., Anderton, R.N., Price, S., Salmon, N.A., Sinclair, G.N., Coward, P.R., Barnes, A.R., Munday, P.D., Moore, M., Lettington, A.H., and Robertson, D.A. (2003). Mechanically scanned real time passive millimetre wave imaging at 94 GHz. In *Passive Millimeter-Wave Imaging Technology VI*, *Proceedings SPIE* 5077:1–6.

Bhartia, P. and Bahl, I. (1984). *Millimeter Wave Engineering and Applications.* New York: John Wiley & Sons.

BIMA. (1993). BIMA observations of the emission from HCN and C4H in IRC+10216. Viewed November 2007. Available at http://bima.astro.umd.edu/images/astronomy/irc10216.gif.html.

CARMA. (2007). Introduction. Viewed November 2007. Available at http://www.mmarray.org/.

Currie, N. and Brown, C. (1987). *Principles and Applications of Millimeter-Wave Radar.* Boston: Artech House.

DERA. (2001). 35 GHz radiometer. Viewed February 2001. Available at http://www.dera.gov.uk.

European Organization for Astronomical Research in the Southern Hemisphere. (2007). The Atacama large millimeter/submillimeter array. Viewed November 2007. Available at http://www.eso.org/projects/alma/.

Klein, L. (1997). *Millimeter-Wave and Infrared Multisensor Design and Signal Processing.* Norwood, MA, Artech House.

Lettington, A., Dunn, D., Rollason, M., Alexander, N., and Yallop, M. (2003). Use of constraints in super-resolution of passive millimeter-wave images. In *Passive Millimeter-Wave Imaging Technology VI*, *Proceedings SPIE* 5077:100–109.

Moffa, P., Yujiri, L., Jordan, K, Chu, R., Agravante, H., and Fornaca, S. (2000). Passive millimeter-wave camera flight tests. *Proceedings SPIE* 4032:14–21.

National Aeronautics and Space Administration. (1997). Passive millimeter wave camera (PMMWC) at TRW. Viewed November 2007. Available at http://grin.hq.nasa.gov/ABSTRACTS/GPN-2000-001305.html.

National Snow and Ice Data Center. (2007). Scanning multi-channel microwave radiometer (SMMR). Viewed November 2007. Available at http://nsidc.org/data/docs/daac/smmr_instrument.gd.html.

Olsen, R., Lovberg, J., Chou, R., and Galliano, J. (1997). Passive millimeter-wave imaging using a sparse phased array antenna. *Proceedings SPIE* 3064:63–70.

Peichl, M., Suss, H., and Dill, S. (2003). High resolution passive millimetre-wave imaging technologies for reconnaissance and surveillance. In *Passive Millimeter-Wave Imaging Technology VI, Proceedings SPIE* 5077:77–86.

Robertson, D. (2004). Compact mm-wave medical imager. In *Passive Millimeter-Wave Imaging Technology VII, Proceedings SPIE* 5410:219–229.

Smith, R., Trott, B.K.D., Sundstrom, B.M., and Ewen, D. (1996). The passive mm-wave scenario. *Microwave Journal* 39(3):22–34.

Smiths Detection. (2007). Smiths Detection Tadar: millimetre-wave people screening system. Viewed November 2007. Available at http://www.tadarvision.com/downloads/Tadar%20Brochure%20May05. pdf.

Suss, H., Gruner, K., and Wilson, W. (1989). Passive Millimeter-Wave Imaging: A tool for remote Sensing. *Alta Frequenza* LVIII(5–6):457–465.

Ulaby, F., Moore, R., and Fung, A. (1987). *Microwave Remote Sensing: Active and Passive*. Boston: Artech House.

5

Active Ranging Sensors

5.1 OVERVIEW

The basic principles of active noncontact range finding are similar for electromagnetic (radar, laser, etc.) and active acoustic sensing. A signal is radiated toward an object or target of interest and the reflected or scattered signal is detected by a receiver and used to determine the range.

As shown in Figure 5.1, a source of radiation is modulated and fed to a transmit antenna, or aperture, which is usually matched to the impedance of the transmission medium to maximize power transfer. This can take the form of a horn for acoustic or radar sensors, or an appropriately coated lens for a laser. The antenna also operates to concentrate the radiated power into a narrow beam so as to maximize the operational range and to minimize the angular ambiguity of the measurement. When the transmitted beam strikes the target, a portion of the signal is reflected or scattered because the target has a different impedance, or refractive index, than the medium through which the signal is propagating. A small percentage of the reflected power travels back to the receiver (which is often collocated with the transmitter), where it is captured by the receiver antenna and converted to an electrical signal that can be filtered to remove extraneous noise before being amplified and detected.

Distance measurement methods can be classified into three categories: interferometry, time of flight, and triangulation. The method used by a particular sensor usually depends on the maximum range and the measurement accuracy required. For example, interferometric methods can be extremely accurate, but are prone to range ambiguity, while time-of-flight methods operate at longer ranges with poorer accuracy.

5.2 TRIANGULATION

Passive triangulation methods have been used for measuring the astronomical distances to the nearest stars using stellar parallax with the Earth's orbit around the Sun as a baseline (Pogge 2007). More mundane applications using active methods include pinpointing the positions of ships at sea, at ranges of a few tens of kilometers, using shorter baselines, and optical range finders that use binocular parallax to measure ranges of 100 m or so.

One of the most common triangulation sensors uses a collimated light-emitting diode (LED) source and a position-sensitive detector (PSD) to measure the direction of arrival of the reflected light beam. As shown in Figure 5.2, the relationship between the range to the target, L_o, and the distance between the transmit and receive apertures, L_B, is a function of the focal length, f, and the displacement, x', from the center of the PSD:

$$L_o = f\frac{L_B}{x'} = f\frac{L_B}{D/2 - x}. \qquad (5.1)$$

An LED or laser source emits a narrow beam of infrared (IR) light in the direction of the target. The small amount of reflected light is focused onto the sensitive surface of the PSD, which generates two output currents, I_A and I_B, that are each proportional to the displacement of the light spot from the center of the device.

Though the current generated by the PSD depends on the intensity of the incident radiation, the ratio of the two currents does not, and so the distance to the target can be determined with good accuracy. The PSD operates by exploiting the change in resistance of a doped silicon semiconductor, which is proportional to the intensity of the incident light. The device is fabricated

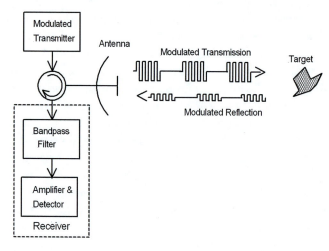

Figure 5.1 Operational principles of a generic time-of-flight sensor.

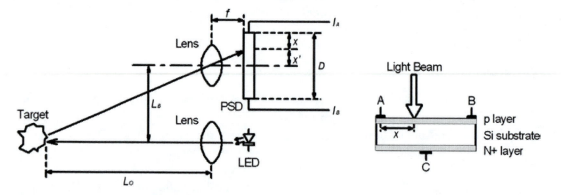

Figure 5.2 Measuring range using a PSD-based sensor.

from a thin slab of high-resistance silicon with the top and bottom layers doped p and n+, respectively. Two electrodes, A and B, are placed at opposite ends of the upper layer and a common electrode, C, covers the bottom. The distance between the upper electrodes is D and the corresponding resistance is R_D.

If the beam strikes the PSD at a distance x from electrode A, the resistance between that point and the electrode is R_x, and a photocurrent, I_o, proportional to the intensity of the light will flow. The amount that flows to the electrodes A and B will be proportional to the relative distances from the incident beam, therefore

$$I_A = I_o \frac{R_D - R_x}{R_D} = I_o \frac{D - x}{D}, \tag{5.2}$$

and

$$I_B = I_o \frac{R_x}{R_D} = I_o \frac{x}{D}. \tag{5.3}$$

Taking the ratio of the two currents I_A and I_B:

$$S = I_A / I_B = (D/x) - 1. \tag{5.4}$$

The distance, x, can then be written as

$$x = D/(S+1). \tag{5.5}$$

Substituting into equation (5.1) gives

$$L_o = f \frac{2L_B}{D} \frac{S+1}{S-1} = k \frac{S+1}{S-1} \tag{5.6}$$

where k is the module geometrical constant. Therefore L_o, the distance from the sensor to the target, can be determined in terms of the ratio of the two currents out of the PSD (Fraden 2003).

A common integrated circuit (IC) that operates using this principle is the Sharp Electronics GP2Y0A21YF, which has an operational range between 10 cm and 80 cm with the characteristics illustrated in Figure 5.3.

In more critical applications, a laser source is used because of its brightness and good spatial coherence and a charge-coupled device (CCD) array is often used as a receiver. On transmit, the width of the laser beam is diffraction limited by the size of the exit aperture. On scattering from the target, the coherence of the laser beam is lost and so, to produce the smallest spot on the CCD array, it must be placed at the focal plane of a lens (Amman 2001).

Triangulation sensors can be used to obtain two-dimensional (2D) or three-dimensional (3D) images of the terrain by scanning the beam. Such scanning mechanisms must be constructed to ensure that both the transmitter and the receiver can see each point, so that as the transmitted spot is swept across the target, a corresponding image is reflected onto the CCD array (Probert-Smith 2001). Figure 5.4 shows a diagram of a sensor developed for 2D scanning (Livingstone and Rioux 1986).

5.3 PULSED TIME-OF-FLIGHT OPERATION

Time of flight is the principle mode of operation for most radar, laser, and active acoustic devices. This technique uses the time between the transmission of a pulse and the reception of an echo

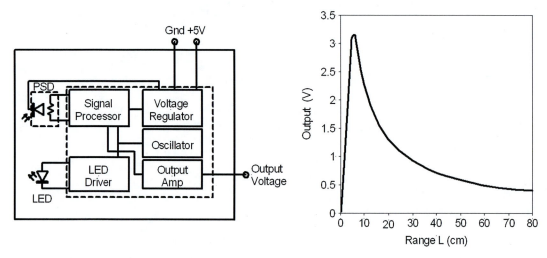

Figure 5.3 Sharp GP2Y0A21YF distance measurement IC schematic block diagram and voltage output as a function of range.

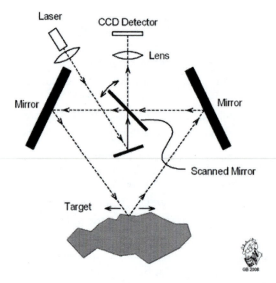

Figure 5.4 Triangulation-based line scanner.

to provide range. Because the round-trip time is measured, there is a factor of two in the formula:

$$R = v\Delta T/2, \tag{5.7}$$

where R is the range (m), v is the propagation velocity (m/s), and ΔT is the round-trip time (s).

In Figure 5.5, which illustrates the basic principle, an aircraft transmits a short radar pulse toward the ground. As the beam intersects with targets which provide good reflectors, a portion of the signal is reflected back toward the sensor. Note that the distance between pairs of pulses is twice the distance between the targets that produced them, and that if the gap is sufficiently large, then the target returns are resolved in range. However, as can be seen in frames (3) and (4), two of the targets are too close and only a single broad pulse is returned.

5.3.1 Sensor Requirements

To operate efficiently, a narrow beam must be formed to concentrate the transmitted energy, the transducer must be matched to the characteristics of the medium to maximize the power transferred, and the receiver must be matched to the transmitted pulse characteristics. In addition, some form of modulation must "mark" the carrier signal so that the round-trip time can be measured. Figure 5.6 shows an amplitude modulated carrier, including the "bang" (transmit) pulse that leaks through from the transmitter into the receiver, and then a pair of echoes after a short delay. The remainder of the trace consists of background noise that is generated by the receiver input stages (see Chapters 2 and 4 for more detail).

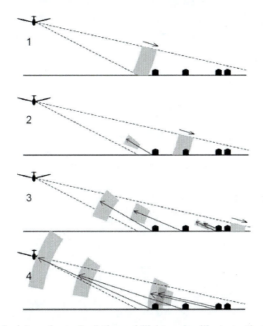

Figure 5.5 Operational principles of a pulsed time-of-flight radar illustrate that the echoes from the nearer houses return to the radar first and that the echoes from the last two houses overlap and appear to the radar as a single return.

It is not convenient to process the modulated carrier because the frequency is often high, and in most cases the carrier frequency carries no information. In general it is detected (demodulated) to produce the envelope shown in Figure 5.7 before the timing information can be extracted.

5.3.2 Speed of Propagation

For electromagnetic radiation, be it light or radio waves, the propagation speed, v (m/s), is a function of the refractive index of the material:

$$v = c/N = c/\sqrt{\varepsilon_r}, \tag{5.8}$$

where c is the speed of light in a vacuum (3×10^8 m/s), N is the refractive index of the material, and ε_r is the relative dielectric constant of the material. It is interesting to note that the refractive

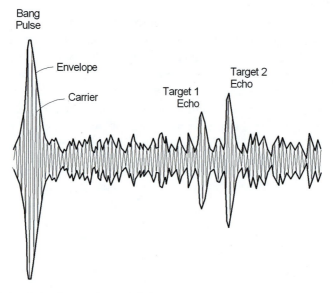

Figure 5.6 Received signal showing carrier modulation.

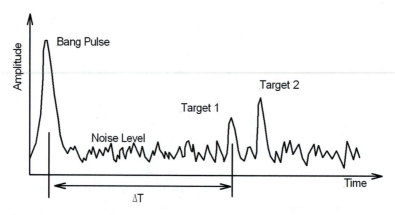

Figure 5.7 Received signal after envelope detection.

index of most materials is also a function of the wavelength, so as a signal propagates through a medium, different frequencies will move at different speeds and the shape of the envelope will evolve.

For acoustic signals, the propagation velocity is a function of the bulk modulus, B, and the density, ρ (kg/m^3), of the material through which the sound is traveling:

$$v = \sqrt{\frac{B}{\rho}} = \sqrt{\frac{1}{K\rho}}, \tag{5.9}$$

where K is its compressibility.

The bulk modulus of the material is defined as the ratio of the applied pressure to the strain:

$$B = \frac{P}{(V_o - V_n)/V_o}, \tag{5.10}$$

where P is the pressure and V is the volume. The bulk modulus of air is equal to $1.4P$, and at sea level $P_{air} = 1.01 \times 10^5$ N/m^2 and $\rho_{air} = 1.23$ kg/m^3. Table 5.1 lists the speed of sound through a number of solids, liquids and gasses.

Note that the temperature, salinity, and density determine the variation in the speed of sound in the sea (see Chapter 7). For underwater sonar, this variation in speed must be considered if accurate measurements are required.

5.3.3 The Antenna

The antenna acts as a transducer to convert electrical signals constrained within wires to those that propagate through space. During transmission it concentrates the radiated energy into a shaped beam that points in the desired direction to illuminate only the selected target, while on reception it collects a small fraction of the energy reflected by the target and delivers it to the receiver. The effectiveness of these two roles is quantified by the transmitting gain and the effective receiving aperture of the antenna. In sonar applications, the gain is known as the directivity index.

For a given input power, the antenna gain is defined as the power density radiated in a particular direction compared to that radiated by an isotropic element (one which radiates equally

Table 5.1 Speed of sound in different materials

Material	Speed (m/s)	Material	Speed (m/s)
Air (STP)	343.2	Sea water	1450–1750
Air 0°C, dry	331.29	Methanol 30°C	1121.2
Nitrogen 0°C	334	Mercury	1451
Oxygen 0°C	316	Aluminum	5000
Carbon dioxide	259	Copper	3750
Helium 0°C	965	Lead	1210
Water 0°C	1402.3	Steel	5250

in all directions). It can be determined from the size of the antenna relative to the wavelength of the radiated signal:

$$G = 4\pi A/\lambda^2, \qquad (5.11)$$

where A is the frontal area of the antenna (m^2) and λ is the wavelength (m).

The antenna beamwidth is generally defined as the total angle between the points on the pattern where the radiated power is equal to 50% of the power at the peak, as shown in Figure 5.8. This is also referred to as the 3 dB beamwidth because the decibel value of 0.5 is −3 dB.

From the definition of the antenna gain, it is obvious that as the beamwidth increases, the gain decreases proportionally. Considering that an isotropic element radiates equally over 4π steradians and that a directional antenna radiates over an area equal to $\theta_{3dB} \times \varphi_{3dB}$, a good approximation for the gain of an antenna is

$$G \approx 4\pi/\theta_{3dB}\varphi_{3dB}, \qquad (5.12)$$

where θ_{3dB} and φ_{3dB} are the beamwidths in orthogonal planes (rad).

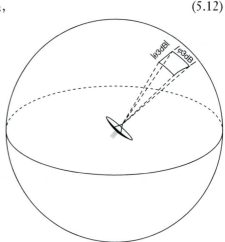

In many antennas, the shape of the beam is determined by a diffraction pattern generated by the radiating aperture. This is known as the diffraction-limited beam, and is common in radar and acoustic antennas where the operational wavelengths are relatively long. In laser systems, where wavelengths of 900 nm or less are common, the antenna pattern is usually tailored by a lens or a mirror for a specific application.

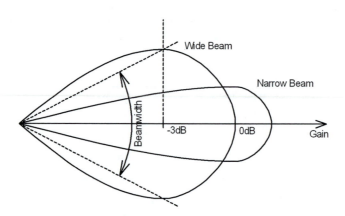

Figure 5.8 Relationship between the gain and beamwidth of an antenna in two dimensions.

For a circular aperture that is uniformly illuminated, the radiated power, P, as a function of the angle off the boresight, θ, is

$$P = \left| \frac{\mathcal{J}_1 \left[(\pi D / \lambda) \sin \theta \right]}{(\pi D / \lambda) \sin \theta} \right|^2, \tag{5.13}$$

where $\mathcal{J}_1$ is the Bessel function of the first kind, D is the diameter of the aperture, and λ is the wavelength of the radiation. This function is plotted in Figure 5.9, which shows that the half-power beamwidth occurs at $u = 1.62$, which corresponds to $\theta_{3\text{dB}} \approx 1.02\lambda/D$. The first minimum occurs at $u = 3.83$, which corresponds to $\theta \approx 1.22\lambda/D$ as expressed by Rayleigh's criterion for resolution defined in Chapter 3. It is interesting to note that, in this ideal case, the theoretical resolution is better than the Rayleigh criterion by a factor of about 1.2.

The pattern for a rectangular aperture is the sync function, and it is also plotted in the figure. The power pattern can be determined by taking the square of this function, and in this case the half-power beamwidth occurs at $u = 1.39$, which corresponds to $\theta_{3\text{dB}} \approx 0.885\lambda/D$:

$$P = \left| \frac{\sin \left[(\pi D / \lambda) \sin \theta \right]}{(\pi D / \lambda) \sin \theta} \right|^2. \tag{5.14}$$

It can be seen that although the majority of the power is radiated by the main lobe, there are significant sidelobes. In the case of the rectangular aperture, the first sidelobe is only a factor of 20 down relative to the peak of the main lobe. Although this is adequate for some applications, there are many instances where the sidelobe level must be reduced further. This can be achieved by weighting the illumination function across the antenna aperture, as is discussed in Chapter 12.

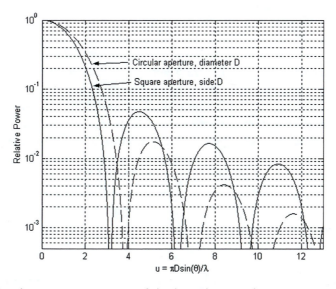

Figure 5.9 Comparison between power patterns of circular and rectangular antenna apertures.

```
% antenna_pattern.m
% generate antenna patterns for circular and square apertures
%
% Arguments:
%
%          k = pi.D*sin(theta)/lamdba
%
% some physical constants:

k=0:0.1:13;

% calculate the power pattern for a square aperture
p_sq = (sin(k)./k).^2;

% calculate the power for a circular aperture
p_cir = (besselj(1,k)./k).^2;
p_cir = p_cir/max(p_cir);

semilogy(k,p_sq,'k-',k,p_cir,'k-');

grid
axis([0,13,5e-04,1])
xlabel('u = \piDsin(\theta)/\lambda')
ylabel('Relative Power')
```

The relationship between the aperture and the antenna beamwidth described above is still valid if the aperture is not square or circular. Many antennas are elliptical or even linear, and will therefore generate antenna patterns that are asymmetrical. A typical example of an asymmetrical antenna would be one used in the ground imaging application illustrated in Figure 5.5. Such an antenna would have a narrow azimuth beamwidth to maximize the cross-range resolution and a wide elevation beamwidth to illuminate a wide swathe of ground. In addition, the antenna may be constructed to concentrate more power at long range and less at short range to compensate for losses.

Examples of two different antennas are shown in Figure 5.10. The circular antenna generates a narrow symmetrical pattern called a pencil beam, and is used for tracking aircraft. The elongated rectangular antenna produces a beam that is narrow in azimuth and wide in elevation, called a fan beam, and is used to search.

5.3.4 The Transmitter

The transmitter generates a carrier signal of the correct frequency that is radiated by the antenna. To tag the carrier so that the time of flight can be measured, it is "marked" using some form

Figure 5.10 Crotale FU showing the fan-beam search antenna and the pencil-beam tracking radar (Army Technology 2007).

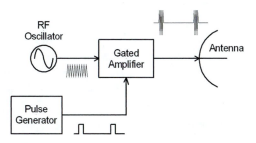

Figure 5.11 On-off amplitude modulation.

of modulation, as discussed in Chapter 2. This can be achieved using any of the following methods:

- Amplitude modulation (AM)
- Frequency modulation (FM)
- Phase modulation (PM)
- Polarization modulation.

For most simple time-of-flight sensors, the carrier is on-off amplitude modulated, as shown in Figure 5.11. In a typical radar application, the oscillator frequency will be in the microwave band, between 1 and 100 GHz, while in a sonar application the frequency will be in the audio or ultrasonic band, depending on the application.

A new range measurement can occur every time a pulse is transmitted. It is therefore useful to maximize this rate, which is known as the pulse repetition frequency (PRF). The relationship between the transmitted and received pulse trains is illustrated in Figure 5.12 from which it is obvious that the time between pulses must be sufficiently long to ensure that echoes, even from the most distant targets of interest, are received before the next pulse is transmitted. Any late-

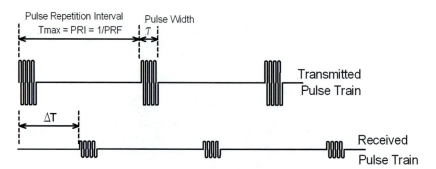

Figure 5.12 Transmitted and received pulse trains.

arriving echoes received after this instant will be ambiguous in range, because they will apparently come from nearby targets. The relationship between maximum unambiguous range, R_{max} (m), and the interval, T_{max} (s), between pulses is

$$R_{max} = vT_{max}/2, \tag{5.15}$$

where v (m/s) is the velocity.

The range at which a target can be detected is proportional to the total energy transmitted by a pulse, and this energy is equal to the product of the peak power and the duration. It is therefore advantageous to make both the power and width as large as possible. There are physical limitations to the power that can be transmitted, whether it is acoustic or electromagnetic in nature, so the maximum range is determined by the pulse width, τ. However, this parameter must also be selected to accommodate the required range resolution, and a compromise is often called for.

In a definition analogous to the Rayleigh criterion for angular resolution, the range resolution, δR (m), is defined as the minimum distance between two targets of equal amplitude, which can still be resolved in range. For a constant frequency, on-off modulated, pulsed sensor, this resolution is proportional to the width of the pulse, τ (s), and the speed of propagation, v (m/s):

$$\delta R = v\tau/2. \tag{5.16}$$

It is possible to use pulse compression techniques that use more complex modulation schemes to improve this resolution beyond that dictated by the pulse width. These options are discussed in detail in Chapter 11.

5.3.4.1 Radar Transmitters   Radar transmitters are made using oscillators that produce extremely high-frequency sinusoidal electrical signals. These pulsed signals are either generated with very high powers directly using microwave tubes such as magnetrons and klystrons, or by lower power continuous-wave oscillators which are subsequently gated and amplified using tubes or solid-state components.

Depending on the application, radar transmitters can be designed to produce powers from less than 1 μW, such as the micropower impulse radar (MIR) sensor (Fraden 2003), to megawatts for long-range surveillance. An example of the latter is the ARSR-1 2D air surveillance radar made by Raytheon (Brookner 1982) that transmits a peak power of 4 MW.

Since their introduction as radar transmitters during World War II, multicavity magnetrons have been the most common of all the high-power sources for radars. They are still used in common commercial applications such as small boat radars because they are both cheap and efficient. Magnetrons are cross-field devices in that they use both an electric and a magnetic field for operation, as illustrated in Figure 5.13. Electrons are released from the hot central cathode and are drawn toward the anode, but the axial magnetic field exerts a force perpendicular to this and forces them to be swept around in a circle. This revolving space charge couples to the resonant cavities and builds in intensity as more direct current (DC) power is converted to an oscillating current. A loop antenna in one of the cavities couples microwave power out of the device.

For low-power applications, transistor oscillators are used. These use the more conventional amplifier and frequency-sensitive feedback configuration shown in Chapter 2 to produce oscillations. With the development of integrated radio frequency (RF) building blocks at chip level, higher frequencies are being generated by multiplying up and amplifying these low-frequency signals. Many 77 GHz automotive radars already use this technique, and as the technology matures it is beginning to compete with the more conventional cavity-based oscillators based on

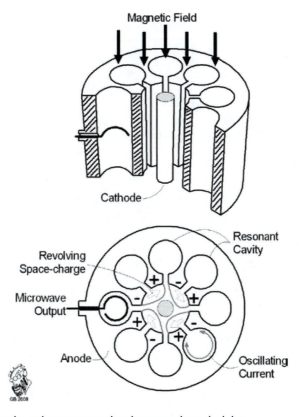

Figure 5.13 Cross section through magnetron showing operating principles.

Gunn or impact ionization avalanche transit-time (IMPATT) diode technology. Figure 5.14 shows the schematic design for one method of constructing oscillators for two-port devices.

Typically a Gunn or IMPATT diode is mounted within a reduced-height waveguide cavity, the height of which has been calculated to match the impedance of the diode. An insulated pin passes through the waveguide wall and biases the diode into its negative resistance region so that oscillations can take place. A sliding short is used to tune the device frequency and a stepped impedance transformer matches the output to the full-height waveguide output to maximize power transfer.

As new technology and improved computer-aided design and simulation software becomes available, increases in both transmit power and frequency have occurred steadily over the past 40 years, as can be seen in Figure 5.15.

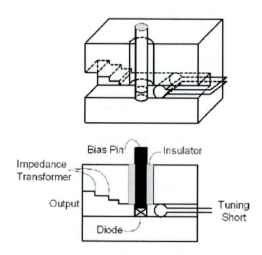

Figure 5.14 Cavity-based IMPATT oscillator schematic (Bhartia and Bahl 1984).

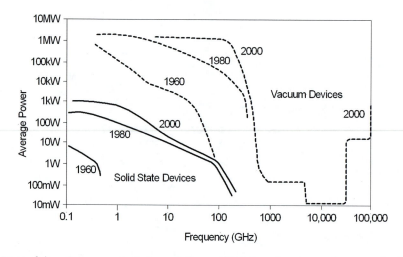

Figure 5.15 State-of-the-art power outputs from various solid-state and vacuum tube transmitters.

5.3.4.2 Underwater Sonar Transmitters Piezoelectric materials possess ferroelectric characteristics, so named by analogy with ferromagnetic materials. Although the piezoelectric effect was discovered in 1880, it was not until 1917, when Langevin used x-cut plates of quartz to generate and detect sound waves in water, that the effect found a practical use.

Some materials like quartz, Rochelle salt (sodium potassium tartarate), and tourmaline are naturally piezoelectric because they contain electric dipoles that are aligned along the crystal axes. In other materials, the dipoles are aligned at random, and the materials need to be poled to develop this characteristic. This involves heating the material to above its Curie temperature, placing it in a strong electric field, and then cooling while the electric field is maintained. The dipoles stay "frozen" in the direction of the applied electric field so long as the material temperature remains below the Curie temperature.

These days, the most common piezoelectric materials are manufactured ceramics, of which the first was barium titanate ($BaTiO_3$). Impurities are now routinely added to lock the polarization in place (Fraden 2003).

Low-frequency, high-power sound sources are conveniently formed using stacks of small transducers. Most of the transducers use piezoelectric materials such as lead titano zirconate because their electromechanical properties are good and their dielectric losses low. Another advantage is the ease with which a variety of shapes can be manufactured using conventional ceramic manufacturing processes. Their main drawback is their relative fragility caused by their low tensile strength. In some cases this can be overcome by prestressing the stack of transducers into compression. Additional problems include variations in the characteristics of individual transducers in the stack and a gradual decay of the polarizing field (and hence efficiency) with time.

A typical transducer is the longitudinally vibrating structure called a tonpilz. It consists of a stack of axially polarized piezoceramic rings and a prestress rod that passes through their center and applies a compressive bias to the transducers and the cement joints, as illustrated in Figure 5.16. Transducers of this type can generate sound intensities of only a few watts per square centimeter, so if high powers are required, they are formed into large arrays.

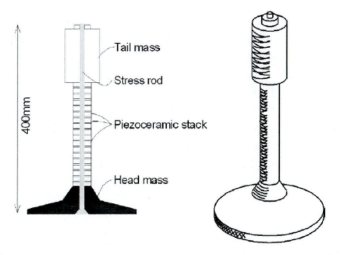

Figure 5.16 Construction of a tonpilz piezoceramic transducer.

Piezoceramic materials have a very high Young's modulus, so for low-frequency operation it is often preferable to use rare earth magnetostrictive alloys, which are less stiff. A magnetostrictive material contains atoms with permanent magnetic moments coupled to the crystalline lattice so that an applied magnetic field can change the lattice spacing, and hence the dimensions of the transducer. Most commonly these transducers consist of a long, thin, nickel ribbon wound tightly into a cylinder. This is surrounded by a solenoid that generates the magnetic field when a current is passed through its windings. The design of most transducers comprises a number of magnetostrictive elements to contain the magnetic flux within the structure, as is common for conventional loudspeakers. A constant field current is applied in addition to an alternating current to linearize the transducer's dynamic behavior.

At even lower frequencies, where a large stroke is required to radiate any useful amount of power into the water, conventional moving coil designs are used. These must include compensation for changes in external pressure with depth to keep the coil centered in the magnetic field (Boucher 1987).

5.3.4.3 Ultrasonic Transmitters   The earliest ultrasonic transmitters were whistles that generated high-frequency oscillations from a jet of compressed air. The Hartmann whistle, which uses a shock wave to excite a resonant cavity, is capable of producing output powers of up to 150 W with an efficiency of about 5%. To generate higher powers, sirens, in which a rotor periodically interrupted the flow of air, were developed. These are capable of producing a few kilowatts of power, but also release a lot of gas.

Conventional electromagnetic transducers can be used to generate ultrasound in the air, but as with loudspeakers, there are problems with maintaining a piston motion without exciting resonances in the radiating plate. Careful design allows transducers to be produced that operate at frequencies up to 20 kHz with an efficiency of 30%. To counteract these resonant modes, shields are sometimes used to block radiation from the out-of-phase sections of the plate, or flexural plates are added. These improvements have been shown to result in efficiencies of up to 81% at low power, with sound-pressure levels (SPLs) in the range of 1 m up to 128 dB (relative to 0.0002 μbar) for 1 W of electrical power.

For good impedance matching and good directivity in air, the size of the radiating plate should be as large as possible, but this generally leads to the issues discussed above. In 1978 a new transducer was developed by J. Gallego that was suitable for high-power applications (Gallego et al. 1978). The structure of this transducer, shown in Figure 5.17, consists of a large circular plate with machined concentric grooves that is driven at its center by a piezoelectric stack through a mechanical amplifier.

The large radiating plate increases the radiation resistance and offers good impedance matching with the air when it flexes in its axis-symmetrical modes. The steps in the plate are one-half wavelength in depth, so when they are flexing in antiphase to the remainder of the plate, the radiated sound is in phase, as can be seen in Figure 5.18. This results in good efficiency and a directivity equal to that of a piston of the same diameter. With careful optimization, including small steps in the back of the radiating plate (visible in Figure 5.18), the original transducer machined from a 6 mm thick titanium plate 335 mm in diameter was 85% efficient and produced a beamwidth of 2° at a frequency of 20 kHz. The SPL obtained for 150 W of input power is about 170 dB, as there is a focusing effect.

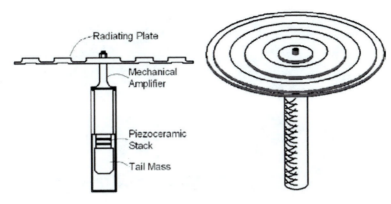

Figure 5.17 High-power stepped-plate ultrasound transducer.

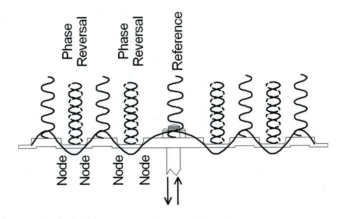

Figure 5.18 Radiation mechanism of the stepped-plate configuration.

One limitation of this design is an extremely narrow bandwidth of ±2 Hz around the central resonant point. As this varies with temperature and the applied load, a feedback mechanism is required to adjust the oscillator frequency (Gallego et al. 1978). This technology is applied by a number of manufacturers, including the Milltronics AiRanger products shown in Figure 5.19.

For short-range applications, most ultrasonic transmitters are made from either thin wafers of piezoelectric material or use electrostatic transducers. A typical piezoelectric transducer similar to the unit shown in Figure 5.20, such as the SensComp 40LT16, has the following characteristics (Senscomp 2007a):

Center frequency: 40 +/−1 kHz,
Bandwidth: 2 kHz,
Transmit SPL: 120 dB min at 40 kHz; 0 dB re 0.0002 μbar per 10 V rms at 30 cm,
Maximum drive voltage: 20 V rms,
Beam angle (6 dB): 55°.

Figure 5.19 Milltronics acoustic measurement system.

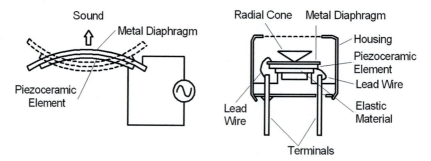

Figure 5.20 Low-power piezoelectric transducer.

Electrostatic transducers are typically larger than the piezoelectric variety, and so have narrower beamwidths. Because they are not resonant, they also exhibit a much wider bandwidth as seen in the abridged specifications for the SensComp 600 series reproduced below (Senscomp 2007b):

- Center frequency: 50 kHz
- Bandwidth: ≈40 kHz
- Transmit SPL: 110 dB minimum at 50 kHz; 0 dB for 20 μPa at 1 m (300 VACpp 150 VDC bias)
- Beam angle (6 dB): 15°.

The output power is specified as an SPL at a fixed distance from the transducer. From the specifications for the two transducers, it can be seen that the units used and the measurement range are different in the two situations. Designers should be aware of this when selecting or comparing transducers.

One of the problems with both low- and high-power transducers is the ring-down period. Because these devices are generally high Q, even after the excitation voltage has been disconnected, they continue to oscillate, as shown in Figure 5.21. During this time the high-amplitude oscillations mask any echoes, so there is a period after transmission during which no target can

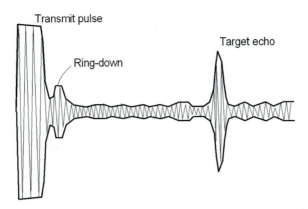

Figure 5.21 Salient features of an acoustic pulse.

be reliably detected. This is known as the blanking time or distance, and is typically between 1 and 10 ms (0.17 to 1.7 m range).

5.3.4.4 Laser Transmitters

Laser (light amplification by stimulated emission of radiation) devices generate a coherent beam of light within a very narrow bandwidth with identical polarization and phase. This is achieved, as the acronym suggests, by stimulated emission in which one photon triggers the release of another with identical characteristics.

When an electron has been excited into a higher energy state by some mechanism, such as an electrical discharge in a gas, it generally does not remain in this state very long, and will return to its original state. This transition results in the release of energy in the form of a photon, a process known as spontaneous emission. If a sufficiently large number of electrons have been elevated to the higher energy state, in a condition known as a population inversion, the presence of a photon with the correct energy can stimulate the electron transition to the lower energy state, with the release of an identical photon in a process known as stimulated emission. These two photons can stimulate another pair, and the cascade of photons grows.

To produce a beam, most lasers are constructed in the form of a resonator comprising two parallel mirrors at opposite ends of the lasing medium. This results in large numbers of photons traveling from end to end coherently and increasing in number with every transition. One of the mirrors is perfectly reflecting and the other is partial, so a small percentage of the light is emitted as a beam. A technique called Q-switching involves the introduction of a shutter that blocks the resonant path and allows the energy stored within the lasing material to build up during the pumping period. When the shutter is opened, most of the energy is released as a single high-intensity pulse.

The most common types of laser transmitters are gas, solid state, and semiconductor. Gas lasers can operate over a wide range of frequencies, from the ultraviolet (UV) through visible and IR right down to the millimeter wave band. Powers range from a few milliwatts up into the kilowatt range. Though the once ubiquitous helium-neon laser has been replaced by semiconductor and solid-state lasers at the red end of the visible spectrum, gas lasers remain the standard sources for most other wavelengths.

The carbon dioxide laser is considered to be the most versatile of the gas lasers, as it can be operated in pulsed or continuous modes and is able to produce the highest output power of any readily available device. These lasers operate using transitions over vibrational-rotational states to produce long-wavelength outputs of between 9 and 11 μm in the IR range. They are commonly used for drilling, welding, and cutting.

Excimer lasers are a family in which light is emitted by short-lived molecules made up of an atom of one of the noble gases (argon, krypton, or xenon) and one halogen (fluorine, chlorine, or bromine). These lasers are important as they can produce high output powers in the UV range which are useful for research and medical applications.

A solid-state laser is one in which the atoms that emit light are fixed in a crystal or glassy material. These include the original ruby laser made by Theodore Maiman, as discussed in Chapter 1, and all operate using the same basic principles. In general, a rod made of the lasing material is manufactured as a resonant cavity by placing mirrors at each end. Light from an external source, typically a flashlamp, excites the light-emitting atoms to form the inverted population needed to generate stimulated emission.

From a range-sensor perspective, neodymium-yttrium aluminum garnet (Nd-YAG) lasers are the most common, as they are used as transmitters for laser range finders. In this formulation, Nd atoms are embedded in YAG to produce an efficient lasing medium that also has good thermal, optical, and mechanical properties. The energy level transitions for this material are shown in Figure 5.22. Typically rods with diameters of between 6 and 9 mm are manufactured in lengths of up to 10 cm. This small size limits the pulsed output power to less than about 0.5 J, but the pulse repetition rate can be high. Average powers of more than 1 kW can be obtained and peak powers can reach hundreds of kilowatts.

The original ruby laser used a spiral flashlamp to pump the laser, but these are inconvenient to make and the energy coupling to the laser rod is low. Most modern lasers use highly polished cylindrical reflectors with elliptical cross sections with a lamp at one focus and the laser at the other, as illustrated in Figure 5.23. Coupling efficiencies of up to 75% can be achieved (Marshall 1968).

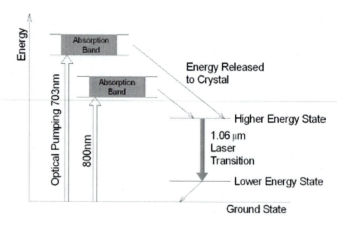

Figure 5.22 Energy levels in neodymium lasers [adapted from (Hecht 1994)].

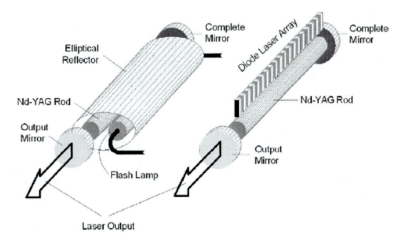

Figure 5.23 Diode and flashlamp pumping of solid-state lasers.

Lamp-pumped Nd-YAG lasers operate at efficiencies of between 0.1% and 1%, but when pumped with semiconductor lasers, efficiencies exceeding 10% can be obtained. This allows small long-range laser range finders to be built (Hecht 1994).

Semiconductor or "diode" laser transmitters rely on the quantum mechanical operation of carefully doped PN junctions to produce coherent radiation in the IR or visible region when biased by a short current pulse. The process occurs when electrons and holes recombine at the junction and drop back from the conduction band into the valence band, with the release of energy as a photon of light. The energy is proportional to the band gap, so in silicon, which has a small band gap, the energy is released as heat, but in gallium arsenide and some other materials, it is released as light in a process that forms the basis for both LEDs and laser diodes.

Important differences between LEDs and laser diodes include the current level. Low currents flow through the former, causing a moderate amount of recombination, with its associated release of photons. However, a much higher current flows through a laser diode junction, sufficient to form a large concentration of electron-hole pairs, enough to dominate over absorption. This generates the required population inversion which is essential for lasing operation.

Spontaneous emission generates photons that travel in any direction, so as with other lasers, a pair of reflective surfaces is used to confine the stimulated emission to a specific direction. This is normally achieved by cleaving the semiconductor crystal to make parallel facets perpendicular to the junction to constrain the stimulated emission to a strip within the junction plane, as shown in Figure 5.24. Because semiconductors have high refractive indices, it is generally not necessary to add to the mirror finish, as at least 30% of the light is reflected. Some lasers do include one reflective face, and others include a photodetector on that face to monitor the output power.

At low current levels, laser diodes generate some spontaneous emission, using the same processes that drive LEDs. However, as the current increases, a threshold is reached above which stimulated emission begins to occur, and the intensity of the output increases dramatically. This threshold current level determines the efficiency of the laser diode, as most of the current below this point is converted to heat and must be dissipated.

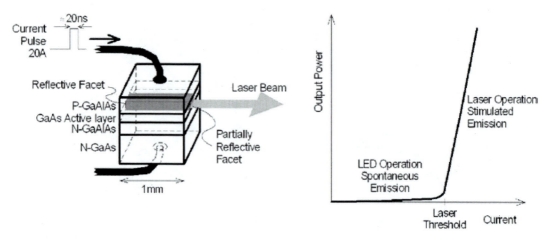

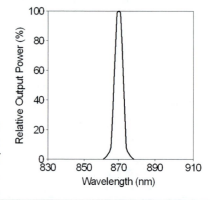

Figure 5.24 Schematic diagram of a laser diode junction and the relationship between the current input and output power.

Most normal diodes are made using the same material with different doping profiles, and are known as homostructures, but laser diodes can be made from different materials on either side of the junction. These different materials have different refractive indices and so constrain the beam within the active junction (rather like a waveguide) to improve the overall efficiency.

The active layer of a laser diode is typically about 1 μm thick and a little wider than that. The beam that exits such a small aperture will be elliptical with a typical divergence of 30° × 10°.

A typical high-power example is the Hamamatsu L7060-2 pulsed-laser diode, which has the following key specifications (Hamamatsu 2007):

- Power output: 30 W
- Wavelength: 870 nm
- Threshold current: 1 A
- Forward operating current: 35 A
- Forward voltage drop: 7 V
- Beam spread: 30° × 9°.

In addition to the development of gallium aluminum arsenide (GaAlAs) for near-IR laser diode material, recent research has focused on mid-IR devices for a number of reasons. This band has a wider range of wavelengths available to match any particular absorption, transmission, or emission feature, which can be useful in research. In addition, at wavelengths greater than 1.4 μm, the vitreous humour of the human eye is nearly opaque and laser light cannot be focused on the sensitive retina. Therefore these devices are relatively eye-safe (Klein 1997).

5.3.5 The Receiver

Each of the different transmitter types discussed in the previous section uses a different type of receiver to convert radiated and reflected energy into an electrical signal that can be processed to extract the relevant range and amplitude information. As discussed earlier in this chapter, microwave radar antennas capture a modulated electromagnetic signal which is then envelope detected. Both sonar and ultrasound systems use reciprocal transducers that act as transmitters by converting electrical signals to sound on transmit, and then convert the reflected sound back to an electrical signal on receive. Laser-based sensors usually use sensitive avalanche or PIN photodiodes, similar to those discussed in Chapter 3, to convert the reflected pulse directly to an electrical signal.

Probably the most critical parts of all of the receivers are the filters that limit the signal entering the device to one that matches the signal transmitted. Systems can include more than one filter to accommodate different characteristics of the signal. For example, a lidar will have an optical filter that blocks all light except the narrow band at the wavelength radiated by the laser. Microwave systems have similar filters that are intrinsic to the narrow-band antennas and waveguide at their inputs, while acoustic receive transducers are often resonant, and hence only sensitive at the radiated signal frequency.

A subsequent tuned band-pass filter ensures that the bandwidth of the received signal is reduced still further so that it matches the characteristics of the transmitted signal to maximize the received signal-to-noise ratio (SNR). This is referred to as a matched filter, and as a rule of thumb, for pulsed sensors, the bandwidth, β, of the transmit signal is determined by its pulse width, τ, according to the relationship

$$\beta = 1/\tau. \tag{5.17}$$

If the bandwidth of the receiver is wide compared to the bandwidth of the transmitted pulse, noise introduced by the extra bandwidth reduces the SNR, while if the receiver bandwidth is too narrow, it will "ring" when it receives a pulse. Figure 5.25 shows the simulation results for an ultrasonic system in which the filter bandwidth is adjusted to illustrate this effect. The rectangular transmit pulse with a duration of 0.5 ms reflects from two targets at ranges of 0.75 m and 0.87 m and the echoes are passed through filters with different bandwidths. For comparison purposes, the unfiltered signal is displayed.

The most obvious result of passing the received signal through all of the filters is improvement in the SNR compared to the unfiltered signal. In both cases the target echo is much easier to discern above the residual noise. The return in the matched case, with a 2 kHz bandwidth, shows two distinct echoes, each roughly triangular in shape and each with a single distinct peak. When the bandwidth is widened to 8 kHz the amount of noise increases, and although two distinct echoes are received, they each have multiple peaks. Finally, in the case where the bandwidth is made too narrow, 0.5 kHz in this case, the filter response is too slow to capture both of the peaks and a single return is observed that continues to ring long after the stimulus has ended.

One additional result that can be observed is that there is an increasing delay through the filter as the bandwidth is reduced. It is therefore important, for time-of-flight measurements, that both the transmit and the receive pulses be passed through the filter if an accurate measure of the range is to be obtained.

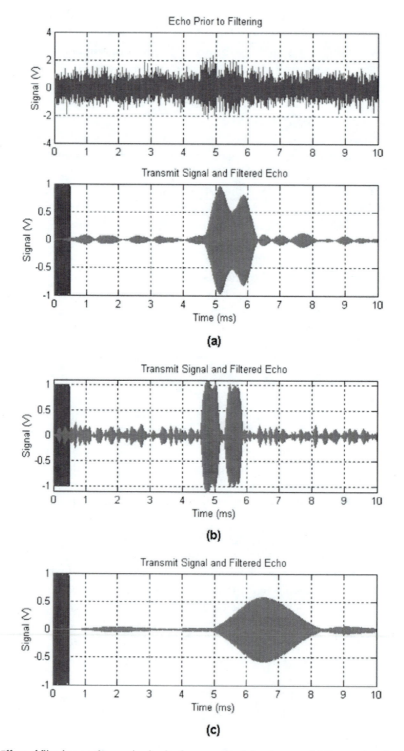

Figure 5.25 Effect of filtering on the received echo from a pair of closely spaced 0.5 ms pulses for (a) a matched filter with a bandwidth of 2 kHz, (b) a filter with a bandwidth of 8 kHz, and (c) a filter with a bandwidth of 0.5 kHz.

```
% Ultrasonic matched filter evaluation
% mat_fil.m

% Variables:
%
%         fcar            carrier frequency (Hz)
%         fmat            matched filter center frequency (Hz)
%         bmat            matched filter bandwidth (Hz)
%         prf             pulse repetition frequency (Hz)
%         tau             pulse-width (s)
%         r1, r2          range of point targets (m)
%         vnoi            noise voltage (V)
%
%
fcar = 40.0e03;                    % center frequency signal (Hz)
fmat = 40.0e03;                    % matched filter center
                                     frequency (Hz)
bmat = 0.5e03;                     % matched filter bandwidth (Hz)
prf = 100;                          % pulse repetition frequency
                                      (pps)
tau = 0.5e-03;                     % pulse period (sec)
r1 = 0.75;                          % range target 1 (m)
r2 = 0.87;                          % range target 2 (m)
vnoi = 0.5;                        % rms value of the noise (V)


% other constants
v = 331.0;                          % speed of sound (m/s)
ts = 1/(20*fcar);                  % 20 samples per carrier
                                      wavelength
ttot = 1/prf;                      % time for one transmit cycle

% generate a continuous transmit signal
t=(0:ts:ttot);
lent=length(t);
sigtx=cos(2*pi*fcar*t);


% generate the transmit pulse modulation
txmod = zeros(size(t));
len = tau/ts;                      % pulse length in samples
dum=(1:len);
x=ones(size(dum));
txmod(1:len)=x;                    % mask for the first pulse
sigtx=sigtx.*txmod;
```

```
% generate the receive signal by convolving the point target
% with the modulated transmit signal
rxmod = zeros(size(t));

del1 = fix(2*r1/(v*ts));              % sample delay to the targets and
                                        back
del2 = fix(2*r2/(v*ts));
rxmod(del1)=1;                        % they are point targets
rxmod(del2)=1;
sigrx=conv(sigtx,rxmod);             % generate the received signal by
                                       convolution
sigrx = sigrx(1:lent);                % truncate to the same length

% add noise to the received signal
noise = vnoi*randn(size(sigrx));
sigrx=sigrx+noise;
% construct the receive band-pass matched filter
wl=2*ts*(fmat-bmat/2);               % lower band limit
wh=2*ts*(fmat+bmat/2);               % upper band limit
wn=[wl,wh];
[b,a]=butter(3,wn);                    % 6th order Butterworth band-
                                         pass filter

% plot the transfer function of the band-pass filter
[h,w]=freqz(b,a,1024);
freq=(0:1023)/(2000*ts*1024);
plot(freq,abs(h));
grid
title('Matched Filter Transfer Function')
xlabel('Frequency (kHz)');
ylabel('|H(f)|')
pause

% filter the received signal
sigrxfil=filter(b,a,sigrx);

% plot the results
subplot(211),plot(t*1000,sigrx)
grid
title('Echo Prior to Filtering')
ylabel('Signal (V)')
```

```
subplot(212),plot(t*1000,sigtx,t*1000,sigrxfil)
grid
title('Transmit Signal and Filtered Echo')
ylabel('Signal (V)')
xlabel('Time (ms)')
```

For other waveforms (pulse compression, etc.), the matched filter form must be calculated for the specific waveform and the type of noise. This is addressed in Chapter 11.

5.4 PULSED RANGE MEASUREMENT

The simplest method for determining the range to a target is to time the interval from the transmission of a pulse until the echo is received. As shown in Figure 5.26, this can be achieved by counting high-speed clock cycles from the leading edge of the bang pulse to the leading edge of the echo pulse. If the clock frequency is selected correctly for the propagation velocity, the counter output can be used to display the range directly.

In this simple example, the echo signal envelope voltage is compared to a threshold that is adjusted so that the probability of a false alarm (noise alone exceeding the threshold) is low, while there is still a good probability that the signal + noise will exceed the threshold. The threshold is set just too low in this case, and a noise pulse triggers a false alarm.

The radar range equation, derived later in this chapter, shows that the amplitude of a fixed cross-section target echo decreases with the fourth power of the range (R^{-4}). It is therefore useful to adjust the detection threshold, V_t, automatically to compensate. This is called sensitivity time control (STC) and its action is illustrated in Figure 5.27. It should not be confused with automatic gain control (AGC), which is addressed in Chapter 13 and which performs a different function.

This technique is ideal for measuring the range of targets using acoustics in air and underwater where the propagation velocity is low, but it cannot easily be used to measure the target range

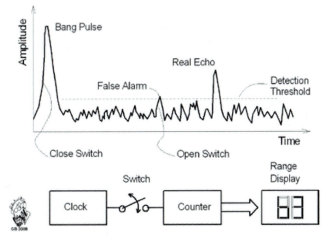

Figure 5.26 Measuring range using a time-of-flight sensor.

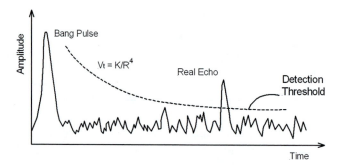

Figure 5.27 Adjustable detection threshold to accommodate propagation losses for targets with a constant cross section.

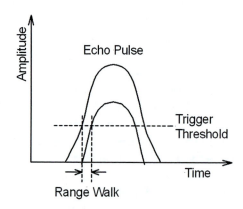

Figure 5.28 Timing errors introduced by variations in the amplitude of the received pulse echo, known as range walk.

to an accuracy of better than about 1 m for radar or laser applications where the propagation velocity is much higher. For example, sound propagation velocity underwater is about 1500 m/s, therefore to measure the range to an accuracy of 10 mm requires a clock speed of only 75 kHz. To achieve the same accuracy using a radar where the electromagnetic pulse travels at 3×10^8 m/s requires a clock speed of 15 GHz, which is impractical.

5.4.1 Timing Discriminators

For electromagnetic applications, the timing process is slightly more complex. The first part involves generating accurate timing pulses to mark the transmission of a pulse and reception of the echo. If a simple threshold-based method is used, then variations in the amplitude of the received echo are converted to timing errors, known as range walk, as shown in Figure 5.28.

The most common technique used to combat range walk is constant fraction discrimination (CFD), which sets a trigger threshold bearing a fixed relationship to the peak amplitude of the received pulse and so reduces errors caused by variations in this value (Amman 2001). This is achieved by splitting the received signal in two, delaying the one path and attenuating the other, and then using the crossover point to trigger a timing mark, as shown in Figure 5.29 (Kilpela

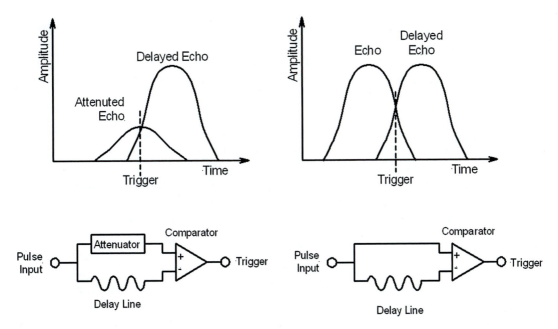

Figure 5.29 Constant fraction discrimination (CFD) techniques: (a) classical, (b) modified.

2004). Optimum values for attenuation and delay are selected according to the shape of the pulse to minimize range walk error.

This technique does not correct for distortion in the pulse shape caused by reflections from oblique targets. Though these errors are seldom a problem with lidar because of its narrow beamwidth, they are an issue with radar systems. A modification of the CFD technique is used in which the pulse amplitudes are maintained to produce an output similar to the "split-gate" trackers discussed in Chapter 13. As before, a timing pulse is generated at the coincidence of the two signals, but in this case the amplitudes of the direct and delayed pulses are the same, as can be seen in Figure 5.29b. When compared to the leading edge timers, this technique is superior because it uses both the leading edge and the trailing edge of the pulse to produce a more accurate result.

Another technique is the crossover timing discriminator in which the pulse is passed through a high-pass filter to produce the derivative. The timing pulse is then triggered at the zero crossing. The trigger point is independent of the pulse amplitude, and it does not require pulse-delay cables, which makes it well suited for integration into application-specific integrated circuits (ASICs) (Kilpela 2004).

To ensure the best accuracy, all of these techniques are used on both the bang pulse and the echo, as this ensures that propagation delays track each other through the circuitry and the changes in characteristics due to variations in temperature and aging.

The time interval between the start and stop pulses is measured using a time-to-digital converter (TDC), which is a fast and accurate timing method. The simplest TDCs comprise a counter, as discussed earlier, that is started by the bang pulse and stopped by the first echo pulse. However, far more sophisticated hybrid techniques are available to measure these periods to subnanosecond accuracies.

The measurement of short time intervals has long been one of the requirements of particle physics for mean lifetime and time-of-flight measurements. A good review of some of these techniques can be found in Porat (1973). These TDCs include the use of complementary clocks running at 500 MHz that can achieve 1 ns resolutions by means of the Vernier principle, which uses two clocks running at slightly different frequencies and a tapped delay-line method to interpolate over one clock cycle, as shown in Figure 5.30.

Most modern lidar and radar systems use hybrid methods. A fast clock is used to measure the synchronous portion of the interval and capacitor-based interpolation is used for the nonsynchronous start and end sections. It is these interpolation methods that have been the focus of much research as more and more accurate measurements are required. At their most basic, time-to-amplitude converters (TACs) consist of a capacitor that is discharged from a fixed value (V_{set}) beginning at the receipt of a trigger pulse and ending at the next clock edge (V_{end}), as shown in Figure 5.31. This value is held constant while it is digitized using a relatively slow 12-bit analog-to-digital converter (ADC) to produce an accurate measure of that short period (T_1 or T_2). This value is added to the final clock count (T_{12}) for the bang pulse or subtracted from it for the echo pulse for reasons that are clear from Figure 5.31 (Maatta and Kostamovaara 1998).

Another common TDC is the dual-slope method in which capacitors are quickly charged during T_1 and T_2 through a constant-current source. They are then discharged very slowly through a separate constant-current source while a counter coupled to the reference clock counts the time taken to discharge. As shown in Figure 5.32, this discharge time is equal to the product of the charge time and the ratio of the charge and discharge currents. This technique is very similar to the circuitry used by conventional dual-slope ADCs.

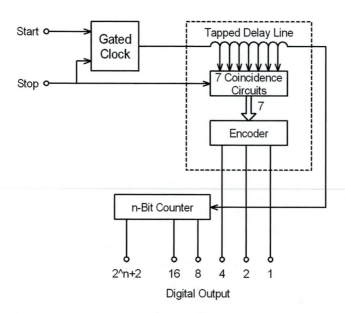

Figure 5.30 Time-to-digital conversion using delay-line interpolation (Porat 1973).

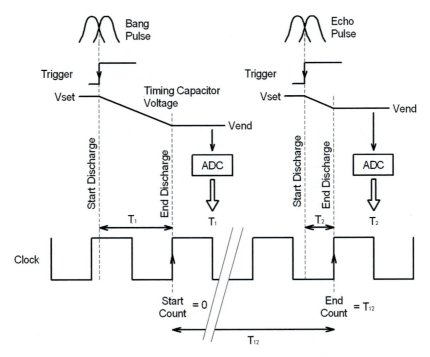

Figure 5.31 Accurate time-to-digital conversion for a lidar using hybrid methods.

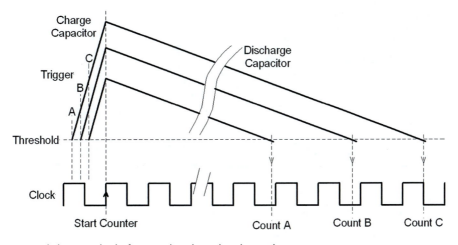

Figure 5.32 Dual-slope method of measuring short time intervals.

5.4.2 Pulse Integration

Pulse integration may be used to decrease the false-alarm rate and to improve measurement accuracy by increasing the effective SNR of the echo. This can be achieved by averaging a number of pulses before thresholding, or by averaging the measured range and discarding outliers. In Figure 5.33, the raw range data are averaged to produce a better estimate. The improvement is proportional to the square root of the number of samples averaged, with submillimeter accuracies possible (Amman 2001).

Integration of pulses to improve the SNR, common in radar systems, requires a storage device that allows the individual echo returns to be averaged. In early radars this was achieved by using a long-persistence phosphor on the cathode ray tube display combined with the integrating properties of the eyes and brain of the operator. However, modern ranging devices often use digital memory and high-speed signal processing techniques to perform the integration function, as shown in Figure 5.34.

Digital processing of radar signals, as illustrated in Figure 5.35, is limited by the sample rate of the required ADCs with sufficient dynamic range. For example, if the pulse width, δR, is 3 m, then the aperture time for the sample-and-hold (S&H) should be

$$\tau = \frac{2\delta R}{c} = 20\,\text{ns},\tag{5.18}$$

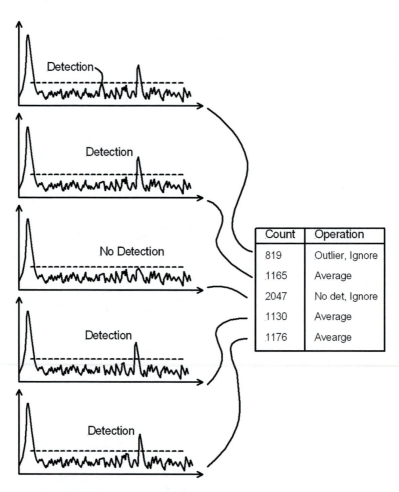

Figure 5.33 Measurement integration to improve range accuracy.

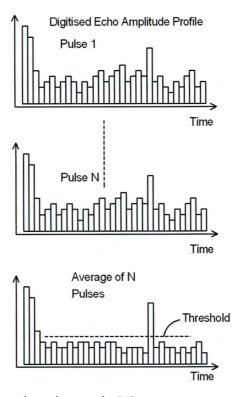

Figure 5.34 Measurement integration to improve the SNR.

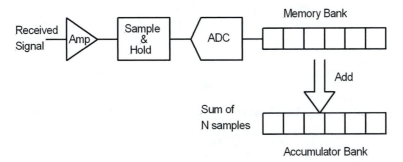

Figure 5.35 Digital pulse integration method.

and the ADC must be clocked at 50 MHz to capture every sample. It is therefore not practical to use this integration technique for very-short-range, high-resolution techniques, as the S&H aperture would have to be too short. It is possible to improve the accuracy of the measured range to much better than the pulse width by interpolation across adjacent range bins that contain the target echo.

5.4.3 Time Transformation

Cost-sensitive applications such as industrial-level sensors and some commercial laser range finders use an analog technique known as "time transformation" to stretch the return time, as this leads to a reduction in clock frequency and ADC requirements. This technique relies on repeated sampling of many pulse cycles. In this technique, two timing ramps are generated, a fast ramp with a period equal to the pulse repetition interval (PRI) and synchronized to it, and a slow ramp that may span thousands of pulses. A comparator triggers a fast S&H every time the fast-ramp voltage exceeds that of the slow ramp, and the sampled echo signal is held constant for just over one complete PRI cycle. From Figure 5.36, it can be seen that this output tracks the echo sequence transformed in time from the original PRI by the ratio of the slow-ramp period to that of the fast-ramp period.

Typical transformation ratios of 1 : 100,000 are common in range-level measurement applications, as these will transform a time-of-flight sensor using electromagnetic (EM) radiation into the equivalent of an acoustic system. Time transformation has the added advantage of integrating repetitive signals automatically because each point on the repeated echo is sampled many times and integrated through a low-pass filter.

It should be noted that this technique is only used for radar and laser devices because of the high speed of EM propagation. This is not required for sonar devices in air (or even in water), as the speed is sufficiently low.

5.5 OTHER METHODS TO MEASURE RANGE

As discussed in the previous sections, it is possible, in theory, to obtain extremely accurate range measurements using pulsed time-of-flight methods and hybrid TDCs. However, in practice, the pulse widths need to be short enough to ensure that the return from a single target is captured. Hence the bandwidth must be very high, which makes achieving a good SNR extremely difficult. A low-bandwidth alternative is to use the relative phase of the received signal to determine the

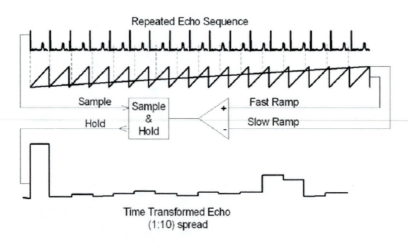

Figure 5.36 Integration by time transformation.

range. An early and extremely accurate sensor using this technology was the microwave tellurometer developed by T. L. Wadley in 1957 (Austin 1992; Bozzolli 1997).

Phase reference measurements are slow and are only suitable for point targets, so if a faster
update rate is required and multiple targets are present in the beam, more conventional wideband
modulation is used.

5.5.1 Ranging Using an Unmodulated Carrier

The most basic ranging method involves measuring the phase shift of the unmodulated carrier
at two or more different frequencies, as shown in Figure 5.37. The equation that defines the
range, R, to the target in terms of the wavelength, λ, of the signal is

$$2R = n\lambda + \Delta\lambda, \tag{5.19}$$

where n is the number of whole wavelengths and $\Delta = \varphi/2\pi$ is fractions of one wavelength.

Because it is difficult to measure phase shifts exactly, an alternative technique is to adjust the
transmitted frequency so that there is a whole number of cycles in $2R$. This can be achieved by
observing the output of the mixer (phase detector) and adjusting the oscillator frequency until it
crosses through zero. By shifting the frequency further until this has been repeated N times, two
equations can be written and solved for the range to the target, R:

$$2R = n_1\lambda_1,$$
$$2R = (n_1 + N)\lambda_2, \tag{5.20}$$

where n_1 is the unknown number of cycles in the round-trip distance, λ_1 is the wavelength at
carrier frequency f_1, and λ_2 is the wavelength at carrier frequency f_2. It is more convenient to
solve these simultaneous equations for the range, R (m), in terms of the frequency (Hz), where
c (m/s) is the speed of light:

$$R = \frac{Nc}{2(f_2 - f_1)}. \tag{5.21}$$

5.5.2 Ranging Using a Modulated Carrier

Most tellurometer systems use a master station and a remote transponder which amplifies
and retransmits the signal to maximize the operating range, but this is not required if a good

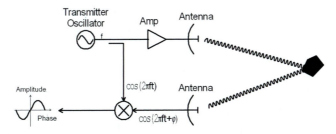

Figure 5.37 Ranging using an unmodulated continuous wave signal.

reflecting surface is available. As before, the round-trip distance between the master and the transponder can be given in terms of wavelengths:

$$2R = n\lambda_m + \Delta\lambda_m,\qquad(5.22)$$

where λ_m is the modulation signal wavelength (m) in this case and the other variables are unchanged. In this equation, both n and R are unknown, and only the phase difference, $\Delta\lambda_m$, can be measured using the principles shown in Figure 5.38.

A wavelength of about 10 m is selected as a compromise between the ability to resolve phase differences, which improves with increasing λ, and the ultimate resolution, which improves as λ is decreased. Electromagnetic radiation with $\lambda = 10$ m equates with a frequency, f, of 30 MHz, for which it is impractical to make narrow-beam antennas. For this reason, the low frequency is modulated onto a higher frequency directional signal to provide accurate measurements.

Resolving the integer ambiguity, n, can be done in a number of ways: Decade modulation involves stepping λ in multiples of 10; for example, 10 km → 1 km → 100 m → 10 m → 1 m, etc. The range must be known to within a few kilometers at the start to avoid ambiguity, and the process is time consuming, as the results must be tabulated and the range calculated for each

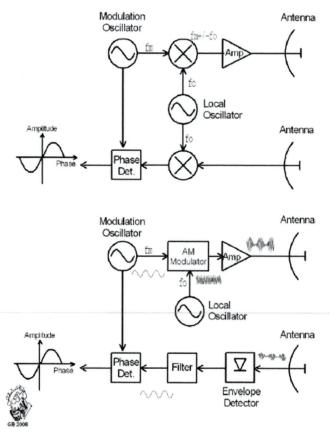

Figure 5.38 Frequency and amplitude modulated carrier-based continuous wave range measurement techniques.

measurement. It also requires the modulator frequency be very accurate and stable. Modern ranging systems including tellurometers, like the MA-200, use a series of similar, stable frequencies. In these, the phase difference is measured for each frequency and the data used to solve a series of simultaneous equations.

This principle is also commonly applied to laser devices which are known as phase-shift range finders (Amman 2001) or amplitude modulated continuous wave (AMCW) lidar (Probert-Smith 2001). These systems are usually reserved for short-range applications where the phase difference is unambiguous and a simple trigonometric relationship exists between the received phase and the range. According to Amman (2001), this is a good method to obtain accuracies of a few millimeters in ranges up to 20 m with natural targets, and if a corner-cube reflector is used, this range accuracy can be improved to better than 50 μm.

In low-cost applications it is possible to use a wideband phase detector IC such as the AD8302 to perform reasonably accurate phase comparisons over a wide range of frequencies from close to DC up to 2.7 GHz and over a wide range of input powers (–60 dBm to 0 dBm). The accuracy of these devices is good, with a calibrated output of 10 mV/deg and a typical nonlinearity of <1° over most of its range, as illustrated in Figure 5.39.

Another method for determining phase is the use of phase-locked loops (PLLs), in which the phase comparator output is a voltage that is directly proportional to the phase difference between the input and the output signals.

5.5.3 Tellurometer Example

In this example, if one modulation is set to a 10 m wavelength, the other would be set to 9.990 m to resolve the distance to the nearest millimeter. This would allow the phase to be resolved without ambiguity out to a range of 5 km ($2R < 10$ km):

$$2R = n_1\lambda_1 + \Delta\lambda_1$$

$$2R = n_2\lambda_2 + \Delta\lambda_2.$$

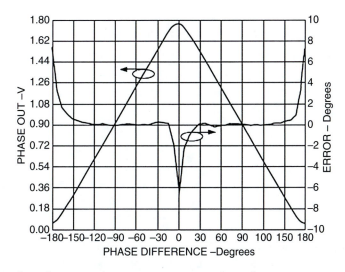

Figure 5.39 AD8302 phase detector performance at 100 MHz and –30 dBm.

If $\lambda_1 = 10$ m and $\lambda_2 = 9.99$ m and the phase difference measurements are respectively $0.0363\lambda_1$ and $0.5989\lambda_2$, what is the measured range?

Because $\Delta\lambda_2 > \Delta\lambda_1$, $n_1 = n_2 = n$, and the equations can be written as follows:

$$2R = 10n + 0.0363 \times 10 \qquad [1]$$

$$2R = 9.99n + 0.5989 \times 9.99. \qquad [2]$$

The following procedure can be used to solve for the two unknowns:

$$[2] - 0.999[1]$$
$$2R(1 - 0.999) = 9.99n - 9.99n + 9.99(0.5989 - 0.0363)$$
$$2R = 5.6203/0.001$$
$$2R = 5620.363\,\text{m}$$
$$R = 2810.18\,\text{m}.$$

To solve these simultaneous equations using MATLAB it is necessary to set up the equations in the form of $A * X = B$:

$$A = \begin{bmatrix} 2 & -10 \\ 2 & -9.99 \end{bmatrix}$$

$$B = \begin{bmatrix} 0.0363 \times 10 \\ 0.5989 \times 9.99 \end{bmatrix}$$

$$X = A \setminus B$$

```
% Solving simple simultaneous equations for tellurometer
% sim_tel.m

A = [2 -10; 2 -9.99];
B = [0.0363*10 ; 0.5989*9.99];
X = A\B
X =

    2810.187
    562.0011
```

The phase difference measurements for the two equations equate to $\varphi_1 = 13.068°$ and $\varphi_2 = 215.604°$ $(-144.396°)$, so assuming that the AD8302 phase detector has been used to determine this, the phase error for φ_1 could be as large as $-1°$ (-7.7%) and that for φ_2 could be up to $0.2°$ (0.14%). Solving for R with these errors gives the following results:

```
X =
    2828.581
    565.682
```

It is obvious that such ICs do not achieve the accuracy required unless the frequency is adjusted for each measurement to maintain a phase shift of close to $90°$ where the phase error is a minimum. If the phase can be measured with an accuracy of $0.1°$, then a worst-case error of just over 3 m occurs.

If $\Delta\lambda_2 < \Delta\lambda_1$, then the phase has wrapped by a complete cycle and $n_2 = n_1 + 1$ and the equations are solved accordingly.

5.6 THE RADAR RANGE EQUATION

The previous sections in this chapter have all assumed that the amplitude of the returned echo was always sufficiently large compared to the noise floor to allow the measurement process. The radar range equation, which is reasonably generic and can be applied to lasers, radars, and acoustic systems, determines the maximum range at which a target can be reliably detected in the presence of noise.

In practice, the range performance obtained using the simple equation described in the following section is optimistic, and in some cases can be out by a factor of two! This is in part because of losses that are not accounted for and higher than expected noise levels. Another important factor that is often overlooked is the fact that the range equation is statistical in nature. The minimum detectable signal, S_{min}, and the target cross section, σ, are both expressed in statistical terms (Skolnik 1980).

5.6.1 Derivation

If the power transmitted by the sensor is P_t through an isotropic radiator (one which radiates uniformly in all directions), then the power density, S_{iso} (W/m^2), at range R from the transmitter will be equal to the transmitted power divided by the surface area, $4\pi R^2$, of an imaginary sphere of radius R:

$$S_{iso} = P_t / 4\pi R^2. \tag{5.23}$$

As discussed earlier in this chapter, the gain, G_t, of an antenna is the measure of the increased power in the direction of the target compared to the isotropic case. Therefore the power density, S_i, as illustrated in Figure 5.40a, that is incident on the target is

$$S_i = P_t G_t / 4\pi R^2. \tag{5.24}$$

The target intercepts a portion of this power and scatters it in all directions. The measure of the proportion of the incident power scattered in the direction of the receiver is denoted as the

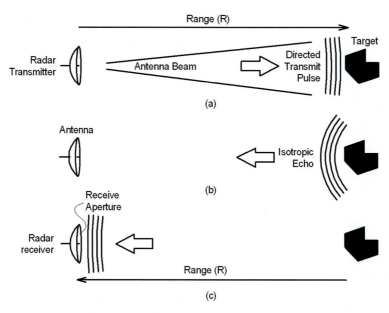

Figure 5.40 Principle for deriving the radar range equation.

target backscatter coefficient, ρ, for a laser radar with a spot size (footprint) smaller than the target, or as the radar cross section, σ (or RCS), for a laser, acoustic, or microwave radar if the beam footprint is larger than the target. The RCS, sometimes referred to as the target cross section, has units of area and is characteristic of a particular target and its orientation. It is discussed in detail in Chapter 8.

Because the target is assumed to scatter isotropically, the power density, S_r, back at the receiver antenna will be

$$S_r = \left(\frac{P_t G_t}{4\pi R^2} \right) \left(\frac{\sigma}{4\pi R^2} \right). \tag{5.25}$$

A small portion of this reflected power will be intercepted by the antenna, as illustrated in Figure 5.40c. The effective area of the receiving antenna, A_e, is defined as the ratio of the received power at the antenna terminals to the power density of the incident wave. The power, S (W), received by the sensor is

$$S = \frac{P_t G_t \sigma}{(4\pi)^2 R^4} A_e. \tag{5.26}$$

It is generally more convenient to describe the receive aperture in terms of its gain rather than having two separate terms to describe the antenna characteristics. The relationship described by equation (5.11) is repeated here. It relates the gain, G_r, of a lossless antenna to its effective aperture and the wavelength, λ (m):

$$G_r = 4\pi A_e / \lambda^2. \tag{5.27}$$

If the aperture of the antenna is large in terms of the wavelength ($d > 20\lambda$) the effective aperture approaches the physical antenna area. Substituting for A_e in equation (5.26) gives

$$S = \frac{P_t G_t \sigma}{(4\pi)^2 R^4} \frac{G_r \lambda^2}{4\pi} = \frac{P_t G_t G_r \lambda^2 \sigma}{(4\pi)^3 R^4}. \tag{5.28}$$

For a monostatic sensor, that is, one that uses the same antenna for receive and transmit, then $G_t = G_r = G$ and the equation can be simplified to

$$S = \frac{P_t G^2 \lambda^2 \sigma}{(4\pi)^3 R^4}. \tag{5.29}$$

All of the known losses can be included by lumping them together and including them as part of the numerator of the equation. They are denoted L, where $L < 1$:

$$S = \frac{P_t G^2 \lambda^2 \sigma L}{(4\pi)^3 R^4}. \tag{5.30}$$

If this equation is considered in isolation, it can be seen that it equates the received power level with the range to the target and a number of parameters specific to the sensor. However, it cannot be used to determine the maximum operational range. To find this, the minimum detectable signal level, S_{min}, must be specified.

If the range equation is rewritten, where the received power $S = S_{min}$, the maximum detection range, R_{max} (m), will be

$$R_{max} = \left[\frac{P_t G^2 \lambda^2 \sigma L}{(4\pi)^3 S_{min}} \right]^{1/4}. \tag{5.31}$$

5.6.2 The Decibel Form

It is very seldom that the range equation is actually used in the form of equation (5.30) or equation (5.31). This is because there is a large variation in the values of the parameters and also because many are specified by their decibel (dB) values as a matter of course. Some of the useful relationships listed below can be used to convert the range equation to decibel form and to manipulate the results:

$$X_{dB} = 10 \log_{10} X,$$
$$X = 10^{X_{dB}/10},$$
$$X_{dB} + Y_{dB} = 10 \log_{10} XY,$$
$$X_{dB} - Y_{dB} = 10 \log_{10} (X/Y),$$
$$10 N X_{dB} = 10 \log_{10} X^N,$$
$$0 = 10 \log_{10} 1,$$
$$-10 = 10 \log_{10} 0.1.$$

The equation for the received power level, S_{dB} (dBW), in decibel form is

$$S_{dB} = P_{tdB} + 2G_{dB} + 10\log_{10}\left(\frac{\lambda^2}{(4\pi)^3}\right) + \sigma_{dB} - L_{dB} - 40\log_{10}R. \tag{5.32}$$

In this case, the received power is relative to 1 W, hence dBW. However, received power levels are often very small and it is convenient to take the reference with respect to 1 mW, in which case the dBm form is used. To convert from dBW to dBm, a factor of 30 is added:

$$1\,W = 0\,dBW = 30\,dBm.$$

5.6.3 Worked Example: Radar Detection Calculation

Determine the maximum detection range of a maritime surveillance radar with the following characteristics:

- Gain: 32 dB
- Transmit peak power: 15 kW ($10\log_{10}15,000 = 41.76$ dBW)
- Frequency: 10 GHz
- Total losses: 5 dB
- Target RCS: 5 sqm ($10\log_{10}5 = 7$ dBsqm)
- Minimum detectable signal: 3.16×10^{-14} W (–135 dBW).

Start by converting everything to decibels as shown and then apply the radar range equation [equation (5.32)] that has been written for decibel values. Plot the received signal, S_{dB}, on a semilog$x$ plot along with the minimum detectable power, as shown in Figure 5.41.

The graphical solution indicates a detection range of 30 km, with the actual crossover point being 30.4 km as determined using the following MATLAB code.

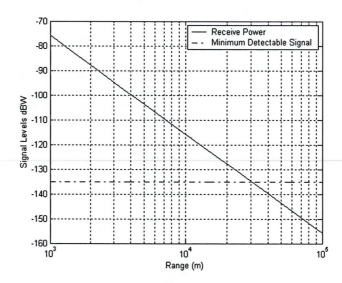

Figure 5.41 Graphical solution to the radar range equation for a maritime search radar.

```
% radar_simple.m
% Generate detection range from the basic radar range equation
% Arguments
%       GdB       antenna gain (dB)
%       Pt        transmit power (W)
%       LdB       losses (dB)
%       freq      frequency (Hz)
%       RCS       radar cross-section (square m)
%       Smin      minimum detectable signal level (W)

GdB = 32;                        % Antenna Gain dB
Pt = 15.0e03;                    % Transmit power (W)
LdB = 5.0;                       % Losses dB
freq = 10.0e09;                  % frequency Hz
RCS = 5.0;                       % radar cross-section (sqm)
Smin = 3.16e-14;                 % minimum detectable signal level (W)

% Calculate wavelength
c = 3.0e+08;
lam = c/freq;

PtdB = 10*log10(Pt);         % transmit power in dBW
RCSdB = 10*log10(RCS);       % radar cross section in dBsqm
SmindB = 10*log10(Smin);     % minimum detectable signal in dBW
kdB = 10*log10(lam^2/(4*pi)^3);

% determine the receive power
R = logspace(3,5,500);

SdB = PtdB + 2*GdB + kdB + RCSdB - LdB - 40*log10(R);
Sminp = ones(size(R))*SmindB;

semilogx(R,SdB,'k-',R,Sminp,'k-.');
grid

xlabel('Range (m)');
ylabel('Signal Levels dBW')
legend('Receive Power','Minimum Detectable Signal')

% Find the intersection range
Rdet = R(find(abs(SdB-SmindB)<0.1))
```

5.6.4 Receiver Noise

The minimum detectable signal level is not an arbitrary value, but one that is determined from the noise level at the input to the radar receiver. Noise can therefore be defined as the unwanted electromagnetic energy that interferes with the ability of the receiver to detect the wanted signal. It may enter the receiver through the antenna along with the desired signal or it may be generated within the receiver.

As discussed earlier, noise is generated by the thermal motion of the conduction electrons in the ohmic portions of the receiver input stages. This is known as thermal or Johnson noise. It is convenient to express the noise power, P_N, in terms of the temperature, T_o, of a matched resistor at the input of the receiver. The power transferred to the receiver is therefore

$$P_N = kT_o\beta, \tag{5.33}$$

where k is Boltzmann's constant (1.38×10^{-23} J/K), β is the receiver noise bandwidth (Hz), and the temperature of the system is T_o, which is usually assumed to be 290 K.

The noise power, N, in practical receivers is always greater than that which can be accounted for by thermal noise alone. This total noise contribution can be described in different ways. One is to consider it to be equal to the noise power output from an ideal receiver multiplied by a factor called the noise figure, NF:

$$N = P_N NF = kT_o\beta NF. \tag{5.34}$$

Another way to describe the noise level is in terms of the effective noise temperature of the receiver, where $T_{eff} = T_o(NF - 1)$. Noise figures for radar systems rise with increasing operational frequency, primarily because of the increasing difficulty in designing low-noise amplifiers. As an example, the X-band (10 GHz) maritime surveillance radar described above would have a noise figure of about 1.78 (2.5 dB), while at E-band (77 GHz), the noise figure of automotive radars is typically greater than 2.8 (4.5 dB). The minimum detectable signal level is most often considered to be the level at which the signal power is equal to the noise power: $S_{min} = N$.

5.6.5 Determining the Required Signal Level

In the process described earlier in this chapter, if the received envelope voltage of the noise exceeds a given threshold, a false alarm is said to have occurred. The probability that this will occur is dependent only on the noise power and the threshold setting. However, if a signal is present with the noise, then the probability that a detection will occur is determined by the powers of both the signal and the noise and also the threshold setting. A relationship must therefore be developed between these two probabilities and the signal and noise levels which determines the maximum operational range of the sensor.

The important quantity in analyzing the detection range is not the signal power itself, but the SNR. The details of the relationship between this ratio, the detection probability, P_d, and the probability of false alarm, P_{fa}, have been analyzed by a number of authors, of whom Blake (1986) is one. These are reproduced in the form of tables or graphs, as shown in Figure 5.42.

Consider the maritime radar example again. Assume that a detection probability, P_d, of 0.9 and a false-alarm probability, P_{fa}, of 10^{-6} are required, then reading from Figure 5.42, a SNR of 13.2 dB is required. This means that the signal level must be 13.2 dB above the noise level defined previously.

Assuming that the transmitted pulse width, τ, is 0.2 μs, the matched filter bandwidth will be $\beta = 1/\tau = 5$ MHz. For a receiver noise figure of 1.78, the noise floor will be

$$N = P_N NF = kT_o\beta NF$$
$$= 1.38 \times 10^{-23} \times 290 \times 5 \times 10^6 \times 1.78$$
$$= 3.56 \times 10^{-14} \text{ W} (-134.5 \text{ dBW}).$$

The required signal level, S_{det}, is determined by adding the 13.2 dB needed to obtain the required detection and false-alarm probabilities:

$$S_{det} = -134.5 + 13.2 = -121.3 \text{ dBW}.$$

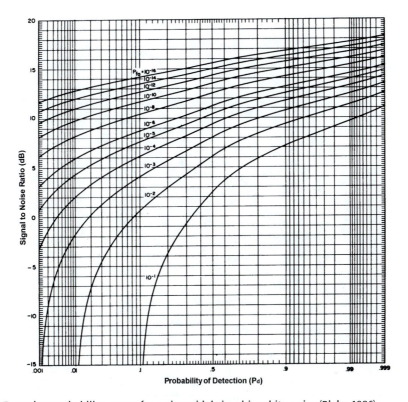

Figure 5.42 Detection probability curves for a sinusoidal signal in white noise (Blake 1986).

5.6.6 Pulse Integration and the Probability of Detection

If the received power from K radar returns is integrated perfectly, the effective SNR is equal to K times the single-pulse SNR for coherent (predetection) integration. However, most radars perform their integration on the video signal after envelope detection, as this is much easier to accomplish. This is known as noncoherent integration, and it is not as effective as coherent integration, with the improvement in effective SNR being only about $K^{0.8}$.

This result assumes that the target amplitude remains constant for the complete integration time and that the noise is white (uncorrelated). If the postdetection noise and signal + noise probability density functions (PDFs) are considered, then the effect of integration is clearly illustrated in Figure 5.43. For the single-pulse case, the tail of the Rayleigh distributed noise envelope overlaps the detection threshold to produce a significant P_{fa}. The signal + noise distribution is Ricean, and it too has a tail which crosses the threshold.

In the integrated case, the means of the PDFs remain unchanged, but their variances decrease, with the result that tails hardly overlap the detection threshold. Therefore, for a given threshold, the probability of detection, P_d, increases and the probability of false alarm, P_{fa}, decreases. If it is required that these two probabilities remain unchanged after integration, then the single-pulse SNR can be reduced, which will result in the signal + noise PDF moving back toward the noise distribution.

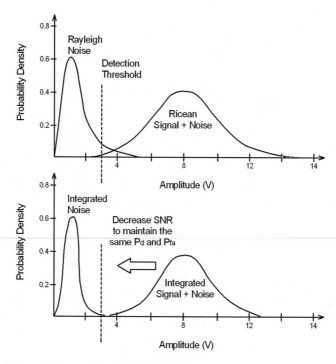

Figure 5.43 Effect of integration on noise and signal PDFs.

Once again, if the maritime radar example is considered with the added assumption that the target is illuminated for sufficient time to integrate 100 pulses, the improvement in SNR, I_{int}, can be calculated as $K^{0.8}$, where $K = 100$:

$$I_{int} = 8\log_{10} 100 = 16 \text{ dB}.$$

To achieve the same P_d and P_{fa} as before, this integration improvement factor allows for the single-pulse SNR to be reduced proportionally. Therefore the signal level for a single pulse, S_{det}, is determined by adding the 13.2 dB to the noise floor as before, then subtracting the integration improvement factor:

$$S_{det} = -134.5 + 13.2 - 16 = -137.3 \text{ dBW}.$$

Note, in this case, that the single-pulse SNR equals $-137.3 + 134.5 = -2.8$ dB. The signal is in fact smaller than the noise, and yet, after integration it is sufficiently large to achieve the specified detection and false-alarm probabilities. In reality, a number of other effects come into play at low SNRs, and also the effects of target amplitude fluctuation must be considered. These are discussed in Chapter 9.

5.7 THE ACOUSTIC RANGE EQUATION

In many respects, the acoustic range equation is similar to the radar version, but because the nomenclature used is unique it can appear to be completely different. This section examines the derivation of this equation to illustrate the similarities.

In acoustic systems, the acoustic power density and SPLs are determined at a reference distance from the transmitter. In air this is usually 30 cm, and in underwater sonar systems it is 1 m.

For an acoustic transmitter radiating a power, P_t, omnidirectionally, the power density, I_{iso} (W/m^2), at a range of 30 cm will be the total radiated power divided by the surface area of a sphere with a radius of 30 cm:

$$I_{iso} = P_t / 4\pi 0.3^2. \tag{5.35}$$

The antenna gain, or directivity index (DI), is defined as the ratio of the solid angle subtended by the surface of the sphere (4π) to the area of the beam:

$$G \approx 4\pi/\theta\varphi, \tag{5.36}$$

where θ, φ (rad) are the two orthogonal 6 dB beamwidths.

The power density in the direction of the beam is the product of this gain and the isotropic power density:

$$I = P_t G / 4\pi 0.3^2. \tag{5.37}$$

Unlike electromagnetic radar, the performance of acoustic systems is seldom determined in terms of the power density, but rather in terms of the SPL. Atmospheric pressure is measured in Pascals (N/m^2), where the air pressure at sea level is defined to be 101,325 Pa. Sound-pressure is the difference between the instantaneous pressure generated by the sound and the air pressure. These differences are very small, with an unbearably loud noise having a sound-pressure of 20 Pa and one that is at the threshold of hearing having a pressure of 20 μPa.

The SPL, L_p (dB), is defined as the square of the ratio of the sound-pressure, P (Pa), to $P_{ref} = 20$ μPa threshold level. It is typically represented in decibel form as

$$L_p = 10\log_{10}(P/P_{ref})^2. \tag{5.38}$$

The relationship between the sound-pressure, P (Pa), and the transmitted power density, I (W/m^2) is

$$P^2 = IZ_{air}, \tag{5.39}$$

where Z_{air} is the acoustic impedance of air. This can be determined by taking the product of the air density, ρ_{air} (kg/m^2) and the speed of sound, c (m/s):

$$Z_{air} = \rho_{air}c = 1.229 \times 340 = 418\,\Omega. \tag{5.40}$$

Rewriting equation (5.38) in terms of equation (5.39) and equation (5.40) gives the SPL, L_p (dB), in terms of the acoustic power density, I (W/m^2), at a given range:

$$\begin{aligned} L_p &= 10\log_{10}\left(IZ_{air}/P_{ref}^2\right) \\ &= 10\log_{10}I + 10\log_{10}Z_{air} - 20\log_{10}P_{ref} \\ &= 10\log_{10}I + 26.21 + 93.98 \\ L_p &= 10\log_{10}I + 120. \end{aligned} \tag{5.41}$$

Assuming that there is no atmospheric attenuation, the SPL drops with the square of the range as the sound propagates out to a target. This is known as the beam-spread loss, H (dB), and is given by

$$H = 20\log_{10}(R/R_{ref}), \tag{5.42}$$

where R is the range to the target (m) and R_{ref} is the range at which the SPL is defined (0.3 m in this case).

The target strength, T (dB), is defined as the ratio of the reflected sound-pressure, P_r, to the incident sound-pressure, P_i, at a nominal range from the center of the target:

$$T = 20\log_{10}(P_r/P_i). \tag{5.43}$$

Further discussion on target strength can be found in Chapter 8 and Chapter 12.

The SPL back at the receiver, L_r (dB), can easily be calculated from equation (5.41), equation (5.42), and equation (5.43):

$$L_r = L_p - 2H + T. \tag{5.44}$$

5.7.1 Example of Using the Acoustic Range Equation

A short-range ultrasonic sonar comprises a pair of SensComp piezoelectric transducers with the following specifications:

- 40LT16: SPL 120 dB at 40 kHz; 0 dB relative to 0.0002 μbar per 10 V RMS at a range of 30 cm
- 40LR16: sensitivity −65 dB at 40 kHz; 0 dB relative to 1 V/μbar
- Beamwidth: 55° (6 dB).

Determine the acoustic power density 30 cm from the transmitter and calculate the received voltage level from a target with $T = -10$ dB at a range of 5 m if the transmitter is driven by 20 V RMS:

$$1\mu\text{bar} = 0.1\,\text{Pa},$$

therefore 0.0002 μbar = 20 μPa, which is the standard reference level for SPL.

The SPL increases with the square of the applied voltage, therefore

$$L_p = 120 + 20\log_{10}(20/10) = 126 \text{ dB}.$$

Calculate the beam spread loss:

$$H = 20\log_{10}(5/0.3) = 24.4 \text{ dB}$$

$$L_r = L_p - 2H + T = 126 - 2\times24.4 - 10 = 67.2 \text{ dB}.$$

Determine the received pressure, P, in microbars:

$$P = 10^{L_r/20} = 10^{3.36} = 2291 \text{ relative to } 0.0002\,\mu\text{bar}.$$

Therefore

$$P = 2291 \times 0.0002 = 0.458\,\mu\text{bar}.$$

The sensitivity response of the receiver is $10^{-65/20} = 5.62 \times 10^{-4}$ V/μbar, therefore the output voltage will be

$$V_{\text{out}} = 0.458 \times 5.6 \times 10^{-4} = 257\,\mu\text{V}.$$

5.8 TOF MEASUREMENT CONSIDERATIONS

Electromagnetic waves travel very fast with $\delta T = 6.6$ ns/m. This means that high data rates are possible, but it is difficult to obtain measurement resolution much better than 1 m. Acoustic signals travel much more slowly, $\delta T = 6$ ms/m (in the air). Data rates are much lower, but good resolution and accuracy (less than 1 mm) are possible with appropriate temperature compensation if the medium is calm.

Using the common pulsed time-of-flight technique, echoes from large targets at long range may be received after the next pulse has been transmitted. To eliminate this range ambiguity requires that the PRF be varied on a pulse-by-pulse or block-by-block basis.

Any object within the beam will return an echo, and as the actual target range may vary considerably across the beamwidth, there may be significant range uncertainty. Targets with large cross sections will dominate the return echo even if they are not in the center of the beam.

Compared to the possible range resolution, angular resolution of typical sensors is poor, as it is determined by the aperture of the antenna relative to the wavelength. Therefore improving the angular resolution requires either a large antenna aperture or operation at a high frequency, neither of which may be practical. Some practical methods to deal with this are discussed in Chapter 13.

5.9 RANGE MEASUREMENT RADAR FOR A CRUISE MISSILE

Under certain conditions the radar cross section of a target may be a function of range. A good example of this is the radio (or radar) altimeter, which has been one of the workhorse sensors in aircraft since its invention in 1928.

Prior to the introduction of the Global Positioning System (GPS), cruise missiles navigated using inertial systems when operating over water, with corrections based on terrain contour mapping when flying over land. Patented in 1958, this technique relied on the simple fact that the altitude of the ground varied as a function of location in a reasonably unambiguous fashion. Therefore, if a radar altimeter measured a height profile it could be correlated to the known terrain contour to pinpoint the missile position.

In the case of a radar altimeter (shown in Figure 5.44), the cross section, σ, is proportional to the area of the beam footprint on the ground, which is a function of the height, R. For a beamwidth θ, the area will be

$$A = \frac{\pi d^2}{4} = \frac{\pi (R\theta)^2}{4}. \tag{5.45}$$

The scale factor σ^o is called the target reflectivity and is defined as the radar cross section per unit area. The RCS is found by taking the product of the reflectivity and the area, as discussed in detail in Chapter 8.

$$\sigma = \sigma^o A = \frac{\pi (R\theta)^2 \sigma^o}{4}. \tag{5.46}$$

Substituting into the basic radar range equation reduces the received signal power to a function of R^{-2}:

$$S = \frac{P_t G^2 \lambda^2 \theta \sigma^o L}{4^4 \pi^2 R^2}. \tag{5.47}$$

Because typical radar cross sections are large, and ranges are short, altimeters can transmit low average powers (100–500 mW), and if required, can operate using low probability of intercept (LPI) spread spectrum waveforms.

As the cruise missile flies into one of the mapped areas, illustrated in Figure 5.45, altitude readings obtained from the on-board radar altimeter are compared with the

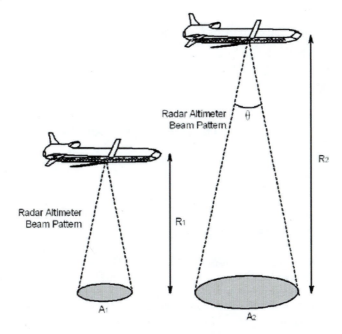

Figure 5.44 Variations in the target cross section for a radar altimeter.

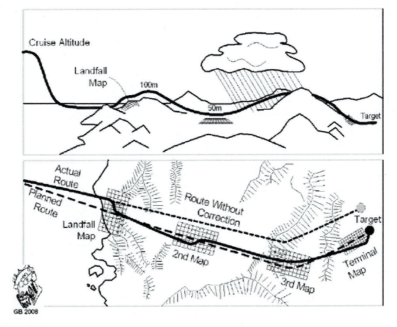

Figure 5.45 Cruise missile map matching [adapted from (Tsipis 1977)].

stored data to determine the missile position to the accuracy of better than one cell. The autopilot is then instructed to make the appropriate correction. Inferences about drifts in the inertial system and prevailing winds, among other factors, can be made at each stage, so subsequent maps may be smaller and have higher resolution, leading to pinpoint targeting.

A terrain altitude map spanning 2 km × 10 km with a cell size of 100 m × 100 m would require only 2000 points. So even if 20 such maps were required, the memory requirements are miniscule compared with image-based systems.

5.10 References

Amman, M. (2001). Laser ranging: a critical review of usual techniques for distance measurement. *Optical Engineering* 40:10–19.

Army Technology. (2007). The Crotale new generation multimission short range air defence missile system. Viewed August 2007. Available at http://www.army-technology.com/projects/crotale/crotale1.html.

Austin, B. (1992). Radar in World War II: the South African contribution. *Engineering Science and Education Journal* 1(3):121–130.

Bhartia, P. and Bahl, I. (1984). *Millimeter Wave Engineering and Applications*. New York: John Wiley & Sons.

Blake, L. (1986). *Radar Range-Performance Analysis*. Norwood, MA: Artech House.

Boucher, D. (1987). Trends and problems in low frequency sonar projectors design. In: *Power Sonic and Ultrasonic Transducers Design*, Lille, France, May 26–27.

Bozzolli, G. (1997). *Forging Ahead*. Johannesburg, South Africa: Wits University Press.

Brookner, E., ed. (1982). *Radar Technology*. Boston: Artech House.

Fraden, J. (2003). *Handbook of Modern Sensors: Physics, Designs and Applications*. Berlin: Springer Verlag.

Gallego, J., Rodriguez, G., and Gaete, L. (1978). An ultrasonic transducer for high power applications in gases. *Ultrasonics* 16(6):267–271.

Hamamatsu. L7060-02 laser diode. Viewed October 2007. Available at http://sales.hamamatsu.com/en/products/laser-group/pulsed-laser-diode/part-l7060-02.php#.

Hecht, J. (1994). *Understanding Lasers: An Entry Level Guide*, 2nd ed. New York: IEEE Press.

Kilpela, A. (2004). Pulsed time-of-flight range finder techniques for fast, high precision measurement applications. PhD dissertation. University of Oulu, Oulu, Finland.

Klein, L. (1997). *Millimeter-Wave and Infrared Multisensor Design and Signal Processing*. Norwood, MA: Artech House.

Livingstone, R. and Rioux, M. (1986). Development of a large field of view 3-d vision system. In: 1986 International Symposium on Optical and Optoelectronic Applied Sciences and Engineering, Quebec, Quebec, Canada, June 2–6, 1986. *SPIE Proceedings* 661:218–223.

Maatta, K. and Kostamovaara, J. (1998). A high-precision time-to-digital converter for pulsed time-of-flight laser radar applications. *IEEE Transactions on Instrumentation and Measurement* 47(2): 521–536.

Marshall, S. (1968). *Laser Technology and Applications*. New York: McGraw-Hill.

Pogge, R. Distances of the stars. Viewed October 2007. Available at http://www.astronomy.ohio-state.edu/~pogge/Ast162/Unit1/distances.html.

Porat, D. (1973). Review of sub-nanosecond time-interval measurements. *IEEE Transactions on Nuclear Science* NS-20(5):36–51.

Probert-Smith, P., ed. (2001). *Active Sensors for Local Planning in Mobile Robotics*. Hackensack, NJ: World Scientific.

Senscomp. (2007a). 40LT16. Viewed October 2007. Available at http://www.senscomp.com/specs/40LT16%20%20spec.pdf.

Senscomp. (2007b). 600 series transducer. Viewed October 2007. Available at http://www.senscomp.com/specs/600%20instrument%20spec.pdf.

Skolnik, M. (1980). *Introduction to Radar Systems*. New York: McGraw-Hill.

Tsipis, K. (1977). Cruise missiles. *Scientific American* 236(2):20–29.

6

Active Imaging Sensors

6.1 IMAGING TECHNIQUES

All time-of-flight sensors produce echoes from targets within their beam. For distributed targets, the amplitude of these returns is determined by a characteristic of the target called reflectivity, which is a normalized (per unit area) measure of how well the target reflects energy. Hence, the magnitude of the reflected power is proportional to the product of the reflectivity and the measurement area. Reflectivity is a complex function of the target material—its effective roughness as a function of the transmitted frequency and the angle at which the surface is illuminated, as shown in Figure 6.1.

In describing the general characteristics of reflectivity, there are three distinct regions that must be considered. In the near-grazing region, the ground may look smooth as determined by Rayleigh's criterion for surface roughness:

$$\Delta h \sin \theta \le \lambda/8, \tag{6.1}$$

where Δh (m) is the root mean square (RMS) height of surface variation, θ (rad) is the grazing angle, and λ (m) is the wavelength.

In the near-grazing region, the reflectivity increases rapidly with increasing grazing angle. In the plateau region, the surface appears rough and the reflectivity increases slowly with increasing grazing angle. In the third region, specular scattering becomes dominant and the reflectivity again increases rapidly with increasing grazing angle.

For targets (such as the ground) in which the dimensions exceed the extent of the antenna footprint, the measurement area is determined by the range gate size and the antenna footprint. Depending on the grazing angle between the radar beam and the target, two different imaging modalities exist. These are range gate (or pulse width)-limited and beamwidth-limited imaging, as illustrated in Figure 6.2 (Brooker et al. 2006).

In the beamwidth-limited case, the echo returns are mostly constrained within a single range gate, with the result that the target area, and hence the image resolution, is determined by the angular extent of the antenna beam. In the range gate-limited case, the image resolution is determined by the azimuth beamwidth of the radar and by the range gate extent.

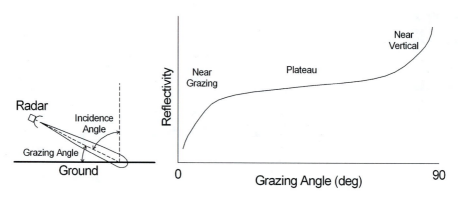

Figure 6.1 The relationship between surface reflectivity and grazing angle.

Figure 6.2 shows that the slope of the target surface and the elevation beamwidth play an important role in determining which of the modes would be most practical for imaging. In most beamwidth-limited sensors, the antenna beam pattern has a circular cross section, and is referred to as a "pencil beam." In contrast, in most range gate-limited sensors, though the azimuth beamwidth is as narrow as possible, the elevation beamwidth is wide, a pattern that is referred to as a "fan beam." These configurations are usually limited to acoustic and radar sensors, while lidar is almost universally beamwidth limited.

It can be seen that in the beamwidth-limited case, the actual reflectivity is not critical as long as a target is detected, because a three-dimensional (3D) surface can be constructed through the detected points. In the range gate-limited case, the terrain surface is assumed to be flat and all of the imaging information is encoded in the intensity of the return, which is directly proportional to the reflectivity of the surface.

6.2 RANGE GATE-LIMITED TWO-DIMENSIONAL IMAGE CONSTRUCTION

In a typical range gate-limited application, a stabilized fan beam antenna is scanned over a limited azimuth angle. During the process, each measurement is time stamped and tagged with the sensor position and the measured antenna azimuth so that it can be converted from polar to Cartesian space and "laid" down onto a rectangular grid pinned to the earth.

The reason for this conversion is because image processing in polar space is generally not practical, particularly if the frame of reference is moving. This transform is most often implemented by selecting a Cartesian grid with a resolution equal to the cross-range resolution at the smallest operational range and then oversampling at longer ranges, as shown in Figure 6.3 (Brooker 2005).

The simplest method of allocating data to the Cartesian image plane is to transform the coordinates (R, θ) of the centroid of each polar pixel into Cartesian coordinates (x, y) and use those to load the nearest Cartesian pixel. However, this method often results in unfilled pixels in the Cartesian plane which are distracting, so the alternative, albeit less efficient method of clocking through all of the Cartesian pixels in the image and allocating to each the magnitude of the nearest polar pixel is performed instead. Partial pixel overlaps are often integrated into their

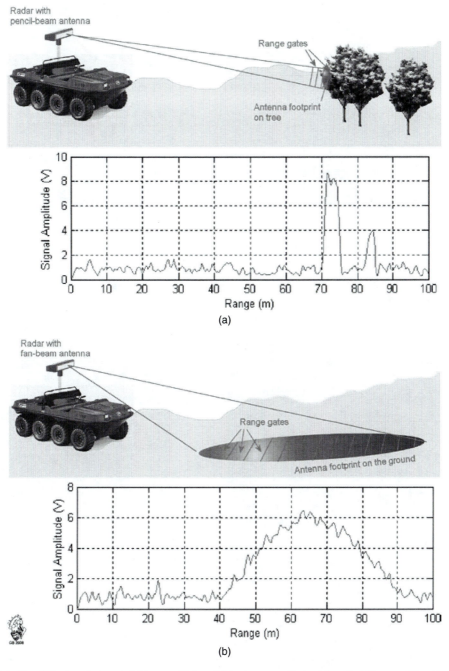

Figure 6.2 The difference between (a) beamwidth and (b) range gate-limited imaging illustrated in this example of a vehicle-mounted radar. The range-amplitude profile for the two modalities is shown below each example.

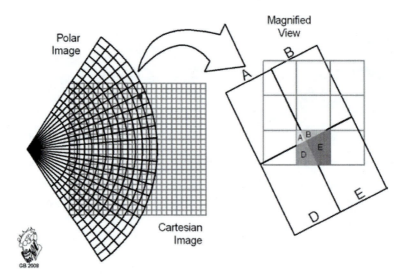

Figure 6.3 Polar to Cartesian image transform shows an example where each Cartesian pixel becomes the weighted average of the polar pixels with which it overlaps.

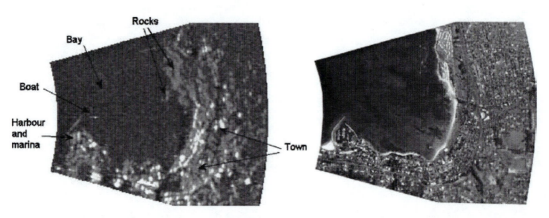

Figure 6.4 Range gate-limited image (a) of the ground transformed from polar to Cartesian space illustrating the good image quality obtained by oversampling and pixel weighting, and (b) an aerial photograph of the same area (Brooker et al. 2006).

respective bins in proportion to the percentage overlap to improve image fidelity, as illustrated in Figure 6.3.

In the example shown, the signal level in the illustrated Cartesian pixel is calculated from the weighted average of the polar pixels A, B, D, and E in proportion to their respective overlapping areas. Obviously the process of oversampling and pixel weighting cannot improve the actual resolution, but it does result in a clear image with a minimum of sampling artifacts, as the example in Figure 6.4a shows.

In this case a 384×128 point polar image has been resampled to produce a 300×200 pixel Cartesian image with a pixel area on the ground of 8 m × 8 m. This is larger than the theoretical range resolution of the radar, which is close to 5 m, but smaller than the cross-range resolution, which is determined by the product of the azimuth beamwidth and the range.

An alternative imaging method produces a reflectivity image by aiming a stabilized fan beam in a direction orthogonal to the direction of flight of an aircraft. This technique, known as side-looking airborne radar (SLAR), produces an image swathe along the direction of travel. The required azimuth resolution can be obtained using a large antenna or by synthesizing an aperture from the forward motion. These processes are discussed in detail in Chapter 12.

6.3 BEAMWIDTH-LIMITED 3D IMAGE CONSTRUCTION

A number of different methods can be used to produce beamwidth-limited images as well. These include using the forward motion of the sensor to supply one of the coordinates, a technique called push-broom scanning, or mechanically scanning in two dimensions across the scene while measuring the range.

6.3.1 Push-Broom Scanning

In the first method, a rotating mirror scans the radar or laser beam at right angles to the direction of travel. These sensors generally produce between 2000 and 8000 pulses every second and each pulse strikes the ground, where because it is rough, some power is reflected back to the receiver. By registering the forward motion of the aircraft using a Global Positioning System/inertial navigation system (GPS/INS) and the beam angle, a two-dimensional (2D) raster is produced, as shown in Figure 6.5. Range or reflected signal amplitude are logged to produce an image of the ground that can be aligned to a conventional image produced by a digital camera.

Digital elevation images made using this technique are commonly referred to as 2½D images because they are sampled onto a fixed Cartesian plane with the z-axis defining the height above some datum. The image shown in Figure 6.6 was generated using only data from a push-broom laser scanner. It combines surface geometry from laser range data and near-infrared surface reflectivity from laser return signal energy. It has also been further manipulated to include shadows cast by trees to make it look more natural.

6.3.2 Mechanical Scanning

Three different mechanical methods, as illustrated in Figure 6.7, are generally used to scan a scene to produce a 3D image. They are the standard pan/tilt unit, the pan-rotating prism/mirror unit, and the 2D mirror unit. This process of using mechanical scanning methods is typically very slow, and high-resolution, high-accuracy images require many minutes or even hours to construct.

- Pan/tilt: This is the conventional method of obtaining an image. A complete time-of-flight sensor is directed across a scene systematically by scanning the beam back and forth quickly in the one axis and slowly in the other. The relative angular rates are selected so that the appropriate overlap between samples is achieved.
- Pan mirror scan: The fast scan axis can be implemented by rotating a flat, multifaceted prismatic mirror. The slow scan axis is implemented by rotating the whole scanner module.

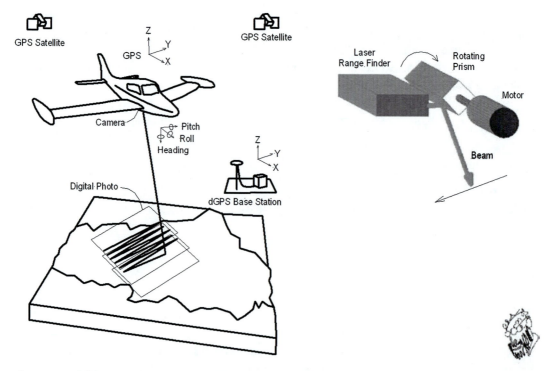

Figure 6.5 Push-broom scan principles showing the generation of an image using a scanning beam and the forward motion of the sensor.

Figure 6.6 Push-broom laser scanner image combining range and reflectivity (Geolas 2007).

As with the previous method, range, amplitude, or both are recorded along with the scan angle to produce an image.

- 2D mirror scan: In this implementation, the time-of-flight sensor is stationary and the beam is scanned across the scene using a mirror that can be tilted in two orthogonal axes.

In the point-cloud image shown in Figure 6.8, where no surface reconstruction is undertaken, only the highest intensity return from each measurement is displayed as a single point in space with height encoded by color (Brooker et al. 2007). This encoding gives no information about

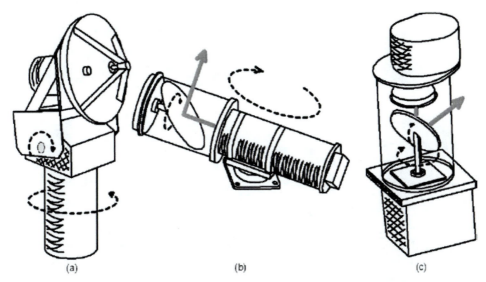

Figure 6.7 Mechanical scanners showing (a) pan/tilt, (b) pan mirror scan, and (c) 2D mirror scan.

(a)   (b)

Figure 6.8 (a,b) A photograph and a 3D point-cloud perspective view of the region adjacent to the ACFR made using a high-resolution, beamwidth-limited millimeter wave radar.

the reflectivity of each return, but it is useful for target volume interpretation and beam penetration analysis.

To obtain this figure, a millimeter wave radar was placed in the center of a grass quadrangle surrounded by buildings, trees, and cars. The image was produced by scanning the pencil beam radar through 360° in azimuth then stepping the beam up by a few tenths of a degree in elevation and repeating the process until a complete 3D sampling of the surroundings had occurred.

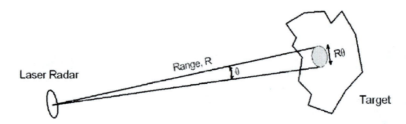

Figure 6.9 Laser radar target area.

6.4 THE LIDAR RANGE EQUATION

The lidar range equation is very similar to the equation that was developed for radar in the previous chapter. There are a few differences in the way that the beam characteristics and the target backscatter are defined, and these lead to an interesting result (Hovanessian 1988).

Referring to the geometry shown in Figure 6.9, the signal power, S (W), received by the laser range finder at range R is

$$S = \frac{P}{4\pi R^2} \cdot \frac{4\pi}{(\pi/4)\theta_{BW}^2} \cdot \frac{\pi}{4}(R\theta_{BW})^2 \cdot \rho \cdot \frac{1}{2\pi R^2} \cdot A \cdot \tau_o, \qquad (6.2)$$
$$\quad\;\; (1) \qquad (2) \qquad\quad (3) \qquad\;\; (4)\;\;(5)\;\;(6)\,(7)$$

where
$\quad$ S = target echo power at the laser radar (W),
$\quad$ P = transmitted pulse power (W),
$\quad$ R = range to the target (m),
$\quad$ θ_{BW} = laser beamwidth (rad),
$\quad$ ρ = target backscattering coefficient (m^2/m^2),
$\quad$ A = lens aperture area (m^2),
$\quad$ τ_o = optical efficiency.

The components that make up the equation are defined as:

1. Point source power over a spherical area $4\pi R^2$. The power density of a point source without the benefit of a lens that focuses the power in a given direction.
2. Measure of the focusing effect of the lens where it is used to direct the radiated power in a given direction. The gain is the ratio of the spread of the power over a sphere of 4π steradians (sr) and the laser beamwidth (in steradians).
3. The area of the target. In this case, the laser spot is smaller than the target area.
4. The backscatter coefficient of the target. This depends on the material reflectivity and the surface roughness.
5. The power from the target is scattered equally over the forward hemisphere of 2π sr. The resulting power density back at the laser will be $1/2\pi R^2$. An alternative is to use the Lambertian scattering assumption, which models the reflected power per solid angle to be proportional to the cosine of the angle between the normal to the surface and the reflection angle. In this case the effective spread is π sr.

6. The target return power that is intercepted by a lens with area A (m^2).
7. The optical efficiency of the laser radar transmission chain from the front aperture of the lens.

For the spherical target power distribution, the target echo power, S (W), can be simplified to

$$S = PA\tau_o\rho/2\pi R^2. \tag{6.3}$$

Note that the increase in the diameter of the illuminated spot on the target and the resultant increase in the amount of reflected power cancels one of the R^2 terms and results in $S \propto R^{-2}$ rather than the more usual R^{-4} associated with the range equation. However, if the laser beam diameter is larger than the target size, term (3) should be substituted for by the target cross section, σ, and the equation becomes

$$S = 2PA\sigma\tau_o/\pi^2 R^4\theta_{BW}^2, \tag{6.4}$$

where the beamwidth, $\theta_{BW} = \lambda/D$ (rad), λ (m) is the wavelength, and D (m) is the lens diameter.

In contrast to the 3 dB half-power beamwidth definition used in microwave radar analysis, the $1/e = 0.367$ level is used in optical systems. This equates to a beamwidth of $1.05\lambda/D$, but the constant 1.05 term is usually ignored.

The beamwidth, θ_{BW} (rad), can be rewritten in terms of the lens area and the radiated wavelength:

$$\theta_{BW}^2 = \lambda^2/D^2 = \lambda^2\pi/4A, \tag{6.5}$$

and the final equation is obtained by substituting equation (6.5) into equation (6.4) and simplifying:

$$S = 8PA^2\sigma\tau_o/\pi^3 R^4\lambda^2. \tag{6.6}$$

6.5 LIDAR SYSTEM PERFORMANCE

To determine the maximum operational range of a lidar system, the minimum detectable signal level must be determined. This depends on the system configuration, of which there are two main types, as shown in Figure 6.10, and the noise floor depends on the type that is implemented.

Direct detection laser receivers convert the echo directly into a voltage or current with PIN or avalanche photodiodes (APDs) using the mechanism discussed briefly in Chapter 3. Heterodyne receivers down-convert the received signal to a lower frequency by mixing it with the output of a stable continuous wave source. The signal can then be amplified and filtered to improve the detection process, and because phase information is maintained, these receivers can be used to measure speed by Doppler processing.

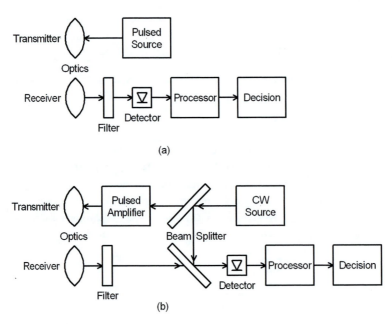

Figure 6.10 Laser receivers using (a) direct detection and (b) heterodyne techniques. (Adapted from Hovanessian 1998.)

Table 6.1 Noise sources in direct detection laser systems

Noise source	Noise current mechanism
Signal photons	Shot noise
Background photons	Shot noise
Bark current	Shot noise
Johnson	Thermal noise
Amplifier	Thermal noise
Generation-recombination	Thermal noise
$1/f$	Surface effects vary with detector

6.5.1 Direct Detection

For direct detection, Table 6.1 lists the main sources of noise (Klein 1997). The detector noise can be determined from the diode specific detectivity, D^*, which includes contributions for all of the noise sources listed in the table (Hovanessian 1988):

$$NEP = (A_d \Delta f)^{1/2} / D^*, \tag{6.7}$$

where

NEP = noise equivalent power level (W),

A_d = detector area (cm^2),

Δf = receiver bandwidth (Hz),

D^* = detectivity (cmHz$^{1/2}$/W).

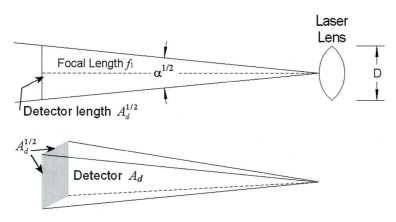

Figure 6.11 Relationship between the detector area and the focal length.

As illustrated in Figure 6.11, the detector area, A_d (cm^2), for a square detector is related to the receiver lens diameter and the focal length:

$$A_d = \left(\alpha_d^{1/2} f_1\right)^2 = \alpha_d f_1^2, \tag{6.8}$$

where α_d (sr) is the instantaneous field of view (IFOV) of the detector, as described in Chapter 3.

The focal length, f_1, can be written as the product of the lens diameter, D, and the lens focal number, f:

$$f_1 = fD. \tag{6.9}$$

Equation (6.8) and equation (6.9) allow the noise equation [equation (6.7)] to be rewritten in terms of the optical parameters if they are known.

6.5.1.1 Direct Detection Photodiodes   One deficiency of conventional PN diodes is their small depletion area (active detection area), which results in many electron-hole pairs recombining before they can create a current in the external circuit. In the PIN photodiode, the depleted region is made as large as possible. A lightly doped intrinsic layer is introduced to separate the more heavily doped p-type and n-type layers. The diode's name comes from the layering of these materials: positive, intrinsic, negative—PIN. Figure 6.12 shows the cross section and operation of a PIN photodiode. The responsivity of these devices is between 0.5 and 1 A/W of incident power.

In the APD, free electrons and holes created by absorbed photons are accelerated by the electric field, gaining several electron-volts of kinetic energy. A collision of these fast carriers with neutral atoms causes them to use some of their own kinetic energy to help the bound electrons break out of the valence shell. Free electron-hole pairs, called secondary carriers, are created by this process, which is called collision ionization. The secondary carriers themselves accelerate and create new carriers, creating an avalanche. Collectively this process is known as photomultiplication, with typical multiplication factors in the tens and hundreds for APDs. For example, a multiplication factor of 80 means that, on average, 80 external electrons flow for every photon of light absorbed.

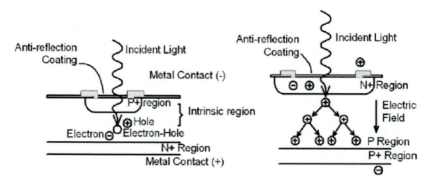

Figure 6.12 Operational principles of (a) a PIN diode and (b) an APD.

Unfortunately APDs require high-voltage power supplies for their operation, with a voltage range from 30 to 70 V for indium gallium arsenide (InGaAs) to more than 300 V for silicon (Si). This and the fact that they are very temperature sensitive further complicates circuit requirements. Because of the added circuit complexity and the high voltages that the parts are subjected to, APDs are always less reliable than PIN detectors. However, because the responsivity of these devices is usually higher than PINs and varies between 0.5 and 100 A/W, there is no option but to use them under certain circumstances.

Because an APD has gain, it must be compared to a PIN photodiode followed by a preamplifier. Preamplifier shot noise increases rapidly with increasing numbers of photons and so, for APDs, the dominant noise at low light levels is the dark current. As the signal level increases, shot noise becomes dominant.

The net result is that PIN diodes tend to have superior signal-to-noise ratios (SNRs) at low frequencies, while at higher frequencies, APDs are superior at low light levels where the dark current dominates. It is interesting to note that laser diodes with pulse lengths of 100 ps are now available, and to detect such short pulses effectively, APDs are essential. Using this technology, systems have been produced with range resolutions of less than 1 mm.

6.5.2 Heterodyne Detection

The noise spectral density of an ideal receiver is a combination of the thermal and photon noise and is given by

$$\psi(f) = \frac{hf}{e^{hf/kT} - 1} + hf,\tag{6.10}$$

where
    $\Psi(f)$ = noise power spectral density (W/Hz),
      h = Plank's constant, 6.6256×10^{-34} (Ws^2),
      f = frequency (Hz),
      k = Boltzmann constant, 1.38×10^{-23} (Ws/K),
      T = absolute temperature (K).

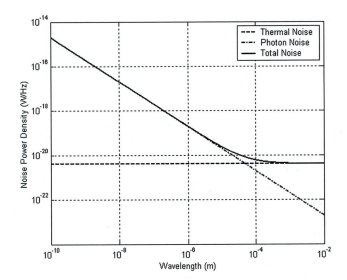

Figure 6.13 Laser radar noise floor as a function of wavelength for heterodyne detection.

The thermal noise power density, $\gamma(f) = kT$, and the photon shot noise power density, $\mu(f) = hf$, contributions are plotted individually and in combination in Figure 6.13.

For microwave radars, the noise power density is determined by the thermal noise floor and is approximately $\gamma(f) = kT$, while in the infrared, the noise power density is dominated by the photon noise, $\mu(f) = hf$. Therefore, if $\lambda < 10\ \mu m$, the noise equivalent power level for a heterodyne receiver can be approximated by

$$NEP = hf\beta/\eta, \tag{6.11}$$

where η is the quantum efficiency (the average number of photons are required to produce one photoelectron), and β (Hz) is the receiver bandwidth.

It is easy to show that at 290 K, the noise floor for a microwave receiver is about 13 dB lower than a photon noise-limited laser receiver operating at 10.6 μm with a quantum efficiency of 0.2. Quantum efficiencies at 850 nm are in the range 0.5 to 0.8, and at the longer wavelength of 10.6 μm they are reduced to between 0.1 and 0.2.

6.5.3 Signal-to-Noise Ratio and Detection Probability

Glint targets represent returns from corner reflectors or normal surfaces (such as the ground) where there is a single dominant scatterer. Returns are normally fairly constant from pulse to pulse and are therefore equivalent to the nonfluctuating case in radar systems. Using the SNR for a glint target calculated using the formulas derived above, the detection probability, P_d, and false alarm probability, P_{fa}, can be determined from Figure 6.14.

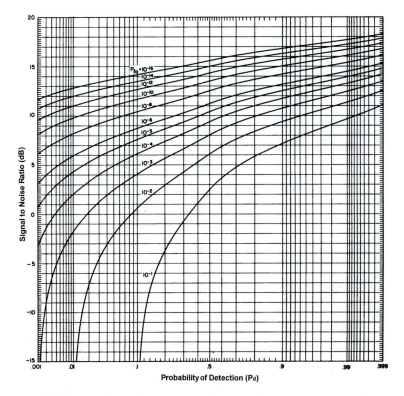

Figure 6.14 Probability of detection for glint (nonfluctuating) targets by lidar (Blake 1986).

6.5.4 Worked Example: Laser Radar Reflection from the Moon

An earth-based carbon dioxide laser operating at a wavelength of 10.6 μm radiates through a collimating lens with a diameter of 500 mm. If it produces 500 W pulses each of duration 0.1 s:

a) What would the diameter of the footprint be on the moon?
b) Ignoring atmospheric effects what would the power density on the moon be in W/m^2?
c) A retroreflector with a diameter of 10 cm and a reflectivity of 0.99 reflects some of the power back to earth. What is the received power density?
d) Is the reflected power density from the moon's surface back on the earth (backscatter coefficient 0.2) larger or smaller than that returned by the retroreflector?
e) If a heterodyne receiver uses the same size lens, what is the single-pulse SNR that we could expect?
f) What is the false alarm probability if a detection probability of 99% is required?

The solutions are determined as follows:

a) The diameter of the footprint on the moon is the product of the beamwidth and the distance to the moon. The mean distance to the moon is 384,400 km and the $1/e$ beamwidth is θ_{BW} (rad) if the system is diffraction limited:

$$\theta_{BW} = 1.05\,\lambda/D = \left(1.05 \times 10.6 \times 10^{-6}\right)/0.5 = 22.3\ \mu\text{rad}.$$

Therefore the diameter of the spot will be

$$d = R\theta_{BW} = 3.844 \times 10^8 \times 22.3 \times 10^{-6} = 8556 \text{ m.}$$

b) The power density on the moon is the total radiated power divided by the area of the footprint. The area of the footprint on the moon is

$$A_{\text{foot}} = \pi d^2/4 = 57.5 \times 10^6 \text{ m}^2 \left[77.6 \text{ dBm}^2\right],$$

and the transmitted power is 500 W, therefore the power density is

$$S_{\text{I}} = P/A_{\text{foot}} = 500/57.5 \times 10^6 = 8.7 \,\mu\text{W}/\text{m}^2.$$

It is often easier to perform calculations in the decibel form. For a transmitted power $P_{\text{dB}} = 10\log_{10}(500) = 27$ dBW, the power density is

$$S_{\text{I}} = P_{\text{dB}} - A_{\text{foot}} = 27 - 77.6 = -50.6 \text{ dBW}/\text{m}^2.$$

c) The received power density from the signal reflected by a retroreflector is determined from the lidar range equation. The effective cross section of the retroreflector is (see Chapter 8):

$$\sigma = 0.99\frac{4\pi D^4}{3\lambda^2} = 0.99\frac{4\pi \times 0.1^4}{3 \times \left(10.6 \times 10^{-6}\right)^2} = 3.7 \times 10^6 \, m^2 = 65.7 \text{ dBm}^2.$$

The power density back on the Earth is then calculated from equation (6.4):

$$S_R = \frac{2P\sigma}{\pi^2 R^4 \theta_{BW}^2} = \frac{2 \times 500 \times 3.7 \times 10^6}{\pi^2 \times \left(3.844 \times 10^8\right)^4 \left(22.3 \times 10^{-6}\right)^2} = 3.45 \times 10^{-17} \text{ W}/\text{m}^2.$$

d) The power density back on Earth is determined from equation (6.3) simplified:

$$S_R = \frac{P\rho}{\pi R^2} = \frac{500 \times 0.2}{\pi\left(3.844 \times 10^8\right)^2} = 2.15 \times 10^{-16} \text{ W}/\text{m}^2,$$

which is 10 times higher than that received from the corner reflector.

e) For a receiver bandwidth matched to the pulse width $\beta = 1/\tau = 10$ Hz, and a quantum efficiency $\eta = 0.2$ (typical at the operational wavelength):

$$NEP = \frac{hf\beta}{\eta} = \frac{hc\beta}{\eta\lambda} = \frac{6.625 \times 10^{-34} \times 3 \times 10^8 \times 10}{0.2 \times 10.6 \times 10^{-6}} = 9.38 \times 10^{-19} \text{ W.}$$

Assuming that the optical efficiency is 100%, the received power is the product of the power density and the receive aperture, where the antenna aperture is $A = \pi D^2/4 = 0.196 \text{ m}^2$, which makes the received power

$$S = S_R A = 2.15 \times 10^{-6} \times 0.196 = 4.21 \times 10^{-17} \text{ W.}$$

The single-pulse SNR is $S/NEP = 45$ [16.5 dB].

f) For an SNR of 16.5 dB and a 99% detection probability, the probability of false alarm lies between 10^{-10} and 10^{-12}, as determined from Figure 6.14.

It should be noted that panels containing corner reflectors were deployed by the Apollo 11, 14, and 15 missions and also by the unmanned Soviet Luna 17 and 21 missions. At the McDonald Observatory of the University of Texas, a ruby laser with a pulse duration of 4 ns was transmitted through a 2.7 m telescope and was able to measure the range to the moon with an accuracy of 15 cm. Data obtained in these experiments have been used to check theories about the lunar orbit, its mass distribution, and to produce a new lunar ephemeris (Ready 1978).

6.6 DIGITAL TERRAIN MODELS

Laser altimeters measure height by transmitting a short duration pulse of IR radiation in a narrow beam toward the ground. Because the ground is rough (compared to the wavelength of the light), the pulse is generally scattered in all directions, with a small fraction of the energy returning to the receiver on the aircraft.

Both the round-trip time and the amplitude of the received echo can be logged, and by using the push-broom technique discussed earlier, strip-map images of the ground can be generated. To produce well-registered, accurate images, the aircraft position is logged using a combination of GPS and INS. Using this technique, the position of a spot on the ground can be determined to an accuracy of about 10 cm (vertically) and 50 cm (horizontally). To allow for subsequent processing, the measured data are resampled onto a Cartesian reference grid "pinned" to the surface of the earth. A weighted average of the nearest neighbors within a given radius of the final coordinate point is used, as illustrated in Figure 6.15. Because the swathe painted on the ground is quite narrow (it depends on the height and the scan angle), it is generally necessary to produce composite images produced by successive passes over adjacent, overlapping areas.

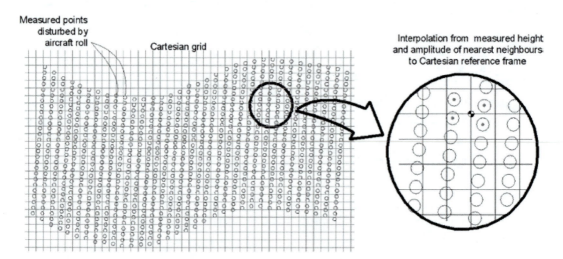

Figure 6.15 Linear interpolation of scanned range and amplitude data onto a Cartesian reference grid.

6.6.1 Surface Models

A digital image is a rectangular array of cells where each cell contains a single value.

- A topological image is produced if the height information is used.
- A reflectivity image is produced if the echo amplitude information is used.

A digital elevation model (DEM) is a digital data file describing the shape of the surface where elevation is a function of longitude and latitude.

- Digital surface models (DSMs) contain elevation values of the air/surface interface, including trees and buildings, etc.
- Digital terrain models (DTMs) reflect pure terrain information as it is represented on contour maps. These are produced by filtering the DSM information.

Figure 6.16 shows an example of a DSM and the DTM that has been obtained by filtering. Depending on the density of the canopy, laser penetration may be limited and the number of samples on the ground will be lower than those available for open ground. Such digital terrain models typically have resolutions between 3 and 7 m, with an absolute vertical accuracy of about 0.5 m (Geolas 2007).

6.6.2 Digital Landscapes

Digital landscapes are DSMs with additional information such as surface color and texture or vegetation types which allows a more realistic representation to be produced. Both DEMs and

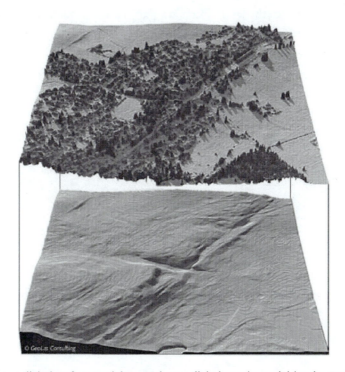

Figure 6.16 Filtering a digital surface model to produce a digital terrain model (Geolas 2007).

DSMs are considered to be 2½D representations of the earth's surface because only a single elevation value is given for each point on the surface, whereas in reality each point may have a multitude of surfaces (tree canopy, building roofs, and ground).

To obtain more detailed information, modern lidar systems no longer record only the first and last pulse range and intensity but produce a complete echo profile by transmitting extremely short pulses (typically 1 ns or less). The received echo is sampled at a matching interval to produce range resolutions of 0.15 m. Using this technique results in each measurement containing information about the density of the tree canopy at different heights as well as the density of the understory vegetation, and even information about the ground surface slope and roughness. This concept is illustrated in Figure 6.17.

The intensity of the echo can be further analyzed to obtain information about the material reflectivity at each level. However, this information is coupled to canopy density and penetration which makes interpretation more difficult.

6.6.3 Thematic Visualization

The detailed information obtained from high-resolution profiling is useful for generating thematic models of the surface. These may include representations of vegetation density and tree heights, or in urban environments, the ratio of sealed (asphalt and concrete) to porous surfaces (sand and vegetation).

For visualization purposes, information about surface roughness and scattering effects may be used to apply realistic texture that differentiate between different forms of cultivated land, as illustrated in Figure 6.16. The usual approach in commercial systems is to use a high-resolution digital camera in parallel with the laser altimeter. This has the advantages of producing a higher resolution image, as well as the ability to image across different spectral bands. However, unlike

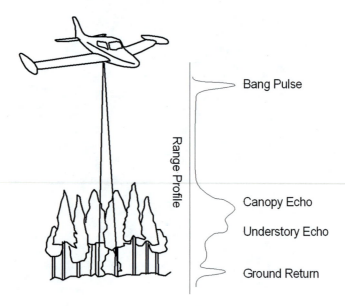

Figure 6.17 High-resolution echo profiling.

the laser images, it requires an external source of illumination, which will probably be at an angle to the camera and thus produce shadows. Integration of lidar scanners and multispectral or thermal scanners along with photogrammatic cameras represents an improved approach to the production of cartographic products, which in the past required significant manual effort to acquire and process.

6.6.3.1 Geographic Information Systems Geographic information systems (GISs) are databases of spatial information that contain a range of specialist information concerning a distributed environment. Such databases can include information on topography, geology, or hydrology, as well as infrastructure and other administrative data. Such data are organized in a manner that facilitates access, retrieval, and display with regard to its geographic location. Imaging lidar can supply elevation and building height information directly into some of the diverse layers of the GIS.

6.6.3.2 3D City Models The good range resolution and high spatial accuracy provided by scanning lidar provides an ideal method for producing 3D models of individual buildings and even complete cities. Laser altimetry is able to produce DSMs of urban areas with resolutions of between 0.5 and 2 m. This is sufficiently accurate to reconstruct heights and footprints, and even the roof shapes of buildings, as shown in Figure 6.18.

The high accuracy available from a DSM can be used to construct accurate road and infrastructure maps. It also allows accurate simulations of flooding to be conducted, a capability which will become more important with changes induced by global warming. Figure 6.19 shows the effect of the Rhine's flooding on the city of Bonn.

Precise lidar data can be used to construct virtual city models. Digital data are converted to a vector format that conforms to the computer-aided design (CAD) standard. These data can then be draped with aerial images to produce texture on roofs and pavements, and images of building

Figure 6.18 Topology image of a cluster of buildings (Geolas 2007).

Figure 6.19 Application of topological images for flood analysis (Geolas 2007).

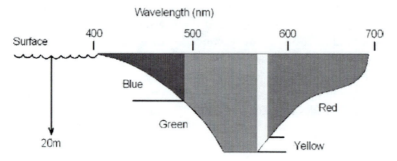

Figure 6.20 Inshore water penetration by light of different colors.

facades taken from the ground can be applied to walls to improve the models still further (Schiewe 2003).

6.7 AIRBORNE LIDAR HYDROGRAPHY

The imaging principles developed for ground mapping can be extended to include subsea mapping if a laser wavelength that penetrates water is selected. Figure 6.20 shows the relative penetration by light of different wavelengths through clear coastal water.

Hydrographic survey systems in use today have been developed by Australian, Swedish, and United States government programs using similar principles with similar performance specifications. The Australian Laser Airborne Depth Sounder (LADS) system operates using a blue-green laser in a push-broom configuration to generate a seafloor profile in relatively shallow coastal water. A conventional laser altimeter measures the height to the water surface, while differential GPS and INS pinpoint the position of the aircraft, as illustrated in Figure 6.21. The LADS

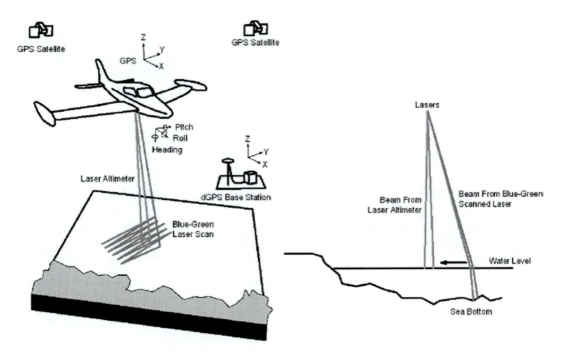

Figure 6.21 Laser airborne depth sounder (LADS) concept.

process can produce depth information at a rate of 64 km²/hr with a spatial resolution of 5 m (Sinclair 2007). The LADS system has the following specifications:

- Sounding rate 900 Hz,
- Water depth between 0.5 and 70 m,
- Swathe width 240 m,
- Resolution normally 5 m × 5 m, but with a 2 m × 2 m option,
- Position accuracy 5 m, CEP 95%.

Areas of cloud cover or high water turbidity that could compromise the performance of the system are identified using satellite images provided by the National Oceanic and Atmospheric Administration (NOAA), and those areas are avoided during the survey.

The LADS system is in use worldwide for shallow-water surveys, and the Royal Australian Navy's unit, mounted on a Fokker F27-500, performed a survey of Sydney Harbor in May 2002, as illustrated in Figure 6.22.

6.8 3D IMAGING

Unlike the processes shown in the previous section that rely on the forward motion of the vehicle to scan one axis, 3D images can be obtained by scanning a range measurement device over two dimensions using one of the mechanical scanning techniques illustrated in Figure 6.7. In most cases the fast axis is scanned using a spinning mirror or prism, while the slow axis is scanned by rotating the whole device (Langer et al. 2000; Probert-Smith 2001).

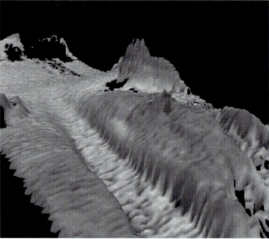

Figure 6.22 LADS survey of Sydney Harbor showing (a) the aircraft and (b) the sow and pigs reef and the western channel (Shallow Survey 2003).

6.8.1 Radar Systems

Under adverse weather conditions or through dust or smoke, the performance of laser systems is degraded due to absorption and scattering (see Chapter 7). An alternative, albeit one with poorer angular resolution, is to use a millimeter wave radar to produce images. Millimeter wave frequencies are used because their wavelength is sufficiently long compared to the sizes of dust, smoke, and mist particles, so that the signal is hardly attenuated as it propagates through the medium.

As illustrated in Figure 6.23, scanners based on standard pan/tilt (slow) or moving mirror (fast) technologies have been developed at the Australian Center for Field Robotics and are used for industrial and mining applications (Brooker et al. 2007). In addition to being relatively immune to weather conditions, because of the relatively wide beamwidth, data produced by these radar scanners can illuminate multiple targets simultaneously. This allows the unit to "see" through trees by exploiting the direct paths through to a target, as well as to obtain foliage density information.

Frequency modulated continuous wave (FMCW) systems with good range resolution, discussed in detail in Chapter 11, produce good 3D images, particularly if the space is oversampled in angle. In a typical implementation, the radar signal is sampled and digitized by a 12-bit analog-to-digital converter (ADC) at 5 MHz for the full duration of each 1 ms chirp. The range amplitude spectrum is produced by a 4096 point fast Fourier transform (FFT) that generates 2048 range bins spanning the range from 0 to 500 m. Thresholding, peak extraction, and interpolation produce a number of targets at each angle.

A 3D image can be built up in a number of ways. In most cases, each observed target, which is defined by its coordinates in polar space (R, θ, φ) and its associated intensity, is converted into Cartesian coordinates and then transformed into a global coordinate frame using the attitude/position data from an associated GPS/INS. This produces a true 3D image that can be passed

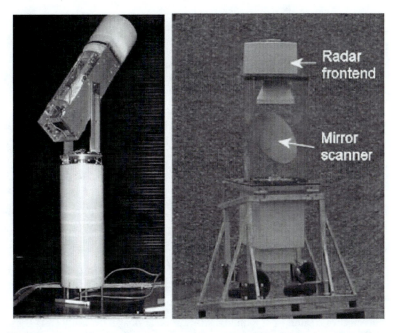

Figure 6.23 Millimeter wave radar imagers based on (a) pan and tilt and (b) mirror scanning.

Figure 6.24 Radar image and photograph of the ACFR and surrounds—plan view.

to a surface reconstruction algorithm (Delaunay triangulation) or displayed unprocessed, as seen in Figure 6.24 and Figure 6.25.

6.8.2 Focused Beam Radar Imaging

The radar images shown in the previous section are produced using collimating antennas operating in the far field where the angular resolution is diffraction limited. For short-range imaging

Figure 6.25 3D perspective created from a millimeter wave radar image. (Courtesy Craig Lobsey)

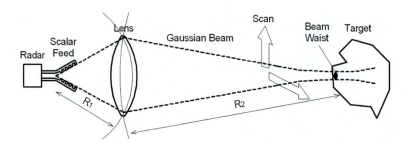

Figure 6.26 Focused beam imaging.

applications, particularly those used in security screening, superior angular resolution can be achieved by focusing the beam onto a spot that is scanned across the target.

In these applications a Gaussian beam is produced by a conical corrugated feed (scalar feed). Quasi-optical processes can be used to manipulate this beam, of which the simplest is a lens to focus it at the required range. This produces a spot with a resolution determined by the beam waist radius, ω_o, as shown in Figure 6.26.

If the lens is thin (one whose focal length is much larger than the width of the beam at the lens) and R_1 and R_2 are the phase-front curvatures of the two beams at the lens, then the focal length, f, of the lens is

$$1/f = 1/R_1 - 1/R_2. \tag{6.12}$$

If the size of the beam emerging from the lens is ω (m), then the waist diameter is given by (Lesurf 1990)

$$\omega_o = \frac{\omega^2}{1 + \hat{z}^2} \tag{6.13}$$

located at a distance

$$z = \frac{R}{1 + (1/\hat{z})^2} \tag{6.14}$$

from the lens, where $\hat{z}$ is defined in terms of the emerging beam diameter, ω (m), the phase front curvature, R (m), and the wavelength of the signal, λ (m):

$$\hat{z} = \frac{\pi \omega^2}{\lambda R}. \tag{6.15}$$

From equation (6.14) and equation (6.15) it can be seen that the beam waist location is a function of R. If $R >> \pi \omega^2 / \lambda$, then $z \approx R/R^2$ and $z \to 0$ as R increases. However, if $R << \pi \omega^2 / \lambda$, then $z \approx R$ and $z \to 0$ as R decreases.

Equation (6.14) and equation (6.15) can be rearranged to form a quadratic equation:

$$R^2 az - R + z = 0, \tag{6.16}$$

where

$$a \equiv \left(\lambda / \pi \omega^2 \right)^2. \tag{6.17}$$

The root of equation (6.16) is

$$R = \frac{1 \pm \sqrt{1 - 4az^2}}{2az}. \tag{6.18}$$

To ensure that R is real, the distance, z, to the beam waist must satisfy the following inequality:

$$z \leq \pi \omega^2 / 2\lambda. \tag{6.19}$$

This means that it is not possible to produce a beam waist at an arbitrary distance from the lens. The maximum range, often called the maximum throw of the lens, is defined where equation (6.19) is set as an equality. If the distance to the waist is less than the maximum, there are two solutions to R, and these produce different waist sizes, ω_o, at the chosen distance, z. These are referred to as the parallel beam waist and the focused beam waist.

Consider Figure 6.26 with the case in which the lens diameter ω is 0.15 m, operating at a frequency of 94 GHz ($\lambda = 0.0032$ m), then the phase front radii and waist diameters will be as shown in Figure 6.27. In the parallel beam case, the phase front radius is very large, implying a very flat lens. The diameter of the beam waist starts out equal to the lens diameter and decreases slightly out to the maximum throw of the lens at 11 m. In the focused beam case, the phase front radius is much smaller and the beam waist diameter increases linearly with distance.

For high-resolution imaging, the focused beam case is of interest, and these results are shown in Figure 6.28. In this case, the phase front radius is almost equal to the distance, z, out to 5 m. In addition, the beam waist diameter also bears an almost linear relationship to the distance. At an imaging range of 3 m, the resolution, as determined by the beam waist diameter, is just over 20 mm.

6.8.3 Lidar Imaging

As illustrated in Figure 6.29, most imaging lidar systems consist of a ranging module and a high-speed rotating prism or mirror housed in a sealed unit. A window allows the scanned beam to exit, and echoes to return, and a pan axis sweeps this complete mechanism across a designated sector to build up a complete 3D image.

It can be seen from Figure 6.30 that laser images are more photorealistic than radar images because the angular resolution is superior. However, this acuity comes at a price in terms of the time taken to produce an image and the sensor's performance in dust, etc. These systems are commonly used in the following fields:

- Topography and mining,
- Architecture and façade measurement,
- As-built surveying,

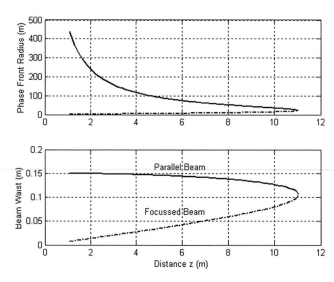

Figure 6.27 Phase front radii and beam waist diameter as a function of the distance to the beam waist for a 150 mm lens.

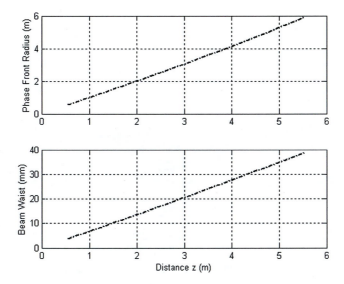

Figure 6.28 Expanded view of the phase front radius and the beam waist diameter for the focused beam case.

Figure 6.29 Commercial 3D laser scanners.

- Archaeology and cultural heritage surveying,
- Civil engineering, and
- City modeling.

Figure 6.30 shows a laser image generated from a number of 3D scans fused into a single point cloud. It shows an image of the same region as those produced by the scanning radar systems reproduced in Figure 6.8 and Figure 6.25. A comparison shows that the superior angular resolution of the laser reproduces architectural details such as walls and windows better than the radar, but that the more amorphous objects such as trees are equally well reproduced, with superior

Figure 6.30 Composite 3D laser scanner color-coded image (Courtesy José Guivant).

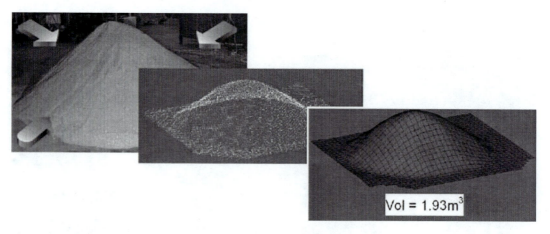

Figure 6.31 Process of estimating volume using a 3D laser scanner (Stone et al. 2000).

foliage penetration exhibited by the radar. The critical difference between the two imaging modalities is that the radar would be equally capable of reproducing the scene if the area was filled with smoke, dust, or fog, while the lidar would be blind.

The availability of low-cost, high-resolution laser scanners has also led to a proliferation of industrial applications from volume estimation and architectural reproduction to CAD and computer-aided manufacturing (CAM). For example, to determine its volume, an object is scanned from a number of carefully surveyed positions to produce individual views that are combined to generate a complete and comprehensive point cloud image, as shown in Figure 6.31. Surface heights on a regular grid are computed from the point cloud data and the volume is calculated by integrating the height of the grid over the surface, as illustrated in Figure 6.31 (Stone et al. 2000).

In the CAD/CAM application, a component is scanned in a similar manner to produce a point cloud image which is then further processed to produce 3D digitized data of the surface. This model can then be used to reproduce the original component using a "rapid prototyping printer" or a computer numerical control (CNC) machine.

6.8.4 Jigsaw Foliage-Penetrating Lidar

An interesting extension to conventional 3D lidar is the Jigsaw system (Marino and Davis 2005). It consists of a novel scanned laser based on a pair of rotating prisms (Risley scanner) and a focal plane array made from 32×32 APDs.

The instantaneous field of view is about 10 mrad $\times$ 10 mrad across the array, making the imaging pixel size just over 300 μrad (5 cm at an operational range of 150 m). A neodymium-yttrium aluminum garnet (Nd:YAG) laser transmits 300 ps long pulses at a pulse repetition frequency (PRF) of 16 kHz, and although this equates to a theoretical range resolution of 45 mm, effects of the APD avalanche time and sampling degrade this to 400 mm. The exceptional angular resolution, high sampling rate, and good range resolution allow this system to reconstruct a complete 3D image of the volume below the tree canopy, as shown in Figure 6.32.

Figure 6.32 Sequence of slices taken by Jigsaw through a thick forest canopy reveal a tank and tree trunks at the lowest level (Marino and Davis 2005).

6.9 ACOUSTIC IMAGING

The use of acoustics for 2D and 3D imaging has a long history and covers myriad applications from medical ultrasound through nondestructive testing to underwater sonar. These applications are discussed elsewhere in this book, with sections on Doppler ultrasound in Chapter 10, sidescan sonar in Chapter 12, and 3D medical ultrasound in Chapter 16. An interesting application is that of acoustic microscopy, and this is discussed briefly in the following section.

6.9.1 Scanning Acoustic Microscopes

The scanning acoustic microscope (SAM) produces images by scanning a focused beam of acoustic energy (sound) across a sample to measure its elastic properties, as illustrated in Figure 6.33. In principle, this process is very similar to that used by the medical ultrasound scanning technology discussed in Chapter 16. The primary difference in this application is the use of mechanical scanning techniques while most medical systems use electronic scanning through one plane.

The principles involved in the formation of an acoustic lens are similar to those used in the objective lens of an optical microscope; that is, different propagation velocities through different media are used to focus a beam so as to resolve small objects. However, unlike its optical counterpart, where the relative velocities between the two media seldom exceed 1.9 (the refractive index difference), with the acoustic microscope the velocity can decrease by a factor of 10 or more when traversing a suitable solid-liquid interface.

A suitable lens such as the one shown in Figure 6.34, manufactured for the original Stanford University microscope (Lemons and Quate 1974; Quate 1979), consists of a concave spherical sapphire lens with a radius of approximately 40 μm immersed in water. At the interface, the sound waves are bent sharply inward as they encounter the liquid, to focus close to the center of curvature of the sphere.

The first step in the imaging process is to convert the high-frequency electrical pulse to an acoustic one by means of a piezoelectric film deposited on the back of the sapphire lens. The acoustic wave travels through the lens material until it strikes the back of the curved section where the focusing takes place. A matching layer of glass improves the coupling from the sapphire

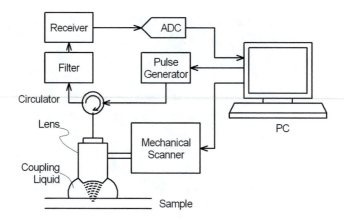

Figure 6.33 Schematic diagram showing the components of a scanning acoustic microscope.

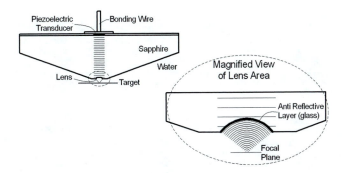

Figure 6.34 Schematic diagram of an acoustic lens.

into the liquid (usually water) in which the target is mounted. The different elastic properties of the target material at the focal point distort the phase of the reflected signal, which then follows the reverse path and is detected by the same piezoelectric film which acts as a microphone.

Most SAMs use short ultrasound pulses at a frequency ranging between 15 and 180 MHz to separate the received and transmitted signals, as well as for time gating to image-specific depths in the medium. An image is formed by scanning the focal point of the lens across the object in a raster fashion, storing the amplitude of the received signal along with the spatial coordinates, and then displaying the complete picture. Modern microscopes are often based on commercial scanning probe microscope (SPM) technology and can produce images in a few seconds.

The resolution, δx (m), that can be obtained using this technique is slightly better than that predicted by the Rayleigh criterion (Briggs 1992). In this case it is given by

$$\delta x = 0.51\lambda/NA, \tag{6.20}$$

where λ is the wavelength of the sound in the medium and NA is the numerical aperture. This is equal to the sine of the subtended semiangle, $\sin\theta_o$.

In a typical application, the operational frequency is 50 MHz and the velocity of sound is 1496.7 m/s in distilled water, making the wavelength 2.99×10^{-2} mm. Estimating θ_o to be between 5° and 10° gives a resolution between 88 μm and 175 μm.

Because the process can take place in water, which is compatible with living biological specimens, it is possible to make images of living cells, as shown in Figure 6.35a (Hildebrand et al. 1981). Such images produce data on the elasticity of individual cells and tissues which can provide useful information on the mechanics of structures such as the cytoskeleton and cell motility.

Most SAMs are used in industrial applications where their ability to image through solid material and identify subsurface cracks and flaws is particularly useful.

6.10 WORKED EXAMPLE: LIDAR LOCUST TRACKER

6.10.1 Requirement

Understanding the behavior of swarms of locusts is a crucial step in the process of minimizing the devastation caused by these pests during their relentless advance across the land. Previous

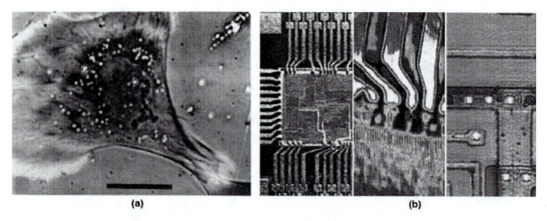

Figure 6.35 Acoustic images of (a) a living cell and (b) the interior of an integrated circuit.

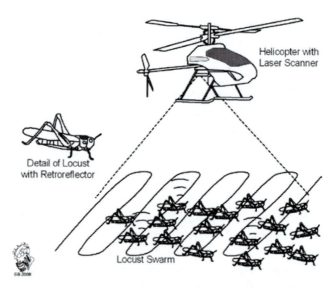

Figure 6.36 Operational principle of the locust tracker showing a push-broom laser scanner mounted on a helicopter that can pinpoint returns from the retroreflectors attached to the backs of locusts.

attempts to track individual insects have been both expensive and time consuming, as they involved tagging individuals with small wireless beacons and then pinpointing their position periodically using radio location devices and GPS.

The following case study is the design of an alternative, less manpower intensive and more effective method of tracking both individual and groups of insects. First, a number of insects are captured and each tagged with a small patch of an efficient retroreflective material. A laser-based push-broom scanner, in conjunction with a helicopter or fixed-wing aircraft fitted with a GPS/INS, can be used to pinpoint the positions of the tagged locusts on a regular basis, as illustrated in Figure 6.36.

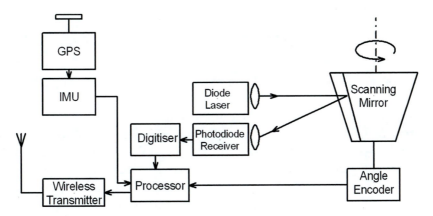

Figure 6.37 Locust tracker system schematic block diagram.

6.10.2 Specifications

- Scanner requirements
  Operational height $h = 1000$ m
  Swathe width $x \geq 1000$ m
- Laser specifications
  Wavelength $\lambda = 905$ nm ± 5 nm (near-infrared)
  Average power $P_{ave} = 2$ mW
  Pulse width $\tau = 20$ ns
  PRF $f_p = 10$ kHz
  Beam divergence $\theta_b = 2$ mrad
  Transmit aperture $d_{tx} = 50$ mm diameter
  Receive aperture $d_{rx} = 50$ mm diameter

The laser peak power can be determined from the average power and the duty cycle:

$$P_p = P_{ave}/\tau f_p = 2 \times 10^{-3}/20 \times 10^{-9} \times 10^4 = 10 \, \text{W}.$$

6.10.3 System Hardware

The block diagram in Figure 6.37 shows the components of the system. A faceted mirror rotates at high speed and scans the laser beam across the ground. Reflections from the retroreflectors on the locusts and returns from the ground are detected by the receiver and digitized. A processor determines the position of each retroreflector from the measured range and angle of the beam in conjunction with the instantaneous position and attitude of the aircraft as measured by the GPS and inertial measurement unit (IMU). The data are then communicated to the ground using a wireless link.

6.10.4 Determining the Required Aircraft Speed

The angle subtended by a swathe width of $x = 1000$ m from a height of $h = 1000$ m is

$$\theta_s = 2\tan^{-1}\frac{x/2}{h} = 2\tan^{-1}\frac{500}{1000} = 53°.$$

Because of the angle doubling for a beam reflected by a rotating mirror, it is convenient to construct the mirror segments at 30° intervals to produce a symmetrical prism with 12 facets. This will produce a beam that scans over 60° before repeating.

To ensure complete coverage of the ground as the beam scans, a sample overlap of at least 50% is required. Hence the beam should scan over 1 mrad (half the beam divergence) between pulses, as shown in Figure 6.38. The maximum angular scan rate is therefore determined by the following:

$$\dot{\theta} = \frac{\theta_b}{2}f_p = \frac{2\times10^{-3}}{2}\times10000 = 10 \ \text{rad/s}$$

The beam scans through 60° (1.05 rad) in 1.05/10 = 0.105 s.

At a height of $h = 1000$ m, the diameter of the footprint on the ground is $x_f = \theta_b h = 2$ m. Therefore, to provide for the same 50% overlap that was achieved for the cross-range scan, the aircraft can advance by $x_f/2 = 1$ m in 0.105 s, which equates to a forward velocity of 9.5 m/s.

6.10.5 Laser Power Density on the Ground

The maximum operational range required if the laser is at an offset angle of 30°, for $h = 1000$ m, corresponds to

$$r_{max} = 1000/\cos 30 = 1155 \ \text{m}.$$

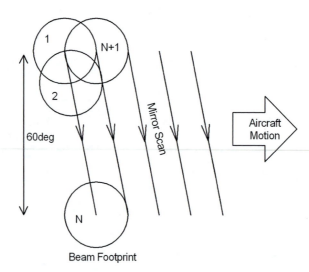

Figure 6.38 The relationship between the beam footprint and the scan pattern.

The spot size on the ground will be slightly elliptical, with a minor axis diameter of

$$d_{min} = r_{max}\theta_b = 1155 \times 2 \times 10^{-3} = 2.31\,m$$

and a major axis diameter of

$$d_{maj} = r_{max}\theta_b/\cos 30 = (1155 \times 2 \times 10^{-3})/0.866 = 2.66\,m,$$

making the total area of the footprint

$$A_f = \pi d_{min} d_{maj}/4 = 4.86\,m^2.$$

The power density of the beam on the ground at the maximum operational range is the peak power divided by the footprint area:

$$S_i = P_p/A_f = 10/4.86 = 2.06\ W/m^2.$$

6.10.6 The Power Density of the Reflected Signals Back at the Laser

The total reflected power is the product of the incident power density, S_i, and the patch area, A_{pat}. Assuming that the reflected light is scattered uniformly over the hemisphere, the power density back at the camera is given by

$$S_r = S_i A_{pat} \cdot \frac{1}{2\pi R^2}.$$

However, because the patch is retroreflective when illuminated, the simplest model is to assume that it becomes an antenna that is diffraction limited by its aperture.

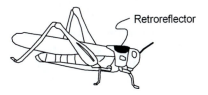

The gain of such an antenna is the ratio of the power radiated in a specific direction relative to the isotropic. The power density back at the laser receiver will be

$$S_r = S_i A_{pat} \cdot \frac{1}{2\pi R^2} G_{pat}.$$

The relationship between the aperture, A_{pat}, and the gain, G_{pat}, of a diffraction-limited antenna is

$$G_{pat} \approx 4\pi A_{pat}/\lambda^2,$$

where G_{pat} is the gain with respect to an isotropic radiator and λ is the wavelength of the radiation. Substituting,

$$S_r = S_i \frac{4\pi A_{pat}^2}{\lambda^2} \frac{1}{2\pi R^2}.$$

For a square retroreflective patch with $d_{pat} = 5$ mm, the formula becomes

$$S_r = S_i \frac{4\pi d_{pat}^4}{\lambda^2} \frac{1}{2\pi R^2}.$$

The optical cross section, σ, is defined as the ratio by which the power density at the receiver exceeds that of an isotropic scatterer. Therefore

$$\sigma = \frac{4\pi d_{pat}^4}{\lambda^2} = \frac{4\pi \times (5\times10^{-3})^4}{(905\times10^{-9})^2} = 9526\, m^2,$$

making the equation

$$S_{rr} = S_i \sigma \frac{1}{2\pi R^2} = 2.06\times9526\times\frac{1}{2\pi\times1155^2} = 2.34\times10^{-3}\, W/m^2.$$

The physical cross section of the footprint on the ground is $A_f = 4.86$ m^2 and the backscatter coefficient, ρ, is 0.1 (see Table 3.1), which makes the backscattered power density at the receiver

$$S_{rg} = S_i A_f \rho \cdot \frac{1}{2\pi R^2} = 2.06\times4.86\times0.1\times\frac{1}{2\pi\times1155^2} = 1.19\times10^{-7}\, W/m^2.$$

A comparison between the previous pair of equations shows that it is not necessary to equate them to determine the relative power densities of the retroreflector and the ground backscatter in order to determine the SNR. The SNR is

$$SNR = 10\log_{10}\frac{S_{rr}}{S_{rg}} = 10\log_{10}\frac{2.34\times10^{-3}}{1.19\times10^{-7}} = 42.9\text{ dB}.$$

This is more than adequate to detect the return from the retroreflector and to differentiate it from backscatter from the ground.

6.10.7 The Effect of the Sun

In addition to competing with laser backscatter from the ground, the receiver must compete with solar backscatter. Figure 6.39 shows the solar spectrum in space and at sea level.

Over the full band from 300 to 2500 nm, the total power density is obtained by determining the integral under the curve. This is approximately 1000 W/m^2. However, to determine the total

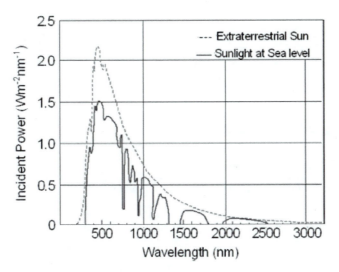

Figure 6.39 Solar spectrum in space and at sea level.

power density for comparison purposes, the bandwidth of any optical filter in the receiver is required.

The specified wavelength for a typical laser range finder is stated as 905 ± 5 nm. Hence an optical filter with a bandwidth of 10 nm would be sufficient. Such filters can be acquired from a number of optics suppliers and have the following specifications:

Full width half max (FWHM) $\lambda_b = 10 \pm 2$ nm
Efficiency $\tau_o = 0.7$

For an incident flux of 0.6 W/m^2/nm at $\lambda = 905$ nm, the total power density will be

$$S_{is} = 0.6\lambda_b = 6 \text{ W}/\text{m}^2.$$

So the total power density back at the laser receiver is, once again, determined by the area of the footprint on the ground, the backscatter coefficient, and the assumption of uniform scattering:

$$S_{rs} = S_{is}A_f\rho \cdot \frac{1}{2\pi R^2} = 6 \times 4.86 \times 0.1 \times \frac{1}{2\pi \times 1155^2} = 3.47 \times 10^{-7} \text{ W}/\text{m}^2.$$

This is another source of noise, so the SNR should again be determined:

$$SNR = 10\log_{10}\frac{S_{rr}}{S_{rg}} = 10\log_{10}\frac{2.34 \times 10^{-3}}{3.47 \times 10^{-7}} = 38.2 \text{ dB}.$$

This is lower than the SNR for the laser backscatter and will therefore define the available SNR at the receiver from external sources.

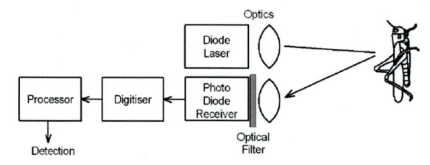

Figure 6.40 Laser range finder configuration showing the position of the optical filter.

6.10.8 The Receiver

The receiver consists of a collimating lens and an optical filter followed by a photodiode to convert the laser echo into an electrical signal, as shown in Figure 6.40. The total power received is equal to the product of the power density at the receiver, S_{rr}, the receive lens aperture, A_{lens}, and the optical efficiency, τ_o.

Assume that the lens diameter is 50 mm, which makes $A_{lens} = 1.96 \times 10^{-3}$ m^2, $S_{rr} = 2.34 \times 10^{-3}$ W/m^2 was determined earlier, and $\tau_o = 0.7$ from the specifications of the optical filter:

$$P_{rec} = S_{rr} A_{lens} \tau_o = 2.34 \times 10^{-3} \times 1.96 \times 10^{-3} \times 0.7 = 3.21 \times 10^{-6} \text{ W}.$$

The characteristics of the Hamamatsu S8223 photodiode, which would be suitable for performing this detection process, are listed in Figure 6.41.

Note that the peak sensitivity (responsivity) occurs at around 900 nm and is $R = 0.53$ A/W. The maximum output current will therefore be

$$I_{rec} = P_{rec} R = 3.21 \times 10^{-6} \times 0.53 = 1.7 \times 10^{-6} \text{ A}.$$

The RMS value for the shot noise is given by

$$I_{shot} = \sqrt{4eI_{rec}\Delta f} = \left(4 \times 1.602 \times 10^{-19} \times 1.7 \times 10^{-6} \times 200 \times 10^{6}\right)^{1/2} = 14.8 \times 10^{-9} \text{ A},$$

where e is the electron charge and Δf is the bandwidth of the photodiode (theoretically from DC up to f_c).

The shot noise is larger than the dark current, therefore the SNR can be determined from the ratio of the received current, I_{rec}, to the shot noise current, I_{shot}:

$$SNR = 20\log_{10}\frac{I_{rec}}{I_{shot}} = 20\log_{10}\frac{1.7 \times 10^{-6}}{14.8 \times 10^{-9}} = 41 \text{ dB}.$$

Because it is more convenient to work with voltage, the output current passes through an op-amp-based current-to-voltage converter, shown in Figure 6.42, before passing through a filter

Part Number	S8223
Package	Plastic
Active Diameter	0.8mm
Peak Wavelength	900nm
Minimum Wavelength	320nm
Maximum wavelength	1060nm
Peak Sensitivity	0.53A/W
Max dark Current	1nA
Cutoff Frequency	200MHz
Capacitance	3pF

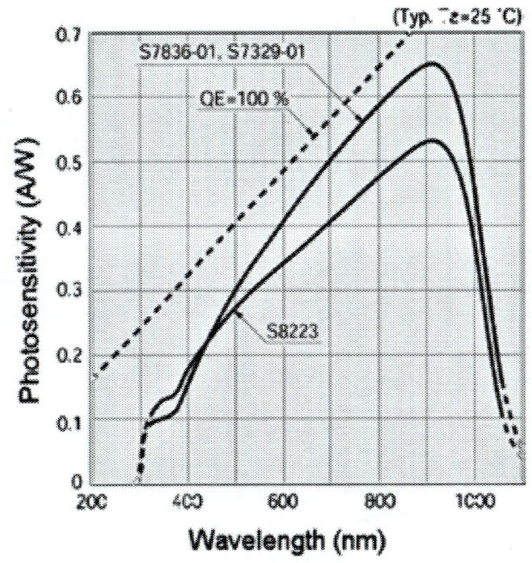

Figure 6.41 Hamamatsu S8223 photodiode characteristics.

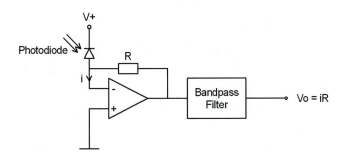

Figure 6.42 Photodiode receiver circuitry.

matched to the laser pulse width. The feedback resistor, R, is selected to produce a reasonable output voltage. For example, by selecting $R = 1\ M\Omega$, a peak voltage of 1.7 V would be produced, for an input current pulse of 1.7 μA.

The transmitted pulse width is $\tau = 20$ ns, so the receiver bandwidth, β, will be the reciprocal of that to a first approximation:

$$\beta = 1/\tau = 1/\left(20 \times 10^{-9}\right) = 50 \text{ MHz.}$$

An appropriately fast op-amp would be required to drive the filter with this short pulse.

If the shot noise current comprises white noise which is uniformly distributed over the 200 MHz bandwidth of the photodiode, placing a matched filter with a bandwidth of 50 MHz at the output improves the SNR by the ratio of the total bandwidth to the filter bandwidth:

$$SNR = 41 + 10\log_{10}\frac{200}{50} = 47 \text{ dB}.$$

It can be shown that the effective SNR out of a square-law detector is also squared, so the SNR of the retroreflector return compared to that from the sun will increase from 38.2 dB to 76.4 dB.

6.10.9 Conclusion

In this example, it can be seen that the retroreflective return will easily be visible above the returns from the backscattered laser signal, the backscatter from the sun, the dark current, and the shot noise. The SNR of 47 dB is limited by the photodiode shot noise.

6.11 References

Blake, L. (1986). *Radar Range-Performance Analysis*. Norwood, MA: Artech House.

Briggs, A. (1992). *Acoustic Microscopy*. Oxford: Clarendon Press.

Brooker, G. (2005). Long-range imaging radar for autonomous navigation. Ph.D. dissertation. University of Sydney, Sydney, NSW Australia.

Brooker, G., Hennessey, R., Bishop, M., Lobsey, C., and Durrant-Whyte, H. (2006). High-resolution millimeter-wave radar systems for visualization of unstructured outdoor environments. *Journal of Field Robotics* 23(10):891–912.

Brooker, G., Hennessy, R., Bishop, M., Lobsey, C., and MacLean, A. (2007). Millimeter wave 3D imaging for industrial application. In: The 2nd International Conference on Wireless Broadband and Ultra Wideband Communications, 2007. AusWireless 2007. Sydney, Australia, August 27–30.

Brooker, G., Widzyk-Capehart, E., Hennessey, R., Bishop, M., and Lobsey, C. (2007). Seeing through dust and water vapor: millimeter wave radar sensors for mining application. *Journal of Field Robotics* 24(7):527–557.

Geolas. (2007). Geolas consulting. Viewed October 2007. Available at http://www.geolas.com/Pages/Hwy.html.

Hildebrand, J., Rugar, D., Johnston, R., and Quate, C. (1981). Acoustic microscopy of living cells. *Proceedings of the National Academy of Sciences* 78(3):1656–1660.

Hovanessian, S. (1988). *Introduction to Sensor Systems*. Norwood, MA: Artech House.

Klein, L. (1997). *Millimeter-Wave and Infrared Multisensor Design and Signal Processing*. Norwood, MA: Artech House.

Langer, D., Mettenleiter, M., Hartl, F., and Frohlich, C. (2000). Imaging ladar for 3-D surveying and CAD modeling of real-world environments. *International Journal of Robotics Research* 19(11):1075–1088.

Lemons, R. and Quate, C. (1974). Acoustic microscope—scanning version. *Applied Physics Letters* 24(4):163–165.

Lesurf, J. (1990). *Millimeter-wave Optics, Devices & Systems*. New York: Taylor & Francis Group.

Marino, R. and Davis, W.J. (2005). Jigsaw: a foliage-penetrating 3D imaging laser radar system. *Lincoln Laboratory Journal* 15(1):23–26.

Probert-Smith, P., ed. (2001). *Active Sensors for Local Planning in Mobile Robotics*. Hackensack, NJ: World Scientific.

Quate, C. (1979). The acoustic microscope. *Scientific American* 241(4):62–70.

Ready, J. (1978). *Industrial Applications of Lasers*. Burlington, MA: Academic Press.

Schiewe, J. (2003). 3D city modeling using ultra high resolution and multi-sensorial remote sensing. In: *Geo-Information Systems GIS 6*. Hüting Verlag, Germany.

Shallow Survey. (2003). Shallow Survey 2003—image gallery. Viewed October 2007. Available at www.shallowsurvey.com/info_files/images3.html.

Sinclair, M. (2007). Laser airborne depth sounder—LADS. Viewed October 2007. Available at http://www.hydrographicsociety.org.nz/reports/report_laser.htm.

Stone, W., Cheok, G., and Lipman, R. (2000). Automated earthmoving status determination. Robotics 2000. ASCE Conference on Robotics for Challenging Environments, February 28–March 2, 2000, Albuquerque, NM.

7

Signal Propagation

7.1 THE SENSING ENVIRONMENT

The environment in which a sensor is expected to operate exerts a strong influence on its performance. These effects, as summarized in Figure 7.1, include interactions of the acoustic or electromagnetic radiation with the target and its surrounds (the background), and particularly with the atmosphere through which the beam must travel between the target and the sensor. These interactions include attenuation by the atmosphere and attenuation and scattering by hydrometeors and other suspended particulates. An additional effect that will only be briefly considered in this book is the effect of interference, be it natural or man made.

As illustrated in Figure 7.2, which shows a radar searching for aircraft targets, returns from the same range gate that includes the target can include returns from the ground surface (through the lower sidelobe in this case) and from particulates suspended in the atmosphere. These are known as surface and volume clutter, respectively, and their magnitude and effect on the detectability of the target are discussed later in this chapter.

In addition, interference can also enter the sensor via the antenna and may come from natural sources such as the sun and local galaxies, or it may be man made, for example, car ignition systems, fluorescent lights, other sensors, etc.

For electromagnetic sensors, clear-air attenuation is mostly caused by molecular absorption and scattering by oxygen and water. Attenuation and scattering are caused by hydrometeors, including rain, hail, clouds, and fog, and also by suspended particulates including dust and smoke.

Interference may be generated in an attempt to deceive the sensor. Smoke can be released to blind infrared (IR) systems and chaff can be released to deceive microwave systems. For ultrasound or underwater sonar sensors, the sources and causes of clutter, interference, and attenuation are different, but their ultimate effect on the performance of the sensor is similar to that for lidar or radar.

7.2 ATTENUATION OF ELECTROMAGNETIC WAVES

As an electromagnetic signal propagates through the atmosphere, molecular interactions with the wave absorb energy and the signal amplitude decreases with range. This process results

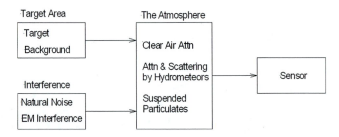

Figure 7.1 The sensing environment.

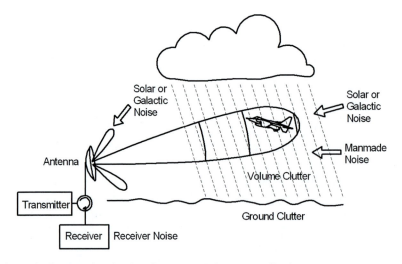

Figure 7.2 Schematic diagram showing interference and clutter contributions.

in the following relationship between the received power, P_r (W), and the transmitted power, P_t (W),

$$P_r = P_t e^{-2\alpha R}, \tag{7.1}$$

where α is the attenuation per meter. It is often referred to as the extinction coefficient.

Using the change of base theorem,

$$\log_a N = \log_b N / \log_b a, \tag{7.2}$$

it can be shown that $\log_{10} N = 0.43429 \log_e N$, which allows the attenuation to be expressed as $\alpha_{dB} = 4342.9\alpha$ (in dB/km) (one way).

The relationship in equation (7.1) is usually expressed in the decibel form:

$$10\log_{10}(P_r) = 10\log_{10}(P_t) - 2\alpha_{dB} R_{km}, \tag{7.3}$$

where

P_r = received power (W),
P_t = transmitted power (W),
α_{dB} = attenuation (dB/km),
R_{km} = range (km).

This nonlinear term must be incorporated into the range equation to obtain accurate estimates of the detection range. Its inclusion makes the equation difficult to solve in a closed form and to overcome this, numerical or graphical methods are normally resorted to.

The radar range equation, including atmospheric attenuation, is used to determine the received power $10\log_{10}(P_r)$ (dBW) for a target with a radar cross section, σ (m²), and various parameters based on the radar characteristics:

$$10\log_{10}(P_r) = 10\log_{10}(P_t) + 10\log_{10}\left(\frac{\lambda^2}{(4\pi)^3}\right) + 20\log_{10}(G) + 10\log_{10}(\sigma)$$
$$- 10\log_{10}(L) - 40\log_{10}(R) - 2\alpha_{dB}R_{km}. \tag{7.4}$$

The attenuation coefficient, α_{dB}, is determined by adding the components from all of the sources of attenuation discussed in the following sections. These include absorption by the atmosphere, as well as suspended particulates and losses due to scattering by particulates in the atmosphere.

7.2.1 Clear Weather Attenuation

Clear weather (or clear air) attenuation, shown graphically in Figure 7.3, is primarily due to absorption by oxygen and water. In this case, the water component is due to the absolute humidity, and it is determined for a representative water content of 7.5 g/m³ at 20°C.

Below 60 GHz there are two main absorption bands: from the 1.35 cm (22.235 GHz) water vapor line and its skirts, and from a series of oxygen lines centered around 0.5 cm (60 GHz). These two peaks bracket the 35 GHz window. The atmosphere becomes progressively more opaque at higher frequencies, with windows at 94, 140, 220, and 360 GHz between the oxygen and water absorption bands.

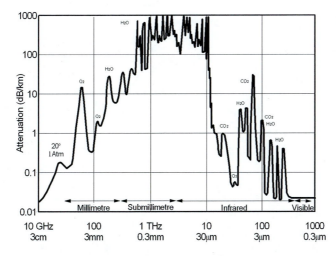

Figure 7.3 Clear-weather attenuation as a function of frequency [adapted from Preissner (1978)].

It is almost completely opaque between 1 THz and 10 THz (0.3 mm and 30 μm) after which windows again appear between absorption lines of water and carbon dioxide in the IR region starting at about 10 μm. Finally, attenuation is very low in the visible region around 500 nm and beyond.

7.2.2 Effect of Atmospheric Pressure (Air Density)

Because total absorption is proportional to air density, Figure 7.3 is adjusted to be representative of the pressure at sea level. However, in many instances it is important to be able to calculate the total attenuation at different altitudes, or through the complete atmosphere. Figure 7.4 shows the difference in the attenuation at sea level and at an altitude of 4 km for horizontal propagation. It can be seen that there is a reasonably consistent decrease in the attenuation across the band at the higher altitude and that the effects of resonances are more pronounced there.

7.2.3 Effect of Rain

The attenuation due to light rain that has a uniform spatial distribution can be predicted with relative ease. However, heavier rain normally comes in squalls with unpredictable intensity and drop size distribution, and this makes accurate estimates of the total path attenuation difficult to determine.

At low frequencies, where the drop size is much smaller that the wavelength, Rayleigh scattering theory is applied. This leads to a simple model for attenuation, α_r (dB/km), as a function of the rain rate, R (mm/hr):

$$\alpha_r = aR^b, \tag{7.5}$$

where a, b are coefficients that are dependent on frequency and temperature. Table 7.1 gives these coefficients for a number of different frequencies (Currie et al. 1992).

If the attenuation at higher frequencies extending into the submillimeter wave band is to be determined, then Mie scattering theory must be used. The theoretical attenuation with frequency

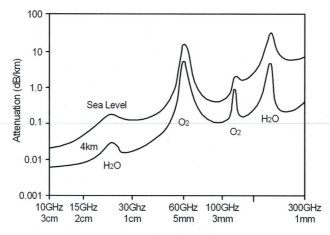

Figure 7.4 Average atmospheric absorption of millimeter waves for horizontal propagation at sea level and at an altitude of 4 km [adapted from Button and Wiltse (1981)].

Table 7.1 Empirical coefficients used to determine attenuation

Frequency (GHz)	a	b
10	0.00919	1.16
35	0.273	0.985
70	0.634	0.868
94	1.6	0.64

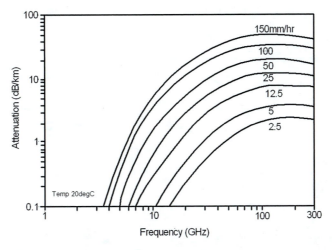

Figure 7.5 Theoretical rainfall attenuation as a function of frequency for different rainfall rates [adapted from Barton (1988)].

up to 300 GHz is plotted in Figure 7.5 at a number of different rainfall rates. Note that the absorption maximum shifts down in frequency as the rain rate increases because the mean drop size increases.

The Laws-Parsons model is used to determine drop-size distributions, if they are required, with the mean raindrop size proportional to rainfall rate and atmospheric conditions. However, it is important to realize that these estimates are inexact and that in reality large variations in attenuation are measured for similar rainfall rates under different environmental conditions. This is illustrated by the large range of drop sizes for different conditions that are listed in Table 7.2.

At IR and visible wavelengths, where lidar operates, the drop size is mostly larger than the wavelength. The scattering and attenuation mechanism is optical and therefore largely independent of drop diameter. It is therefore directly proportional to the liquid water content (LWC) (in g/m^3) which is, in turn, directly proportional to the rain rate:

$$\alpha_r = k_r R, \tag{7.6}$$

where k_r is the scaling coefficient and R is the rain rate (mm/hr).

Table 7.2 Typical drop size for different conditions (Bhartia and Bahl 1984)

Condition	Drop diameter (μm)
Haze	0.01–3
Fog	0.01–100
Clouds	1–50
Drizzle	3–800
Moderate rain (4 mm/hr)	3–1500
Heavy rain (16 mm/hr)	3–3000

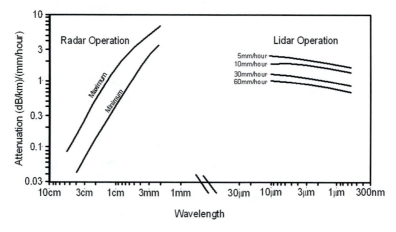

Figure 7.6 Specific attenuation (dB/km/mm/hr) in rain for radar and lidar operation [adapted from Brookner (1982)].

Figure 7.6 provides a comparison between the specific attenuation in the microwave band for wavelengths between 3 mm and 3 cm and for lidar in the band between 500 nm and 10 μm. As an example, compare the attenuation at a wavelength of 3 mm and 3 μm for a rainfall rate of 5 mm/hr. In the lidar case, the intersect on the y-axis is about 2.2 (dB/km/mm/hr), making the total attenuation 11 dB/km. In the radar case, the specific attenuations are a minimum of 2.7 and a maximum of 5.3, making the total attenuation something between 13.5 dB/km and 26.5 dB/km. This can be compared to the theoretical value of about 9 dB/km obtained from Figure 7.5. Such large discrepancies in the measurement and estimation of attenuation parameters are not uncommon because of the uncertainty in defining or measuring drop size distributions, etc.

7.2.4 Effect of Fog and Clouds

Clouds and fog are caused by the condensation of atmospheric water vapor into water droplets that remain suspended in the air. Fog is defined as having a water droplet density that restricts visibility to less than 1 km. It can be generated either by advection or radiation mechanisms that result in different droplet sizes, as shown in Table 7.3. Advection fog is caused by a horizontal

Table 7.3 Fog characteristics (Bhartia et al. 1984)

	Radiation fog	Advection fog
Average drop diameter	10 μm	20 μm
Typical drop size range	5–35 μm	7–65 μm
Liquid water content	0.11 g/m^3	0.17 g/m^3
Droplet concentration	200/cm^3	40/cm^3
Visibility	100 m	200 m

movement of a warm air mass over cold water, and radiation fog is caused by the cooling of air overnight, and often occurs over rivers and swamps.

Because it is difficult to determine the actual water content of fog in the field, its density is quantified in terms of visibility. This is still a fairly subjective measure, however, as it is defined as the range at which a high-contrast target can just be seen. That notwithstanding, the following empirical relationship has been determined which relates the visibility to LWC for an advection fog:

$$M_w = 308V^{-1.43}, \tag{7.7}$$

where M_w (g/m^3) is the LWC and V (m) is the visibility. For example, the LWC of a reasonably heavy fog with a visibility, V, of 125 m will be $M_w = 0.31$ g/m^3.

Because the drop diameter of both fog and clouds is less than 100 μm, the Rayleigh (low-frequency) approximation can be used to evaluate the attenuation at microwave and millimeter wave frequencies. Therefore, the attenuation, α_f (dB/km), will bear a linear relationship to the LWC:

$$\alpha_f = k_f M_w, \tag{7.8}$$

where k_f is the attenuation coefficient (dB/km/g/m^3) at a specific frequency. Since the complex index of refraction for water is very temperature dependent, this will also be a function of temperature.

Using equation (7.8) and taking k_f from Figure 7.7, for $f = 100$ GHz, gives an attenuation of 4 dB/km/g/m^3 at 20°C, making the attenuation in fog with a visibility of 125 m

$$\alpha_f = 4 \times 0.31 = 1.24 \text{ dB/km}.$$

As with the rain information, these empirical results are not particularly accurate due to variations in the characteristics of the medium.

The structure of clouds is similar to that of fog, with drop size diameters of less than 100 μm. As shown in Table 7.4, the LWC can vary from 0.15 g/m^3 in stratus clouds, right up to 1 g/m^3 in heavy cumulus clouds.

As the frequency increases, the wavelength becomes an appreciable percentage of the circumference of the drop and resonance occurs. Because there is such a large variation in drop size and

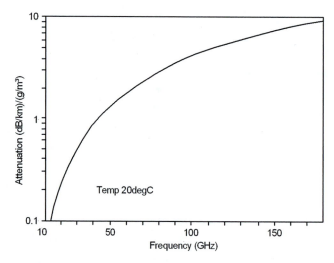

Figure 7.7 Specific attenuation through clouds and fog as a function of frequency [adapted from Falcone and Abreu (1979)].

Table 7.4 Liquid water content of clouds (Bhartia et al. 1984)

Cloud type	Liquid water content (g/m³)
Heavy fog	0.15–0.4
Moderate fog	0.01–0.1
Cumulus	0.5–1
Altostratus	0.35–0.45
Stratocumulus	0.3–0.6
Nimbostratus	0.6–0.7
Stratus	0.15–0.3

LWC for mist and fog, the actual attenuation that will occur in the IR region around this resonance is difficult to determine accurately.

Figure 7.8 shows the specific laser attenuation, k_f (dB/km/mg/m³) for various drop radii, a (μm), and wavelengths. The total attenuation, α_f, can then be determined by multiplying this specific attenuation coefficient by the LWC (mg/m³):

$$\alpha_f = k_f M_w. \tag{7.9}$$

For radiation fog with a drop diameter of 8 μm ($a = 4\,\mu$m) and an LWC of 0.1 g/m³ (100 mg/m³) the attenuation is

$$\alpha_f = 0.2 \times 100 = 20\,\text{dB/km (one way)}.$$

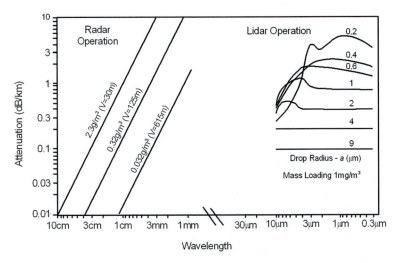

Figure 7.8 Comparison of radar and laser fog attenuation [adapted from Brookner (1982)].

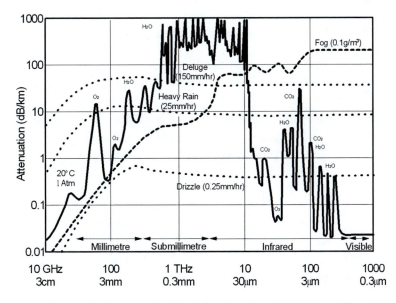

Figure 7.9 Summary of atmospheric attenuation for clear air, rain, and fog [adapted from Preissner (1978)].

For a heavy advection fog (visibility 50 m), the LWC was calculated to be 1.2 g/m³ (1200 mg/m³) for the same drop diameter (8 μm), which makes the attenuation 0.2 × 1200 = 240 dB/km.

7.2.5 Overall Attenuation

As an indication of the relative magnitudes of the various contributions to the total attenuation, Figure 7.9 summarizes the results presented in the earlier sections. Note that resonance effects due to fog particle size occur at wavelengths between 30 μm and 3 μm and those for rain occur at wavelengths between 3 mm and 1 mm, depending on the rainfall rate.

7.2.6 Attenuation Through Dust and Smoke

An interesting characteristic of almost all measured dust clouds is that the distribution of particle sizes is bimodal, as illustrated in Figure 7.10. This is apparently independent of the source of the dust, as it has been measured for wind-blown dust (Patterson 1977; Pinnick et al. 1985), vehicle-generated dust, and dust generated by explosions (Pinnick et al. 1983). Unexpectedly, the distribution holds irrespective of the density of the dust cloud, though the proportions may change. Typical mass mean radii for the two peaks are between 1 μm and 10 μm for fine dust and between 10 μm and 50 μm for larger particles, depending on the source of the dust.

The distribution of particle sizes hardly affects the attenuation in the microwave and millimeter wave bands because the wavelength is much larger than even the largest dust particle, so scattering can be accurately determined using the Rayleigh model. However, in the IR and visible bands, the relative particle sizes are close to the wavelength and Mie scattering must be applied to model the attenuation.

As with water vapor, the density of dust and other particulates is often characterized according to the visibility of targets within that environment. Using this definition, it has been found that in wind-blown dust the visibility can be related to the mass of dust per cubic meter of air (mass loading) (Gillett 1979):

$$M_d = 37.3V^{-1.07}, \tag{7.10}$$

where M_d (g/m^3) is the mass loading of dust, V (m) is the visibility. The coefficients are dependent on the particle type and the meteorological conditions, and these have been selected as being typical for sand storms in the desert (Ghobrial and Sharief 1987).

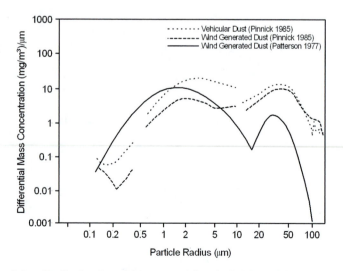

Figure 7.10 Measured size distribution for vehicle-generated and wind-blown dust (Pinnick et al. 1985).

According to Pinnick et al. (1985), dust is mostly made up of clumps of finer particles, so the individual density is lower than that of the actual rock material and seldom exceeds 1.5 g/cm^3. The density of the actual rock material varies from about 2.75 g/cm^3 for limestone (Nelson 2001) down to 2.44 g/cm^3 for quartz dust particles collected around Khartoum (Ghobrial et al. 1987) and 1.8 g/cm^3 for sand with a 8% water content (Pinnick et al. 1985). The characteristics of coal are of importance in some industrial measurement applications, and it has a density of 1.48 g/cm^3 (Nelson 2001).

7.2.6.1 Attenuation of Radar Signals   Two mechanisms determine the attenuation of radar signals. They are α_a (dB/km), the attenuation due to absorption, and α_s (dB/km), the attenuation due to scattering (Currie et al. 1992):

$$\alpha_a = 81.86 \times 10^{-3} \frac{\text{Im}(-K) M_d}{\lambda \rho}, \tag{7.11}$$

$$\alpha_s = 1.353 \times 10^{-19} \frac{M_d a^3 K^2}{\lambda^4 \rho}, \tag{7.12}$$

where M_d (g/m^3) is the mass loading, ρ (g/cm^3) is the effective particle density, a (μm) is the particle radius, and λ (m) is the wavelength. K is a constant determined by the dielectric properties of the material:

$$|K|^2 = \left| \frac{m^2 - 1}{m^2 + 2} \right|^2 = \left| \frac{\varepsilon_r - 1}{\varepsilon_r + 2} \right|^2, \tag{7.13}$$

where m is the complex index of refraction of the material.

In most cases the attenuation due to scattering is so small that it can be ignored, and the attenuation due to absorption can be determined in terms of the real component, ε', and the imaginary component, ε'', of the permittivity. These are also known respectively as the dielectric constant and the loss factor because the imaginary part is related to the rate at which energy is absorbed by the material.

For a material density of 2.44 g/cm^3 and using the relationship between the mass loading and the visibility from equation (7.10), the attenuation can be rewritten in terms of the visibility (Ghobrial et al. 1987; Goldhirsh 2001):

$$\alpha_a = \frac{3.76 \varepsilon''}{\left[(\varepsilon' + 2)^2 + \varepsilon''^2 \right] \lambda} V^{-1.07}. \tag{7.14}$$

Substituting into equation (7.14) for the permittivity of rock dust, which has been measured to be $\varepsilon_{\text{dust}} = 5.23 - j0.26$ (Ghobrial et al. 1987), produces the result shown in Figure 7.11 at a wavelength of 3.2×10^{-3} m (94 GHz). If the dust is replaced by water with a density of 1 g/cm^3 and a permittivity $\varepsilon_w = 8.35 - j15.45$ (measured at 89 GHz), the losses are much higher, as shown.

What is obvious from these results is that the total attenuation is extremely sensitive to the water content of the dust particles. As it is uncommon to find completely dry dust, significant variations in the attenuation can be expected in any measurements.

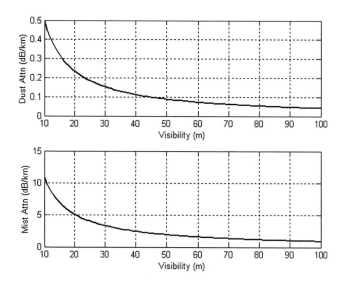

Figure 7.11 Relationship between visibility and attenuation for dust and mist droplets at 94 GHz.

7.2.6.2 Attenuation of Laser Signals   Lasers are the ideal sensors for monitoring air quality because measurements are noncontact and, in fact, can operate at quite a long distance from the sample volume to be monitored. The basic method involves sending a beam of laser radiation through the sample of atmosphere to be investigated and then monitoring variations in the transmission of the light.

The relationship between the radiated intensity, I_o (W), and that received, I (W), after traveling through a range R is (Ready 1978)

$$I = I_o e^{-\alpha R}, \tag{7.15}$$

where α is the extinction coefficient arising from absorption and scattering. The extinction coefficient can be expressed as the sum of the absorption component, α_{abs}, a component due to scattering from molecules in the atmosphere, α_{ray}, and a component due to scattering from larger particles, α_{mie}:

$$\alpha = \alpha_{abs} + \alpha_{ray} + \alpha_{mie}. \tag{7.16}$$

Mie scattering in this case includes interactions in the optical region, where the size of the dust particles is much larger than the laser wavelength.

Of most interest from a range performance perspective is the relationship between the extinction coefficient and the mass loading. This is generally expressed as the ratio α/M, commonly called the mass extinction coefficient. Pinnick et al. (1985) derive the equation for this parameter on the assumption that the particle size distribution is bimodal, as shown in Figure 7.10, and that the particles are spheroidal in shape. This is an extension to the original work by Trabert (1901), Mie theory, and the complex angular momentum (CAM) theory.

To give an indication of the complexity of the models, the Mie theory formula is shown for illustrative purposes:

$$\frac{\alpha}{M} = \frac{\int \pi a^2 Q_E n(a)\,da}{\rho \int \frac{4\pi}{3} a^3 n(a)\,da} = \frac{4}{3\rho}\frac{\langle r^2 Q_E\rangle}{\langle r^3\rangle}, \tag{7.17}$$

where Q_E is the Mie extinction efficiency and is a function of the wavelength and the particle size,

$$Q_E = 2 + 1.992386\left(\frac{2\pi a}{\lambda}\right)^{-2/3}, \tag{7.18}$$

a is the radius of the particles, $n(a)$ is the size distribution, $\langle a^2\rangle$ and $\langle a^3\rangle$ are the second and third moments of that size distribution, and ρ is the density of the dust.

According to Pinnick et al. (1985), Trabert's formula takes into account only the size distribution of the particles, while the CAM theory does the same in a more sophisticated way, but still ignores refractive index effects. Mie theory includes both the distribution and refractive index, while the spheroid model includes all of those and the particle shape.

It is very difficult to determine the extinction and mass loading properties of a dust cloud accurately because of spatial and temporal fluctuations in the dust concentration. One way to improve measurement accuracy is to integrate the extinction coefficient and the total dust concentration with time, as shown in Figure 7.12. The mass extinction coefficient can then be determined from the gradient of the graph.

It can be seen that the scatter in the measurements makes it difficult to decide which of the extinction models is more accurate. However, as a first approximation, the mass extinction

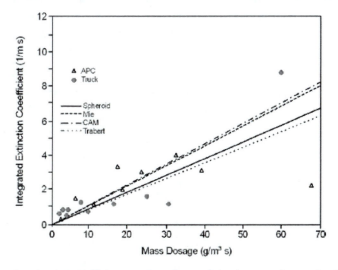

Figure 7.12 Integrated extinction coefficient at 9.5 μm versus the integrated mass loading for vehicle dust derived from sandy soil (Pinnick et al. 1985).

coefficient at 9.5 μm is $\alpha/M_d \approx 0.1/\text{m/g/m}^3$ for near source dust derived from sandy soil. Other tests, conducted in the lower absorption region at 10.59 μm, show a much smaller value for the mass extinction coefficient. In this case the measured value is $\alpha/M_d \approx 0.03/\text{m/g/m}^3$ (Pinnick et al. 1985).

To make this representation consistent with the results for attenuation in fog, it is necessary to convert these values to an attenuation in decibels per kilometer (dB/km) for a normalized mass loading of 1 mg/m^3. This is performed by utilizing the change of base theorem described in equation (7.2) and normalizing the mass extinction coefficients to milligrams per cubic meter (mg/m^3):

$$\alpha_{dB} = 4.3429(\alpha/M_d). \tag{7.19}$$

Applying this formula to the measured data results in an attenuation of $\alpha_{dB} = 0.43$ dB/km/mg/ m^3 for the higher mass extinction coefficient, and $\alpha_{dB} = 0.13$ dB/km/mg/m^3 for the lower one. This compares reasonably well with the results shown in Figure 7.8 for fog with larger drop sizes.

As an example, consider a dust cloud with an optical visibility of 50 m. This corresponds to a mass loading of 0.57 g/m^3 (570 mg/m^3), as read from Table 7.5, and the attenuation, using the lower value of mass extinction for particle sizes of 9.5 μm, is

$$\alpha_{dB} = 570 \times 0.13 = 74 \text{ dB/km}.$$

Measurements made at 0.55 μm and 10.4 μm are reproduced in Figure 7.13 and show that the measured transmissions in the visible and IR wavelengths are similar through explosion-generated dust. This means that most of the attenuation is caused by large particles operating in the optical region rather than smaller particles in the Mie region (Pinnick et al. 1983). This relationship is confirmed (Patterson 1977), where measurements at 0.55 μm and 10.6 μm show no wavelength dependence for soil-derived dust either. This relationship confirms the well known rule of thumb that states that "if you can see through it, a laser range finder can probably measure the range."

Table 7.5 Relationship between visibility and mass loading

Visibility (m)	Mass loading (g/m^3)
1	37
2*	17.8
5	8.5
50	0.57
100	0.27
1000	23 mg/m^3

*Minimum visibility for heavy sand storm in Khartoum (Goldhirsh 2001).

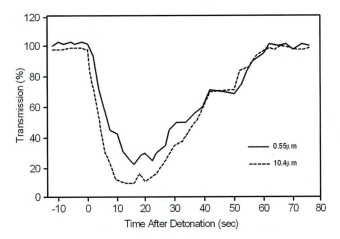

Figure 7.13 Measured transmission through explosion-generated dust showing good correlation between visible and IR wavelengths (Pinnick et al. 1983).

Table 7.6 Calculated attenuation through different aerosol types (Adapted from Bhartia 1984)

Source	Material	Density (g/cm^3)	Mass concentration (g/m^3)	Visibility (m)	Attenuation (dB/km) $\lambda = 3\ mm$	Attenuation (dB/km) $\lambda = 1–10\ \mu m$
Desert dust	Quartz	2.6	10	3.4	1.56	1300
Stack: stone mill	Quartz	2.6	80	0.5	12.5	10400
Stack: steel mill	Coal	1.5	10	3.4	2.2	1300
Volcanic dust	Quartz	2.6	5	6.5	0.8	650
Light fog	Water	1	0.001	7 km	0.004	0.2
Heavy fog	Water	1	1	55	4	200

Table 7.6 compares the typical attenuation characteristics of a number of different materials in the millimeter wave and IR bands. From this table and the remainder of the section, it can be seen that in contrast to the poor transmission through dust at the shorter wavelength, the attenuation in the millimeter wave band is fairly insignificant.

7.2.7 Effect of Atmosphere Composition

In industrial applications where extremely accurate range measurement is often required, a correction factor must be applied to compensate for changes in propagation velocity caused by differences in the pressure, temperature, and composition of the medium. Essentially the variation is caused by differences in the relative dielectric constant, ε_r,

$$v = c/\sqrt{\varepsilon_r}\,, \qquad (7.20)$$

where v is the velocity (m/s) and c is the speed of light in air, 2.997925×10^8 (m/s). This effect is tabulated in Table 7.7 for a number of common gases and vapors at different temperatures.

Table 7.7 Gas makeup effect on velocity

Vapor content	Temperature (°C)	Relative dielectric constant	Velocity ($\times 10^8$ m/s)	Error at 30 m (m)
Air	0	1.000590	2.997925	0
Helium	140	1.000068	2.997823	+0.00104
Hydrogen	100	1.000264	2.997529	+0.00403
Oxygen	100	1.000523	2.997141	+0.00797
Nitrogen	100	1.000580	2.997055	+0.00885
Ammonia	0	1.007200	2.9871904	+0.10953
Benzene	400	1.002800	2.993736	+0.04265
Carbon dioxide	100	1.000985	2.996449	+0.01501
Water	100	1.007850	2.986226	+0.11940

7.2.8 Electromagnetic Propagation Through Solid Materials

Absorption of electromagnetic radiation by solids and liquids is generally determined by the real component of the relative dielectric constant, ε_r, and loss tangent, $\tan \delta$. In a dielectric material, the loss tangent describes the phase difference between the electric field component due to the real component, ε', and the displacement field due to the imaginary component, ε''. The loss tangent can therefore be written as

$$\tan \delta = \varepsilon'' / \varepsilon'. \tag{7.21}$$

The attenuation coefficient, α (dB/m), for propagation through an unbounded dielectric material is

$$\alpha = \frac{27.3}{\lambda_o} \sqrt{\varepsilon_r} \tan \delta, \tag{7.22}$$

where ε_r is the relative dielectric constant and λ_o is the wavelength (m) in the material.

It should be remembered that a portion of the electromagnetic signal will be reflected from the surface of the materials due to the mismatch in dielectric constant (or characteristic impedance). For normal incidence, the reflection coefficient is

$$\Gamma = \frac{Z_L - Z_o}{Z_L + Z_o} = \frac{\sqrt{\varepsilon_r} - 1}{\sqrt{\varepsilon_r} + 1} = \frac{n-1}{n+1}, \tag{7.23}$$

where Z_o is the characteristic impedance of free space $\sqrt{\mu_o / \varepsilon_o}$, and Z_L is the impedance of the material $\sqrt{\mu_o / \varepsilon_o \varepsilon_r}$, ε_r is the relative dielectric constant (permittivity), and n is the refractive index. The portion, T, that passes through the boundary is $T = 1 - \Gamma$:

$$T = 2Z_o / (Z_L + Z_o). \tag{7.24}$$

Solid material characteristics are important for the design of radomes and lens antennas or to determine the propagation characteristics for ground penetrating radar (GPR), as displayed in Figure 7.14. Lists of dielectric constants and loss tangents for various organic and inorganic materials that might be used as radomes or lenses for microwave radar applications can be found in a number of textbooks. Some examples include Teflon with $\varepsilon_r = 2.10$ and $\tan \delta = 0.0005$, Lexan with $\varepsilon_r = 2.86$ and $\tan \delta = 0.006$, and aluminum oxide with $\varepsilon_r = 7.85$ and $\tan \delta = 0.0005$ (Currie and Brown 1987).

7.3 REFRACTION OF ELECTROMAGNETIC WAVES

Refraction effects are independent of wavelength at frequencies below 100 GHz. The decrease in air density, and hence refractive index with height, results in a downward curvature of horizontally launched radar beams as illustrated in Figure 7.15. Bending is equal to about one-quarter of the curvature of the earth. To plot ray paths as straight lines, the effective earth radius is four-thirds of the actual value.

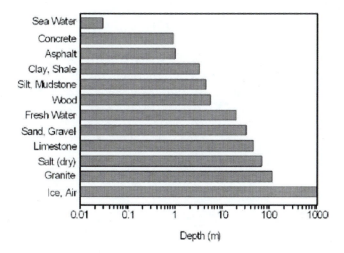

Figure 7.14 Potential GPR penetration depth.

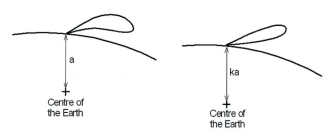

Figure 7.15 Bending of the beam due to refraction by the earth's atmosphere requires an increase in the effective earth radius to produce an apparently unrefracted beam.

The radar horizon can be approximated by

$$d = \sqrt{2kah},$$ (7.25)

where
d = radar horizon (m),
k = 4/3,
a = actual radius of the earth (6378×10^3 m),
h = radar height (m).

This can be simplified to the following for the both horizon range and radar height in kilometers:

$$d_{km} = 130\sqrt{h_{km}}.$$ (7.26)

Refraction effects can also occur closer to the earth's surface, where heating of the ground can generate a thermal gradient or a temperature inversion in the lower atmosphere. In conjunction with the water vapor gradient, this effect can result in pronounced curving or even ducting of the propagation path.

7.4 ACOUSTICS AND VIBRATION

Unlike electromagnetic radiation which can propagate through space, acoustic waves are transmitted through alternate compression and expansion of the medium (liquids and gases). Solids support shear, so acoustic signals can propagate as compression or as shear waves in that medium.

The literature abounds with different units that are used to measure pressure in the atmosphere. Some of these are listed in Table 7.8.

Table 7.8 Common units of pressure

Units	Atmospheric pressure
Atmospheres (atm)	1
Bar	1.01325
Pascal (Pa)	1.01325×10^5
Pounds/in^2 (psi)	14.6960
Inches Hg	29.9213
mm of Hg (torr)	760
Inches of water	406.8
Dynes/cm^2	1.01325×10^6

7.4.1 Characteristic Impedance and Sound Pressure

As sound propagates through a medium, each small volume element of that medium oscillates about its equilibrium position. For pure harmonic motion, the displacement is

$$y = y_m \cos \frac{2\pi}{\lambda}(x - vt). \tag{7.27}$$

It can be shown that the pressure exerted by the sound wave is

$$p = \left(k\rho_0 v^2 y_m\right)\sin\left(kx - \omega t\right), \tag{7.28}$$

where $k = 2\pi/\lambda$ is the wave number and ω (rad/s) is the angular frequency.

The amplitude of the wave is the sound pressure, $p_m = k\rho_0 v^2 y_m$, so a sound wave may be considered to be a pressure wave with a phase offset of 90° to that of the displacement wave. The pressure at any point changes sinusoidally with time around a mean value. The root mean square (RMS) (effective) value of this fluctuating component is known as the acoustic pressure, P. This can be measured using an appropriately calibrated microphone or hydrophone.

The characteristic impedance is defined as the ratio of the acoustic pressure and the RMS volume velocity, ξ. Note that this is not the wave velocity:

$$Z = P/\xi. \tag{7.29}$$

This is a complex quantity. However, for an idealized medium (with no loss), Z is real and reduces to a proportionality factor that equates intensity to the square of the sound pressure:

$$Z = \rho_0 c, \tag{7.30}$$

where ρ_0 (kg/m^3) is the density and c (m/s) is the speed of propagation. For air with $\rho_0 = 1.3$ kg/m^3 and $c = 332$ m/s, $Z_{air} \approx 400$ acoustic ohms, while in water with $\rho_0 = 1000$ kg/m^3 and $c = 1400$ m/s, $Z_w = 1.4 \times 10^6$ acoustic ohms.

7.4.2 Sound Intensity

Intensity of a sound is defined as the power transferred per unit area. This is also known as the sound pressure level (SPL), and it can be expressed through the acoustic impedance (similar to Ohms law for electrical circuits):

$$I = P\xi = P^2 Z, \tag{7.31}$$

where I (W/m^2) is the sound intensity (SPL), P (N/m^2 or Pa) is the acoustic pressure, and ξ (m^3/s) is the volume velocity.

It is common to specify the sound not as an intensity, but as a ratio, β (dB), relative to the minimum audible threshold, where $P_0 = 2 \times 10^{-5}$ N/m^2 (2.9×10^{-9} psi) is the minimum audible threshold. The reference intensity, I_0, can be determined from P_0 using equation (7.31):

$$I_0 = P_0^2/Z = \left(2 \times 10^{-5}\right)^2/400 = 10^{-12} \text{ W/m}^2.$$

The ratio is therefore

$$\beta = 10 \log^{10}(I/I_o) = 20 \log_{10}(P/P_o). \tag{7.32}$$

To give some insight into the relative levels of some common sounds, Table 7.9 has been compiled. It shows that the normal range of auditory input extends from 0 dB (the minimum audible threshold) up to about 120 dB, with the pain threshold 10 dB higher than this. It is incredible that the range of human hearing spans 12 orders of magnitude.

7.4.3 Sound Propagation in Gases

The speed of sound, c (m/s), in an ideal gas can be written in terms of the equations of state ($PV = nRT$) for ideal gases (Cook 1972):

$$c = \sqrt{\frac{RT\gamma}{M}} = \sqrt{\frac{P\gamma}{\rho_o}} = \sqrt{\frac{K}{\rho_o}}, \tag{7.33}$$

where

R = universal gas constant (8134.3 J/kmol),
T = temperature (K),
M = molecular weight of the gas (g/mole),
γ = ratio of specific heats (adiabatic exponent) C_p/C_v for the gas,
P = pressure (1 atm = 1.013×10^5 Pa),
ρ_o = density (kg/m^3),
K = bulk modulus ($K = \Delta P/\Delta V$).

Table 7.9 Sound levels

Sound	Level β (dB)
Stream flow, rustling leaves	15
Watch ticking, soft whisper	20–30
Quiet street noises	40
Normal conversation	45–60
Normal city or freeway traffic	70
Vacuum cleaner	75
Hair dryer	80
Motorcycle, electric shaver	85
Lawn mower, heavy equipment	90
Garbage truck	100
Screaming baby	115
Racing car, loud thunder, rock band	120–130
Jet airplane takeoff from 120 feet	120
Pain threshold	130
Rocket launch from 150 feet	180

The adiabatic exponent, γ, can be estimated as follows:

- 1.66 for monatomic gases (helium, neon, argon),
- 1.40 for diatomic gases (hydrogen, oxygen, nitrogen),
- 1.33 for triatomic and more complex gases (ammonia, methane, toluene),
- 1.286 for very long molecules.

Table 7.11 gives a more accurate measure of the adiabatic exponent for various gases.

7.4.3.1 Worked Example: Effect of Molecular Weight on the Speed of Sound   The speed of sound in any gas mixture can be calculated by using the molecular weight to determine the density. Air is made up of oxygen, nitrogen, and carbon dioxide in proportions (21:78:1), therefore the molecular weights are 16 + 16 = 32, 14 + 14 = 28, 12 + 16 + 16 = 44. The molecular mass, M, of the mixture is therefore

$$M = \frac{21 \times 32}{100} + \frac{78 \times 28}{100} + \frac{1 \times 44}{100} = 29.$$

Remembering that at atmospheric pressure and at 0°C the volume of 1 mole of gas is 22.414 L, the density can be determined by

$$\rho_0 = \frac{29 \times 10^{-3}}{22.4 \times 10^{-3}} = 1.29 \, \text{kg/m}^3,$$

and finally, the speed of sound through the air can be calculated as

$$c_{air} = \sqrt{\frac{1.013 \times 10^5 \times 1.4}{1.29}} = 331.6 \, \text{m/s}.$$

The speed of sound in helium can be calculated in a similar way. In this case $M = 4$, making ρ_0 equal to 0.18 kg/m^3:

$$c_{He} = \sqrt{\frac{1.013 \times 10^5 \times 1.66}{0.18}} = 970.7 \, \text{m/s}.$$

7.4.3.2 Effect of Temperature and Pressure   The density of the gas at any temperature and pressure is related to that at STP for the pressure in millimeters of mercury and the temperature in Kelvin (Cook 1972):

$$\rho = \rho_0 \left(\frac{P}{760} \right) \left(\frac{273.16}{T} \right). \tag{7.34}$$

For rough calculations of the speed of sound (ignoring changes in C_v), the changes in density are assumed to be the only variable affecting the velocity. Table 7.10 lists the speed of sound in a number of different gases at a pressure of 1 atm and a temperature of 0°C. Of the common

Table 7.10 Velocity of sound in gases (Cook 1972)

Gas	Formula	Speed (m/s) at 0°C
Air		331.45
Ammonia	NH_2	415
Argon	A	307.8
Carbon monoxide	CO	337.1
Carbon dioxide	CO_2	258.0
Carbon disulfide	CS_2	189
Chlorine	Cl_2	205.3
Ethylene	C_2H_4	314
Helium	He	970
Hydrogen	H_2	1269.5
Methane	CH_4	432
Neon	Ne	435
Nitric oxide	NO	325
Nitrogen	N_2	337
Nitrous oxide	N_2O	261.8
Oxygen	O_2	317.2
Steam (100°C)	H_2O	404.8

gases, carbon dioxide is the slowest, with a speed of only 258 m/s, and hydrogen the fastest, with a speed of nearly 1270 m/s.

Industrial range sensing using ultrasound requires an accurate method of determining the speed of sound, so more accurate values for the adiabatic components C_p and C_v are required. These are listed for a number of gases in Table 7.11.

7.4.4 Sound Propagation in Water

As with gases, the speed, c (m/s), is equal to the bulk modulus of elasticity divided by the density of the liquid:

$$c = \sqrt{\frac{B}{\rho}}. \tag{7.35}$$

However, because the modulus of elasticity and the density are functions of the pressure (depth), temperature, and salinity, an empirical equation (Wilson 1960) is usually used to calculate the speed:

$$c = 1449.2 + 4.62T - 0.055T^2 + 1.4(S - 35) + 0.017z, \tag{7.36}$$

where T (°C) is the temperature, S (parts per thousand) is the salinity, and z (m) is the depth.

Table 7.11 Specific heat at constant pressure (C_p) and the ratio, γ, of C_p to the specific heat at constant volume C_v^* (Cook 1972)

Gas	Temperature (°C)	C_p (cal/g deg)	γ (C_p/C_v)
Air	−120 (10 atm)	0.2719	1.415 (−118°C, 1 atm)
	−50 (10 atm)	0.2440	1.408 (−78°C, 1 atm)
	0	0.2398	1.403
	50 (10 atm)	0.2480	1.403 (17°C, 1 atm)
	100	0.2404	1.401
	400	0.2430	1.393
	1000	0.2570	1.365
	1400	0.2699	1.341
Ammonia	15	0.5232	1.310
Argon	15	0.1253	1.668
Carbon dioxide	15	0.1989	1.304
Carbon monoxide	15	0.2478	1.404
Chlorine	15	0.1149	1.355
Ethylene	15	0.3592	1.255
Helium	−180	1.25	1.660
Hydrogen	15	3.389	1.410
Hydrogen sulfide	15	0.2533	1.32
Methane	15	0.5284	1.31
Neon			1.64
Nitric oxide	15	0.2329	1.400
Nitrogen	15	0.2477	1.404
Nitrous oxide	15	0.2004	1.303
Oxygen	15	0.2178	1.401
Propane			1.13
Steam	100	0.4820	1.324
Sulfur dioxide	15	0.1516	1.29

Table 7.12 lists the speed of sound in fresh water as a function of temperature and pressure. It is interesting to note that the speed changes by almost 10% for a temperature increase from 0°C to 100°C, while to achieve a similar increase, the pressure must be increased to about 3000 psi (200 atm).

Table 7.13 and Table 7.14 list the speed of sound in seawater at different pressures, temperatures, and salinities. Note that the speed of sound at the sea surface (35 ppt salt) at 0°C is 1449.2 m/s, as contrasted with 1403 m/s for fresh water, a fact that might be useful in determining the salinity of a water sample. The variation in velocity, which equates to differences in acoustic impedance, is instrumental in causing sound rays to be refracted away from a straight path, as discussed later in this chapter.

Table 7.12 Speed of sound in fresh water (Cook 1972)

Temperature (°C)	Speed, $P = 1$ atm (m/s)	Pressure (psi)	Speed, $T = 30°C$ (m/s)
0	1403.0	14.7	1,509.7
10	1447.8	2,000	1,532.9
20	1482.9	4,000	1,556.2
30	1509.7	6,000	1,579.4
40	1529.3	8,000	1,602.4
50	1542.9	10,000	1,625.3
60	1551.3	12,000	1,648.1
70	1555.1	14,000	1,670.9
80	1554.7		
90	1550.5		
100	1542.5		

14.7 lbf/in.^{2} = 101 kN/m^2 = 1.01 bar.
1000 lbf/in^2 = 6.89 MN/m^2 = 68.9 bar.
1 dyne = 10 μN.
1 kgf = 9.807 N.
1 Pa = 1 N/m^2.
10 m depth = 1 bar increase in pressure.

Table 7.13 Speed of sound in seawater at $P = 1$ atm (Cook 1972)

Temperature (°C)	Speed (m/s)		
	$S = 3.3\%$	$S = 3.5\%$	$S = 3.7\%$
−3	1431.9	1435.0	1437.6
0	1446.3	1449.4	1451.9
5	1468.2	1471.2	1473.5
10	1487.6	1490.4	1492.7
15	1504.6	1507.4	1509.5
20	1519.6	1522.2	1524.2
25	1531.9	1535.1	1536.9
30	1543.8	1546.2	1547.9

7.4.5 Sound Propagation in Solids

In an isotropic solid for which the wave front is a large number of wavelengths, longitudinal and shear waves can exist which have velocities (m/s) (Cook 1972)

$$c_{long} = \sqrt{\frac{\lambda + 2\mu}{\rho_o}}, \tag{7.37}$$

$$c_{shear} = \sqrt{\frac{\mu}{\rho_o}}, \tag{7.38}$$

where μ, λ (N/m^2) are the Lamé elastic moduli for the material.

Table 7.14 Speed of sound in seawater at T = 20°C
(Cook 1972)

Pressure (kg/cm²)	Speed (m/s)		
	S = 3.3%	S = 3.5%	S = 3.7%
1.033	1519.6	1522.2	1524.1
100	1535.4	1537.1	1540.1
200	1551.5	1554.2	1556.3
300	1567.7	1571.5	1572.5
400	1584.0	1586.8	1588.9
500	1602.0	1603.1	1605.2
600	1616.8	1619.5	1621.6
700	1633.1	1635.8	1637.9
800	1649.4	1652.1	1654.2
900	1665.5	1668.2	1670.3

For a rod, whose diameter is a small fraction of the wavelength, extensional and torsion waves can also propagate:

$$c_{tor} = \sqrt{\frac{\mu}{\rho_o}}, \tag{7.39}$$

$$c_{ext} = \sqrt{\frac{Y_o}{\rho_o}}, \tag{7.40}$$

where Y_o (N/m²) is Young's modulus and can be determined from the Lamé elastic moduli:

$$Y_o = \mu \frac{3\lambda + 2\mu}{\lambda + \mu}. \tag{7.41}$$

The salient characteristics that affect sound propagation of a number of solids are listed in Table 7.15.

The characteristic impedances, Z (kg/m²s) for longitudinal and shear wave propagation can be calculated using

$$Z_{long} = \sqrt{\rho(\lambda + 2\mu)} \tag{7.42}$$

$$Z_{shear} = \sqrt{\rho\mu}. \tag{7.43}$$

Table 7.15 Elastic constants, wave velocities, and characteristic impedances of various materials

Material	Y_o (N/m²)	μ (N/m²)	λ (N/m²)	Poisson's ratio (σ)	c_{long} (m/s)	c_{shear} (m/s)	c_{ext} (m/s)	Z_l (kg/m²s)	Z_s (kg/m²s)
Aluminum	70×10^9	25×10^9	61×10^9	0.355	6420	3040	5000	17.3×10^6	8.2×10^6
Brass	104×10^9	38×10^9	113	0.374	4700	2110	3480	40.6×10^6	18.3×10^6
Copper	125×10^9	46×10^9	131	0.37	5010	2270	3750	44.6×10^6	20.2×10^6
Iron	152×10^9	59.9×10^9	69.2	0.27	4994	2809	4480	37.8×10^6	21.35×10^6
Lead	16×10^9	5.4×10^9	33	0.43	1960	690	1210	22.4×10^6	7.85×10^6
Nickel	214×10^9	80×10^9	164	0.336	6040	3000	4900	53.5×10^6	26.6×10^6
Steel	216×10^9	82.9×10^9	100.2	0.276	5941	3251	5250	46.5×10^6	25.4×10^6
Titanium	116×10^9	44×10^9	77.9	0.32	6070	3125	5090	27.3×10^6	14.1×10^6
Pyrex glass	62×10^9	25×10^9	23	0.24	5640	3280	5170	13.1×10^6	7.6×10^6
Lucite	4×10^9	1.43×10^9	5.62	0.4	2680	1100	1840	3.16×10^6	1.3×10^6
Nylon 6-6	3.55×10^9	1.22×10^9	5.11	0.4	2620	1070	1800	2.86×10^6	1.18×10^6
Polyethylene	0.76×10^9	0.26×10^9	2.88	0.458	1950	540	920	1.75×10^6	0.48×10^6

7.4.6 Attenuation of Sound in Air

For a plane acoustic wave propagating through the atmosphere, there are three primary mechanisms that interact with the sound to cause attenuation. These are molecular absorption, viscosity, and thermal effects.

A reasonably simple model for the attenuation, α (dB/m), which includes the effects of both temperature and humidity has been derived from measured data and is accurate for relative humidities greater than 30% (Knudsen and Harris 1950):

$$\alpha_c = \left(\frac{f}{1000} \right)^{3/2} \frac{0.283}{20 + \varphi_t}, \tag{7.44}$$

where

f = frequency (Hz),

$\varphi_t = \varphi_{20}(1 + 0.067\Delta t)$,

φ_{20} = relative humidity (%) at 20°C,

Δt = the temperature difference from 20°C.

Consider the attenuation, α_c (dB/m), determined using this model at a frequency of 30 kHz for a temperature of 20°C and a relative humidity of 30%:

$$\alpha_c = \left(\frac{30 \times 10^3}{1000} \right)^{3/2} \frac{0.283}{20 + 30} = 0.93 \, \text{dB/m}.$$

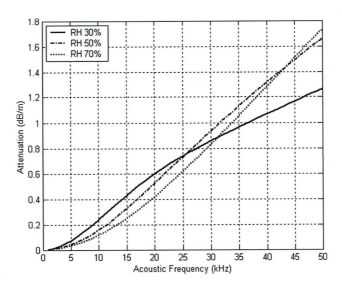

Figure 7.16 Sound attenuation as a function of frequency and relative humidity in air at 20°C.

A far more comprehensive model developed (Burnside 2004) and implemented in MATLAB (listing below) conforms to the ISO-9613 standard and has been used to generate the results shown in Figure 7.16. These results give an attenuation of 0.88 dB at 20°C and a relative humidity of 30%, which is in reasonably good agreement with the results from equation (7.43).

```
function [a] = atmAtten(Tin,Psin,hrin,dist,f)
% A function to return the atmospheric attenuation of sound due to the
vibrational relaxation times of oxygen and nitrogen.
% NOTE:  This function does not account for spherical spreading!
%
% Usage:  [a] = atmAtten(T,P,RH,d,f)
%                 a - attenuation of sound for input parameters in dB
%                 T - temperature in deg C
%                 P - static pressure in inHg (29.92inHg = 1 atm)
%                 RH - relative humidity in %
%                 d - distance of sound propagation
%                 f - frequency of sound (may be a vector)
%
% Nathan Burnside 10/5/04
% AerospaceComputing Inc.
% nburnside@mail.arc.nasa.gov
%
```

```
% Refs: Bass, et al., "Journal of Acoustical Society of America", (97)
p680, Jan 95.
%          Bass, et al., "Journal of Acoustical Society of America",
(99) p1259, Feb 96.
%          Kinsler, et al., "Fundamentals of Acoustics", 4th ed.,
p214, John Wiley 2000.


T = Tin + 273.15;           % temp input in K
To1 = 273.15;               % triple point in K
To = 293.15;                % ref temp in K


Ps = Psin/29.9212598;       % static pressure in atm
Pso = 1;                    % reference static pressure


F = f./Ps;                  % frequency per atm

% calculate saturation pressure
Psat = 10^(10.79586*(1-(To1/T))-5.02808*log10(T/To1)+1.50474e-4*
(1-10^(-8.29692*((T/To1)-1)))-4.2873e-4*(1-10^(-4.76955*((To1/T)-1)))-2.2195983);

h = hrin*Psat/Ps;           % calculate the absolute humidity

% Scaled relaxation frequency for Nitrogen
FrN = (To/T)^(1/2)*(9+280*h*exp(-4.17*((To/T)^(1/3)-1)));

% scaled relaxation frequency for Oxygen
FrO = (24+4.04e4*h*(.02+h)/(.391+h));

% attenuation coefficient in nepers/m
alpha = Ps.*F.^2.*(1.84e-11*(T/To)^(1/2) + (T/To)^(-5/2)*(1.275e-2*exp(-
2239.1/T)./(FrO+F.^2/FrO) + 1.068e-1*exp(-3352/T)./(FrN+F.^2/FrN)));

a = 10*log10(exp(2*alpha))*dist;
```

7.5 ATTENUATION OF SOUND IN WATER

Attenuation of sound through seawater is caused by viscosity and molecular relaxation. The viscosity effects are proportional to the square of the frequency and occur in both fresh and salt water. The mechanism of molecular relaxation involves the reduction of molecules to ions caused by the sound pressure (Waite 2002).

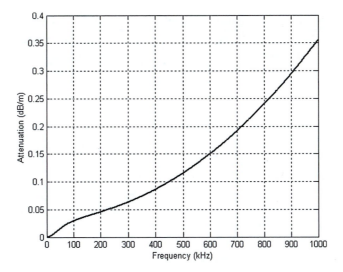

Figure 7.17 Sound attenuation in seawater at 5°C and 1 atm.

Figure 7.17 shows a plot of an empirically derived equation commonly used to determine the attenuation coefficient of sound in seawater (Urick 1975):

$$\alpha_{dB} = \frac{0.036f^2}{f^2 + 3600} + 3.2 \times 10^{-7} f^2, \tag{7.45}$$

where α_{dB} (dB/m) is the attenuation coefficient and f (kHz) is the frequency.

Because of the strong frequency dependence, tables are often constructed to exclude the f^2 characteristic. For Table 7.16, the extinction coefficient, α, can be converted to an attenuation coefficient, α_{dB}, by applying the change of base theorem:

$$\alpha_{dB} = 8.686\alpha. \tag{7.46}$$

For example, to determine the attenuation coefficient at a temperature of 5°C and a frequency of 500 kHz using these tables,

$$\alpha = 44.1 \times \left(500 \times 10^3\right)^2 \big/ 10^{15} = 0.011$$

$$\alpha_{dB} = 0.011 \times 8.686 = 0.0958 \text{ dBm}.$$

This result is in reasonable agreement with the result shown in Figure 7.17, with the small differences in attenuation attributable to the effects of the dissolved salt.

Figure 7.18 gives a comparison between the attenuation in salt and fresh water. When plotted on log-log axes, the freshwater response is linear, with a slope of two confirming that the viscosity losses are proportional to the square of the frequency. The difference between this curve and the

Table 7.16 Attenuation of sound in fresh water at 1 atm (Pinkerton 1947; Litovitz and Carnevale 1955)

Temperature (°C)	Attenuation (1 atm) $10^{15}\alpha/$ f^2 (sec^2/m)	Pressure (atm)	Attenuation (30°C) $10^{15}\alpha/$ f^2 (sec^2/m)
0	56.9	0	18.5
5	44.1	500	15.4
10	36.1	1000	12.7
15	29.6	1500	11.1
20	25.3	2000	9.9
30	19.1		
40	14.6		
50	12.0		
60	10.2		
70	8.7		
80	7.9		

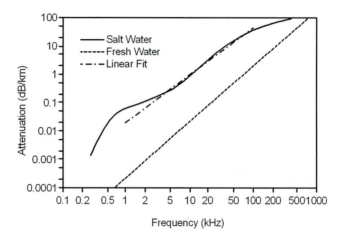

Figure 7.18 Sound attenuation in water as a function of frequency.

one for salt water is the additional loss due to molecular relaxation. This loss is greatest at low frequency, but by the time the frequency reaches 500 kHz, the pressure changes are too rapid for relaxation to occur and therefore no additional energy is absorbed.

An approximation for the attenuation coefficient (dB/km) that is valid for f between 0.5 kHz and 100 kHz in standard sea water is (Waite 2002)

$$\alpha_{dB} = 0.05 f^{1.4}. \tag{7.47}$$

The accuracy with which this equation fits the saltwater attenuation coefficient can be seen in Figure 7.18, where it is labeled "Linear Fit."

7.6 REFLECTION AND REFRACTION OF SOUND

Reflection and refraction of acoustic signals occur in regions where there are changes in the impedance of the medium. These include transitions from one material type to another or the more gradual changes introduced by temperature or pressure gradients.

7.6.1 Waves Normal to the Interface

If a wave with intensity I_o is propagating through medium 1 with a characteristic impedance of Z_1 and passes through a boundary normal to the direction of travel into medium 2 with characteristic impedance Z_2, the transmitted, I_{trans}, and reflected, I_{ref}, sound intensities are

$$I_{ref} = I_o \left(\frac{|Z_1 - Z_2|}{|Z_1 + Z_2|} \right), \tag{7.48}$$

$$I_{trans} = I_o \left(\frac{|2Z_2|}{|Z_1 + Z_2|} \right). \tag{7.49}$$

It can be seen that the proportion of the signal reflected increases as a function of the difference in characteristic impedance between the two media.

7.6.2 Waves at an Angle to the Interface

When the incident wave strikes the interface at an angle θ_i to the normal, a reflected wave moves away from the interface at an angle $\theta_r = \theta_i$, while the transmitted wave is refracted at an angle θ_t to the normal in the second medium. The angle of refraction is given by Snell's law:

$$\theta_t = \sin^{-1}(c_2 \sin \theta_i / c_1), \tag{7.50}$$

where c_1 and c_2 are the speeds of propagation in the two media. In this case, the formulas describing the relative intensities of the reflected and transmitted signals include the angles of incidence and refraction:

$$I_{ref} = I_o \left(\frac{|Z_1 \cos \theta_t - Z_2 \cos \theta_i|}{|Z_1 \cos \theta_t + Z_2 \cos \theta_i|} \right), \tag{7.51}$$

$$I_{trans} = I_o \left(\frac{|2Z_2 \cos \theta_i|}{|Z_1 + Z_2|} \right). \tag{7.52}$$

The effects of reflection and refraction are generally visualized using a technique called ray acoustics that is analogous to standard electromagnetic ray tracing techniques. As illustrated in Figure 7.19. For example, if a wave is propagating through n horizontal layers in each of which the speed is considered to be constant, Snell's law can be rewritten in terms of the speed, c, and φ_n, the angle made with the horizontal at that point:

$$c_1 / \cos \varphi_1 = c_2 / \cos \varphi_2 = c_n / \cos \varphi_n. \tag{7.53}$$

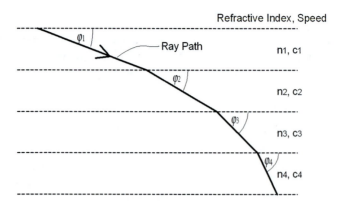

Figure 7.19 Path taken by an acoustic signal propagating through water layers of increasing refractive index.

Note than φ_n is the complement of the angle usually associated with Snell's law. It is called the grazing angle.

In practice, the temperature does not change abruptly, but rather the gradient will increase or decrease at a measurable rate,

$$c(z) = c_0 + gz, \tag{7.54}$$

where c_0 is the speed at the surface (or the transducer depth) and g is the gradient dc/dz between the surface and depth z. The net result of this is that the rays appear to be curved rather than traveling in straight lines, with a radius of curvature ρ at any point:

$$\rho = c_0/g = c/g\cos\varphi. \tag{7.55}$$

In general, the gradient is not constant due to variations in temperature, pressure, and salinity, so the path followed by sound can be quite convoluted, with some possibilities discussed in the following section.

7.6.3 Refraction and Reflection

Understanding sound propagation in water requires a good knowledge of local conditions. This includes the pressure, salinity, and temperature-induced velocity profile, the conditions of the sea surface and the seafloor, and the prevailing currents. Some examples of the different ways that sound propagates using both refraction and reflection are illustrated in Figure 7.20.

In general, however, many of the common effects are primarily due to refractive effects because reflection from the surface or the bottom is determined by the characteristics of those surfaces, which must be reasonably flat to become effective reflectors of short-wavelength acoustic energy. The definition of a smooth surface that will reflect energy in a specular manner is discussed in Chapter 8.

A number of examples showing the large variety of sound propagation patterns by refraction alone are shown in Figure 7.21, with one of the more interesting effects occurring at depths of approximately 1000 m. At this depth, pressure becomes the important factor, where it combines with temperature and salinity to produce a zone of minimum sound speed. This zone has been

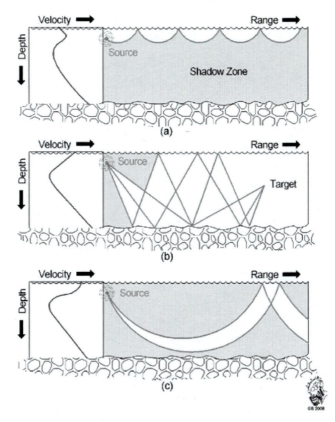

Figure 7.20 Sound propagation effects due to refraction and reflection: (a) surface duct, (b) bottom bounce, and (c) convergence zone.

named the sound fixing and ranging (SOFAR) channel. If a sound is generated by a point source in the SOFAR zone, it becomes trapped by refraction, as shown in the figure, and dispersed horizontally rather than in three directions, allowing it to travel for great distances.

7.7 MULTIPATH EFFECTS

7.7.1 Mechanism

Specular reflection of electromagnetic signals from the ground or sonar signals reflecting from the surface or the seafloor can have a number of effects on target detection probability and tracking accuracy. Figure 7.22 shows that the signal radiating from the antenna can reach the target via two separate paths. One is the direct line-of-sight path from A to B and the other is the path reflected from the flat reflecting surface.

The echo reflected by the target arrives back at the radar along the same two paths and the magnitude of the resultant signal will depend on the amplitudes and relative phase differences between the direct and the reflected path. Modifications to the field strength (volts per meter in the electromagnetic case, and sound pressure in the sonar case) at the target caused by the presence of the surface are expressed by the ratio η (Meeks 1982).

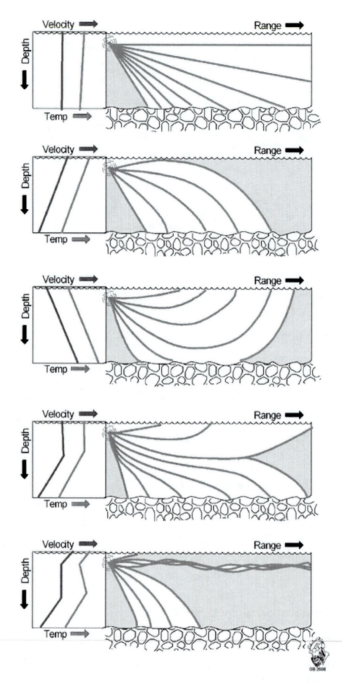

Figure 7.21 Sound propagation effects due to refraction: (a) isovelocity—straight line transmission; (b) negative gradient—rays curve downward; (c) positive gradient—rays curve upward; (d) isovelocity over negative gradient—split beam pattern; and (e) negative gradient over positive—sound channel.

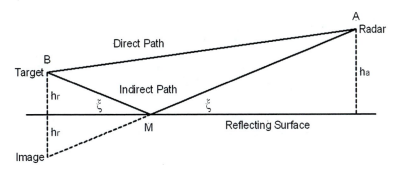

Figure 7.22 Multipath propagation over a flat surface.

The reflection coefficient from the surface may be considered as a complex quantity, where ρ is the real part of the reflection coefficient and φ is the phase shift on reflection:

$$\Gamma = \rho e^{j\varphi}. \tag{7.56}$$

In this example it is assumed that $\Gamma = -1$, that is, the reflected wave suffers no change in amplitude, but its phase is shifted by 180°.

The difference between the reflected path AMB and the direct path AB for $R \gg h_a$ is given by (Skolnik 1980)

$$\Delta = 2h_a \sin \xi. \tag{7.57}$$

For small ξ, $\sin(\xi)$ may be replaced by $(h_a + h_t)/R$ to produce

$$\Delta = \frac{2h_a(h_a + h_t)}{R}. \tag{7.58}$$

This can be further simplified if $h_t \gg h_a$:

$$\Delta \approx 2h_a h_t / R. \tag{7.59}$$

The total phase difference (including the 180°) between the direct and reflected signals as they combine at the target is therefore

$$\psi = \psi_d + \psi_r = \frac{2\pi}{\lambda} \frac{2h_a h_t}{R} + \pi. \tag{7.60}$$

The vector sum of the two signals with the same amplitude, but a phase difference Ψ can be written as $[2(1 - \cos \Psi)]^{1/2}$. As the ratio of the direct and the reflected powers at the target is the square of the amplitude, this can be written as

$$\eta^2 = 2\left(1 - \cos\frac{4\pi h_a h_t}{\lambda R}\right) = 4\sin^2\frac{2\pi h_a h_t}{\lambda R}. \tag{7.61}$$

By reciprocity, the path from the target back to the radar follows the same two paths, as shown in Figure 7.23, so the power ratio back at the radar is

$$\eta^4 = 16\sin^4\frac{2\pi h_a h_t}{\lambda R}. \tag{7.62}$$

The range equation describing the received echo power in a multipath environment must be modified by the propagation factor η^4 (often also called F^4) (Meeks 1982).

7.7.2 Multipath Lobing

Since sine varies from 0 to 1, the factor η^4 varies between 0 and 16. Therefore, taking into account the fourth-power relationship between the range and the echo signal amplitude, the radar detection range for a constant radar cross section target varies between 0 and 2 times the range of the same radar in free space.

The field strength is a maximum when the argument in the sine term satisfies $(2n + 1)\pi/2$, where n is an integer. Therefore

$$4h_a h_t/\lambda R = 2n + 1 \tag{7.63}$$

are maxima, and

$$4h_a h_t/\lambda R = n \tag{7.64}$$

are minima.

The presence of a perfectly conducting flat ground plane causes the continuous elevation beam coverage to break up into a lobed structure, as shown in Figure 7.24. Hence, if a target is in the

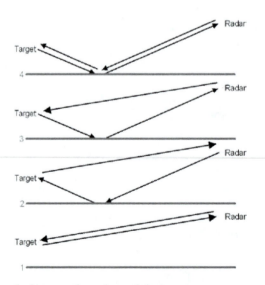

Figure 7.23 Possible signal paths between the radar and the target.

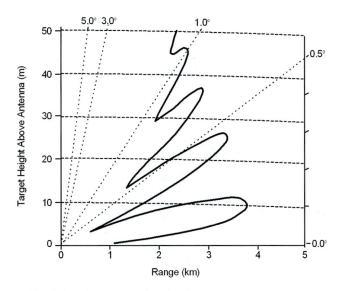

Figure 7.24 Effect of multipath interference on elevation beam pattern.

lobe, it will be detected at twice the range that it would in free space, but if it is between lobes, then theoretically it will not be detected at all. There is also an angle below which the signal level drops off rapidly and where there is minimal coverage. This is the region below the first lobe which occurs at an angle of $\lambda/4h_a$ (rad).

The curvature of the earth cannot be neglected when predicting radar performance at low elevation angles near the horizon. In this case the lobing is not as severe as it is for the flat earth approximation because the beam is more divergent, so the minima are not so deep, nor the maxima so large.

Predicting range height coverage can be managed by computer using packages such as VCCALC (Fielding and Reynolds 1987), or using purpose written MATLAB code (Mahafza 2000). Figure 7.24 shows a typical plot originally generated by VCCALC for an antenna height of 3 m and a wavelength of 3.1 cm (X-band).

Because the positions of the lobes is a function of both the geometry and the transmitted frequency, a frequency agile radar can be constructed that moves the lobes by altering its frequency. A range of frequencies is selected to obtain uniform coverage over the region of interest. However, it is not possible to fill in the region below the lowest lobe.

7.7.3 Multipath Fading

From Figure 7.25 it can be seen that an aircraft approaching the radar at a constant height of 10,000 ft would fly through a number of lobes once it appeared over the horizon at a range of 240 km. In reality, the surface of the earth is never totally flat, so in addition to a specular component, a diffuse component also exists. This diffuse component includes reflections from an elliptical "glistening" surface that extends from the ground below the radar to the ground below

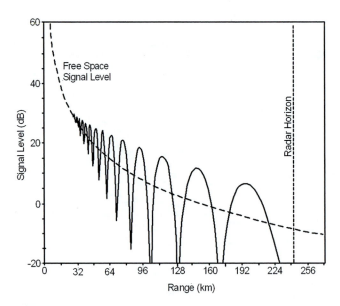

Figure 7.25 Multipath fading as a function of range compared to the free space signal for an aircraft approaching the radar at a height of 10,000 feet.

the target, and will fill in the nulls to a certain extent. However, that notwithstanding, fades will occur, and it is often possible to detect aircraft targets at some ranges and not at others during an approach. In addition to the possibility of losing the target because of signal fading, in the case of a tracking radar, multipath fading has a major effect on the elevation tracking accuracy, as discussed in the following section.

7.7.4 Multipath Tracking

Tracking a moving target in a multipath environment can result in some interesting behavior, as shown in Figure 7.26. In this example, a radar is tracking an aircraft flying away at a constant height (Skolnik 1980).

- Region A: At close range the target elevation angle is large and the antenna beam does not illuminate the surface, as shown in Figure 7.27, so the tracking is smooth.
- Region B: At intermediate range, where the elevation angle varies from as much as 6 down to 0.8 beamwidths, the surface reflected signal enters the radar through the antenna near-in sidelobes. The surface-reflected signal is small, so the radar makes small oscillations around some mean position.
- Region C: At greater ranges, where the main beam illuminates the surface, the interference between direct and reflected signals can result in large errors in elevation angle. As with the glint error, these angular excursions can be many times the angular separation between the target and its image.

Particularly over smooth water, the elevation tracking errors can generate such severe nodding in a mechanically tracked antenna that it can cause the radar to break lock. Over the years, a number of methods have been derived to mitigate against this problem, of which the surest is to

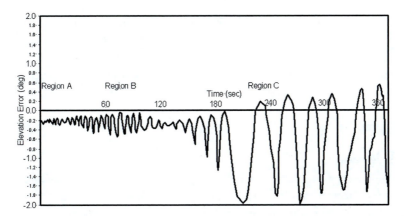

Figure 7.26 Elevation tracking errors caused by multipath interference on a phased array tracking radar with a beamwidth of 2.7°. The aircraft flew out at a constant altitude.

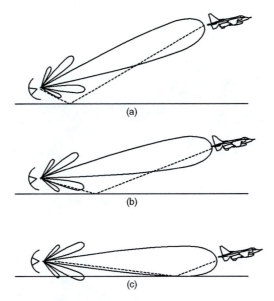

Figure 7.27 Illustration of the three multipath regions.

use an antenna with a sufficiently narrow beamwidth that it does not illuminate the surface. This requires either a large antenna or a high operating frequency.

A second method is to fix the antenna's elevation (once the target has reached the 0.8 beamwidths region) and to continue with closed loop tracking. This holds the antenna still so that it cannot nod the beam off the target even though fades still occur.

For very short pulses it is possible to separate the direct return and the delayed multipath return. However, this is often not practical, as the range resolution required to separate the signals is very fine:

$$\Delta R = 2h_a h_t / R. \tag{7.65}$$

For a radar height of 30 m, a target height of 100 m, and a range of 10 km, the range resolution must be 0.6 m, which corresponds to a pulse width of 4 ns, which is unusually short for a tracking radar.

Finally, frequency agility can also be effective in reducing multipath error, as a change in frequency changes the phase relationship between the direct and the reflected signals. However, it turns out that the bandwidth required to achieve this fix is equivalent to that required for the short-pulse option.

7.7.5 Effects on Imaging

For short-range imaging applications which are typical for robotic guidance and collision avoidance, the primary effect of multipath is to displace targets from their correct bearing and elevation. In the elevation plane, the grazing angles are very shallow and specular reflection occurs. This results in targets appearing below the ground surface which could lead the autonomous vehicle, shown in Figure 7.28, to conclude that the gradient has changed, or more catastrophi-

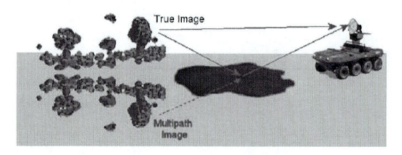

Figure 7.28 Effect of multipath on the interpretation of the ground level.

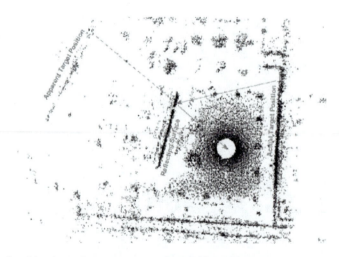

Figure 7.29 Effect of multipath on the apparent angular position of features.

cally, to allow it to drive into a dam under the misguided assumption that it has identified some trees in that direction.

Multipath in azimuth generally caused by the flat walls of buildings has the similar effect of displacing and doubling up of the number of features visible to the radar, as shown in Figure 7.29. This can result in severe navigation problems (Brooker 2005).

7.8 References

Barton, D. (1988). *Modern Radar Systems Analysis*. Norwood, MA: Artech House.

Bhartia, P. and Bahl, I. (1984). *Millimeter Wave Engineering and Applications*. New York: John Wiley & Sons.

Brooker, G. (2005). Long-range imaging radar for autonomous navigation. Ph.D. dissertation. University of Sydney, Sydney, NSW, Australia.

Brookner, E., ed. (1982). *Radar Technology*. Norwood, MA: Artech House.

Burnside, N. (2004). A function that returns the atmospheric attenuation of sound. MATLAB Central. Available at http://www.mathworks.com/matlabcentral/fileexchange/loadFile.do?objected= 6000&objectType=FILE.

Button, K. and Wiltse, J., eds. (1981). *Infrared and Millimeter Waves*, Vol. 4, *Millimeter Systems*. Burlington, MA: Academic Press.

Cook, R., ed. (1972). *American Institute of Physics Handbook, Section 3: Acoustics*. New York: McGraw Hill.

Currie, N. and Brown, C. (1987). *Principles and Applications of Millimeter-Wave Radar*. Norwood, MA: Artech House.

Currie, N., Hayes, R., and Trebits, R. (1992). *Millimeter-Wave Radar Clutter*. Norwood, MA: Artech House.

Falcone, V. and Abreu, L. (1979). Atmospheric attenuation of millimeter and submillimeter waves. IEEE EASCON-79, Electronics and Aerospace Systems Conference, Arlington VA, October 9–11.

Fielding, J. and Reynolds, G. (1987). *VCCALC: Vertical Coverage Calculation Software and Users Manual*. Boston: Artech House.

Ghobrial, S. and Sharief, S. (1987). Microwave attenuation and cross polarization in dust storms. *IEEE Transactions on Antennas and Propagation* AP-35(4):418–425.

Gillett, D. (1979). Environmental factors affecting dust emission by wind erosion. In: *Saharan Dust*. C. Morales, ed. New York: John Wiley & Sons; pp. 71–94.

Goldhirsh, J. (2001). Attenuation and backscatter from a derived two-dimensional duststorm model. *IEEE Transactions on Antennas and Propagation* 49(12):1703–1711.

Knudsen, V. and Harris, C. (1950). *Acoustical Designing in Architecture*. New York: John Wiley & Sons.

Litovitz, E. and Carnevale, J. (1955). Effect of pressure on sound propagation in water. *Journal of Applied Physics* 26(7):816–820.

Mahafza, B. (2000). *Radar Systems Analysis and Design Using MATLAB*. Boca Raton, FL: CRC Press.

Meeks, M. (1982). *Radar Propagation at Low Altitudes*. Boston: Artech House.

Nelson, S. (2001). Measurement and calculation of powdered mixture permittivities. *IEEE Transactions on Instrumentation and Measurement* 50(5):1066–1070.

Patterson, E. (1977). Atmospheric extinction between 0.55 μm and 10.6 μm due to soil-derived aerosols. *Applied Optics* 16(9):2414–2418.

Pinkerton, J. (1947). A pulse method for the measurement of ultrasonic absorption in liquids: results for water. *Nature* 160:128–129.

Pinnick, R., Fernandez, G., and Hinds, B. (1983). Explosion dust particle measurements. *Applied Optics* 22(1):95–102.

Pinnick, R.G., Fernandez, G., Hinds, B.D., Bruce, C.W., Schaefer, R.W., and Pendleton, J.D. (1985). Dust generated by vehicular traffic on unpaved roadways: sizes and infrared extinction characteristics. *Aerosol Science and Technology* 4(1):99–121.

Preissner, J. (1978). The Influence of the Atmosphere on Passive Radiometric Measurements. AGARD Conference Reprint No. 245: Millimeter and Submillimeter Wave Propagation and Circuits.

Ready, J. (1978). *Industrial Applications of Lasers*. Burlington, MA: Academic Press.

Skolnik, M. (1980). *Introduction to Radar Systems*. Tokyo: McGraw-Hill Kogakusha.

Trabert, W. (1901). Die extinction des lichtes in einem turben medium. *Meteor* Z(18):518–525.

Urick, R. (1975). *Principles of Underwater Sound*. New York: McGraw-Hill.

Waite, A. (2002). *Sonar for Practicing Engineers*. New York: John Wiley & Sons.

Wilson, W.D. (1960). Speed of sound in seawater as a function of temperature, pressure and salinity. *Journal of the Acoustical Society of America* 32:641–644.

8

Target and Clutter Characteristics

8.1 INTRODUCTION

The operation of all active sensors requires that some of the energy transmitted by the device be reflected off objects within the beam. The characteristics of the reflected signal can then be analyzed in various ways to extract information about the target. It should be stressed that the word "target" does not necessarily convey the idea that the object is being hunted, though that was almost certainly the root of the word. In this book, target is defined as the object of interest to the sensor, in contrast to clutter, which is defined as any object whose echoes may interfere with the investigation of the target.

8.2 TARGET CROSS SECTION

The "cross section" qualitatively relates the amount of power that strikes the target to the amount of power that is reflected into the receiver. Assuming that the power density of a plane-wave incident on the target is S_i (W/m^2), and the amount of power scattered isotropically is P_r (W), which is defined in terms of the cross section, σ (m^2), as follows,

$$P_r = \sigma S_i, \tag{8.1}$$

then the power density, S_r (W/m^2) of the scattered wave for an isotropic scatterer at the receiving antenna is

$$S_r = P_r / 4\pi R^2. \tag{8.2}$$

This allows the cross section to be defined in terms of the ratio of the power density at the receiver to that incident on the target:

$$\sigma = 4\pi R^2 (S_r / S_i). \tag{8.3}$$

To ensure some rigor in the definition, in order to ensure that the receiving antenna is in the far field and that the waves are planar, equation (8.3) is written as

$$\sigma = \lim_{R \to \infty} 4\pi R^2 \left| \frac{S_r}{S_i} \right|. \tag{8.4}$$

The definition holds for acoustic and the electric and magnetic components of electromagnetic waves as follows:

$$\sigma = \lim_{R \to \infty} 4\pi R^2 \left| \frac{E_r^2}{E_i^2} \right|$$

$$\sigma = \lim_{R \to \infty} 4\pi R^2 \left| \frac{H_r^2}{H_i^2} \right|$$

$$\sigma = \lim_{R \to \infty} 4\pi R^2 \left| \frac{P_r^2}{P_i^2} \right| \tag{8.5}$$

where E_r and E_I are the electric field magnitude at the receiver and incident on the target, H is the magnetic field, and P is the acoustic pressure.

By definition, if a target were to scatter power uniformly over all angles its cross section would be equal to the area from which power was extracted from the incident wave. Since only a sphere has the ability to scatter isotropically, it is convenient to interpret cross section in terms of the projected area of an equivalent sphere.

8.2.1 Cross Section and the Equivalent Sphere

A sphere with radius $a \gg \lambda$ will intercept power contained in πa^2 of the incident wave. It will scatter power uniformly over 4π steradians of solid angle. The cross section of a sphere is therefore (by definition) equal to the projected area, πa^2, even though the actual area that returns power to the receiver is from a very small area where the surface lies nearly parallel to the incident wave front.

The cross section of a sphere is therefore equal to one-quarter of its surface area, and by extension it can be shown that the average cross section of any large object that consists of continuous curved surfaces will be equal to one-quarter of its total surface area.

8.2.2 Cross Section of Real Targets

In reality, very few targets are spherical and the cross sections of most targets are complicated functions of the viewing aspect, the frequency, and the polarization of the incident signal. Though it is possible to derive the equations describing the cross sections of some common shapes such as cylinders and flat plates, in general, sophisticated modeling software is used to obtain estimates of target cross sections. Ideally, however, the cross section is obtained by measurement, but even this method is fraught with difficulty, particularly if the target is large, as it is often impractical to measure the cross section over all aspect angles.

The target cross section is often related to the physical size of the item, but under certain circumstances it may be much larger or much smaller. For example, a corner reflector (retroreflec-

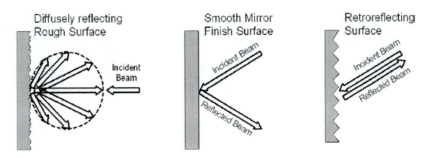

Figure 8.1 Different modes of reflection depend on the surface characteristics of the target.

tor) has an extremely large target cross section in relation to its size, whereas a B-2B stealth bomber has a very small cross section.

Because all but the simplest targets are made up of many scattering surfaces, the cross section will also be made up of reflections from a large number of scatterers. This means that even very small changes in the aspect angle of the target will result in relative phase changes between the scatterers and an altered cross section. The effective surface roughness of a target (as a function of λ) also plays an important role in determining its cross section.

Three mechanisms, shown in Figure 8.1, determine individually or in combination the target reflection characteristics. They are diffuse reflection, which occurs when the surface is rough compared to the wavelength of the incident radiation, specular reflection (like a mirror) if the surface is flat, and retroreflection if the surface is made up of right-angled sections.

There are a number of significant differences between the cross sections of targets measured using electromagnetic waves and those measured using acoustic waves, although the underlying theory of reflection is the same in the two cases. One of the differences occurs because the nature of electromagnetic waves supports the concept of polarization, while acoustic waves propagating in air or water cannot because shear waves are not supported in those media.

8.3 RADAR CROSS SECTIONS

Qualitatively, the radar cross section (RCS) of an object is a measure of its "size" as seen at a particular radar wavelength and polarization. The RCS has units of square meters (m^2), but is often expressed in decibels relative to 1 m^2 (dBm^2) as

$$\sigma(dBm^2) = 10\log_{10}\left[\sigma(m^2)\right].$$ (8.6)

To account for the polarization dependency of RCS, the relationship between the transmitted and received electric fields must be considered in terms of their orthogonal linear polarization components, E_H and E_V, and the proportionality constants that relate them

$$\begin{bmatrix} E_H^r \\ E_V^r \end{bmatrix} = \begin{bmatrix} a_{HH} & a_{VH} \\ a_{HV} & a_{VV} \end{bmatrix} \begin{bmatrix} E_H^t \\ E_V^t \end{bmatrix}.$$ (8.7)

This matrix of constants is referred to as the target scattering matrix and it can be used to justify the definition of a radar cross-section scattering matrix with the same form:

$$\sigma = \begin{bmatrix} \sigma_{HH} & \sigma_{VH} \\ \sigma_{HV} & \sigma_{VV} \end{bmatrix}. \tag{8.8}$$

The relationship between the target and RCS scattering matrices is as follows (for each of the four terms):

$$a_{HH} = \sqrt{\sigma_{HH}} \, e^{j\varphi_{HH}}, \tag{8.9}$$

where $\sqrt{\sigma_{ij}}$ is the magnitude of each matrix component and φ_{ij} is the phase of the associated element.

For most applications it is sufficient to define the target RCS in terms of one polarization only. Therefore, henceforward it will be assumed to have a single value that corresponds to either σ_{HH} or σ_{VV}, depending on the incident polarization.

8.4 THE RCS OF SIMPLE SHAPES

A number of simple reflectors are useful, particularly for use as calibration targets. The flat plate offers the highest RCS for a given size, but it is specular, which makes it difficult to align so that the reflected power returns to the receiver. An alternative is to use a retroreflector, of which the trihedral corner reflector is the most common. It also produces a high RCS that remains reasonably constant over a wide angle and is therefore easy to align. It is the reflector of choice for use as a calibration target if signal level is a consideration. The sphere is also often used as a reference for moving targets, as its RCS is invariant with the observed angle. Because the RCS is very small for its size, a sphere can only be used if the incident power is large. In addition, compared to the trihedral reflector, large conductive perfect spheres are difficult to manufacture.

8.4.1 Flat Plate

The flat plate has the largest peak RCS for its size of any target. It is specular, and so the RCS falls off sharply with changes in incidence angle from 0°. The length, a, of the plate determines the width of the specular lobe and the interval between peaks as defined by

$$\sigma(\theta) = \sigma_o \left[\frac{\sin\left(\dfrac{2\pi}{\lambda} a \sin\theta\right)}{\dfrac{2\pi}{\lambda} a \sin\theta} \right]^2 \cos^2\theta \tag{8.10}$$

$$\sigma_o = \frac{4\pi a^2 b^2}{\lambda^2}, \tag{8.11}$$

where $\sigma(\theta)$ (m^2) is the RCS as a function of the angle, a,b (m) are the sides of the plate, and θ (rad) defines the plate rotation around the axis.

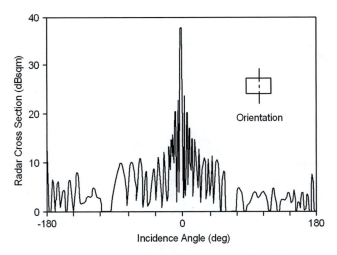

Figure 8.2 Measured RCS of a flat, square plate as a function of orientation [adapted from Trebits (1978)].

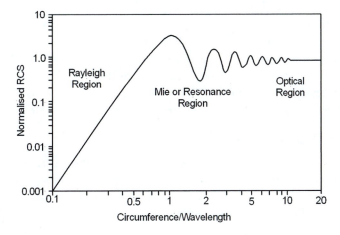

Figure 8.3 The RCS of a sphere as a function of the circumference normalized by the wavelength of the incident radiation.

At incidence angles close to 90°, complex edge-diffraction effects become dominant and small irregularities in the surface can destroy the symmetry of the data, as shown in the measured data (Figure 8.2).

8.4.2 The Sphere

A perfectly conducting sphere is the simplest shape whose RCS can be determined exactly. With its three-dimensional (3D) symmetry the RCS is aspect independent, so it is often used as a calibration target if the size selected is appropriate to the operating frequency of the measurement system. Depending on the relative differences between the size of the sphere and the wavelength of the electromagnetic signal, different interactions dominate, as is apparent from Figure 8.3:

- Rayleigh region ($2\pi a/\lambda < 1$): The RCS is inversely proportional to the fourth power of the wavelength, as indicated by the slope of the curve in that region.

- Mie region ($1 < 2\pi a/\lambda < 10$): In this resonance region, a creeping wave travels around the sphere and back toward the receiver where it interferes constructively or destructively with the specular backscatter to produce a cyclical variation in the RCS.
- Optical region ($2\pi a/\lambda > 10$): The RCS of the sphere approaches its geometric projected area πa^2.

8.4.3 Trihedral Reflector

Trihedral corner reflectors have large nonspecular radar cross sections and so make ideal test targets, as alignment is not critical. Their behavior for linear polarizations is good, but they cannot be used for circularly polarized measurements, as the triple bounce reverses the sense of the polarization. For a symmetrical reflector having each vertex with length a (m), the peak radar cross section (m^2) is

$$\sigma = 4\pi a^4/3\lambda^2. \tag{8.12}$$

From Figure 8.4 it can be seen that azimuth and elevation misalignments of up to about 10° can be tolerated without a significant reduction in RCS. Note that the RCS is proportional to the square of the projected area and inversely proportional to the square of the wavelength. Because some natural and many man-made objects are made up of corners, these become dominant as the radar frequency increases. For example, if the RCS of a car is measured, at low frequency the curved body is dominant, but as the frequency increases, small angular objects like wing mirrors and grille work can become the major contributors to the return amplitude.

8.4.4 Other Simple Calibration Reflectors

A number of other simple reflectors have cross sections that can be determined analytically and are listed in Table 8.1. These are most often useful in their own right for calibration, or because they can be used to determine the characteristics of more complex shapes. In these examples it is assumed that the surface finish is sufficiently smooth to ensure specular properties.

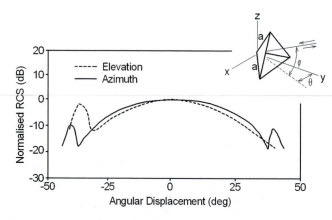

Figure 8.4 The RCS of a trihedral reflector as a function of orientation [adapted from Currie (1984)].

Table 8.1 Maximum RCS for typical calibration targets [adapted from Currie and Brown (1987)]

Target		Maximum Cross-section	Advantages	Disadvantages
	Cylinder	$\sigma = \dfrac{2\pi a b^2}{\lambda}$	Nonspecular along the radial axis	Low RCS for size, specular along axis
	Sphere	$\sigma = \pi a^2$	Nonspecular	Lowest RCS for size, radiates isotropically
	Diplane	$\sigma = \dfrac{8\pi a^2 b^2}{\lambda^2}$	Large RCS for size, nonspecular along one axis. Useful for testing polarisation	Specular along one axis
	Triangular Trihedral	$\sigma = \dfrac{4\pi a^4}{3\lambda^2}$	Nonspecular	Cannot be used for cross polarised measurements
	Square Trihedral	$\sigma = \dfrac{12\pi a^4}{\lambda^2}$	Large RCS for size, nonspecular	Cannot be used for cross polarised measurements
	Circular Trihedral	$\sigma = \dfrac{0.507\pi^3 a^4}{\lambda^2}$	Large RCS for size, nonspecular	Cannot be used for cross polarised measurements
	Flat Rectangular Plate	$\sigma = \dfrac{4\pi a^2 b^2}{\lambda^2}$	Largest RCS for size	Specular along both axes, difficult to align
	Top Hat	$\sigma = \dfrac{2\pi a b^2}{\lambda \cos^3 \varphi}$	Low RCS for size	Difficult to align rotated seam
	Bruderhedral	φ is the elevation angle to the cylinder, $\varphi=0°$ is perpendicular to the cylinder c>b to be effective for 90° rotation of polarisation $\varphi = 45°$	Large RCS, easier to align for rotation than Top Hat	Moderately specular along one axis

8.5 RADAR CROSS SECTION OF COMPLEX TARGETS

Reasonably complex targets can be analyzed by breaking the structure into simpler elements. For example, a cannon shell or missile could be modeled as a cone on a cylinder. However, in most cases direct measurement or modeling using commercial electromagnetic software are the only ways to obtain reasonably realistic estimates of the RCS.

8.5.1 Aircraft

The RCS of a full-size aircraft can be measured in an outdoor range by placing the aircraft on a pedestal mounted on a turntable. To reduce the mass of such targets they are often gutted. However, this can introduce inaccuracies in the results as jet engines and cockpit instrumentation often add a significant contribution to the overall RCS.

Figure 8.5 shows the now classic radar cross section of a B-26 bomber measured at a wavelength of 10 cm and plotted on polar axes as a function of azimuth angle. Note that in the original measurements the axes were in dBft^2, and these have been altered to suit the metric notation. It can be seen that, even at this low frequency, the RCS exceeds 25 dBm^2 (316 m^2) from certain aspect angles. In contrast, the RCS of the B-2 stealth bomber is about –40 dBm^2.

A more modern aircraft, the C29 cargo plane, shows a more dramatic variation in the RCS, with strong peaks at ±90° generated by the aircraft fuselage, as can be seen in Figure 8.6. From the tail-on aspect, this aircraft also shows a reasonably large RCS that may be generated by the engine outlets.

8.5.2 Ships

Measuring the RCS of a ship involves sailing in a circle while measuring the return from a fixed point and then compensating for variations in range, as is illustrated in Figure 8.7. The median

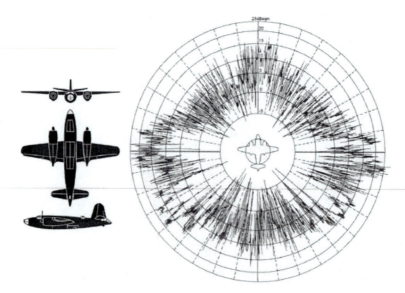

Figure 8.5 Diagram and RCS of a B-26 bomber as a function of aspect angle at a wavelength of 10 cm [adapted from Ridenour (1947)]. Courtesy McGraw-Hill Book Company.

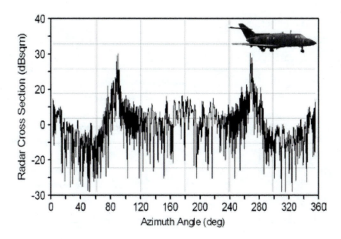

Figure 8.6 The RCS of a C29 cargo plane with aspect angle.

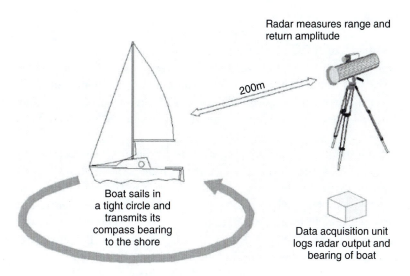

Figure 8.7 Polar RCS measurement setup for a ship.

RCS, σ (m^2), of a typical ship at low grazing angles (excluding the specular broadside return) is related to its size by the empirical formula (Skolnik 1970)

$$\sigma = 52 f^{1/2} D^{3/2}, \tag{8.13}$$

where f (MHz) is the frequency of the incident signal and D (kilotons) is the displacement of the vessel.

Table 8.2 Median RCS of a number of ship types

Type	Length (m)	Mass (tons)	Median RCS (dBsqm)
Fishing Vessel	10	5	Q at ~10
Small Coaster	50	225	S ■■■ B/Q (~12–30)
Coaster	65	500	nS ■■■ B/Q (~17–33)
Large Coaster	80	900	BW ■■■ Q (~28–40)
Collier	85	1570	nB ■■■ Q (~18–33)
Frigate	120	2000	BW ■■■ Q (~33–50)
Cargo Liner	135	8000	BW ■ Q (~37–42)
Bulk Carrier	200	8200	nB ■■■ B/Q (~25–40)
Ore Carrier	240	25400	■■■ nB (~30–45)
Container Ship	250	26500	BW ■■■ B/BW (~35–50)
Medium Tanker	260	35000	nB ■■■ Q (~43–58)

Scale markers: 10 20 30 40 50 60

S - Stern On, Q - Quarter, B - Broadside, BW - Bow, n - Near

This empirical relationship shows reasonable agreement with the measured data shown in Table 8.2, although it is not considered to be accurate at high frequencies due to the quadratic increase in the RCS of simple targets (flat plates and corners) with increasing frequency. At higher grazing angles, the median RCS is approximately equal to the displacement tonnage (in m^2).

The definition of the RCS of a target is the composite sum of the returns from the complete object. However, it is often both practical and illuminating to take measurements in which the range resolution is much finer than the dimensions of the target. Such measurements can go some way to pinpointing the position of scatterers (Brooker et al. 2008; Moon and Bawden 1991).

The RCS of ships does, by definition, include interactions with the medium. For example, the corner made by the water surface and the near-vertical sides of the ship add significantly to the amplitude of returns at higher grazing angles. In Figure 8.8, the polar RCS of a naval auxiliary ship is shown for horizontal incident polarization at two different frequencies. These results, in contrast to those presented for the aircraft, have been processed over 2° intervals to show the 80, 50, and 20 percentile levels. Note that RCS values around the 50 dBm^2 mark are common at all aspect angles, and these can exceed 70 dBm^2 at broadside.

High-resolution measurements of the RCS of a number of small boats have been made at 94 GHz from a shore-mounted radar that tracked the craft in angles as they sailed slowly in a tight circle at an appropriate range. The radar measured and logged the returns from the target, while the boat angle, measured by an electronic compass on board, was communicated back to the shore where it was also logged. The data were then used to produce a polar RCS measurement such as the one shown in Figure 8.9.

It is interesting to note that the median RCS of this boat is about 10 dBm^2, and the RCS, calculated using equation (8.13) for a 13 ton displacement and a frequency of 94 GHz, is 13 dBm^2. It is possible that the "corner reflector" effect is not evident in this measurement because the boat is made primarily from fiberglass.

The RCS measurements of scale models in the millimeter and submillimeter wave bands have been used to predict the RCS of ships at lower frequencies. This is a particularly useful technique,

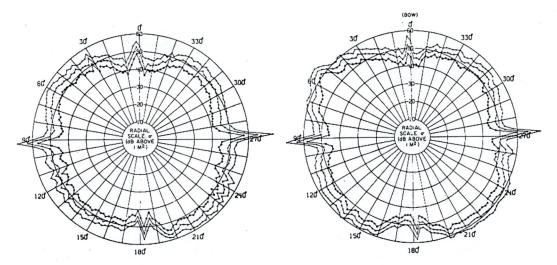

Figure 8.8 The RCS of a naval auxiliary ship with aspect angle at (a) 2.8 GHz and (b) 9.225 GHz (Skolnik 1980). Courtesy McGraw-Hill Book Company.

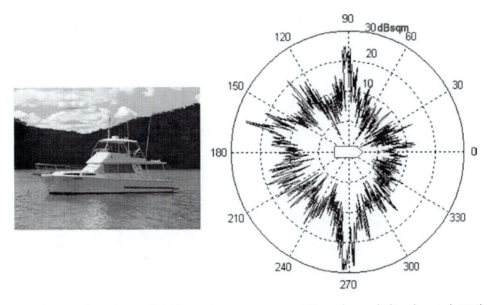

Figure 8.9 The RCS of a Steber 42 flybridge cruiser at 94 GHz at a 2.5° grazing angle (Brooker et al. 2008).

as it is a relatively simple matter to alter the shape of a model to tune the signature of the vessel (Paddison et al. 1978).

8.5.3 Ground Vehicles

Measurement techniques for ground vehicles generally involve measuring the RCS through 360° in azimuth by rotating the vehicle on a turntable. These are normally made from a reasonably short range so that it is possible to produce plots at various grazing angles by adjusting the height of the measurement radar. An example of such measurements, made for a Toyota pickup truck at 35 GHz, is shown in Figure 8.10.

It is possible to produce reasonably good RCS measurements without a turntable. For the measurements shown in Figure 8.11, the vehicle was driven slowly in a tight circle while the radar measurement system was directed manually and automatic range tracking kept the measurement gates centered on the target.

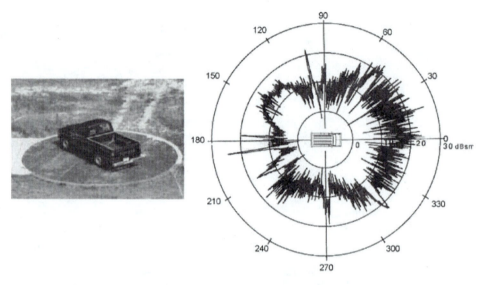

Figure 8.10 The RCS of a Toyota utility vehicle at 35 GHz from a 0° grazing angle [adapted from Currie (1989)].

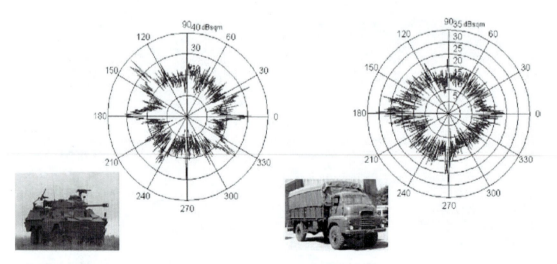

Figure 8.11 The RCS of large military vehicles at 94 GHz: (a) Ratel armored personnel carrier and (b) Bedford truck.

The RCSs of most military vehicles are larger than those of military aircraft because the latter are generally more rounded for aerodynamic reasons and because of stealth concerns (as discussed later in this chapter). Ground vehicles, on the other hand, are often made up of flat armor plates and lots of brackets, antennas, etc., as can be seen in Figure 8.11a.

8.6 EFFECT OF TARGET MATERIAL

Of course, all targets are not made of metal, and therefore it cannot be assumed that perfect reflection will occur from the surface. In industrial applications, in particular, radars are used to measure range to a large variety of solids and liquids, all of which will have different dielectric constants and therefore different reflectivities, as listed in Figure 8.12.

- The radar reflectivity characteristic is inversely related to its relative dielectric constant.
- Reduced reflection from low dielectric materials allows the radar to penetrate foam layers above liquids. It also allows the tracking of water levels in tanks containing hydrocarbons.
- As with acoustics, for solid targets, the particle size and angle of repose will have an effect on the echo strength.
- Liquid-level radars often rely on the fact that only one smooth high-reflectivity target will be visible to measure ranges to submillimeter accuracy. This is useful in custody transfer applications (petrol and oil).
- In industrial applications, low-power pulsed radars are good for high dielectric constant materials, $\varepsilon_r > 8$, but often do not have the sensitivity to detect lower dielectric materials.
- Frequency modulated continuous wave (FMCW) radars can operate where the dielectric constants are much lower, as their sensitivity can be better because the average power transmitted is higher.

8.7 THE RCS OF LIVING CREATURES

8.7.1 Human Beings

If it can be assumed that the human body is roughly elliptical and perfectly conductive, then the RCS should increase with frequency in the Rayleigh region. In the Mie region, it should oscillate before settling to a reasonably constant value in the optical region. However, as shown in Table 8.3, this is not the case. First, at low frequencies, significant penetration of the body occurs and some resonance effects may occur with internal organs. As the frequency increases, penetration depth decreases and Mie scattering occurs from the surface of the body. Finally, as the optical region is approached, the magnitude of the measured variations does decrease. In this region, because the human body is convex, with a surface area approaching 2 m^2, the RCS should equal the projected area, which is approximately 0.5 m^2.

Measurements made of the human torso with a radar footprint diameter of 1.3 m made at 94 GHz are reproduced in Figure 8.13. These include a reference 1 m^2 reflector in the beam about a meter off the ground that is eclipsed as the human subject moves into the beam. Because the measurements are made in decibels, the difference between the signal level returned by the reflector and that returned by the body gives a direct measure of the RCS. It can be seen that

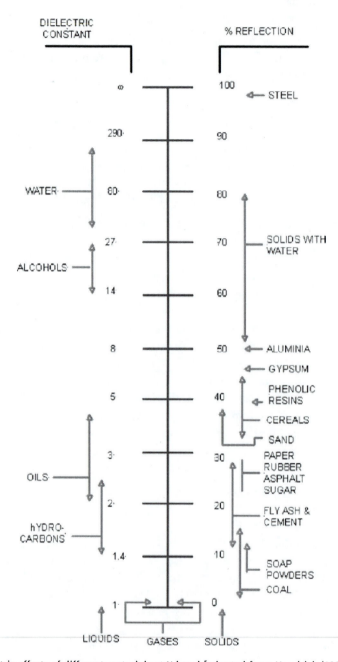

Figure 8.12 Dielectric effects of different materials at X-band [adapted from Hendrick (1992)].

Table 8.3 RCS of a human being at different frequencies (Skolnik 1980)

Frequency (MHz)	Wavelength (m)	RCS (m^2)	Scattering region
410	0.7	0.033–2.33	Rayleigh
1120	0.27	0.098–0.997	Mie
2890	0.1	0.140–1.05	Mie
4800	0.06	0.368–1.88	Mie
7375	0.04	0.495–1.22	Optical

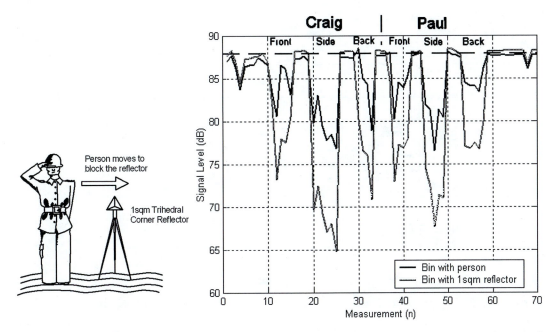

Figure 8.13 Human RCS at 94 GHz is determined by having a person block the path from the radar to the 0 dBm^2 corner reflector. The RCS is determined by the decrease in the signal level.

the RCS of the two human beings varies between −3 dBm^2 and −8 dBm^2. This corresponds to an RCS of between 0.5 m^2 and 0.15 m^2.

8.7.2 Birds

Radar returns are often returned from areas that appear clear. These are called ghosts or angels and are often returns from flocks of birds or swarms of insects. Because birds can fly at up to 50 knots, their returns are not rejected by Doppler processing or moving target indicator (MTI).

An excellent and comprehensive review by Charles Vaughn on the radar characteristics of birds is a good introduction to the subject (Vaughn 1985). In Table 8.4, the mean and median values of the RCS of a number of bird types are reported at UHF, S-band, and X-band. It can be seen that resonance effects play a large role in the measured RCS.

Table 8.4 RCS of birds at three different frequencies

Bird type	Frequency	Mean RCS (cm^2)	Median RCS (cm^2)
Grackle (170 g)	X	16	6.9
	S	25	12
	UHF	0.57	0.45
Sparrow (25 g)	X	1.6	0.8
	S	14	11
	UHF	0.02	0.02
Pigeon (450 g)	X	15	6.4
	S	80	32
	UHF	11	8.0

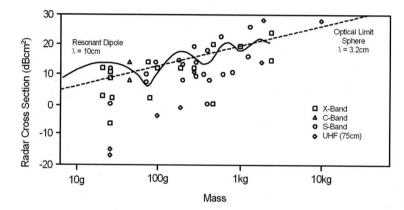

Figure 8.14 Bird RCS as a function of mass at various frequencies [adapted from Vaughn (1985)].

One way to estimate the RCS of birds is to assume that they are either spherical or elliptical and made of water. As shown in Figure 8.14, this is not a particularly accurate prediction method, with errors of 10 dB being common.

Accurate measurements of individual birds are hard to obtain because they are generally not very cooperative in regard to remaining still, keeping a constant posture with wings open or folded, as required. Many measurements have been made of birds in flight because they make excellent targets of opportunity for aircraft tracking and surveillance radars (Diehl and Larkin 2005; Gauthreaux 2003).

Fluctuations in the RCS of a single bird in flight have been measured to have a log-normal distribution, with variations synchronized with wing beat. Variations in excess of 30 dB from the minimum to the maximum observed RCS are common (Vaughn 1985). Other useful information gleaned from these observations includes the relationship between the wing beat frequency, f (Hz), and the length of the bird, l (mm), which is

$$fl^{0.827} = 572. \tag{8.14}$$

8.7.3 Insects

Appreciable echoes are only obtained from insects if their body length exceeds $\lambda/3$, otherwise they are too low down in the Rayleigh scattering region to be visible. Insects viewed broadside typically have RCS values between 10 and 1000 times larger than when viewed head on. At X-band, the RCS of a variety of insects shows a variation from 0.02 to 9.6 cm^2 with the polarization axis along the body, and between 0.01 and 0.95 cm^2 for the transverse case. Figure 8.15 shows the RCSs for a number of insects in comparison to that of a water droplet with the same mass.

A bee would have a broadside RCS of about 1 cm^2 at X-band. This would not increase significantly up to W-band as the scattering mechanism moved from the Mie to the optical region.

8.8 FLUCTUATIONS IN RCS

So far in this chapter the RCS has been given as a single value which corresponds to the mean or median of the measured values. However, if the polar patterns shown in the previous figures are considered, it is clear that even small variations in the observation angle often result in large changes in the measured amplitude. It is this variation that translates into both temporal and spatial fluctuations of the RCS.

8.8.1 Temporal Fluctuations

If the target is made up of many scattering elements of comparable size and with phase relationships that vary in a random fashion as a function of time, the distribution of cross section is exponential. This is known as the Rayleigh amplitude distribution and is shown in Figure 8.16 (Barton 1976).

The RCS of a target such as the Steber 42 flybridge cruiser, as shown in Figure 8.9, can be displayed as a probability density function based on the full 360° azimuth coverage, or as a smaller sector if required. Though these are in effect spatial variations in RCS, if the vehicle is moving

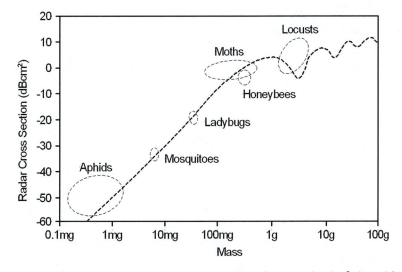

Figure 8.15 The RCS of insects at 9.4 GHz compared to the RCS of a water droplet [adapted from Riley (1985)].

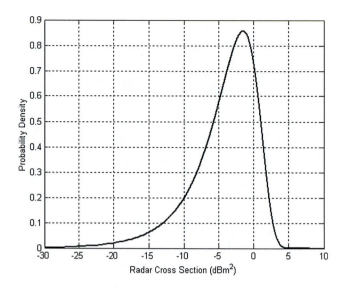

Figure 8.16 Log plot of the Rayleigh distribution.

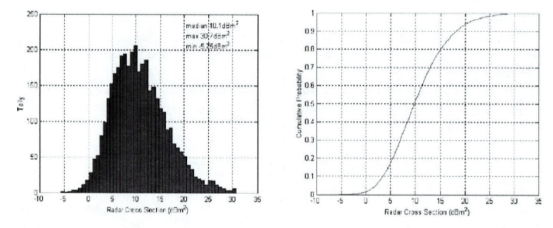

Figure 8.17 The PDF and cumulative probability distributions of the RCS of a Steber 42 flybridge cruiser at 94 GHz.

and showing a changing aspect to the radar, they map into temporal variations in the echo amplitude.

Because the target is made up of a lot of independent scatterers, it could be expected that the probability distribution function (PDF) would be Rayleigh. However, an examination of the PDF displayed in Figure 8.17 shows a distribution that is basically log-normal with a long tail on the high side. The reason for this different distribution is twofold. First, the high-side tails are generated by the large specular returns from the sides of the boat, and second, the log-normal form is due to the high RCS contribution of small flat plates and corners because the operational frequency is so high.

8.8.2 Spatial Distribution of Cross Section

The various reflecting bodies across a target interfere constructively and destructively as it moves. This results in a physical displacement of the effective target position that varies with time. The angular position of the target measured by a radar system is determined by the direction of arrival of the phase front of the reflected radar signal from all scatterers, and so it is possible that the angular error can extend beyond the physical boundaries of the target. These variations in direction of arrival are known as angle glint. Figure 8.18 shows a sample of the angular tracking error measured for an aircraft flying toward the radar, which gives an indication of the size of the glint error.

Assuming that the aircraft path is undisturbed, and can be approximated by the mean value of this error signal, if the extent of the tracking error is compared to the physical extent of the aircraft, it is found that the PDF extends past the wingtips, as shown in Figure 8.19.

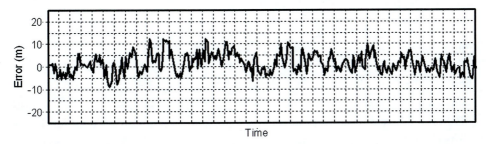

Figure 8.18 Displacement of the tracked centroid of an aircraft echo with time illustrates a phenomenon known as glint.

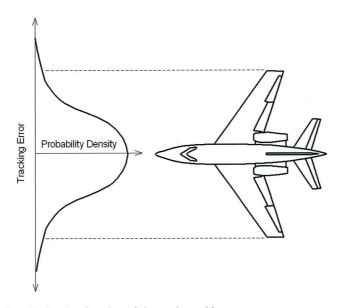

Figure 8.19 Probability distribution function of the angle tracking error.

Measured glint spectra show that a significant portion of the return occurs at frequencies below 3 Hz. This is within the bandwidth of most tracking filters, and so glint can cause serious tracking errors if the appropriate precautions are not taken (see Chapter 13).

In the same way, because all of the scatterers that make up the target are not at the same range, the centroid of the composite return can move back and forth along the body of the aircraft to produce range glint. As with angle glint, much of this variation occurs at low frequency and can induce errors in estimates of both the range and the range rate of the target.

8.9 RADAR STEALTH

8.9.1 Minimizing Detectability

Stealth in this context is the ability of an aircraft to evade radar. However, that is only one of several factors that must be considered in keeping aircraft undetected. The others can be almost as important and include the following:

- Sound,
- Sight,
- Heat,
- Leaking electronic signals.

The drawing of a typical jet fighter, shown in Figure 8.20, indicates the major contributions to the RCS. It can be seen that they include many of the structures that are intrinsic to the design of the aircraft, including engine components, control surfaces, and even the cockpit. Other large contributions also include external structures like drop tanks and ordnance. It is therefore difficult to apply stealth retroactively to an existing design.

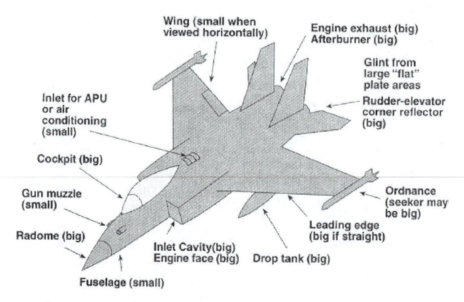

Figure 8.20 Major contributors to RCS of a jet fighter (Fulghum 2001a). Courtesy AIAA.

The basics of stealth have been known since the 1950s and the computational power required to design stealth aircraft has been available since the 1970s. These requirements dictate the shape of modern stealthy aircraft and are summarized in Figure 8.21. Critical to the design are blended components and curved edges, as well as shielded inlet and outlet nozzles among other things.

The first stealth aircraft relied on faceting to reflect power away from its source, as depicted in the graphic of the F-117 shown in Figure 8.22. This technique is effective against monostatic systems, but not against bistatic or multistatic systems where the transmitter and receiver are spatially separated.

In addition to a design shape that minimizes returns, the construction material is important. Multilayered composite structures are made to match the impedance to free space (ensuring that very little energy is reflected). They then progressively absorb it using materials that are loaded with resistive carbon, which is known as radar-absorbing material (RAM).

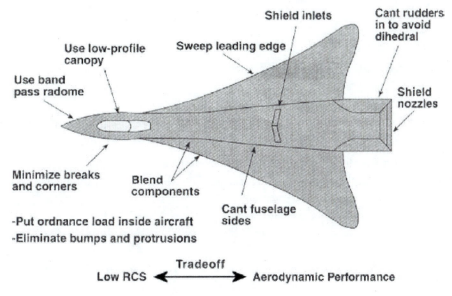

Figure 8.21 Designing aircraft for stealth (Fulghum 2001a). Courtesy AIAA.

Figure 8.22 F-117 faceted stealth design.

Windows are covered with conductive materials to minimize penetration into the cockpit and subsequent reflection from the instruments which would be difficult to make stealthy. Air intakes are covered with mesh, or follow convoluted paths designed to ensure that the radar signal is reflected many times and mostly absorbed before it exits.

Research is being conducted into ways of creating a plasma that can cover the aircraft. Plasma makes an excellent RAM, as is not reflective, but offers high attenuation. Under certain circumstances, a 10 mm thick plasma could reduce the radar reflectivity of the underlying surface by 20 dB.

That the existing monostatic stealth designs are effective is borne out by the table of values in Figure 8.23. Excellent examples are the F-22 and the B-2, both of which have RCS values smaller than those of individual insects.

Hand in hand with stealth development is that of electronic countermeasures. Broadband noise-like sources increase the noise floor and can reduce the effectiveness of a radar, and these and other electronic jammers form a major part of the defensive strategy of modern military aircraft (see Chapter 9 and 13 for more information).

Stealth aircraft cannot operate in total electronic silence, so communications and radar systems must be low probability of intercept (LPI) and the radars must have extremely low sidelobes to reduce the probability of detection still further.

8.9.2 Antistealth Technology

As with most systems developed for military applications, as soon as a new technology appears, countermeasures are developed. Antistealth technology includes the following techniques:

- Radar with wavelengths longer than the aircraft.
- Bistatic and multistatic radar configurations. These can use dedicated transmitters or existing FM or mobile phone broadcasts.
- Wideband radar, as it is difficult to make good wideband RAM.
- Wake and exhaust detection and tracking, as neither of these can be completely eliminated.
- Wingtip vortex detection, as vortices generate turbulence that changes the refractive index of the air, which then reflects radar signals.

Multistatic radar-like systems based on mobile phone technology have been shown to be capable of detecting and pinpointing all modern stealth aircraft to an accuracy of better than

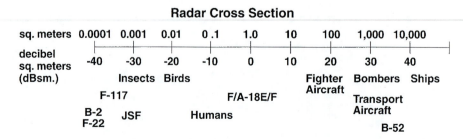

Figure 8.23 Comparison of the monostatic RCS of various objects (Fulghum 2001a).

10 m. The existing mobile infrastructure operates to flood the area with electromagnetic radiation and a series of antennas are deployed to detect it. As a stealth aircraft flies through this environment it alters the received phase of the signal because of multipath, and these fluctuations can be detected by the receive array and exploited to track the aircraft with almost pinpoint accuracy.

One further consideration that is likely to render stealth technology less effective has been the introduction of Ka-band satellite television. This is the band used by the B-2 radar (Figure 8.24), so its radiation would be detected by the ubiquitous network of commercial receivers in people's homes (Fulghum 2001a).

8.10 TARGET CROSS SECTION IN THE INFRARED

Much of the discussion on RCS is directly applicable to the shorter wavelength radiated by lidar, though there are a number of significant differences:

- Laser targets are generally larger than the beam footprint.
- Targets which appear smooth at microwave frequencies will appear rough at the shorter wavelength.
- A diffuse surface in the microwave band may appear as a collection of specular scatterers to a laser.

For a laser, the scattering cross section is

$$\sigma = \rho G A, \tag{8.15}$$

where
σ = cross section (m^2),
ρ = reflectivity of the surface,
G = gain of the target,
A = projected physical area (m^2).

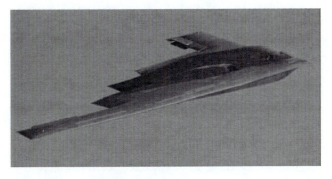

Figure 8.24 B-2 stealth bomber (Courtesy Richard Seaman).

The gain of the target is determined by its area of coherence, A_c (m^2), and the wavelength, λ (m):

$$G = 4\pi A_c / \lambda^2, \tag{8.16}$$

where

$$A_c = \lambda^2 / \Omega, \tag{8.17}$$

and Ω is the solid angle spread of the reflected light around the direction between the laser and the target. Combining equation (8.16) and equation (8.17):

$$G = 4\pi / \Omega. \tag{8.18}$$

On a scattering surface of physical area A (m^2), there are M independent coherence areas such that $M = A/A_c$. For coherent scattering, $M = 1$ or $M \gg 1$. For intermediate values of M, interference between coherence areas must be considered, and that is an effect which is beyond the scope of this book.

For $M = 1$, a specular scatterer, the target cross section is

$$\sigma = 4\pi \rho A^2 / \lambda^2. \tag{8.19}$$

This is similar to the equation for the target cross section of a flat plate described by equation (8.11) with an added term to account for the reflectivity of the material.

For $M \gg 1$, the surfaces become diffuse, and from this, according to Lamberts law, the scattered power density decreases proportional to $\cos(\theta)$, where θ is the angle from the surface normal, as illustrated in Figure 8.25. In this case, the solid angle spread, $\Omega = \pi$, and the target cross section is

$$\sigma = 4\pi \rho A \cos\theta. \tag{8.20}$$

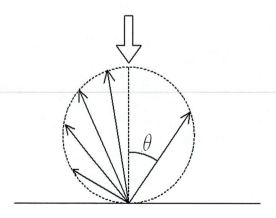

Figure 8.25 Lambertian scatterer.

Instead of using cross section to define the characteristics of a laser target, distributed targets are often characterized by their "effective" reflection coefficient, ρ (as discussed in Chapter 6), and physical cross section. The reflection coefficient is a function of frequency, so the tables reproduced earlier for microwave and 10 μm infrared will not be the same as those for 0.9 μm infrared, shown in Table 8.5. For a diffuse scatterer, the effective reflection coefficient cannot exceed 100%, but for a specular scatterer, the reflection coefficient can be many times this value as it is the product of the actual reflectivity and the gain of the target.

Because of the short wavelengths involved and because a large number of different materials absorb light in the infrared, the drive to produce stealthy aircraft in this band has not been as focused. It can be seen from Table 8.5 that it is only necessary to coat the object with a rough

Table 8.5 Infrared reflectivity of various materials at 0.9 μm

Diffusely reflecting material	Reflectivity (%)
White paper	Up to 100
Cut clean dry pine	94
Snow	80–90
Beer foam	88
White masonry	85
Limestone, clay	Up to 75
Newspaper with print	69
Tissue paper two-ply	60
Deciduous trees	Typ 60
Coniferous trees	Typ 30
Carbonate sand (dry)	57
Carbonate sand (wet)	41
Beach sand and bare desert	Typ 50
Rough wood pallet (clean)	25
Smooth concrete	24
Asphalt with pebbles	17
Lava	8
Black neoprene	5
Black rubber tire wall	2
Specular reflecting material	
Reflecting foil 3M2000X	1250
Opaque white plastic[1]	110
Opaque black plastic[1]	17
Clear plastic[1]	50

[1] Measured with the beam perpendicular to the surface to achieve maximum reflection.

layer of black rubber or neoprene to reduce the reflectivity to between 2% and 5%. Manufacturers of laser range finders generally specify their performance for a target with 80% diffuse reflectivity, therefore a reduction down to 2% reduces the operational range of the sensor to $\sqrt{2/80} = 0.16$ of the specified performance.

8.11 ACOUSTIC TARGET CROSS SECTION

8.11.1 Target Composition

As with electromagnetic radiation, in the acoustic domain all materials will partially reflect, partially absorb, and partially transmit the incident wave. This is quantified by the coefficient of reflectivity, K_r, of the target:

$$K_r = \frac{I_r}{I_i} = \left(\frac{Z_a - Z_o}{Z_a + Z_o} \right)^2, \tag{8.21}$$

where
    I_r = reflected intensity of the sound (W/m^2),
    I_i = incident intensity of the sound (W/m^2),
    Z_a = acoustic impedance of the medium, air or water (acoustic ohms),
    Z_o = acoustic impedance of object, target (acoustic ohms).

As the acoustic impedance, Z_o, is related to the propagation velocity, hard, dense targets tend to reflect well (as their propagation velocity is high), while soft, light targets tend to transmit or absorb. Z_o is not only a function of the target material, but also the material texture. Lower frequencies are often absorbed by porous targets so, to maximize detection, it is generally best to operate at the highest frequency that will propagate effectively through the medium.

8.11.2 Target Properties

While material properties are important at a microscopic level, the acoustic pulse interacts with a relatively large area of the target, so the strength and quality of an echo from the target will also depend on its geometry. For short-range applications where the insonified footprint is smaller than the target extent, shape is important because many targets are specular in nature. As the frequency increases, the effective surface roughness, as a function of the acoustic wavelength, increases and targets become less specular.

One of the major issues with short-range ultrasound sonar applications, particularly in structured environments like the interiors of buildings, is that the walls are specular and the corners are retroreflectors. These characteristics, in conjunction with a significant multipath problem, result in extremely confusing returns. It can be seen in Figure 8.26 that most of the returns are from corners, the walls at normal incidence, or the result of multiple bounce echoes from corners via the walls at oblique angles.

From these results it is obvious that the use of sonar technology for indoor navigation is fraught with difficulty if the observation is made from a single position. However, by fusing the results of multiple scans made from different positions within the room, a better match between the measured and true interior is possible (Probert-Smith 2001).

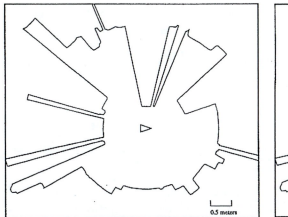 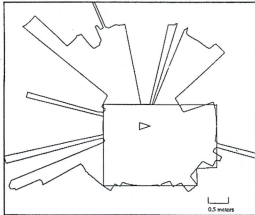

Figure 8.26 Scanned range-azimuth image of a room: (a) sonar data only, (b) room dimensions drawn on the sonar image (Leonard and Durrant-Whyte 1992).

8.11.3 Particulate Targets

Many acoustic systems are used for industrial applications such as level measurement in bins or silos. The target in these instances is often particulate in nature and it has two characteristics that are important regarding the echo strength:

- Small-scale granularity, and
- Large-scale angle of repose and undulations.

Granular particles scatter the reflected wave in all directions, which is essential for an echo return if the material is lying at an angle to the normal. If, however, the particle size is comparable to $\lambda/2$, significant cancellations can occur. As a rule of thumb, the acoustic wavelength should be chosen to exceed the grain size by a factor of four.

If the material surface lies at an angle to the incident acoustic wave, the echo can be reflected away from the transducer toward the walls of the vessel. This can result in the echo return following a zigzag path back to the receiver and an incorrect range reading. With an undulating surface in which the period of the undulations is shorter than the beam footprint, the echo can be directed to follow multiple paths back to the transducer, which can spread the pulse and result in a lower probability of detection.

8.11.4 Underwater Targets

The target strength, TS, is defined as the ratio of the reflected acoustic intensity, I_r (W/m^2), at 1 m from the acoustic center of the target to the incident intensity, I_i (W/m^2), at the same distance

$$TS = 10\log_{10}(I_r/I_i). \tag{8.22}$$

According to Waite (2002), *TS* can be calculated using either peak pressures of the incident, P_i (Pa), and reflected, P_r (Pa), waves or their total integrated energies:

$$TS_{peak} = 20 \log_{10}(P_r/P_i), \tag{8.23}$$

$$TS_{ave} = 10 \log_{10} \frac{\int_0^{T_t} p_r^2(t)\,dt}{\int_0^{T_p} p_i^2(t)\,dt}, \tag{8.23}$$

where $p_i(t)$ and $p_r(t)$ are the time-varying incident and reflected pressures across the acoustic pulse, and where T_t (s) is the extent of the target and T_p (s) is the extent of the pulse.

8.11.4.1 Target Strength of a Sphere   A large sphere with a radius a (m) will intercept $\pi a^2 I_i$ (W) from a plane acoustic wave with an intensity I_i. It can be assumed that the sphere scatters isotropically and so the power density at a radius, r (m), from the center of the sphere will be

$$I_r = \pi a^2 I_i / 4\pi r^2,$$

which reduces to $I_r = a^2 I_i/4$ at the reference distance of 1 m. Therefore, using the definition in equation (8.22), the *TS* (dB) of a sphere can be calculated as

$$TS_{sphere} = 10 \log_{10}(a^2/4). \tag{8.24}$$

8.11.4.2 Target Strength of Other Shapes   If any shape is larger than about 5λ, its *TS* can be determined from a composite of the shapes listed in Table 8.6. To accommodate changes in the incidence angle, θ (rad), the following corrections can be added to the formulas for the rectangular plate or the cylinder:

$$TS_{ang} = 10 \log_{10}\left(\frac{\sin x}{x}\right)^2 + 10 \log_{10} \cos^2 \theta,$$

where

$$x = (2\pi L/\lambda) \sin \theta.$$

Note how similar these are to the RCS equations for the flat plate and the cylinder determined earlier in this chapter.

Table 8.6 Target strengths of some common shapes (Waite 2002)

Shape	TS (dB)	Incidence	Comment
Sphere	$10 \log_{10}(a^2/4)$	Any	a is the radius
Convex surface	$10 \log_{10}(ab/4)$	Normal to surface	a,b are principal radii
Plate of any shape	$20 \log_{10}(A/\lambda)$	Normal to surface	A is the area
Rectangular plate	$20 \log_{10}(aL/\lambda)$	Normal to surface	a,L are sides
Circular plate	$20 \log_{10}(\pi a^2/\lambda)$	Normal to surface	a is the radius
Cylinder	$10 \log_{10}(aL^2/2\lambda)$	Normal to surface	a is radius. L is length

Consider a sea mine that can be approximated by a 2 m long cylinder with a diameter of 1 m and hemispherical ends. The *TS* at a wavelength of $\lambda = 0.15$ m ($f = 10$ kHz) can be determined by the sum of the two contributions shown in Figure 8.27. It can be seen that the specular contribution from the cylindrical section dominates the return around an incidence angle of 0°, and then the constant *TS* of the cylindrical end cap dominates at larger angles.

The *TS* of torpedoes and submarines can be approximated using the same techniques, although modern submarines are designed with some stealth features. For example, a moderately large Kilo class submarine can have a displacement of 3000 tonnes with dimensions of 74 m × 10 m × 6.6 m and will have a maximum *TS* in excess of 20 dB. However, if the submarine is clad with an appropriate sound-absorbing material, this value can be reduced by 10 to 15 dB. Table 8.7 lists typical *TS* values for a number of other man-made and natural targets.

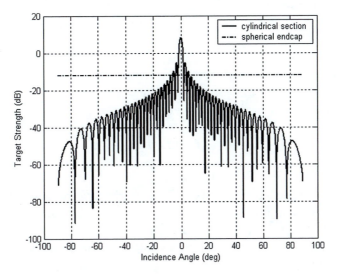

Figure 8.27 Theoretical target strength of a cylindrical sea mine with hemispherical ends as a function of incidence angle.

Table 8.7 Typical values of TS for other targets (Waite 2002)

Target	Aspect	TS (dB)
Surface ship	Beam	+25
	Off beam	+15
Mines	Beam	0
	Off beam	−10 to −25
Torpedoes	Random	−15
Towed array	Beam	<0
Whale 30 m	Dorsal	+5
Shark 10 m	Dorsal	−4
Iceberg	Any	>+10

8.12 CLUTTER

One complicating factor in the study of clutter is that it means different things in different situations. For example, to an engineer developing a missile to detect and track a tank, the return from vegetation and other natural objects would be considered to be "clutter." However, a remote sensing scientist would consider the return from natural vegetation to be the primary target. Clutter is thus defined as return from a physical object or a group of objects that is undesired for a specific application. Clutter may be divided into sources distributed over a surface (land or sea), within a volume (weather or chaff), or concentrated at discrete points (structures, birds, or vehicles).

The magnitude of the signal reflected from the surface back to the receiver is a function of the material, the roughness, and the grazing angle. There are three primary scattering types into which clutter is generally classified. These are specular, retro, and diffuse, as shown in Figure 8.28.

8.12.1 Ground Clutter

Because of the statistical nature of clutter, the mean reflectivity is most often quoted. A convenient mathematical way to describe this mean value for surface clutter is the constant γ model (Barton 1988) in which the surface reflectivity is modeled as

$$\sigma^\circ = \gamma \sin \Psi, \tag{8.25}$$

where σ° is the reflectivity (cross section per unit area m^2/m^2), Ψ is the grazing angle at the surface (deg), and γ is a parameter describing the scattering effectiveness.

Figure 8.29 shows that at low grazing angles the measurements fall below the model because of propagation factor effects, and at high grazing angles the measured reflectivity rises above the value predicted by the model because of quasi-specular reflections from surface facets. However, in the region between 5° and 60°, the model is reasonably accurate.

For different surface types, the following values are typical:

- Values for γ between −10 and −15 dB are widely applicable to land covered by crops, bushes, and trees.

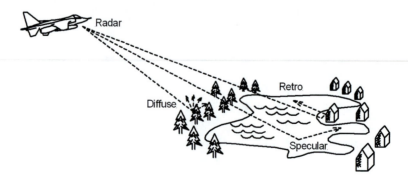

Figure 8.28 Three types of clutter: diffuse, specular, and retro.

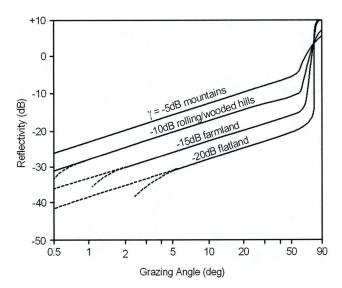

Figure 8.29 Effect of grazing angle on clutter reflectivity for different clutter types [adapted from Barton (1988)].

- Desert, grassland, and marsh are more likely to have γ near −20 dB.
- Urban or mountainous regions will have γ near −5 dB.

These values are almost independent of wavelength and polarization, but they only apply to modeling of mean clutter reflectivity and not fluctuation levels.

8.12.2 Spatial Variations

At microwave frequencies, the spatial distributions of reflectivity within uniform areas are Rayleigh distributed except for elevation angles of less than 5° (Boothe 1969). Where the statistics of the radar return from a number of terrain types are considered, the distribution appears more nearly log-normal and the standard deviation is greater. To confirm this hypothesis, a custom suite of programs was written to examine millimeter wave radar images in minute detail with the specific goal of determining the differences in reflectivity of various forms of clutter at low grazing angles.

The following measurements were made from a radar system mounted at a height of 3 m operating across relatively flat ground near an airfield. This equates to a grazing angle of about 1.1° at a range of 150 m. Figure 8.30 shows a section of runway and taxiway being analyzed (Brooker 2005).

A crude representation of the raw statistics of the highlighted rectangular section of the image is shown in the histogram at the top right of the figure. It can be seen that the return levels from this particular piece of clutter are roughly Gaussian with a mean of −60 dBm. The cursor arrow marks a single image pixel on the tarmac intersection of two taxiways. The small image on the right shows a magnified view of this area, and the details below it are for the individual pixel being measured.

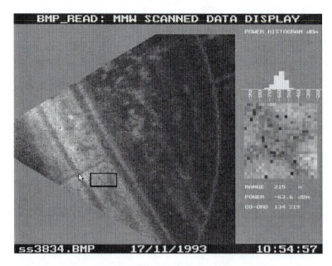

Figure 8.30 Image reflectivity analysis tool output showing the captured file ss3834.bmp which consists of a section of runway with two intersecting taxi ways and the surrounding scrub.

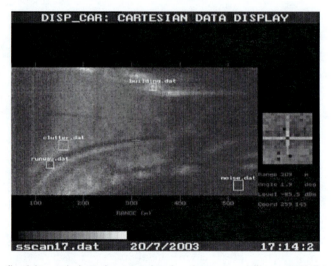

Figure 8.31 Image reflectivity analysis tool output showing the captured file sscan17.dat with various sections marked for storage and further processing.

Using another program, the outputs of which are shown in Figure 8.31, selected areas of the image can be highlighted and then stored (along with their range) for further processing in MATLAB. This allows for a more comprehensive contrast analysis to be made between the reflectivities of different target types.

The received power is a function of the target reflectivity, the antenna gain, the beamwidth, and the receiver bandwidth (gate size). To calibrate the system, a reference target reflector with an RCS of 10 m^2 was placed at a range of 140 m, aligned perfectly using a telescopic sight, and the magnitude of the return logged. Because the terrain between the radar and the reflector was rough grass and scrub, it was assumed that multipath interference would be insignificant and so no additional precautions were taken. From the received power level of −26 dBm, it is simple to produce the appropriate formula to convert from received power to reflectivity.

For a range gate size of 3 m, corresponding to the 3 kHz radio frequency (RF) bandwidth of the spectrum analyzer, and a 3 dB beamwidth, as stated previously, of 0.85°, the conversion is

$$10\log_{10}\sigma^\circ = 10\log_{10}P_r - 36 + 30\log_{10}R, \qquad (8.26)$$

where σ° (m²/m²) is the reflectivity, P_r (dBm) is the received power, and R (m) is the range. This can be confirmed using the relationship between the radar parameters and the area of each clutter cell using the relationship defined in Figure 8.39.

The reflectivity data PDFs for these boxed areas are shown, corrected for range, in Figure 8.32. In this example, the median value for the runway reflectivity is −47 dB, with the clutter some 10 dB higher at −37 dB and the building higher still at −22 dB. The standard deviations of the runway and clutter reflectivity are similar at about 5 dB, with the building a bit larger at 7 dB.

The spatial distributions of the target reflectivity for both the runway (tarmac) and the clutter (rough grass and low scrub) are log-normal in shape, an observation which has been confirmed by other researchers (Foessel 2002). The return from the building, which is characterized as a reflectivity value for purposes of comparison even though it is more likely to be the return from a few high RCS points, has a less well-defined distribution that appears to be made up from two distinct reflectors with different cross sections.

The statistical data from Currie et al. (1992) shown in Figure 8.33 for grass and crops confirm the variation in reflectivity measured above, but models the mean reflectivity to be more than

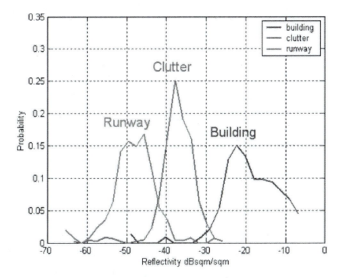

Figure 8.32 Range-corrected histograms of reflectivity distribution for the different target areas show the typical log-normal distributions obtained for millimeter wave clutter in the specular, diffuse, and retroreflective regions.

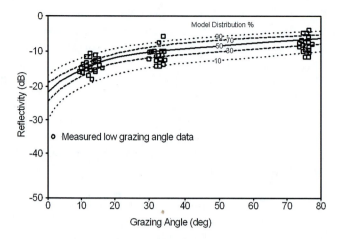

Figure 8.33 Comparison between the measured reflectivity of grass and crops at 95 GHz with the Georgia Tech statistical model predictions. The mean value of the measured low gazing angle data is also presented [adapted from Currie et al. (1992)].

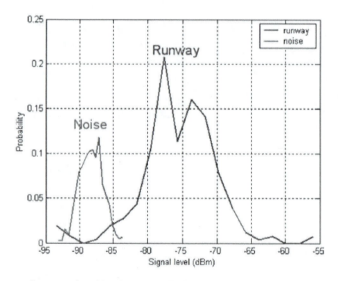

Figure 8.34 Histograms of runway signal level compared with the noise floor confirm that the runway reflectivity statistics are not noise limited.

10 dB higher than the data shown in Figure 8.32. This is probably due to the fact that the data for low grazing angles are extrapolated from measurements made above 10°, whereas the data presented in the previous section are measured at a grazing angle of about 1°.

The noise floor is not a function of the range and so it cannot be plotted on the reflectivity data axes. As it defines the minimum level below which measurements cannot be made, its PDF is plotted along with that of the runway. It can be seen from Figure 8.34 that the returns from the runway are still more than 10 dB above the noise floor and thus the reflectivity measured for this and all of the other target types is valid.

8.12.3 Temporal Variations

Temporal fluctuations are those variations that occur within a single resolution cell with time. They are described by amplitude statistics and frequency spectra or decorrelation times. Measurements made on the same areas at 10, 16, 35, and 95 GHz show that both the standard deviation and the shape of the distribution changes with increasing frequency. Whereas at 10 GHz the distribution is generally Rayleigh in shape, at 95 GHz it more closely approximates log-normal, with its associated longer tails (Currie et al. 1987). It should be remembered that these statistics are approximations only and in reality the data seldom fit a specific distribution exactly. Variations will be a function of a number of parameters including processing bandwidth, radar resolution, scene content, geometry, etc.

Temporal analysis of a typical area that might be traversed by an autonomous vehicle was undertaken. The clutter area chosen was predominantly grassland with the occasional bush. The pattern of the grass growth was irregular, with some clumps and strips reaching a height of 500 mm and other areas much shorter (100 mm).

A section of clutter with a reasonably small spatial variation at a distance between 250 and 350 m was used to determine the extent of temporal variations in reflectivity. Returns from this area were logged from a radar overlooking the scene and the results displayed as shown in Figure 8.35.

The pseudo-image in the figure is a representation of the same clutter slice measured at intervals of just over 2 s for 1 min. These data, converted to a reflectivity, are averaged to produce the mean value at each range, which is displayed in decibels. Minimum and maximum values are also determined and displayed in the adjacent graph.

The temporal variation is measured to have a standard deviation of between 2 and 3 dB with a peak-to-peak variation on either side of the mean of between −6 dB and 4 dB. Because the data

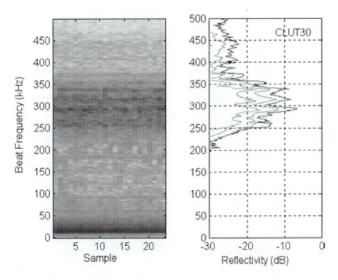

Figure 8.35 Measured clutter reflectivity from grassland and scrub showing (a) the time-range pseudo-image and (b) the variation in reflectivity levels (minimum, median, and maximum).

were only sampled every 2.5 s, the temporal clutter correlation period cannot be determined from these measurements.

At low frequencies, Doppler-shift spectra due to clutter movement (Currie et al. 1992) can be represented by a Gaussian function of the form

$$W(f) = W_o \exp\left(-af^2 / f_t^2\right), \tag{8.27}$$

where

W = power density (W/Hz),
W_o = power density at zero velocity (W/Hz),
f_t = cutoff frequency (Hz),
a = experimentally determined coefficient.

For many clutter types, particularly trees, a physical mechanism exists that generates two different spectral signatures (a low-frequency component due to limb movement and a high-frequency component due to leaf flutter), as shown in Figure 8.36. The low-frequency distribution is modeled by the Gaussian function described in equation (8.27), while the high-frequency component is modeled using the following Lorentzian function:

$$W(f) = \frac{A}{1 + (f/f_c)^n}, \tag{8.28}$$

where A is an empirically derived constant, f_c (Hz) is the cutoff frequency, and n is also an empirically derived factor (dependent on the frequency). At 94 GHz, typical values for $f_t = 3.5$ Hz, $f_c = 35$ Hz, $n = 2$, and the breakpoint is 18 dB for high wind speeds. At low wind speeds, the low-frequency component remains unchanged but the high-frequency cutoff reduces to 6 Hz.

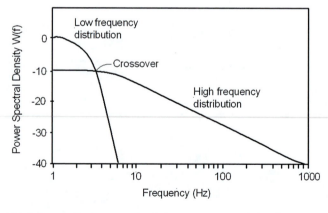

Figure 8.36 Low- and high-frequency components of the power spectral density returns for wind-blown trees showing differences in the spectra that can be attributed to leaves and branches [adapted from Currie et al. (1992)].

8.12.4 Sea Clutter

When the constant γ model is applied to sea clutter (Barton 1988), averaging over all wind directions it is found that γ depends on the Beaufort wind scale, K_B (see Table 8.8), and the wavelength, according to an empirical relationship,

$$10\log\gamma = 6K_B - 10\log\lambda - 64. \tag{8.29}$$

At millimeter wavelengths, even tiny structures become important, and it is no longer possible to model gravity wave effects while neglecting surface ripples and other fine structures. Reflectivity models become complicated functions of frequency, sea state, incidence angle, wind direction, polarization, and other factors as discussed in Currie et al. (1987).

Figure 8.37 shows a composite of data derived from a number of measurements. It does not correspond to any particular set of data, but represents the reflectivity trends at X- and L-band for all directions and in wind speeds between 10 and 20 knots (Skolnik 1970). The Ka-band data are from a more limited set of data (Currie et al. 1992). In this figure, the subscript VV means that both the transmitted and received signals are vertically polarized. The same applies to HH, but for the horizontal. It is possible to measure the cross-polar RCS, HV or VH, as well.

At low grazing angles, typically less than 10°, sea clutter takes on an interesting characteristic. Sharp clutter peaks, known as sea spikes, with reflectivities 10 to 20 dB higher than the mean begin to appear at moderate to high wind speeds.

Below a critical angle ($\approx 1°$) researchers have noted a sharp drop in reflectivity, probably due to the interference of direct and reflected signals. This interaction can account for an R^{-7} decrease in reflectivity with range which is sometimes observed. However, at other times, no critical angle occurs, and sometimes the power relationship is different (Long 1983; Skolnik 1990). From a

Table 8.8 The Beaufort scale

Beaufort no.	Description	Wind speed (knots)
0	Calm	<1
1	Light air	1–3
2	Light breeze	4–6
3	Gentle breeze	7–10
4	Moderate breeze	11–16
5	Fresh breeze	17–21
6	Strong breeze	22–27
7	Near gale	28–33
8	Gale	34–40
9	Strong gale	41–47
10	Storm	48–55
11	Violent storm	56–63
12	Hurricane	>64

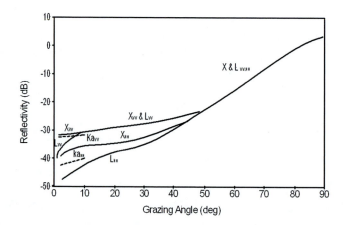

Figure 8.37 Effect of grazing angle on sea clutter reflectivity in a moderate sea [adapted from Skolnik (1970) and Currie et al. (1992)].

temporal perspective, at low grazing angles, sea clutter spectra are normally distributed, with the peak frequency of the upwind spectrum determined primarily by the orbital velocity of the largest sea waves and a wind-induced component.

8.13 CALCULATING SURFACE CLUTTER BACKSCATTER

Target detection in a clutter environment depends on identifying the differences in the target return from those of the surrounding clutter. These differences may simply be due to differences in the magnitudes of the two signals as determined by their relative RCS values, or by differences in polarization, Doppler signature, or a combination of effects.

The cross section, σ, of the surface clutter is the product of the reflectivity, σ^o, and the illuminated area, A (m^2):

$$\sigma = \sigma^o A. \tag{8.30}$$

The accurate measurement of surface reflectivity is fraught with difficulty. To begin with, two possible geometries must be considered when calculating the measurement area, depending on whether the whole ground echo falls within one range cell or many.

The beamwidth-limited resolution case shown in Figure 8.38 occurs when

$$\tan \varphi > \frac{2\pi R \tan(\theta_{EL}/2)}{\alpha c\tau/2}. \tag{8.31}$$

It can be seen that the illuminated area is defined by the intersection of the conical beam with the surface of the target. Due to the geometry and the length of the transmitted pulse, it is assumed that all of the power reflected from this surface contributes to the RCS of the target, and so the area can be approximated by an elliptical footprint,

$$A = \pi r_1 r_2. \tag{8.32}$$

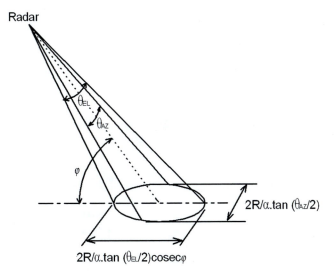

Figure 8.38 Definition of a beamwidth-limited resolution cell.

From the geometry of the figure, this can be expanded to

$$A = \left(\frac{\pi R^2}{\alpha^2}\right)\left(\tan\frac{\theta_{AZ}}{2}\right)\left(\tan\frac{\theta_{EL}}{2}\right)\csc\varphi, \tag{8.33}$$

where

A = area illuminated on the surface (m²),
R = slant range to the surface (m),
α = beam shape factor = 1.33 for a Gaussian-shaped beam,
θ_{AZ} = azimuth for a 3 dB two-way beamwidth (rad),
θ_{EL} = elevation for a 3 dB two-way beamwidth (rad),
φ = grazing angle (rad).

This operational mode can be extended to encompass the measurement of the returns from walls and trees if the appropriate geometry is applied, and the tree surface is assumed to be impenetrable by the radar signal.

One alternative is to gate the radar signal in range, but once again, this is not satisfactory unless a well-defined surface is being measured, as shown in the range gate-limited case of Figure 8.39. If the projected pulse width (gate size) is sufficiently short compared to the length of the elliptical footprint defined by the elevation beamwidth, then the area can be approximated by a rectangle and the formula for the area, A (m²), is

$$A = \frac{Rc\tau}{\alpha}\left(\tan\frac{\theta_{AZ}}{2}\right)\sec\varphi, \tag{8.34}$$

where c is the speed of light (3×10^8 m/s) and τ (s) is the transmitted pulse width.

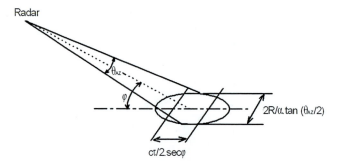

Figure 8.39 Definition of a pulse-width (range-gate)-limited resolution cell.

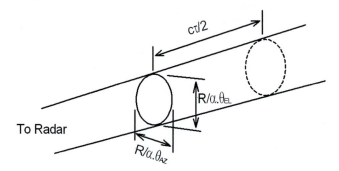

Figure 8.40 Definition of a volume clutter resolution cell.

8.14 CALCULATING VOLUME BACKSCATTER

For the measurement of foliage, rain, or dust where some penetration occurs, the volume clutter formulation is appropriate. In this formulation, a normalization parameter analogous to $\sigma°$ called η exists for volume scatterers. This volume reflectivity is defined as the RCS per unit illuminated volume (m^2/m^3).

As with the surface cases, the illuminated volume can be calculated from the geometry of the beam and the range gate or pulse-width characteristics shown in Figure 8.40. A good approximation at long range is the volume, V (m^2), of an elliptical cylinder with diameters $R\theta_{AZ}$ (m) and $R\theta_{EL}$ (m) and with length $c\tau/2$ (m). As with the previous cases, a correction factor, α, is applied for a Gaussian beam shape:

$$V = \pi R^2 \theta_{AZ} \theta_{EL} c\tau/8\alpha^2. \tag{8.35}$$

The cross section, σ (m^2), is then the product of the volume reflectivity, η (m^2/m^3), and the volume, V (m^3):

$$\sigma = \eta V. \tag{8.36}$$

Because the values of η are generally very small, it is often convenient to work with the decibel form: $10\log_{10}\eta$.

This model assumes that there is minimal attenuation of the radar signal over the length of the cell. This may be a valid assumption at lower frequencies, where foliage and rain penetration is good, but it is not so in the millimeter wave band, where the measured two-way attenuation in dry foliage exceeds 4 dB/m and is even higher in dense green foliage. In these instances, the backscatter must take into account the attenuation of the signal as it propagates through the medium.

8.14.1 Rain

Figure 8.41 shows the theoretical values for the backscatter as a function of rainfall rate at different frequencies (Currie et al. 1987). It is interesting to note that at $\lambda = 1$ mm, the reflectivity hardly increases with increasing rain rate; it must therefore be independent of liquid water content.

These data are determined using the relationship between the reflected and incident power on small spherical targets, as discussed earlier in the section on the RCS of a sphere. Although a given rainfall rate does not imply a specific drop size distribution, the trend for the drops to get bigger as the rainfall rate increases generally holds true.

8.14.2 Dust and Mist Backscatter

The mass loading of dust or mist that can be supported in the atmosphere is extremely small and so the backscatter can often be neglected for electromagnetic radiation with wavelengths of 3 mm or more. However, under certain circumstances, if the dust density is very high (such as in rock crushers) or if the propagation path through the medium is very long (in dust storms), then the total backscatter can be significant.

Figure 8.42 shows the measured backscatter through dust at 10 GHz and 35 GHz generated by an explosion. In theory, the backscatter is proportional to λ^{-4}, therefore the difference between the two graphs should be $40 \log_{10}(35/10) = 21.8$ dB. This is true after about 15 s, when most of

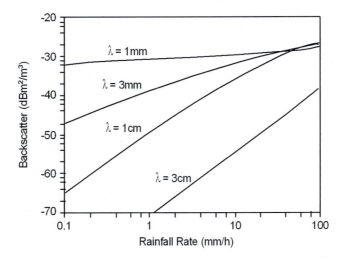

Figure 8.41 Theoretical raindrop backscatter as a function of rainfall rate using Marshall-Palmer drop size distribution.

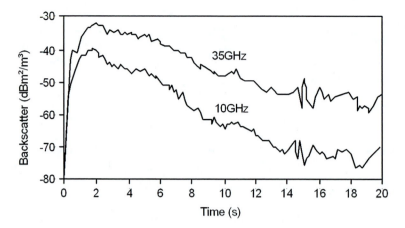

Figure 8.42 Backscatter from dust after explosion at 10 GHz and 35 GHz [adapted from Martin (1980)].

the larger particles have settled, but just after the explosion the difference is less than 10 dB, implying that the dominant scattering mechanism is not Rayleigh, but Mie or even optical.

Under most circumstances the relationship between the particle radius, a, and the wavelength, λ, is such that the Rayleigh approximation can be applied ($2\pi a/\lambda < 1$). In this case the backscatter for a cloud of dust particles or water droplets can be approximated to be (Ghobrial and Sharief 1987; Goldhirsh 2001)

$$\eta = \frac{\pi^5}{\lambda^4}|K_o|^2 Z, \tag{8.37}$$

where K_o is related to the relative dielectric constant of the particle and the Z factor is determined from the number and size of the particles:

$$|K_0|^2 = \left|\frac{\varepsilon - 1}{\varepsilon + 2}\right|^2, \tag{8.38}$$

$$Z = \sum_i D_i^6 N_i \Delta D, \tag{8.39}$$

where $N_i(D)\Delta D$ is the number of particles whose diameters are between D_i and $D_i + \Delta D$, and ε is the complex dielectric constant of the particle.

The equation for Z can be rewritten in terms of the total number of airborne particles per unit volume, N_T, and P_i, which is the probability that a particle has a diameter between D_i and $D_i + \Delta D$, per unit volume:[1]

$$Z = N_T \sum_i D_i^6 P_i. \tag{8.40}$$

---

1 Also called the number fraction distribution.

The total number of particles per unit volume can also be expressed in terms of the mass loading

$$M = \frac{4}{3}\pi\rho\sum_i N_i r_i^3 = \frac{4}{3}\pi\rho N_T \sum_i P_i r_i^3. \qquad (8.41)$$

As discussed in Chapter 7, in a rain or dust storm the visibility can be related to the mass loading or the liquid water content (g/m^3) in the air by

$$M = C/V^\gamma, \qquad (8.42)$$

where M (g/m^3) is the mass loading of dust/water, V (m) is the visibility, and C and γ are constants that depend on the particle type and the meteorological conditions. Typical values for these constants are $C = 37.3$ and $\gamma = 1.07$ (Ghobrial et al. 1987; Goldhirsh 2001).

Using the mass loading formulation in equation (8.42) modified to include the density, ρ (kg/m^3), of the material and solving for N_T, the following is obtained:

$$N_T = \frac{3}{4\pi}\cdot\frac{C}{\rho}\cdot\frac{1}{V^\gamma}\cdot\frac{1}{\sum_i P_i r_i^3} = \frac{2.25\times10^{-9}}{V^{1.07}\sum_i P_i r_i^3}. \qquad (8.43)$$

Substituting into equation (8.40):

$$Z = \frac{2.25\times10^{-9}}{V^{1.07}}\frac{\sum_i P_i D_i^6}{\sum_i P_i r_i^3}. \qquad (8.44)$$

Estimates are made for the two summation terms based on typical size distributions determined for sandstorms (Ghobrial et al. 1987):

$$\sum_i P_i r_i^3 = 4\times10^{-14}\ (m^3),$$

$$\sum_i P_i D_i^6 = 2\times10^{-24}\ (m^6).$$

Substituting back into equation (8.37) to obtain a formula for the backscatter as a function of the wavelength, the dielectric constant, and the visibility becomes

$$\eta = \frac{\pi^5}{\lambda^4}|K_o|^2\cdot\frac{1.125\times10^{-19}}{V^{1.07}}. \qquad (8.45)$$

Figure 8.43 shows the backscatter at 35 GHz plotted for rock dust with a density of 2.44 g/cm^3 and a dielectric constant, ε_r, of $5.23 - j0.26$ as a function of the visibility. The second graph shows

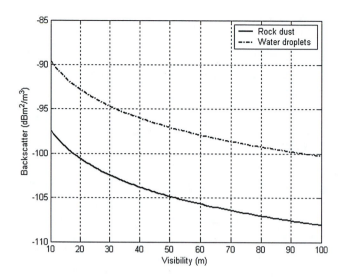

Figure 8.43 Backscatter from rock dust and water with identical particle size distributions as a function of the visibility at 35 GHz.

the backscatter for water spray with a density of 1 g/cm^3 and a complex dielectric constant, ε_r, of $8.45 - j15.45$. In each case the particle/droplet size distributions are assumed to be the same.

A comparison of these results, and the measured data presented in Figure 8.42, point to a much higher backscatter for the explosion-generated dust than is predicted from the theoretical results derived for dust storms. These differences can be explained by the fact that the dust particles carried in a dust storm are much smaller than those generated by an explosion.

As the frequency is increased beyond the millimeter wave into the infrared and the backscatter mechanism moves out of the Rayleigh region into the Mie region, total backscatter increases significantly and the medium becomes more opaque. This is the reason that clouds look white at optical frequencies, whereas even heavy rain is more transparent even though the mass loading of rain is often higher than that of clouds.

8.15 SONAR CLUTTER AND REVERBERATION

8.15.1 Backscatter

The scattering strength from the sea bottom is a function of the acoustic frequency, the grazing angle, roughness, and the material make-up. The characteristics encountered are similar to those shown in Figure 8.28, being diffuse, specular, and retroreflective. In most cases, at frequencies below 10 kHz, specular scattering is associated with mud or sand, while diffuse scattering is associated with a rocky bottom. With increasing frequency, however, particle size plays a large role and a sea bottom comprising sand or pebbles becomes a diffuse scatterer.

Backscatter from the sea surface is a function of frequency and the sea state, which is to a large extent determined by the wind speed. The backscatter strength, S_b (dB), is defined as the ratio of the scattered intensity to the incident intensity per unit area:

$$S_b = 10\log_{10}(I_{scat}/I_i). \tag{8.46}$$

Table 8.9 Typical backscatter strengths at low grazing angles for low-frequency sonar (from Waite 2002)

Material	Backscatter strength, S_b (dB)
Mud	−40 to −45
Sand and shingle	−32 to −40
Pebbles and rock	−24 to −32
Sea surface (sea state 6)	−33 to −43
Sea surface (sea state 2)	−45 to −58

Typical backscatter strengths at frequencies below 10 kHz and grazing angles of less than 10° are given in Table 8.9 (Waite 2002).

The target strength due to backscatter is determined by converting the illuminated area to a decibel value and adding it to the backscatter strength, as shown in Figure 8.38 and Figure 8.39. For range gate-limited imaging this can be simplified to

$$TS_b = 10\log_{10}\frac{c\tau}{2}R\theta_{AZ} + S_b, \qquad (8.47)$$

where c (m/s) is the speed of sound in water, τ (s) is the pulse width, R (m) is the range, and θ_{AZ} (rad) is the azimuth beamwidth.

8.15.2 Volume Reverberation

This is identical in form to the volume backscatter formulation defined for radar systems, with the exception that the beam shape correction is seldom applied. For a narrow-beam, short-pulse sonar, the target strength will be approximately

$$TS_v = 10\log_{10}\frac{c\tau}{2}\frac{\pi R^2\theta_{AZ}\theta_{EL}}{4} + S_v, \qquad (8.48)$$

where S_v (dB) is the volume backscatter strength, and is primarily a function of the amount of solid material insonified by the beam. Typically $S_v \approx -80$ dB.

8.16 WORKED EXAMPLE: ORE-PASS RADAR DEVELOPMENT

The development of a device to determine the depth of rock in an ore pass requires an intimate knowledge of propagation mechanisms through dust and the reflectivity characteristics of rock.

8.16.1 Requirement

- To measure the range from 10 m to the bottom of a 300 m deep, 6 m diameter ore pass so that an estimate (accurate to 1%) can be made of the amount of ore available.

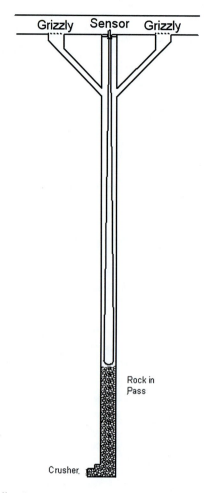

Figure 8.44 Ore pass schematic diagram.

- A typical pass configuration is shown in Figure 8.44.
- The pass will be filled with loose rock, which may be dry or wet.
- There will be lots of dust and the visibility may be as low as 20 m at times.
- A grizzly (coarse grid) at the top of the pass ensures that rocks do not exceed 1 m in diameter.
- The sensor should be capable of operating while rock is being tipped into the pass.
- The range measurement update rate should be sufficiently high to monitor the progress of the rock as it falls down the pass.
- Blasting takes place within 50 m of the radar and the concussion wave that travels through the area is intense.

8.16.2 Selection of a Sensor

To be effective, the sensor must be capable of measuring the range to the target with as little ambiguity as possible. This requires that only the target of interest is illuminated and that a good signal-to-noise ratio (SNR) is maintained. In this example the target is 6 m in diameter and the maximum range is 300 m. This requires that the 3 dB beamwidth of the sensor be less than 1.1° for reasons that are demonstrated later in this example.

At first glance it would seem that a laser range finder is the ideal sensor, but the dust issue is a problem. The mass loading for a visibility of 20 m can be determined from equation (7.10):

$$M_d = 37.3V^{-1.07} = 37.3 \times 20^{-1.07} = 1.5\,\text{g}/\text{m}^3.$$

The extinction coefficient, α (dB/km) is between 0.1 and 0.43 dB/km/mg/m^3. It will therefore be between 150 dB/km and 645 dB/km one way for this dust density. This equates to a minimum two-way attenuation of 90 dB, which makes the use of a laser range finder impractical for this application.

The second low-cost alternative is to use ultrasound. Consider that the maximum-size transducer that could possibly be used has a diameter of 1 m. The longest wavelength that could be used to produce a beamwidth of 1.1° is

$$\lambda = \theta_{3dB}/70 = 0.016\,\text{m}.$$

For a sound velocity of 340 m/s, this equates to a minimum frequency of 21.6 kHz, which corresponds to an attenuation coefficient of about 0.6 dB/m one way as read from the graphs in Figure 7.16. Over a range of 300 m, this results in a two-way attenuation of 360 dB, which again is completely impractical. By default, the only sensor that could possibly perform this function is radar.

8.16.3 Range Resolution

From the size of the grizzly, it can be deduced that large rocks may be dropped into the pass, therefore the target surface will not be regular. Large rock diameters and the angle of repose of the rock surface will result in reflections occurring over at least 1.5 m in range. To obtain a measurement accuracy of 1% over a 300 m deep pass requires an accuracy of 3 m or better. To achieve this, a range resolution of about 2 m is selected. This is quite well matched to the target size (to maximize the RCS) and is also less than the required measurement accuracy. To obtain a range resolution of 2 m, the transmitted pulse width, $R = c\tau/2$, and the range gate size ΔR must both be 2 m.

8.16.4 Target Characteristics

The pile of rock may be wet or dry. It can be shown that the RCS, σ, is a function of the relative dielectric constant, ε_r:

$$\sigma = k\left|\frac{\varepsilon_r - 1}{\varepsilon_r + 2}\right|^2.$$

For the rock, $\varepsilon_r = 5.23$, and for water it is 8.45. The ratio of the RCS for wet and dry rock targets is $\sigma_{water}/\sigma_{rock} = 0.51/0.34 = 1.5$ (1.8 dB).

The pile of rock can be described as a number of facets of various sizes and facing in different directions. Scattering from the various facets may add constructively or destructively and thus a large variation in the reflectivity can be expected. From Figure 8.12 it can be seen that the reflection coefficient for rock is about 0.5. If the surface is considered to be completely rough, then it will scatter isotropically on average, making the overall reflectivity, σ°, $1 \times 0.5 = 0.5$ (−3 dB) when the rock is dry. Because both deep fades and large specular returns can occur, variations in the reflectivity will be modeled as a log-normal distribution with the tails extending 15 dB on either side of the mean.

8.16.5 Clutter Characteristics

The walls of the pass are made of the same material as the target. They are also very rough, so we can assume the same variation in reflectivity, but because the grazing angle is much lower, a reduction in the mean reflectivity to −10 dB can be expected.

8.16.6 Target Signal-to-Clutter Ratio

For adequate detection probability, the target signal-to-clutter ratio (SCR) requirements can be determined in a similar manner as the SNR requirements. Assume that at least 13 dB is required for adequate P_d and P_{fa}.

The maximum mean target cross section is the product of the mean reflectivity and the beam footprint, $\sigma = \sigma^\circ A$. This occurs when the beam fills the pass, as shown in Figure 8.45a.

To simplify the calculations, convert everything to decibels. The target area in decibels is just $10\log_{10}(A) = 10\log_{10}(\pi d^2/4) = 14.5$ dBm^2. The mean target RCS is therefore the sum of the area in decibels and the mean reflectivity in decibels, $\sigma_{tar} = 14.5 - 3 = 11.5$ dBm^2.

The clutter area within the same gate as the target echo is a cylinder of the pass with diameter, d, and height equal to the gate size, ΔR. The clutter area is therefore $10\log_{10}(\pi d \Delta R) = 10\log_{10}(37.7)$ $= 15.8$ dBm^2, making the mean clutter RCS $\sigma_{clut} = 15.8 - 10 = 5.8$ dBm^2.

The target SCR is $11.5 - 5.8 = 5.7$ dB, which is too low for a good probability of detection. It is not possible to use integration to improve the effective SCR because the target returns are

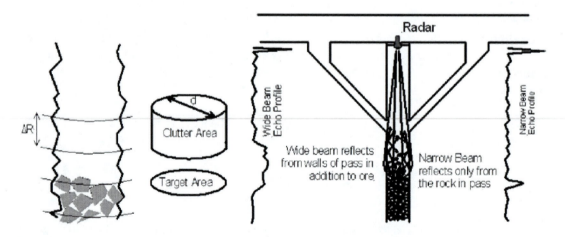

Figure 8.45 Diagrams showing (a) target and clutter areas, and (b) beamwidth effect on echo.

correlated in the same way as the signal returns. Therefore the logical alternative is to ensure that the beamwidth is sufficiently narrow that no reflections are returned from the walls of the pass.

8.16.7 Antenna Size and Radar Frequency

The beamwidth, θ_{3dB} (deg), and the antenna diameter, d (m) (for a circular aperture), are related by

$$d = 70\lambda/\theta_{3dB}.$$

The appropriate operational frequency can be determined, given the following constraints:

- The smaller the antenna the easier it is to mount and align the radar.
- Components costs are proportional to frequency.
- Propagation losses increase proportional to frequency.

It can be seen from Table 8.10 that a frequency of 77 or 94 GHz would be satisfactory.

8.16.8 Radar Configuration

Assuming that the simplest radar possible must be constructed to operate at either 77 or 94 GHz, then for this application a simple noncoherent pulsed radar is appropriate. The schematic diagram for such a system is shown in Figure 8.46.

8.16.9 Component Selection

8.16.9.1 Antenna Options Commercial antennas are available with diameters of 200, 250, and 300 mm. A 250 mm diameter antenna for operation at 94 GHz would be a good choice.

Select a 250 mm diameter Cassegrain antenna from Millitech or Elva or a 250 mm horn lens from Flann Microwave as shown in Figure 8.47. At 94 GHz, the characteristics of the two antenna types are similar, although the Cassegrain antenna sidelobes will be slightly higher because of aperture blockage.

The specifications supplied by the manufacturers can be confirmed by calculation using equation (5.11). For an aperture efficiency of $\rho_A = 0.7$ (typical for a Cassegrain antenna),

$$G = 4\pi\rho_A A/\lambda^2 = 42{,}432\,(46.2\,\text{dB})$$

$$\theta = \varphi = 70\lambda/d = 0.89°.$$

Table 8.10 Antenna diameter as a function of operational frequency

f (GHz)	λ (m)	d (m)	Comment
10	0.03	2.1	Much too large
35	0.0086	0.6	Probably too large
77	0.0039	0.27	Okay
94	0.0032	0.22	Okay

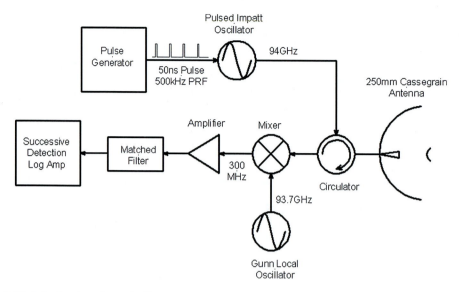

Figure 8.46 Pulsed radar schematic diagram.

Figure 8.47 Millimeter wave parabolic Cassegrain and horn lens antennas.

8.16.9.2 Radar Transmitter The transmitter needs to accommodate an uncompressed pulse width of 2 m which corresponds to, from equation (5.16), a pulse duration of

$$\tau = 2\Delta R / c = 13.3 \text{ ns.}$$

The lowest-cost option at this time is a pulsed radar based on a noncoherent solid-state Gunn or IMPATT diode-based transmitter. The off-the-shelf options from Millitech are as follows:

- Pulsed Gunn: $\tau = 20$ ns to $1000\,\mu$s with a maximum duty cycle of 50% and $P_t = 0.1$ W (20 dBm). Typical chirp 100 MHz.
- Pulsed IMPATT: $\tau = 50$ ns or 100 ns with a pulse repetition frequency (PRF) between 10 and 75 kHz and $P_t = 12$ W (40.8 dBm). Typical chirp 100 MHz.

Neither transmitter meets the 13.3 ns pulse-width requirement. Select the Gunn option as being the closest at 20 ns (3 m), which is still equal to the specified 1% without using interpolation methods to improve the measurement resolution. This would degrade the SCR still further if the beamwidth were wider, but because a sufficiently narrow beam was chosen, increasing the pulse duration slightly has little effect.

8.16.9.3 Receiver Options The receiver configuration could be one of the following:

- RF amp – mixer – intermediate frequency (IF) amp – matched filter ($G = 20$ dB, DSB $NF = 6$ dB)
- Mixer – IF amp – matched filter ($L = 8$ dB, DSB $NF = 7$ dB).

Amplifiers at 94 GHz are still extremely expensive ($\approx$ $15,000 each), so the small noise figure advantage is not justified and the second option will be used.

Local Oscillator The local oscillator requires sufficient power to drive the down-conversion mixer, so at least 13 dBm of drive power is required at 93.7 GHz. A mechanically tuned Gunn oscillator with an output power, P_{out}, of 40 mW (16 dBm) is adequate.

Duplexer For a radar that uses a single antenna, the duplexer switches between the transmitter and the receiver, while protecting the receive port from the high transmit powers. Options include the following:

- 3 dB directional coupler, 20 dB directivity, 1.6 dB transmit insertion loss, and 4.6 dB receive insertion loss.
- Junction circulator, 20 dB isolation, 0.8 dB insertion loss for both transmit and receive paths.

From both insertion loss and isolation (directivity) the circulator is either superior or equal to the coupler. The coupler can handle higher powers, but the circulator is good to 5 W peak, which is sufficient for this application. The circulator is also smaller and lighter than the coupler.

Matched Filter Assuming that the matched filter requirement is satisfied by $\beta\tau = 1$ as described by equation (5.17), for $\tau = 20$ ns, the optimum bandwidth, β, is 50 MHz. Because the transmitter chirps about 100 MHz during the pulse period, using a filter with a bandwidth of 50 MHz would result in a significant loss of received power, $10\log_{10}(50/100) = 3$ dB, if the chirp were linear. However, increasing the bandwidth to 100 MHz to accommodate the full chirp increases the noise floor by 3 dB, so not much is gained. It is very difficult to make a matched filter for the uncontrolled transmitter chirp as it is extremely nonlinear and is a function of a number of factors that are difficult to control, therefore the filter bandwidth of 50 MHz will be maintained.

The Intermediate Frequency   The IF is selected according to the following:

- Amplifier components easy to obtain and low cost.
- The matched filter with a bandwidth of 50 MHz is easy to construct.
- Detectors are available at that frequency.

A typical amplifier would have the following specifications:

- Band 200–400 MHz
- Gain 30 dB
- Noise figure 1.5 dB.

The Transmit and Local Oscillator Frequencies   For the selected IF center frequency of 300 MHz, the transmitter is tuned to operate at 94 GHz and the local oscillator (LO) operates at 93.7 GHz. There is no image filter, so the transmitter could just as well operate at 93.4 GHz, which would result in the same IF.

Dynamic Range Requirements   The system dynamic range requirements are as follows:

- Target RCS variation 30 dB due to physical characteristics.
- Target RCS variation 2 dB due to wet/dry surface.
- Because the area illuminated (and hence the RCS) is proportional to R^2, the range-dependent change in signal level, S_{rec}, as predicted using the radar range equation is a function of R^{-2}.
- Dynamic range = $20\log_{10}(R_{max}/R_{min})$ = 30 dB.

The total echo dynamic range is therefore 30 + 2 + 30 = 62 dB.

Detector Options   The options are shown in Figure 8.48.

- Envelope detector with an STC-controlled variable gain amplifier to minimize the dynamic range requirements of the rest of the system.

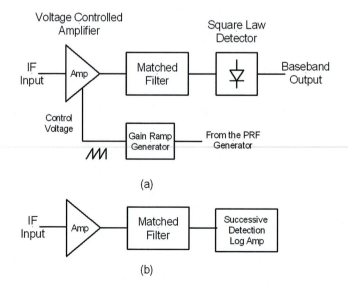

Figure 8.48 Detector options: (a) STC and square-law or (b) successive detection log amplifier.

- Successive detection log amplifier (SDLA) with an instantaneous dynamic range of more than 70 dB and no STC requirements.

Because of the uncertainties in the overall design (RCS levels, etc), the SDLA is selected because its performance is more robust than the detector. It is also easier to interface to the postdetection electronics. A Pascal SDLA has a direct current (DC) voltage output proportional to the input power, as can be seen from its transfer function reproduced in Figure 8.49.

The specifications are as follows:

- Dynamic range: >70 dB
- Tangential sensitivity: −75 dBm
- Pulse rise time: 3 ns
- Pulse decay time: 6 ns
- Transfer function: 25 mV/dB
- Output level: 2 V for a 0 dBm input signal

8.16.10 Signal-to-Noise Ratio

The transmitted power is $P_{tx} = P_{osc} - L_{line} - L_{circ} = 20 - 0.4 - 0.8 = 18.8$ dBm. For the SSB noise figure, if we use the formula that includes the mixer loss $L_m = 8$ dB and an IF amplifier with a noise figure of 1.5 dB as well as line losses, $L_{rec} = L_{line} + L_{circ} = 0.4 + 0.8 = 1.2$ dB:

$$NF_{rec} = L_{rec} + L_m + NF_{IF} = 1.2 + 8 + 1.5 = 10.7 \text{ dB}.$$

Matched filter loss, $L_{match} = 3$ dB, is added to the noise figure, making the total noise figure $NF_{tot} = 13.7$ dB.

8.16.11 Output Signal-to-Noise Ratio

The received power, P_r (dBm), is calculated using the radar range equation which is rewritten in decibel terms:

$$P_r = P_t + 2G + 10\log_{10}\frac{\lambda^2}{(4\pi)^3} + \sigma - 40\log_{10} R.$$

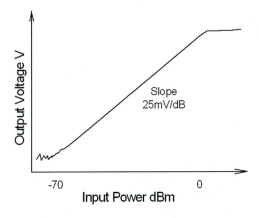

Figure 8.49 The SDLA transfer function.

At the maximum operational range of 300 m, and using the mean RCS of 11.5 dBm^2, the received power, P_r (dBm), is

$$P_r = 18.8 + 2 \times 46 - 82.9 + 11.5 - 99 = -59.6 \text{ dBm.}$$

The noise power, P_n (dBm), for a bandwidth of 50 MHz is

$$P_n = 10\log_{10}(kT\beta) + NF_{\text{tot}} = -127 + 13.7 + 30 = -83.2 \text{ dBm.}$$

The SNR is then $-59.6 - (-83.2) = 23.6$ dB. However, because of fluctuations in the target RCS, the minimum predicted single-pulse SNR may be 15 dB lower than this at

$$SNR_{\text{min}} = 23.6 - 15 = 8.6 \text{ dB.}$$

8.16.12 Required IF Gain

For a good detection probability, the signal must be at least 13 dB above the noise floor, therefore a tangential sensitivity (TSS) of -75 dBm, the lowest signal level, should be $-75 + 13 = -62$ dBm. Given that the actual signal power after down-conversion for the minimum predicted RCS at the longest range would be

$$P_{\text{if}} = P_r - L_{\text{rec}} - L_{\text{m}} - 15 = -59.6 - 1.2 - 8 - 15 = -83.8 \text{ dBm,}$$

a minimum IF gain of at least 21.8 dB would be required.

8.16.13 Detection Probability and Pulses Integrated

Assuming that a detection probability, P_d, of 0.95, and a very low false alarm probability, P_{fa}, of 10^{-12} is needed, then an effective SNR of 16.3 dB must be achieved. To achieve a postdetection integration gain of $16.3 - 8.6 = 7.7$ dB, N pulses must be integrated, where $N = 10^{(7.7/8)} = 9$ pulses.

  Note that this is not altogether true, as the formula was derived for a square-law detector and an SDLA is being used. To compensate, integrate an additional 7 pulses ($N = 16$).

8.16.14 Measurement Update Rate

For a maximum unambiguous range of 300 m, the radar can be operated at a maximum PRF of $c/2R_{\text{max}} = 500$ kHz. With 16 pulses integrated, the update rate for measurement output is reduced to 31.25 kHz.

8.16.15 Monitoring Rock Falling Down the Pass

Assume that the rock that enters the pass accelerates due to gravity until it hits the bottom.

  • There is no terminal velocity due to air resistance.
  • There is no terminal velocity due to friction from the walls of the pass.

By the time the rock reaches 300 m down it will be traveling at 76 m/s. At an update rate of 31.25 kHz, the rock will hardly have moved at all between samples. The Doppler shift will be

$f_d = 2v/\lambda = 39$ kHz, which is a very small fraction of the 50 MHz IF bandwidth, so it can be ignored.

8.16.16 Prototype Build and Test

A prototype pulsed radar unit was built as described and is shown, without its housing, in Figure 8.50. The radar was installed, along with its data acquisition hardware, over an ore pass where it produced strong echoes from the rock, as can be seen from the snapshot shown in Figure 8.51.

Figure 8.50 The prototype ore pass radar.

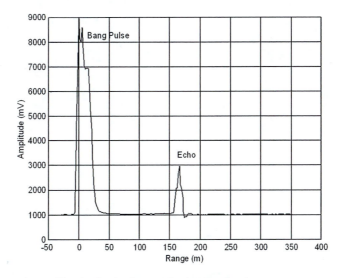

Figure 8.51 Ore pass echo profile obtained using a pulsed W-band radar.

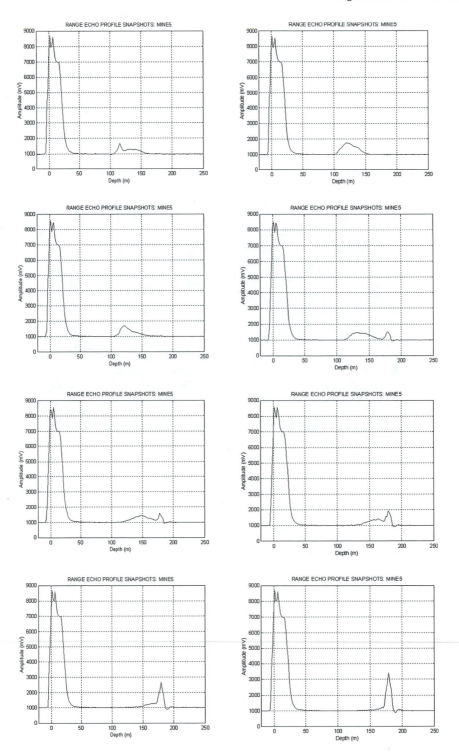

Figure 8.52 Pulsed radar snapshots of rock falling down a pass.

Although it was not possible to log the data continuously at 31 kHz, snapshots of the radar echo profile were stored every few seconds so that the process of ore falling down the pass could be documented. The sequence of images shown in Figure 8.52 clearly shows this process.

It can be seen from the distributed nature of the echoes that the rock is sufficiently spread out for detections to be made from regions within the falling mass. However, the bulk of the ore is sufficiently thick to block the beam completely and sometimes no echo at all is returned from the bottom.

8.17 References

Barton, D. (1976). *Radar Systems Analysis*. Norwood, MA: Artech House.

Barton, D. (1988). *Modern Radar Systems Analysis*. Norwood, MA: Artech House.

Boothe, R. (1969). *The Weibull Distribution Applied to Ground Clutter Backscatter Coefficient*. Huntsville, AL: U.S. Army Missile Command.

Brooker, G. (2005). Long-range imaging radar for autonomous navigation. Ph.D. dissertation. University of Sydney, SYDNEY, NSW, Australia.

Brooker, G., Lobsey, C., and Hennessey, R. (2008). Radar Cross Sections of Small Boats at 94 GHz. Radar-Con 2008, Rome, Italy, May 26–30.

Currie, N. (1984). *Techniques of Radar Reflectivity Measurement*. Norwood, MA: Artech House.

Currie, N. (1989). *Radar Reflectivity Measurement: Techniques and Applications*. Norwood, MA: Artech House.

Currie, N. and Brown, C. (1987). *Principles and Applications of Millimeter-Wave Radar*. Norwood, MA: Artech House.

Currie, N., Hayes, R., and Trebits, R. (1992). *Millimeter-Wave Radar Clutter*. Norwood, MA: Artech House.

Diehl, R. and Larkin, R. (2005). Introduction to the WSR-88D (NEXRAD) for ornithological research. Technical Report PSW-GTR-191, USDA Forest Service, Washington, DC; pp. 876–888.

Foessel, A. (2002). Scene modeling from motion-free radar sensing. PhD dissertation. Robotics Institute, Carnegie Mellon University.

Fulghum, D. (2001a). Counterstealth tackles U.S. aerial dominance. *Aviation Week and Space Technology* February 5:55–57.

Fulghum, D. (2001b). Stealth retains value, but its monopoly wanes. *Aviation Week and Space Technology* February 5:53–55.

Gauthreaux, S. (2003). Radar ornithology and biological conservation. *Auk* 120(2):266–277.

Ghobrial, S. and Sharief, S. (1987). Microwave attenuation and cross polarization in dust storms. *IEEE Transactions on Antennas and Propagation* AP-35(4):418–425.

Goldhirsh, J. (2001). Attenuation and backscatter from a derived two-dimensional dust storm model. *IEEE Transactions on Antennas and Propagation* 49(12):1703–1711.

Hendrick, W. (1992, Spring). Industrial applications of radar technology for continuous level measurement. ISA/92. Proceedings of the 32nd Symposium: Instrumentation in the Pulp and Paper Industry, Canada.

Leonard, J. and Durrant-Whyte, H. (1992). *Directed Sonar Sensing for Mobile Robot Navigation*. Boston: Kluwer.

Long, M. (1983). *Radar Reflectivity of Land and Sea*. Norwood, MA: Artech House.

Martin, E. (1980). Radar propagation through dust clouds lofted by high explosive tests. MISERS BLUFF Phase II: Final Technical Report on Project A2465 for SRI International. Atlanta: Georgia Institute of Technology.

Moon, T. and Bawden, P. (1991). High resolution RCS measurements of boats. *IEEE Proceedings F Radar and Signal Processing* 138(3):218–222.

Paddison, F., Shipley, C., Maffett, A., and Dawson, M. (1978). Radar cross section of ships. *IEEE Transactions on Aerospace and Electronic Systems* AES-14(1):27–33.

Probert-Smith, P., ed. (2001). *Active Sensors for Local Planning in Mobile Robotics*. Hackensack, NJ: World Scientific.

Ridenour, L., ed. (1947). *Radar Systems Engineering: MIT Radiation Laboratory Series*. New York: McGraw-Hill.

Riley, J. (1985). Radar cross section of insects. *Proceedings of the IEEE* 73(2):208–232.

Skolnik, M. (1970). *Radar Handbook*. New York: McGraw-Hill.

Skolnik, M. (1980). *Introduction to Radar Systems*. Tokyo: McGraw-Hill Kogakusha.

Skolnik, M. (1990). *Radar Handbook*. New York: McGraw-Hill.

Trebits, R., Hayes, R., and Bomar, L. (1978). MM-wave reflectivity of land and sea. *Microwave Journal* 21(8):47–49.

Vaughn, C. (1985). Birds and insects as radar targets: a review. *Proceedings of the IEEE* 73(2):205–227.

Waite, A. (2002). *Sonar for Practicing Engineers*. New York: John Wiley & Sons.

9

Detection of Signals in Noise

9.1 RECEIVER NOISE

Noise is the unwanted energy that interferes with the ability of the receiver to detect the wanted signal. It may enter through the antenna along with the desired signal or it may be generated within the receiver itself. In underwater sonar systems, external acoustic noise is generated by waves and wind on the water surface, by biological agents (fish, prawns, etc.), and by man-made sources such as engine noise. In radar and lidar sensors the external electromagnetic noise is generated by various natural mechanisms such as the sun and lightning, among others. Man-made sources of electromagnetic noise are myriad, from car ignition systems and fluorescent lights to other broadcast signals. The deliberate transmission of noise in an attempt to mask a target echo or otherwise deceive a sensor is known as jamming.

9.1.1 Radar Noise

As discussed earlier, noise within the sensor is generated by the thermal motion of the conduction electrons in the ohmic portions of the receiver input stages. This is known as thermal or Johnson noise.

Noise power can be expressed in terms of the temperature of a matched resistor at the input of the receiver. The amount of power, P_N (W), transferred from the resistor into the receiver is

$$P_N = kT_o\beta, \tag{9.1}$$

where k is Boltzmann's constant (1.38×10^{-23} J/K), T_o (K) is the system temperature (usually 290 K), and β (Hz) is the receiver noise bandwidth.

The signal-to-noise ratio (SNR) in practical receivers is always worse than that which can be accounted for by thermal noise alone. This is either because additional noise is introduced by the active input stages of the receiver, or because the input is attenuated due to losses in the signal path.

The total noise at the output of the receiver, N (W), can be considered to be equal to the noise power output from an ideal receiver multiplied by a factor called the noise figure, NF:

$$N = P_N F_N = kT_o\beta NF. \tag{9.2}$$

It can be shown that the noise figure, NF, for a cascaded receiver chain made up of a number of stages, each with gain and individual noise figures (Barton 1976) is

$$NF = NF_1 + \frac{NF_2 - 1}{G_1} + \frac{NF_3 - 1}{G_1 G_2}. \tag{9.3}$$

For the receiver shown schematically in Figure 9.1, the noise figure is determined primarily by the losses between the antenna and the first amplifier, and the contribution of that amplifier. For example, if this initial loss is 0.5 dB, followed by a low-noise amplifier with a noise figure of 1.5 dB and a gain of 20 dB, the third stage is a mixer with a conversion loss of 6 dB.

Because the noise figure of a lossy section is equal to the attenuation, and the gain is the negative of the loss, $NF_1 = 0.5$ dB and $G_1 = -0.5$ dB. For the amplifier, $NF_2 = 1.5$ dB and $G_2 = 20$ dB, and finally, for the mixer, $NF_3 = 6$ dB. These are converted to ratios as follows:

$$NF_1 = 10^{0.5/10} = 1.122$$

$$G_1 = 10^{-0.5/10} = 0.8913$$

$$NF_2 = 10^{1.5/10} = 1.4125$$

$$G_2 = 10^{20/10} = 100$$

$$NF_3 = 10^{6/10} = 3.98$$

$$NF = 1.122 + \frac{1.4125 - 1}{0.8913} + \frac{3.98}{0.8913 \times 100} + \ldots = 1.122 + 0.4628 + 0.0447 + \ldots = 1.6295.$$

The noise figure in this case is determined almost entirely by the first two terms of the calculation, and all subsequent terms are reduced to negligible proportions because of the high gain of that first amplifier. This result should be compared to the one described in Chapter 4, where no amplifier separates the antenna and the mixer.

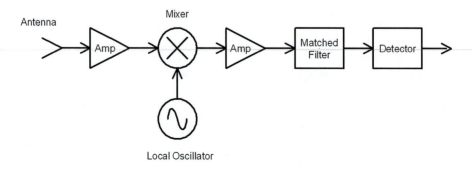

Figure 9.1 Block diagram of a typical heterodyne receiver used in radar systems.

9.1.2 Noise Probability Density Functions

Consider the radar receiver shown in Figure 9.1 which consists of an antenna followed by a wideband amplifier and a mixer that down-converts the signal to an intermediate frequency (IF), where it is further amplified and filtered (bandwidth β_{IF}). This is followed by an envelope detector and further filtering (bandwidth $\beta_V = \beta_{IF}/2$).

The noise entering the IF filter is assumed to be Gaussian, as it is thermal in nature (Skolnik 1980), with a probability density function (PDF) given by

$$p(v) = \frac{1}{\sqrt{2\pi\psi_o}} \exp\frac{-v^2}{2\psi_o}, \tag{9.4}$$

where $p(v)dv$ is the probability of finding the noise voltage v between v and $v + dv$, and ψ_o is the variance of the noise voltage.

If Gaussian noise is passed through a narrow-band filter (one whose bandwidth is small compared to the center frequency), then the PDF of the postdetection envelope of the noise voltage output can be shown to be

$$p(R) = \frac{R}{\psi_o} \exp\frac{-R^2}{2\psi_o}, \tag{9.5}$$

where R is the amplitude of the envelope of the filter output. This has the form of the Rayleigh PDF.

Figure 9.2 shows the PDFs for a simulation of 10^6 samples of Gaussian noise generated in MATLAB, followed by a band-pass filter with a relative bandwidth of $\beta/f_c = 0.2$, detection, and subsequent filtering. Most of the noise generated within the receiver that contributes to the noise factor is also Gaussian.

9.1.3 Infrared Detection and Lidar Noise

As discussed in Chapter 2, there are four common sources of noise that affect infrared (IR) systems, irrespective of whether they are associated with passive or active sensors. They are thermal noise, shot noise, avalanche noise, and $1/f$ noise. In the detector shown in Figure 9.3, it is convenient to determine the magnitude of these sources in terms of the noise current generated, as these individual contributions can be summed to obtain the overall noise contribution.

9.1.3.1 Thermal Noise Thermal noise is generated by the random thermal motion of current carriers (electrons and holes). It is broadband and therefore the root mean square (RMS) noise current, i_{therm} (A) is

$$i_{therm} = \sqrt{\frac{4kT\Delta f}{R_{shunt}}}, \tag{9.6}$$

where k is Boltzmann's constant, T (K) is the temperature, Δf (Hz) is the receiver bandwidth, and R_{shunt} (Ω) is the shunt resistance of the detector.

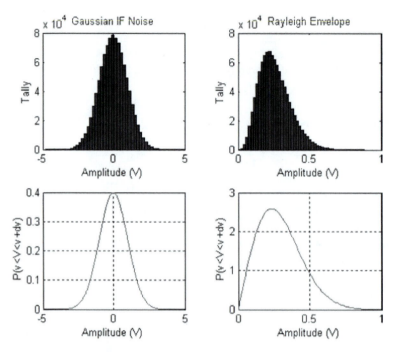

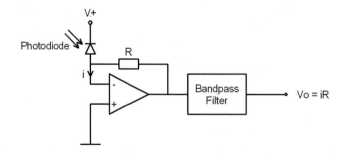

Figure 9.2 Amplitude distributions of thermal noise pre- and postdetection.

Figure 9.3 Block diagram in a typical direct detection receiver for light.

9.1.3.2 Shot Noise   Shot noise is generated from fluctuations in the rates of generation and recombination of carriers. It is also broadband and therefore the RMS current, i_{shot} (A), is

$$i_{shot} = \sqrt{2e\left(i_{photo} + i_{dark}\right)\Delta f},$$ (9.7)

where i_{photo} and i_{dark} are the mean photo and dark currents, respectively, and e is the charge on an electron. It is interesting to note that the magnitude of the shot noise increases with increasing photo current.

It is shown in Chapter 2 that event-driven noise conforms to a Poisson distribution in which the mean value and the variance are equal. However, as the number of events per unit time, γ, increases, the distribution becomes more normal. For $\gamma > 1000$, a normal distribution with equal mean and variance becomes an excellent approximation of the Poisson distribution. Therefore, for most IR systems, it is practical to model the shot noise as Gaussian.

9.1.3.3 Avalanche Noise Avalanche noise occurs in reverse-biased PN junctions at breakdown. It is therefore an important source of noise in avalanche photodiode detectors (APDs). These are discussed in Chapter 6.

9.1.3.4 1/f Noise Generated by surface and contact effects, $1/f$ noise is not well understood and is usually determined empirically for different detector families. It is dominant only at frequencies below 100 Hz.

9.1.3.5 Total Noise Contribution Manufacturers of photodiodes often provide some specifications for the device, including the responsivity, R (A/W), the dark current, i_{dark} (nA), the shunt resistance, R_{shunt} (Ω), and the cutoff frequency. In this case, i_{photo} can be determined from the input power, the responsivity and shot noise can be calculated using equation (9.7), and the thermal noise can be derived from equation (9.6). The output SNR is the square of the ratio of the photocurrent and the total noise current.

As discussed in Chapter 6, the total detector noise contribution can also be calculated from the specific detectivity, D^* (cmHz$^{1/2}$/W), which is often also provided by the photodiode manufacturer. This is used to determine noise equivalent power, NEP (W):

$$NEP = (A_d \Delta f)^{1/2} / D^*,$$ (9.8)

where A_d (cm^2) is the detector area and Δf (Hz) is the receiver bandwidth. The output SNR in this case is the received signal power divided by the NEP.

9.1.4 Sonar Noise

The primary sources of noise in sonar systems are thermal, noise from the sea, and noise generated by vessels.

9.1.4.1 Thermal Noise In common with other sensors, the sonar receiver adds its own noise to the signal that it receives. In this application it is the noise voltage, v_{therm} (V), which is usually calculated (Waite 2002) as

$$v_{therm} = \sqrt{4kTR\Delta f}.$$ (9.9)

In an underwater transducer, thermal noise is generated from the thermal agitation of the molecules of water, which produce pressure fluctuations on the face of the receiver transducer. In a completely "dead" sea this is (Waite 2002)

$$N_{\text{therm}} = -15 + 20\log_{10} f,$$ (9.10)

where N_{therm} (dB at 1 μPa) is the noise pressure and f (kHz) is the frequency.

9.1.4.2 Noise from the Sea At frequencies below 30 kHz, agitations much larger than those produced by thermal effects add to the noise level. These are a function of the sea state and the local rainfall rate. In addition, the shipping density has an effect at low frequencies, as summarized in Figure 9.4.

9.2 EFFECTS OF SNR

The noise levels described in the previous section set the minimum threshold below which a target signal cannot be detected. Because both the noise and the target signals are statistical in nature, target detection and false alarm are both defined in terms of probability of occurrence.

9.2.1 Probability of False Alarm

A false alarm occurs whenever the noise voltage exceeds a defined threshold voltage, V_t, as illustrated in Figure 9.5. Notice from the figure, that the average time interval between crossings of the threshold, called the false alarm time, T_{fa} (s), can be written as

$$T_{fa} = \lim_{N \to \infty} \frac{1}{N} \sum_{k=1}^{N} T_k, \tag{9.11}$$

where T_k (s) is the time between crossings of the threshold, V_t, by the noise envelope (when the slope of the crossing is positive).

The false alarm probability can be defined as the ratio of the average amount of time that the envelope is above the threshold, $<t_k>_{ave}$, to the total time between false alarms,

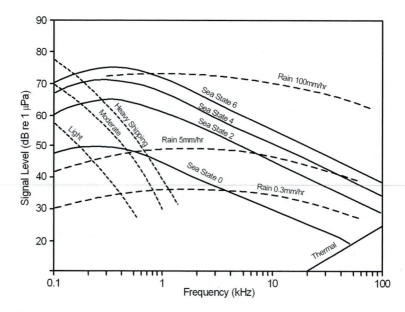

Figure 9.4 Ambient sea noise [adapted from Waite (2002)].

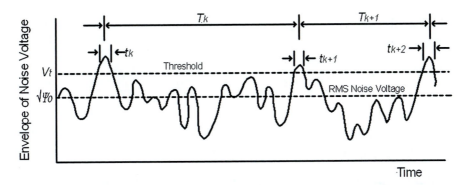

Figure 9.5 Receiver output voltage illustrating false alarms due to noise [adapted from Skolnik (1980)].

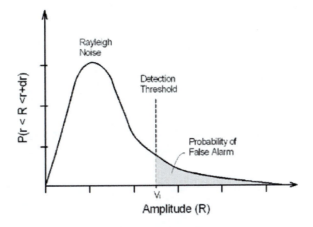

Figure 9.6 Probability of false alarm is determined from the PDF of the noise envelope voltage.

$<T_k>_{ave}$, if the duration of the false alarms is small compared to the total time, $<t_k>_{ave} << <T_k>_{ave}$, then

$$P_{fa} = \frac{\sum\limits_{k=1}^{N} t_k}{\sum\limits_{k=1}^{N} T_k} = \frac{\langle t_k \rangle_{ave}}{\langle T_k \rangle_{ave}}. \tag{9.12}$$

If the average duration of a noise pulse is the reciprocal of the receiver bandwidth, β (Hz), and $<T_k>_{ave} = T_{fa}$, then equation (9.12) can be rewritten as

$$P_{fa} = 1/T_{fa}\beta. \tag{9.13}$$

The postdetection envelope of the noise voltage for a radar signal can be described by the Rayleigh distribution in equation (9.5). Therefore the probability of a false alarm occurring is equal to the shaded area shown in Figure 9.6, and this can be determined analytically by integrating the noise power PDF:

$$P_{fa} = \text{Pr}\,ob\,(V_t < R < \infty) = \int_{V_t}^{\infty} \frac{R}{\psi_o} \exp \frac{-R^2}{2\psi_o} \, dR$$

$$= \exp \frac{-V_t^2}{2\psi_o} \tag{9.14}$$

For a bandwidth $\beta = \beta_{IF}$, the false alarm time, T_{fa} (s), can be determined from equation (9.13) and equation (9.14) to be

$$T_{fa} = \frac{1}{\beta_{IF}} \exp \frac{V_t^2}{2\psi_o}. \tag{9.15}$$

The false alarm times of practical sensors must be very large (usually a couple of hours), so the probability of false alarm must be very small, typically $P_{fa} < 10^{-6}$.

9.2.2 Example

The noise output variance of a surveillance radar is $1\ V^2$. If it has a pulse width of $0.3\ \mu s$, what will the detection threshold be to achieve a probability of false alarm of 10^{-10}, and what is the false alarm time.

Matching the bandwidth to the pulse width:

$$\beta_{if} = \frac{1}{\tau} = \frac{1}{0.3 \times 10^{-6}} = 3.33\ \text{MHz}$$

$$T_{fa} = \frac{1}{P_{fa} \beta_{if}} = \frac{1}{10^{-10} \times 3.33 \times 10^6} = 3 \times 10^3\ \text{s.}$$

Note that this is only 50 min between false alarms!

From equation (9.14), the relationship between the threshold voltage and the probability of false alarm is

$$P_{fa} = \exp \frac{-V_t^2}{2\psi_o}.$$

Therefore

$$V_t = \sqrt{-2\psi_o \ln P_{fa}} = \sqrt{-2 \times 1 \times \ln 10^{-10}} = 6.79\ \text{V.}$$

9.2.3 Probability of Detection

Consider that a sine wave with amplitude, A, is present along with the noise at the input to the IF filter, and that the frequency of the sine wave is equal to the center frequency of the IF filter.

It was shown by Rice (Skolnik 1980) that the signal at the output of the envelope detector will have the following PDF, known as a Rice distribution,

$$P_s(R) = \frac{R}{\psi_o} \exp\left(-\frac{R^2 + A^2}{2\psi_o}\right) I_o\left(\frac{RA}{\psi_o}\right), \tag{9.16}$$

where $I_o(Z)$ is a modified Bessel function of order zero and argument Z.

It can be shown that for large Z, an asymptotic expansion for $I_o(Z)$ is

$$I_o(Z) \approx \frac{e^Z}{\sqrt{2\pi Z}}\left(1 + \frac{1}{8Z} + \dots\right). \tag{9.17}$$

The probability that the signal will be detected is the same as the probability that the envelope, R, will exceed the threshold, V_t, and is

$$p_d = \int_{V_t}^{\infty} p_s(R)\, dR = \int_{V_t}^{\infty} \frac{R}{\psi_o} \exp\left(-\frac{R^2 + A^2}{2\psi_o}\right) I_o\left(\frac{RA}{\psi_o}\right) dR. \tag{9.18}$$

Unfortunately this cannot be evaluated in a closed form and so numerical techniques or a series approximation must be used. Fortunately this has already been done, and tables and a series of curves have been produced by various researchers (Blake 1986; Skolnik 1970).

In terms of the PDFs for the noise and the signal + noise voltages, the detection and false alarm process is shown graphically in Figure 9.7. The lightly shaded area represents the P_d.

From Figure 9.8, for a SNR of 15 dB and a probability of false alarm, P_{fa}, of 10^{-10}, the detection probability, P_d, is 0.9. Remember that this is only true for a sinusoidal signal with a fixed amplitude (nonfluctuating) and no additional losses.

The following MATLAB script was generated to confirm that the processes described in the previous section do in fact produce the Rayleigh and Rice distributions. Figure 9.9 shows these distributions for one million signal and noise samples, and they look correct.

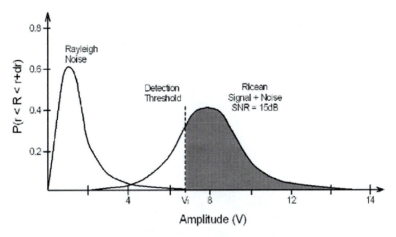

Figure 9.7 The PDFs of noise and signal + noise for $P_{fa} = 10^{-10}$ and $\psi_0 = 1$.

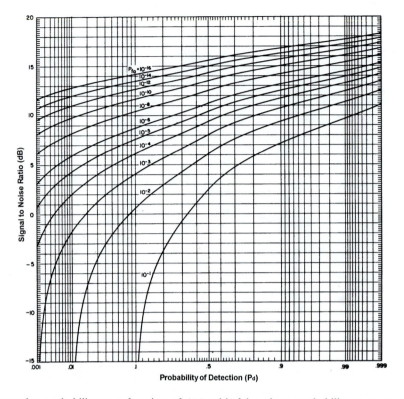

Figure 9.8 Detection probability as a function of SNR with false alarm probability as a parameter (Blake 1986).

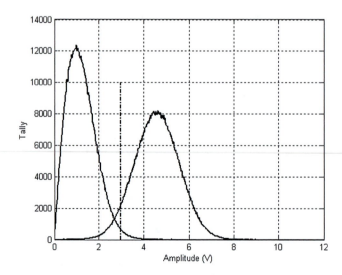

Figure 9.9 Signal and signal + noise PDFs generated by the time domain MATLAB simulation.

```
% Pd and Pfa determined numerically by running a time domain simulation
% detection.m

% Arguments
SNR = 10;                      % signal-to-noise ratio (dB)
vt = 3;                        % threshold voltage (V)

% Determine the amplitude of a sine wave to achieve the required SNR
amp = sqrt(2)*10.^(SNR/20);

% Generate the noise and signal+noise sequences for a noise variance
of 1
a = (1:1000000);
x = randn(size(a));
y = randn(size(a));
sigi = amp*sin(a/1000);
sigq = amp*cos(a/1000);

% Determine the envelope of the signal and signal+noise sequences
noise = sqrt(x.^2+y.^2);
sig_noise = sqrt((x+sigi).*(x+sigi)+(y+sigq).*(y+sigq));

%Plot the distributions over a voltage range 0 to 12V
edges=(0:0.02:12);
n=histc(noise,edges);
ns=hist(sig_noise,edges);
plot(edges,n,'k',edges,ns,'k');
hold

plot([vt,vt],[0,10000],'k-.');
grid
xlabel('Amplitude (V)')
ylabel('Tally')

% Look for noise peaks above the threshold and determine the Pfa
nfa = find(c>vt);
pfa = length(nfa)/1000000

% Look for S+N peaks above the threshold and determine the Pd
nd = find(csig>vt);
pd = length(nd)/1000000
```

Further confirmation that this model is correct is obtained by examining the detection and false alarm probabilities for a specific SNR and threshold. For SNR = 10 dB and a detection threshold $V_t = 3$ V shown in Figure 9.9, the measured probabilities are $P_{fa} = 0.0111$ and $P_d = 0.9462$. These results are identical to those obtained from equation (9.14) and Figure 9.8. Note that this MATLAB simulation is not accurate far down the tails of the distribution, as the number of samples required to achieve such small probabilities becomes too large.

9.2.4 Detector Loss Relative to an Ideal System

An envelope detector is used by a radar system when the phase of the received pulse is unknown. This is called noncoherent detection, and it results in a slightly higher SNR requirement than the curves shown in Figure 9.8. This loss in SNR is known as the detector loss and it becomes significant for low single-pulse SNRs, as can be seen in Figure 9.10.

The detector loss factor, C_1 (dB), is approximately (Barton 1988)

$$C_1 \approx 10 \log_{10} \frac{SNR_1 + 2.3}{SNR_1}, \qquad (9.19)$$

where C_1 is the loss in SNR and SNR_1 (ratio) is the predetector single-pulse SNR required to achieve a particular P_d and P_{fa}.

The graph shows that for a good SNR, the detector loss is very small. For example, in the case where $P_d = 0.9$ and $P_{fa} = 10^{-6}$, the SNR is 13.2 dB and the loss is only about 0.4 dB. This effect is known as small signal suppression, as it becomes much more pronounced as the SNR decreases.

The received IF signal consists of two orthogonal signals, generally referred to as the in-phase and quadrature components. One of the advantages of coherent detection is that it has zero response to the quadrature noise component. In contrast, this component is converted into phase modulation after envelope detection, with the result that the effective SNR is degraded. However,

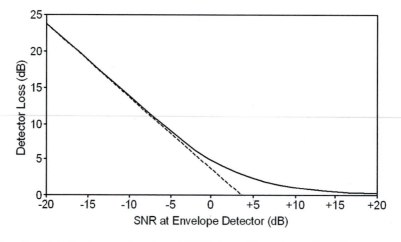

Figure 9.10 Envelope detector loss as a function of SNR [adapted from Barton (1988)].

the phase modulation is very small for high SNRs, and few advantages can be gained by coherent detection for SNRs greater than 10 dB (C_1 = 1 dB) over envelope detection.

9.3 THE MATCHED FILTER

To achieve the best possible SNR, the characteristics of the IF filter must be matched to those of the signal pulse. It must be stressed that this in not a match for maximum power transfer in the electrical sense, but a match in which the output SNR is maximized.

The peak signal-to-(average) noise power ratio of the output response of the matched filter is equal to twice the received signal energy, E (J), divided by the single-sided noise power density, N_o (W/Hz):

$$\left(\frac{\hat{S}}{N}\right)_{out} = \frac{2E}{N_o},\tag{9.20}$$

where $\hat{S}$ (W) is the peak instantaneous signal power seen during the matched filter response to a pulse and N (W) is the average noise power.

The received energy, E (J), is the product of the received power, S (W), as determined by the range equation and the pulse duration, τ (s):

$$E = S\tau,\tag{9.21}$$

and the noise power density, N_o (W/Hz), is the received noise power, N (W), divided by the bandwidth, β_{IF} (Hz):

$$N_o = N/\beta_{IF}.\tag{9.22}$$

Substituting equation (9.21) and equation (9.22) into equation (9.20),

$$\left(\frac{\hat{S}}{N}\right)_{out} = \frac{2S\tau}{N/\beta_{IF}} = \left(\frac{S}{N}\right)_{in} 2\beta_{IF}\tau.\tag{9.23}$$

When the bandwidth of the signal at IF is small compared to the center frequency, the peak power is approximately twice the average power in the received pulse. So the output SNR is

$$\left(\frac{S}{N}\right)_{out} \approx \left(\frac{S}{N}\right)_{in} \beta_{IF}\tau.\tag{9.24}$$

This is a general result and can be applied to pulsed systems, pulse compression, or continuous wave systems. In practice, true matched filter implementations are often not easily achieved and compromises are made. For example, in pulsed systems, analog filters are often constructed from tuned circuits and these incur a loss in effective SNR, as shown in Table 9.1. In this book, the matched filter is often approximated using a sixth-order Butterworth band-pass filter with $\beta\tau = 1$, as discussed in Chapter 5.

Table 9.1 Efficiency of matched filter approximations (Skolnik 1970)

Input signal shape	Matched filter characteristic	Optimum $\beta\tau$	Loss in SNR compared to matched filter (dB)
Rectangular pulse	Rectangular	1.37	0.85
Rectangular pulse	Gaussian	0.72	0.49
Gaussian pulse	Rectangular	0.72	0.39
Gaussian pulse	Gaussian	0.44	0 (matched)
Rectangular pulse	Single tuned circuit	0.4	0.88
Rectangular pulse	Two cascaded tuned circuits	0.613	0.56
Rectangular pulse	Five cascaded tuned circuits	0.672	0.5

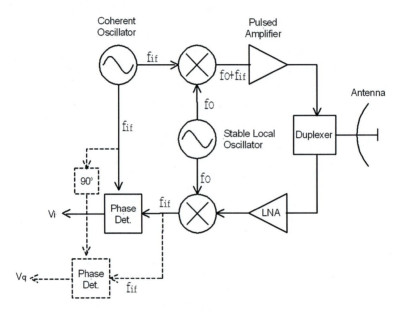

Figure 9.11 Block diagram of a simple coherent radar sensor.

9.4 COHERENT DETECTION

Under some circumstances, noncoherent, or envelope, detection is not an effective method of identifying a target. This could be because the SNR is very low or because it is important to utilize the fact that the target is moving to discriminate its echo from those of static targets in the same range gate.

Coherent detection relies on the use of a phase detector that compares the echo signal with a phase reference generated by the transmitter, as shown in Figure 9.11. The amplitude of the bipolar video output is proportional to the product of the amplitude of the received echo and the cosine of the phase difference between the reference signal and echo.

One of the problems with the simple configuration using a single-phase detector is that the output signal reduces to zero if the phase difference between the reference and echo signals approaches 90°. This is usually addressed by including a second reference phase signal, shifted by 90°, and providing in-phase, V_i, and quadrature, V_q, outputs (shown dashed) that can be further processed. For example, if the two signals are combined in quadrature, a unipolar output similar to that provided by the envelope detector is produced:

$$V_o = \sqrt{V_i^2 + V_q^2}. \tag{9.25}$$

An alternative is to pass the signals through a moving target indicator (MTI) circuit to suppress static returns and enhance the echoes produced by moving targets. This is discussed in detail in Chapter 14.

The provision of a second channel and the application of coherent detection improves the effective SNR of the received signal by 3 dB and eliminates the envelope detector losses discussed in the previous section.

9.5 INTEGRATION OF PULSE TRAINS

The relationships developed earlier between SNR, P_d, and P_{fa} apply to a single pulse only. However, it is possible to improve the detection performance of a pulsed sensor by averaging a number of returns. For example, as a search radar beam scans, the target will remain in the beam sufficiently long for more than one pulse to hit it (Skolnik 1980). This number, known as hits per scan, can be calculated as

$$n_b = \frac{\theta_b f_p}{\dot{\theta}_s} = \frac{\theta_b f_p}{6\omega_m}, \tag{9.26}$$

where

n_b = hits per scan,
θ_b = azimuth beamwidth (deg),
$\dot{\theta}_s$ = azimuth scan rate (deg/s),
ω_m = azimuth scan rate (rpm).

For a typical long-range surveillance radar with an azimuth beamwidth of 1.5°, a scan rate of 5 rpm, and a pulse repetition frequency of 30 Hz, the number of pulses returned from a single point target (hits per scan) is 15.

The process of summing all these hits is called integration, and it can be achieved in many ways, some of which were discussed in Chapter 5. If integration is performed prior to the envelope detector, it is called predetection or coherent integration, while if integration occurs after the detector, it is called postdetection or noncoherent integration.

Predetection integration requires that the phase of the signal be preserved if the full benefit of the summing process is to be achieved. Because phase information is destroyed by the envelope detector, postdetection integration, though easier to achieve, is not as efficient. For example, if n pulses are perfectly integrated by a coherent integration process, the integrated SNR will be exactly n times that of a single pulse in white noise. However, in the noncoherent case, although the integration process is as efficient, there are the detector losses (discussed earlier) that reduce the effective SNR at the output of the envelope detector.

Integration improves the P_d and reduces the P_{fa} by reducing the noise variance, thus narrowing the noise and signal + noise PDFs, as shown in Figure 9.12 (Currie and Brown 1987). For n pulses integrated, the single-pulse SNR required to achieve a given P_d and P_{fa} will be reduced. However, this results in increased detector losses, and hence a reduced effective integration efficiency (Skolnik 1980).

The integration efficiency may be defined as

$$E_n = \text{SNR}_1/n\text{SNR}_n, \qquad (9.27)$$

where

E_n = integration efficiency for n pulses integrated,

SNR_1 = single-pulse SNR required to produce a specific P_d if there is no integration,

SNR_1 = single-pulse SNR required to produce a specific P_d if n pulses are integrated perfectly.

The improvement in SNR if n pulses are integrated postdetection is thus nE_n. This is the integration improvement factor, or the effective number of pulses integrated, and can be represented by a single curve, shown in Figure 9.13, for typical values of $P_d > 0.7$ and $P_{fa} < 10^{-4}$.

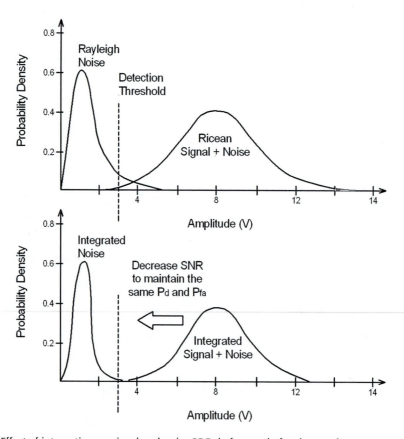

Figure 9.12 Effect of integration on signal and noise PDFs before and after integration.

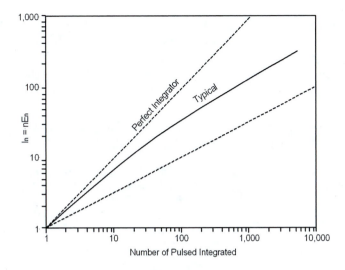

Figure 9.13 Integration improvement factor as a function of pulses integrated [adapted from Skolnik (1980)].

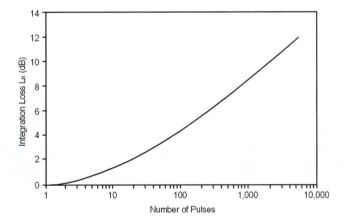

Figure 9.14 Integration loss as a function of the number of pulses integrated [adapted from Skolnik (1980)].

The integration loss in decibels is defined as

$$L_n = 10\log_{10}(1/E_n).\qquad(9.28)$$

The integration improvement factor is not a sensitive function of either the P_d or P_{fa} and can also be represented by a single curve for $P_d > 0.7$ and $P_{fa} < 10^{-4}$, as shown in Figure 9.14.

9.6 DETECTION OF FLUCTUATING SIGNALS

The analysis in the previous section assumes that signal amplitude does not vary from pulse to pulse during the integration period. However, from the discussion on target cross section in

Chapter 8, it is obvious that the RCS of any moving target (with the exception of a sphere) will fluctuate with time as the target aspect as seen by the radar changes.

To properly account for these fluctuations, both the PDF and the correlation properties with time must be known for a particular target and trajectory. Ideally these characteristics should be measured for a target, but this is often impractical. An alternative is to postulate a reasonable model for the target fluctuations and to analyze the effects mathematically.

Four fluctuation models proposed by Swerling, and shown graphically in Figure 9.15, are often used to describe a wide range of target types (Currie et al. 1987):

- Swerling 1: Echo pulses received from the target on any one scan are of constant amplitude throughout the scan, but uncorrelated from scan to scan. The Rayleigh PDF is given by

$$p(\sigma) = \frac{1}{\sigma_{av}} \exp \frac{-\sigma}{\sigma_{av}}, \tag{9.29}$$

where σ_{av} is the average cross section over all target fluctuations.
- Swerling 2: The PDF is as for case 1, but the fluctuations are much faster and are taken to be independent from pulse to pulse.
- Swerling 3: The fluctuations are independent from scan to scan, but the PDF is chi-squared (4 degrees of freedom) and is given by

$$p(\sigma) = \frac{4\sigma}{\sigma_{av}^2} \exp \frac{-2\sigma}{\sigma_{av}}. \tag{9.30}$$

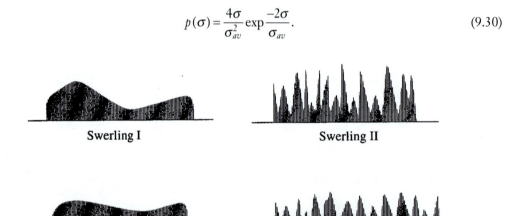

Swerling I   Swerling II

Swerling III   Swerling IV

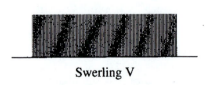

Swerling V

Figure 9.15 Radar returns for different Swerling fluctuations (Mahafza 2000).

- Swerling 4: The PDF is as for case 3, but the fluctuations are independent from pulse to pulse.
- Swerling 5: Nonfluctuating.

The PDFs for cases 1 and 2 are indicative of a target with many (more than five) scatterers of equal amplitude. These are typical for complicated targets like aircraft, while the PDFs for cases 3 and 4 are indicative of a target with one large scatterer and many small scatterers.

As can be seen in Figure 9.16, the single-pulse SNR required to achieve a particular P_d (for $P_d > 0.4$) will be higher for a fluctuating target than for a constant amplitude signal. This is known as the fluctuation loss, L_{fluct}. However, for $P_d < 0.4$, the detection process takes advantage of the fact that a fluctuating target will occasionally present echo signals larger than the average, and so the required SNR is lower.

To account for fluctuating targets and integration, a further set of curves describing the integration improvement factor have been developed and are shown in Figure 9.17. If these curves are examined in isolation, it would appear that the integration efficiency is $E_n > 1$ under certain conditions. One is not getting something for nothing, as the single-pulse SNR is much higher in these cases than it would be for the single-pulse case.

The procedure for using the range equation when one of these Swerling targets is used is as follows:

1. Find the SNR for the single-pulse, nonfluctuating case that corresponds to the P_d and P_{fa} required using the curves in Figure 9.8.
2. For the specific Swerling target, find the additional SNR needed for the required P_d using the curves in Figure 9.16.
3. If n pulses are to be integrated, the integration improvement factor $I_n = nE_n$ is then determined using the curves in Figure 9.17.
4. The SNR_n and nE_n are substituted into the range equation along with σ_{av} and the detection range is found.

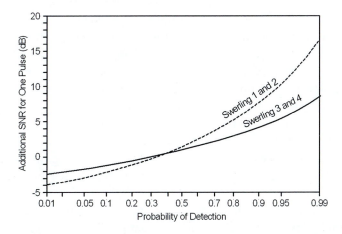

Figure 9.16 Effect of target fluctuation on required SNR [adapted from Skolnik (1980)].

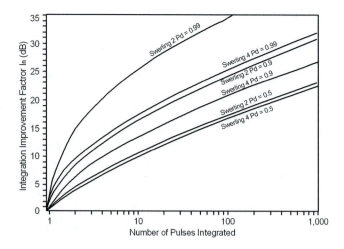

Figure 9.17 Integration improvement factor as a function of the number of pulses integrated for fluctuating targets [adapted from Skolnik (1980)].

9.7 DETECTING TARGETS IN CLUTTER

In many applications the target return is accompanied by echoes from the surrounding clutter. This includes ground clutter in applications where low-flying aircraft or ground vehicles are to be detected by radar, or a ship sonar detecting submarines close to the sea floor. It also includes volumetric clutter in the case where the sensor is required to detect airborne targets in the rain or submarines through schools of fish or krill.

The primary difference between this problem and the detection of targets in noise is that the statistics of the clutter returns are often correlated over a number of measurement periods. It is therefore not possible to improve the effective SNR, and hence the probability of detection, by integrating a number of pulses.

All is not lost, however, as a number of other techniques, apart from integration, are available that can be used to improve the signal-to-clutter (SCR) ratio. The most common method is to rely on target motion to discriminate between the target and clutter returns. The classical MTI process relies on the clutter return remaining stationary and correlated over a number of pulses so that its coherently detected output remains constant. Because the target is moving, the relative phase of the echo and the reference changes, with the result that the output of the coherent detector changes, allowing detection to take place. This is discussed at length in the chapters on Doppler (Chapter 10) and on tracking moving targets (Chapter 14).

An alternative is to improve the resolution of the sensor by narrowing the beam or improving the range resolution to the extent that the amplitude of the clutter returns are lower than those from the target. This technique can be further extended by examining the two orthogonal polarization returns (in the radar case), as discussed briefly in the chapter on high range resolution (Chapter 11).

One of the techniques available in rain is to operate using circular polarization. Because raindrops are nearly spherical, a circularly polarized radar signal incident on one will be reflected with the opposite sense of rotation. A complex target, such as an aircraft, reflects energy that is more uniformly divided between the two senses of rotation, so some target-to-clutter discrimination can be achieved by using the copolar return only.

9.8 CONSTANT FALSE ALARM RATE PROCESSORS

As can be seen from the tail in the Rayleigh distributed noise power density function, the false alarm time is very sensitive to the setting of the detection threshold voltage. Changes in radar characteristics with time (aging) and changes in the target background characteristics mean that a fixed detection threshold is not an effective way of maintaining a constant false alarm time. To do this, adaptive mechanisms called constant false alarm rate (CFAR) processors must be used.

The CFAR process works by sampling the region around the cell under test and determining the noise statistics from the ensemble, as shown in Figure 9.18. These statistics are then used to set a detection threshold which should maintain a constant false alarm rate. This is known as a cell averaging CFAR (CA-CFAR) (Currie et al. 1987).

Needless to say, target detection using the CFAR process introduces additional losses as the statistics are incompletely characterized. For example, a cell averaging process will exhibit a 3.5 dB loss (compared to an ideal single-pulse detector) if 10 cells are used in broadband noise or clutter with a Rayleigh PDF. This decreases to 1.5 dB for 20 cells and 0.7 dB for 40 cells. The loss decreases with an increasing number of integrated pulses. For a 10-cell CFAR with 10 pulses integrated, the loss is only 0.7 dB and is reduced to only 0.3 dB for 100 pulses integrated.

For aircraft detection, obtaining good background noise statistics is not a problem, as the area around the craft is generally empty. However, for ground targets, the CFAR threshold must be determined from the clutter statistics, which may not be homogeneous. This is made worse by the fact that targets (tanks, etc.) often hide at the edges of clutter boundaries to reduce the probability of detection.

The CFAR can operate along range cells, cross-range cells, or both, as shown in Figure 9.19. In general, however, along-range is the most common, as this allows a new threshold to be generated with every measurement, while the other methods require that the sensor scan an area before the process can occur.

In the following examples, two targets are present, the larger in range bin 500 and the smaller in range bin 750. In Figure 9.20, a 25 + 25 point moving average is used to determine the mean value of the signal, from which the CFAR algorithm creates an adaptive threshold that detects

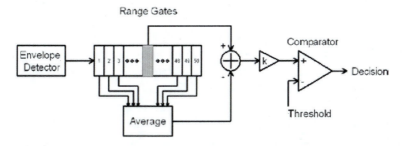

Figure 9.18 Multiple cell averaging CFAR.

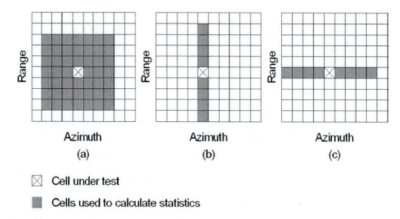

Azimuth   Azimuth   Azimuth
(a)   (b)   (c)

⊠ Cell under test

▧ Cells used to calculate statistics

Figure 9.19 Constant false alarm rate options: (a) area CFAR, (b) range-only CFAR, and (c) azimuth angle-only CFAR [adapted from Currie et al. (1987)].

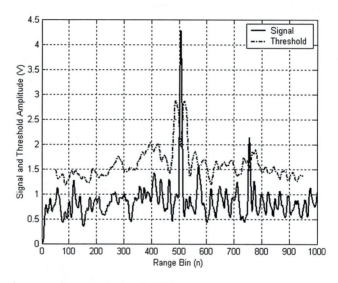

Figure 9.20 CFAR performance for a constant noise floor.

the two targets without detecting the noise. However, a fixed threshold of about 1.6 would perform just as well.

In many radar systems, the noise floor varies with range, as shown in Figure 9.21. In this instance, a fixed threshold does not perform adequately, as any threshold that could detect the target in bin 750 would also detect the short-range noise. The CFAR-generated thresholds easily detect the two targets with no danger of detecting the noise.

The conventional CFAR is ineffective in detecting targets sitting at the interface between clutter types because the statistics are skewed by the transition. In the example shown in Figure 9.22, the smaller target is placed just before this transition and three different CFAR algorithms are implemented. The first generates its threshold based on the statistics of the bins before the

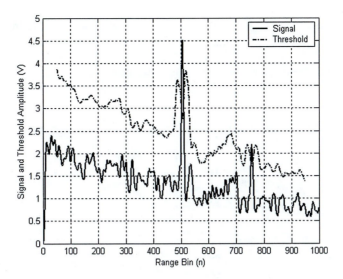

Figure 9.21 CFAR performance for a sloping noise floor.

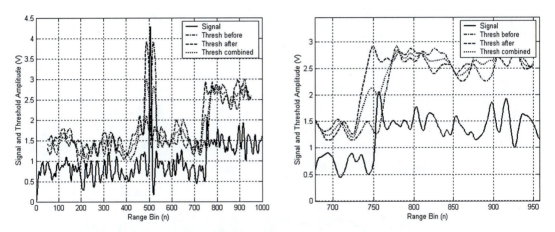

Figure 9.22 Performance of a modified CFAR algorithm designed to detect targets at a transition between two clutter types: (a) complete trace, (b) expanded view of transition.

cell under test, while the second uses the bins after the cell, and the third uses both in the conventional manner. From the expanded view around the transition (Figure 9.22b), it can be seen that although the conventional CFAR algorithm almost fails to detect the target, the threshold-before algorithm detects it easily.

9.9 TARGET DETECTION ANALYSIS

The process of searching for and detecting a target usually involves scanning the sensor beam in azimuth and elevation while monitoring the range echo for returns that exceed a particular threshold. As discussed in this chapter, it is important to set the detection threshold to achieve acceptable false alarm rates while maintaining a good detection probability. In the following example, this process is analyzed in detail.

9.9.1 Worked Example: Target Detection with Air Surveillance Radar

Figure 9.23 shows a photograph of an air surveillance radar antenna illuminated by a pair of horns to produce the lower and upper beams. Mounted above it is the identification friend or foe (IFF) antenna, often referred to as the secondary surveillance radar (SSR) antenna.

The specifications of this radar are:

Type:	Two-dimensional (2D) air surveillance radar
Band:	L
Frequency:	1250 to 1350 MHz
Peak power:	5 MW
Antenna size:	12.8×6.7 m
Antenna gain:	36 dB (lower beam)
	34.5 dB (upper beam)
Beam shape:	cosec^2
Elevation beamwidth:	$4° \, \mathrm{cosec}^2$ to $40°$
Azimuth beamwidth:	$1.25°$
Scan:	Mechanical
Scan rate:	6 rpm
PRF:	360 pps
PRF stagger:	Quadruple
Pulse width:	$2 \, \mu s$
Noise figure:	4 dB

Calculate the theoretical detection range for a 1 m^2 aircraft target if the detection probability, P_d, is 0.9 and the mean time between false alarms is 9 hr.

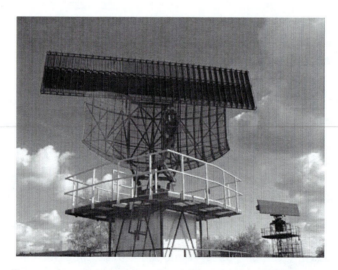

Figure 9.23 Air surveillance radar active antenna and IFF antenna (Courtesy ELDIS Radar Systems).

9.9.1.1 Determine Receiver Parameters
In the absence of any specifications, reasonable assumptions can be made.

Matched Filter It is assumed that the radar transmits a rectangular pulse, being the most common type, and that the matched filter is approximated by two cascaded tuned circuits. From Table 9.1, the specifications of the filter are

$$\text{Loss} = 0.56 \text{ dB}$$
$$\beta\tau = 0.613.$$

Hits per Scan From the pulse repetition frequency, f_p, rotation rate, ω_m, and the azimuth beamwidth, θ_b, it is possible to calculate how many pulses will hit the target as the beam scans past it:

$$n_b = \frac{\theta_b f_p}{6\omega_m} = \frac{1.25 \times 360}{6 \times 6} = 12.5 \left[\text{use } n_b = 10 \right].$$

False Alarm Probability This can be determined from the matched filter bandwidth, β, and the false alarm time, T_{fa}:

$$P_{fa} = 1/T_{fa}\beta.$$

From the matched filter assumptions, $\beta\tau = 0.613$, and the pulse width $\tau = 2$ μs, the IF bandwidth $\beta = 306$ kHz. For the specified false alarm time, $T_{fa} = 9$ hr $= 32,400$ s,

$$P_{fa} = 1/\left(32.4 \times 10^3 \times 306 \times 10^3\right) = 10^{-10}.$$

Single-Pulse SNR Assuming a detection probability $P_d = 0.9$ and the calculated probability of false alarm, the required single-pulse SNR can be obtained from Figure 9.8:

$$SNR_1 = 15.2 \text{ dB}.$$

Effect of Fluctuating Target The radar is designed to detect aircraft which can be modeled as fluctuating targets with Swerling 2 statistics. Using the curves in Figure 9.16, the additional SNR required to maintain the required detection and false alarm probabilities is

$$L_{fluct} = 8 \text{ dB}.$$

Effect of Pulse Integration Integration of pulses reduces the required SNR to maintain the specified detection and false alarm probabilities. Using the curves in Figure 9.17 for 10 pulses integrated, a Swerling 2 target, and $P_d = 0.9$ to obtain an improvement factor

$$I_{10} = 15 \text{ dB}.$$

Total n Pulse SNR Required This is obtained by adding up all of the contributions

$$SNR_1 + L_{fluct} + I_{10} = SNR_{10}$$

$$15.2 + 8 - 15 = 8.2 \text{ dB}.$$

9.9.1.2 Radar Range Equation Apply the standard radar range equation converted to decibels:

$$P_r = \frac{P_t G^2 \lambda^2 \sigma}{(4\lambda)^3 R^4 L}$$

$$P_r = P_t + 2G + 10\log_{10}\frac{\lambda^2}{(4\pi)^3} + \sigma - 40\log_{10} R - L.$$

Tabulate the Losses

Transmitter line = L_{tx} = 2 dB (incorporated into transmitter power)
Receiver line = L_{rec} = 2 dB (incorporated into receiver noise figure)
1D scanning loss = 1.6 dB
Matched filter = 0.56 dB
CFAR loss = 0.7 dB
Miscellaneous loss = 1.3 dB
Total L = 1.6 + 0.56 + 0.7 + 1.3 = 4.16 dB.

Transmitter Power (dBW)

$10\log_{10}(5 \times 10^6)$ = 67 dBW
Less L_{tx} = 2 dB
Radiated peak power P_t = 65 dBW

Antenna gain (dB) The lower beam is used in this analysis: G = 36 dB.

Radar Cross Section (dBm²)

$$\sigma = 10\log_{10}(1) = 0 \text{ dBm}^2$$

Propagation constant (dB)

$$10\log_{10}\frac{\lambda^2}{(4\pi)^3} = -45.7 \text{ dB}$$

Received Power (dBW) This is calculated from the range equation:

$$P_r = 65 + 2 \times 36 - 45.7 + 0 - 4.16 - 40\log_{10} R$$
$$= 87.14 - 40\log_{10} R.$$

9.9.1.3 Determine the Receiver Noise and SNR The receiver noise is a combination of the thermal noise and the receiver noise factor:

$$N = P_N F_N = kT_{sys}\beta F_N,$$

where

k = Boltzmann's constant, 1.38×10^{-23} J/deg,
T_{sys} = temperature, 290 K,
$\beta = 306 \times 10^3$ Hz.

We are given the receiver noise figure in decibels:

$$10\log_{10}(F_N) = 4 \text{ dB}.$$

The receiver waveguide loss is incorporated into the noise figure to give the total receiver noise level:

$$\begin{aligned} N &= 10\log_{10}(kT_{sys}\beta) + 10\log_{10}(F_N) + L_{rec} \\ &= -149 + 4 + 2 \\ &= -143\,\text{dBW}. \end{aligned}$$

Received SNR (dB)

$$\begin{aligned} SNR_{rec} &= P_r - N \\ &= 87.14 - 40\log_{10}R + 143.0 \\ &= 230.1 - 40\log_{10}R. \end{aligned}$$

This must equal the required single-pulse SNR if 10 pulses are integrated, SNR_{10}, to achieve the specified P_d and P_{fa}:

$$8.2 = 230.1 - 40\log_{10}R.$$

9.9.1.4 Solve for the Detection Range   In the absence of any atmospheric attenuation or other propagation effects (like multipath), the detection range is

$$\begin{aligned} R &= 10^{(230.1-8.2)/40} \\ &= 352,776.8\,\text{m}\,[352.8\,\text{km}]. \end{aligned}$$

Effect of Atmospheric Attenuation   The atmospheric attenuation, α_{dB}, at L-band is about 0.003 dB/km (one way), so the total two-way attenuation over 353 km is 2.1 dB, which will result in a significant reduction of the detection range. The equation becomes nonlinear because there are two terms that include the range, and therefore it is best solved graphically using MATLAB:

$$P_r = P_t + 2G + 10\log_{10}\frac{\lambda^2}{(4\pi)^3} + \sigma - L - 40\log_{10}R - 2\alpha R_{km}.$$

Figure 9.24 shows two received power curves, one that does not take into account the atmospheric attenuation which crosses the minimum detectable signal level threshold at 353 km. The other graph, which takes attenuation into account, crosses the threshold at 320 km.

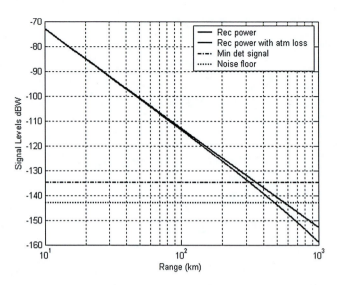

Figure 9.24 Graphical solution to the radar range equation including the atmospheric loss.

```
% Generate detection range from the basic radar range equation
% Arguments

GdB = 36;                    % Antenna Gain (dB)
Pt = 5.0e06;                 % Transmit power (W)
LdB = 4.16;                  % Losses (dB)
Lrx = 2;                     % receiver line loss (dB)
Ltx = 2;                     % transmit line loss (dB)
freq = 1.3e09;               % frequency (Hz)
RCS = 1.0;                   % radar cross-section (sqm)
beta = 306e03;               % receiver bandwidth (Hz)
Fn = 4;                      % receiver noise figure (dB)
SNR10 = 8.2;                 % signal-to-noise ratio rqd for detection (dB)
alpha = 0.003;               % clear air attenuation (dB/km)


% Other constants
k = 1.38e-23;                % Boltzmann's constant
T = 290;                     % Temperature (K)


% Calculate wavelength
c = 3.0e+08;
lam = c/freq;


PtdB = 10*log10(Pt)-Ltx; % transmit power in dBW
RCSdB = 10*log10(RCS);   % radar cross section in dBsqm
```

```
kdB = 10*log10(lam^2/(4*pi)^3);

% determine the noise power
N = 10*log10(kbeta)+Fn+Lrx;

% determine the receive power
R = logspace(4,6,500);
Rkm = logspace(1,3,500);

SdB = PtdB + 2*GdB + kdB + RCSdB - LdB - 40*log10(R);
SdBatm = SdB-2*alpha*Rkm;
Np = ones(size(R))*N;
Tdet = Np+SNR10;

semilogx(Rkm,SdB,'k',Rkm,SdBatm,'k',Rkm,Tdet,'k-.',Rkm,Np,'k:');
grid

xlabel('Range (km)');
ylabel('Signal Levels dBW')
legend('Rec power','Rec power with atm loss','Min det signal','Noise
floor')
```

9.9.2 Range Analysis Software Packages

Radar performance analysis software (Barton 1988; Blake 1986) has been available for many years. In this case, a package called RGCALC (Fielding and Reynolds 1987) is used. For the 2D radar, it provides the results shown in Figure 9.25.

Note that the predicted performance using RGCALC for a Swerling 2 target is only 285 km, whereas the MATLAB analysis shown above predicts a detection range of 320 km. This lower range is because the package makes the assumption that the optimum matched filter has $\beta\tau = 1$, and hence calculates a wider IF bandwidth. This adds 2.1 dB to the noise floor, which accounts for the difference.

9.9.3 Detection Range in Rain

Should detection performance in rain be required, then the MATLAB software can easily be modified to include the additional attenuation. However, some caution is needed, because even if the SNR is sufficiently high to detect the target, volumetric backscatter from the rain may reduce the SCR to less than that required for detection.

If the radar is operating through heavy rain (50 mm/hr), the attenuation increases to about $\alpha = 0.02$ dB/km and the backscatter coefficient is about $\eta = 10^{-7}$ m^2/m^3. For an antenna beamwidth

```
Radar Name or Description -- Enroute 2D Air Surveillance Radar


    Radar and Target Parameters (inputs) --

    Peak Pulse Power (kilowatts) ..................    5000.0
    Pulse Duration (usec) ........................    2.0000
    Transmit Antenna Gain (dB) ...................      36.0
    Receive Antenna Gain (dB) ....................      36.0
    Frequency (MHz) ..............................    1300.0
    Receiver Noise Factor (dB) ...................       4.0
    Bandwidth Correction Factor (dB) .............        .6
    Antenna Ohmic Loss (dB) ......................        .0
    Transmit Transmission Line Loss (dB) .........       2.0
    Receive  Transmission Line Loss (dB) .........       2.0
    Scanning-Antenna Pattern Loss (dB) ...........       1.6
    Miscellaneous Loss (dB) ......................       2.0
    Number of Pulses Integrated ..................        10
    Probability of Detection .....................      .900
    False-Alarm Probability (Negative Power of Ten)   10.0
    Target Cross Section (Square Meters) .........    1.0000
    Target Elevation Angle (Degrees) .............       .40
    Average Solar and Galactic Noise Assumed
    Pattern-Propagation Factors Assumed = 1


         ***********************************


    Calculated Quantities (Outputs) --

    Noise Temperatures, Degrees Kelvin --
         Antenna (TA) ............................     117.2
         Receiving Transmission Line (TR) .......      169.6
         Receiver (TE) ..........................      438.4
         TE X Line-Loss Factor = TEI ............      694.9
         System (TA + TR + TEI) .................      981.7
    Two-Way Attenuation Through Entire Troposphere (dB)   3.1
```

Swerling Fluctuation Case	Signal-to-Noise Ratio, dB	Tropospheric Attenuation, Decibels	Range, Nautical Miles	Range, Kilometers
0	6.76	2.77	163.7	303.2
1	15.28	2.13	104.0	192.6
2	7.91	2.70	153.9	285.0
3	11.33	2.45	128.2	237.4
4	7.38	2.73	158.3	293.2

Figure 9.25 Output from RGCALC radar range performance software.

of $1.25° \times 4°$ (0.02 rad $\times$ 0.07 rad) and a pulse width of $2~\mu s$, the rain clutter RCS is a function of range:

$$\sigma_{clut} = \eta \frac{\pi R^2 \theta_{AZ} \theta_{EL} c\tau}{8\alpha^2}$$
$$= 10^{-7} \times \frac{\pi \times 0.02 \times 0.07 \times 3 \times 10^8 \times 2 \times 10^{-6}}{8 \times 1.33^2} R^2$$
$$= 1.86 \times 10^{-8} R^2.$$

This can be rewritten in decibel form as

$$\sigma_{\text{clut}} = -77.3 + 20 \log_{10} R.$$

For the sake of simplicity, assume that the target will be detectable if the signal level exceeds the clutter level by the single-pulse SNR, $SNR_1 = 15.2$ dB. In that case, the maximum detection range can be determined from

$$\sigma_{\text{clut}} = \sigma_{\text{tar}} - 15.2$$
$$-77.3 + 20 \log_{10} R = \sigma_{\text{tar}} - 15.2$$
$$R = 1274 \, \text{m}.$$

From an atmospheric attenuation perspective, the detection range is only reduced by a small amount, to 216 km. It is obvious from these calculations that the SNR has very little to do with the detection range unless the backscatter from the rain can be decreased significantly. As discussed earlier in this chapter, moving target discrimination or using circular polarization may go some way to providing the answer.

9.10 NOISE JAMMING

The most common form of electronic countermeasures (ECM) is noise jamming, which is intended to prevent detection of an echo by raising the receiver noise level. There are two main types of noise jammers: self-screening and standoff. In the first case, the aircraft carrying the jammer is also the target and is assumed to be flying toward the radar. This configuration works to the advantage of the jammer, as its power increases along with the target echo amplitude and all of the noise from the jammer enters the main lobe of the radar. In the standoff jammer case, the azimuth angle is generally different from the target, and the noise generally enters the radar through its sidelobes. However, because standoff jammers are usually fitted to dedicated electronic warfare (EW) aircraft, their output power can be much higher (Blake 1986).

The radar equation can be used to determine the detection (or burn-through) range in the presence of a jammer by replacing the receiver noise power density, $N_r = kT_oNF$, with the jammer noise power density. The jammer noise power density, N_j (W/Hz), at the radar is (Skolnik 1980)

$$N_j = P_j G_j G_r \lambda^2 / (4\pi)^2 R^2 \beta_j, \tag{9.31}$$

where P_j (W) is the jammer power, G_j and G_r are the line-of-sight gains of the jammer and radar antennas, respectively, and β_j (Hz) is the jammer bandwidth.

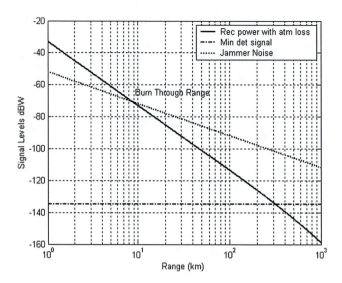

Figure 9.26 Performance of a self-screening jammer against an air surveillance radar.

9.10.1 Noise Jamming Example

Consider an L-band self-screening jammer with the following specifications:

Jammer power	P_j = 2 kW (33 dBW)
Gain	G_j = 6 dB
Wavelength	λ = 0.2308 m
Bandwidth	β_j = 500 MHz (87 dBHz)

This is in operation against the air surveillance radar illustrated in Figure 9.23. Rewriting equation (9.31) in the decibel form gives

$$N_j = P_j + G_j + G_r + 20\log_{10}\frac{\lambda}{4\pi} - \beta_j - 20\log_{10}R$$
$$= 33 + 6 + 36 - 34.7 - 87 - 20\log_{10}R$$
$$= -46.7 - 20\log R.$$

The results of this simulation showing the burn-through range are shown in Figure 9.26.

9.11 References

Barton, D. (1976). *Radar Systems Analysis*. Norwood, MA: Artech House.

Barton, D. (1988). *Modern Radar Systems Analysis*. Norwood, MA: Artech House.

Blake, L. (1986). *Radar Range-Performance Analysis*. Norwood, MA: Artech House.

Currie, N. and Brown, C. (1987). *Principles and Applications of Millimeter-Wave Radar*. Norwood, MA: Artech House.

Fielding, J. and Reynolds, D. (1987). *RGCALC: Radar Range Detection Software and User's Manual*. Norwood, MA: Artech House.

Mahafza, B. (2000). *Radar Systems Analysis and Design Using MATLAB*. Boca Raton, FL: CRC Press.

Skolnik, M. (1970). *Radar Handbook*. New York: McGraw-Hill.

Skolnik, M. (1980). *Introduction to Radar Systems*. Tokyo: McGraw-Hill Kogakusha.

Waite, A. (2002). *Sonar for Practicing Engineers*. New York: John Wiley & Sons.

10

Doppler Measurement

10.1 THE DOPPLER SHIFT

The Doppler shift is the apparent difference between the frequency at which sound or light waves leave a source and that at which they reach an observer, caused by the relative motion of the observer and the wave source. This phenomenon is used in astronomical measurements, in radar, and in modern navigation sensors. It was first described in 1842 by Austrian physicist Christian Doppler.

Examples of the Doppler effect, as illustrated in Figure 10.1, include the following:

- As a blowing horn is approached at speed, the perceived pitch is higher until the horn is reached and then it becomes lower as the horn is passed.
- The light from a star, observed from the Earth, shifts toward the red end of the spectrum (lower frequency or longer wavelength) if the Earth and star are receding from each other and toward the violet (higher frequency or shorter wavelength) if they are approaching each other. The Doppler effect is used in studying the motion of stars.
- Radar velocity measurement, in which the shift in frequency of the echo compared to the transmit frequency is used to determine the radial speed of the target and to discriminate moving targets from stationary ones.

10.1.1 Doppler Shift Derivation

Consider the relationship between the frequency of sound produced by a source moving with velocity v_s (m/s) and the frequency received by a receiver moving with velocity v_r (m/s). For simplicity, assume that both the source and the receiver are moving in a straight line in the same direction, as shown in the figure (Liley 1998). At time $t = 0$, the source, S, and receiver, R, are separated by a distance d (m):

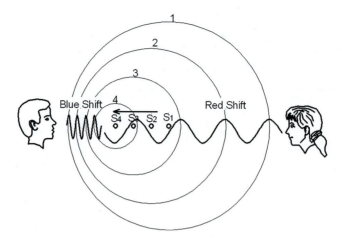

Figure 10.1 Illustration of Doppler shift for a sound source moving from S_1 to S_4.

The source emits a wave that propagates at velocity c (m/s) and reaches the receiver after time t (s).

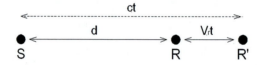

As the receiver has moved $v_r t$ (m):

$$ct = d + v_t t, \tag{10.1}$$

or

$$t = d/(c - v_r). \tag{10.2}$$

At time τ, the source, S, would have moved τv_s (m). Let the wave emitted at that instant be received at time t' by R. In this time, R would have moved $v_r t'$ (m),

$$c(t' - \tau) = (d - v_s \tau) + v_r t', \tag{10.3}$$

making

$$t' = \frac{d + (c - v_s)\tau}{c - v_r}.$$ (10.4)

Thus, for the receiver, the interval between the waves has been τ' (s):

$$\tau' = t' - t = \frac{c - v_s}{c - v_r}\tau.$$ (10.5)

Whereas for the source, the interval between waves has been τ (s). Now the number of waves emitted in τ (s) by the source must equal the number of waves received by the receiver in τ' (s),

$$f_r t' = f_s \tau,$$ (10.6)

making

$$f_r = \frac{c - v_r}{c - v_s} f_s.$$ (10.7)

This is the general result and can be used in most cases so long as the source and receiver speeds are nonrelativistic (applicable for electromagnetic radiation only).

For v_s and $v_r \ll c$, which is the case for electromagnetic radiation,

$$f_r = \frac{1 - v_r/c}{1 - v_s/c} f_s = \left[1 - \frac{v_r}{c}\right]\left[1 - \frac{v_s}{c}\right]^{-1}.$$ (10.8)

Expanding the last term using the binomial expansion,

$$(1 + x)^n = 1 + nx + \frac{n(n-1)}{2!}x^2 + \dots.$$ (10.9)

For $x \ll 1$, the higher order terms can be ignored and the received frequency becomes

$$f_r \approx \left[1 - \frac{v_r}{c}\right]\left[1 + \frac{v_s}{c}\right] = \left(1 - \frac{v_{rs}}{c}\right)f_s,$$ (10.10)

where $v_{rs} = v_r - v_s$ (m/s) is the velocity of the receiver relative to the source.

The Doppler shift, f_d (Hz), is thus

$$f_d = f_r - f_s = (-v_{rs}/c)f_s.$$ (10.11)

The frequency moving away from the source will be less than the frequency measured at the source, whereas the frequency measured at a receiver moving toward the source will be greater than the frequency measured at the source.

Figure 10.2 Doppler geometry: separated transducers [adapted from Liley (1998)].

10.2 DOPPLER GEOMETRY

In most Doppler sensors both the transmitter and receiver are stationary and they illuminate a moving target. In addition, for some applications (Doppler ultrasound imaging) they may not be collocated, as shown in Figure 10.2.

The velocity of the target relative to the source will be $v\cos\theta_s$ (m/s) and the velocity of the target relative to the receiver will be $v\cos\theta_r$ (m/s). The Doppler shift arising under these circumstances can be calculated assuming the following:

- The target is a receiver moving away from the source with a velocity $v\cos\theta_s$ (m/s).
- The receiver is moving away from the target (source) with a velocity $v\cos\theta_r$ (m/s).

This is equivalent to the receiver moving away from the source with velocity $v\cos\theta_t + v\cos\theta_r$, as shown in the figure, even though both are stationary.

$$V_s \longleftarrow \quad \underset{S\ R}{\bullet\ \bullet} \quad \longrightarrow V_r$$

10.2.1 Targets Moving at Low Velocities ($v \ll c$)

The Doppler frequency for separated transducers is

$$f_d = \frac{f_s v}{c}\left(\cos\theta_s + \cos\theta_r\right), \tag{10.12}$$

and using one of the trigonometric identities, it can be rewritten as

$$f_d = -\frac{2f_s v}{c}\cos\left(\frac{\theta_r + \theta_s}{2}\right)\cos\left(\frac{\theta_r - \theta_s}{2}\right). \tag{10.13}$$

If the transmit and receiver transducers are collocated, then $\theta_r \approx \theta_s = \theta$, and the formula for the Doppler frequency reduces to

$$f_d = -\frac{2f_s v}{c}\cos\theta = -\frac{2v}{\lambda_s}\cos\theta. \tag{10.14}$$

10.2.2 Targets Moving at High Speed ($v < c$)

If the target is moving at a speed which is a large fraction of the speed of propagation, then it is not possible to use the electromagnetic radiation approximation, and equation (10.7) must be used for separated transducers. Note that the sign of the source component is reversed because it is effectively moving away from the receiver and the original equation was derived for the source and receiver moving in the same direction. This is almost always the equation that should be used with ultrasound measurements in air:

$$f_r = \frac{c - v\cos\theta_r}{c + v\cos\theta_s} f_s,$$ (10.15)

and the Doppler frequency is

$$f_d = f_r - f_s.$$ (10.16)

For collocated transducers, $\theta_r \approx \theta_t = \theta$, just substitute for a common angle.

10.3 DOPPLER SHIFT EXTRACTION

As illustrated in Figure 10.3, the simplest way of extracting the Doppler shift involves mixing the received signal with a portion of the transmitted signal to obtain a beat frequency, f_d (Hz), which is the difference between the two. The accuracy with which the Doppler shift can be measured is determined by $1/T_d$ (Hz), the reciprocal of the observation time.

Assuming that an ultrasound transducer operating at a frequency $f_s = 40$ kHz produces a signal of the form

$$x_s(t) = \xi_s \cos(\omega_s t).$$ (10.17)

The corresponding received signal from a single moving target will be

$$x_r(t) = \xi_r \cos([\omega_s + \omega_d]t + \varphi),$$ (10.18)

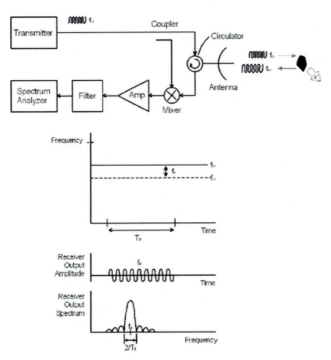

Figure 10.3 Configuration of a generic Doppler sensor and the output waveform.

where φ (rad) is the phase term dependent on the distance to the target, the source frequency, $\omega_s = 2\pi f_s$ (rad/s), and the Doppler shift, $\omega_d = 2\pi f_d$ (rad/s).

The transmit and received signals are mixed (multiplied together) to produce

$$x_s(t)x_r(t) = \xi_r\xi_s \cos(\omega_s t)\cos([\omega_s + \omega_d]t + \varphi)$$
$$= \frac{\xi_s\xi_r}{2}\{\cos(\omega_d t + \varphi) + \cos([2\omega_s + \omega_d]t + \varphi)\}. \tag{10.19}$$

This resulting output signal is low-pass filtered to remove the component at $2f_s$, leaving only the Doppler signal and the phase shift,

$$x_d(t) = \frac{\xi_s\xi_r}{2}\cos(\omega_d t + \varphi). \tag{10.20}$$

In a typical application, for an ultrasound signal at 40 kHz, the component at $2f_s$ (80 kHz) is filtered out using a low-pass filter. The Butterworth filter response is shown in Figure 10.4 for reasons discussed in Chapter 2, but it is possible to use any of the other configurations. For example, the elliptical filter offers a much steeper cutoff if required. In addition, the reflected signal amplitude from nonmoving objects in the beam can be 40 to 50 dB larger than the Doppler signal, and so additional high-pass filtering is often required to remove this direct current (DC) component as well.

10.3.1 Direction Discrimination

The Doppler process discussed above can provide only an absolute difference frequency; it contains no information regarding the direction of motion (Liley 1998). A number of techniques can be applied to preserve this directional information:

- Sideband filtering,
- Offset carrier demodulation,
- In-phase/quadrature demodulation.

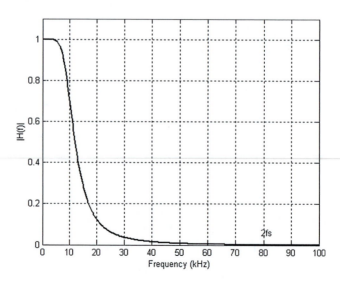

Figure 10.4 Low-pass filter to remove the component at $2f_s$.

In the descriptions that follow, it should be remembered that $\omega_d > 0$ means that the target is traveling toward the sensor and $\omega_d < 0$ means that it is traveling away.

10.3.1.1 Sideband Filtering   As shown in Figure 10.5, the received signal is split and passed through two band-pass filters, one passing signals over the frequency band $\omega_s < \omega < \omega_s + \omega_m$ and the other passing signals over the frequency band $\omega_s - \omega_m < \omega < \omega_s$. The output of each filter passes through a mixer and filter as usual.

If the target is approaching, the signal appears in the first output, and if it is receding it appears in the second, as illustrated in Figure 10.6. If the signal falls exactly halfway between the two, it appears with equal amplitudes in both outputs.

10.3.1.2 Offset Carrier Demodulation   This process involves heterodyning (mixing) the received signal by a reference signal $\omega_1 + \omega_s$, as shown in Figure 10.7. The received signal is

$$x_r(t) = \xi_r \cos([\omega_s + \omega_d]t + \varphi), \qquad (10.21)$$

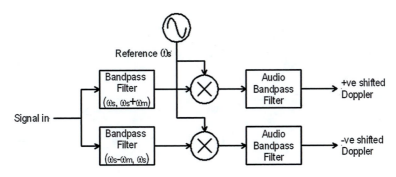

Figure 10.5 Sideband filtering [adapted from Liley (1998)].

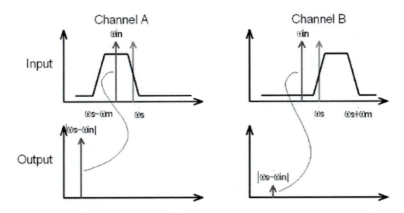

Figure 10.6 Sideband filtering spectra in the two channels.

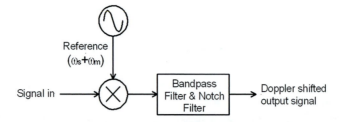

Figure 10.7 Offset carrier demodulation [adapted from Liley (1998)].

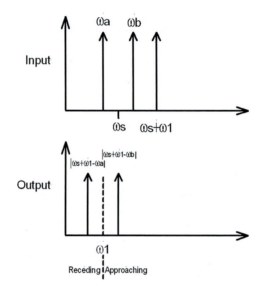

Figure 10.8 Offset carrier spectra.

and the reference signal is

$$x_1(t) = \xi_1 \cos([\omega_s + \omega_1]t). \tag{10.22}$$

Mixing the two signals of equation (10.21) and equation (10.22) gives

$$x_1(t)x_r(t) = \frac{\xi_1\xi_r}{2}\{\cos([\omega_1 + \omega_d]t + \varphi)\cos([2\omega_s + \omega_1 + \omega_d]t + \varphi)\}, \tag{10.23}$$

where ω_1 is chosen so that $\omega_1 > |\omega_{dmax}|$, and, as usual, the mixed signal is filtered to remove the component at $2\omega_s$ to produce the output spectra shown in Figure 10.8. In this case, $\omega_1 + \omega_d > \omega_1$ occurs for positive Doppler shift and $\omega_1 + \omega_d < \omega_1$ for negative Doppler shift.

10.3.1.3 In-Phase/Quadrature Demodulation   In this implementation, the received signal is split into two channels. In the in-phase channel it is mixed with the transmitted signal, and in the quadrature channel it is mixed with the transmitted signal that has been phase shifted by $\pi/2$, as shown in Figure 10.9.

Mixing and filtering as before results in two output signals, an in-phase signal, $i(t)$, and a quadrature signal, $q(t)$:

$$i(t) = \cos(\omega_d t + \varphi)$$
$$q(t) = \sin(\omega_d t + \varphi).$$

(10.24)

The direction of the Doppler shift, and hence the direction of motion, is determined by noting the phase relationship between $i(t)$ and $q(t)$:

- $\omega_d > 0$, $q(t)$ is $\pi/2$ retarded with respect to $i(t)$.
- $\omega_d < 0$, $q(t)$ is $\pi/2$ advanced with respect to $i(t)$.

As can be seen in the illustration in Figure 10.10, if the $i(t)$ and $q(t)$ are fed into the x and y inputs of an oscilloscope, respectively, a circular Lissajous figure is produced that will rotate clockwise for a positive Doppler shift and counterclockwise if the Doppler shift is negative.

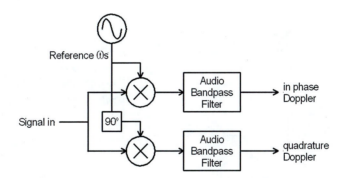

Figure 10.9 In-phase/quadrature demodulation [adapted from Liley (1998)].

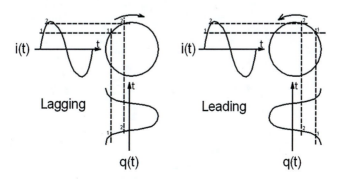

Figure 10.10 Lissajous figure generated by applying $i(t)$ and $q(t)$ to the x and y inputs of an oscilloscope can be used to determine the sign of the Doppler shift.

10.4 PULSED DOPPLER

Pulsed Doppler is identical to the continuous wave version except that the transmit signal is interrupted periodically. This allows the technique to measure range as well as velocity. As shown in Figure 10.11, a continuously running reference version of the transmitted signal must be maintained so that the received signal can be synchronously detected.

The received echo from a pulsed Doppler sensor is shown in Figure 10.12a. The Doppler shift is determined by mixing this pulse train with the transmitter reference signal. Depending on the pulse width, τ (s), and the Doppler shift, f_d (Hz), each processed pulse may include a number of cycles, as shown in Figure 10.12b for $f_d > 1/\tau$, or may be a fraction of a cycle, as illustrated in Figure 10.12c for $f_d < 1/\tau$.

The results shown in Figure 10.12c serve to illustrate why the detection process must be coherent. In the event that this was not the case, the output pulse amplitudes would be random, and no Doppler information could be extracted. However, because the output can be thought of

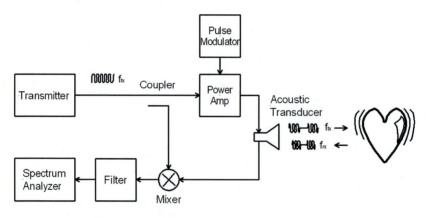

Figure 10.11 Pulsed Doppler ultrasound schematic diagram.

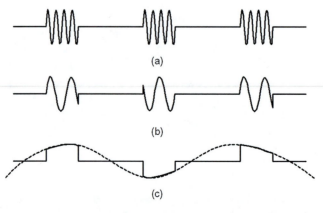

Figure 10.12 Pulsed Doppler waveforms: (a) RF echo pulse train, (b) video pulse train for $f_d > 1/\tau$, and (c) video pulse train for $f_d < 1/\tau$ [adapted from Skolnik (1980)].

as the equivalent of a continuous wave system that it sampled after the mixing (phase detection) process, a sine wave can be reconstructed irrespective of how low the Doppler frequency is.

An alternative approach, which has been exploited in the manufacture of low-cost speed radars, is to ensure that the start phase of each transmitted pulse remains the same. In this case it is easiest to think of the phase shift of the received signal as being proportional to the round-trip time to the target, and that will change as the target moves.

The waveforms in Figure 10.13 show simulation results, generated by the MATLAB script, of the basic pulsed Doppler technique extended to incorporate in-phase and quadrature detection

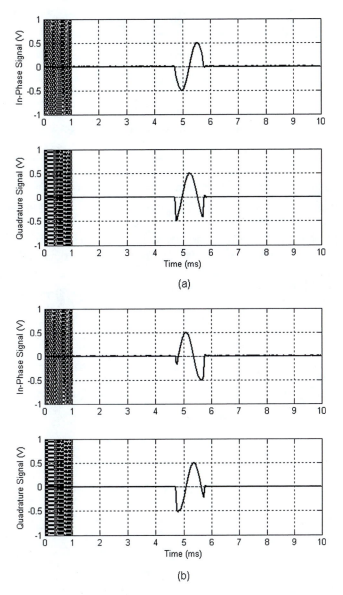

Figure 10.13 In-phase and quadrature Doppler outputs for (a) a receding and (b) an approaching target with a velocity $v_{tar} = 3.9$ m/s.

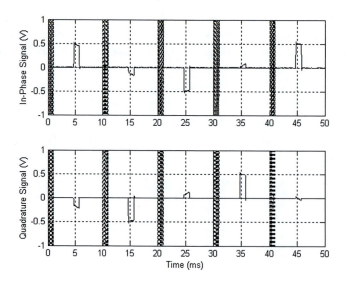

Figure 10.14 In-phase and quadrature Doppler outputs for a receding target velocity v_{tar} = 0.1 m/s.

so that the direction of travel can be determined. In this figure it can be seen that the quadrature signal leads the in-phase signal for a receding target (Figure 10.13a), and that it lags the in-phase signal for an approaching target (Figure 10.13b).

In Figure 10.14, the target velocity is slowed down to 0.1 m/s and more pulses are shown. In this case, it is clear that the Doppler shift can be determined by constructing a sinusoidal signal through the peaks of the detected pulses. A careful examination of these results shows that the quadrature output is still phase shifted by 90° with respect to the in-phase signal.

```
% ultrasonic doppler detection simulation
% dop_filter.m
%
% Arguments
%
fcar = 40.0e03;                    % center frequency signal (Hz)
bdop = 10.0e03;                    % doppler filter bandwidth (Hz)
prf = 100;                         % pulse repetition frequency (pps)
tau = 1.0e-03;                     % pulse period (sec)
rtar = 0.8;                        % target range (m)
vtar = 5;                          % target velocity m/s
v = 340.0;                         % speed of sound (m/s)
ts = 1/(21*fcar);                  % 21 samples per carrier wavelength
ttot = 5/prf;                      % time for five transmit cycles
```

```
% generate the reference i and q transmit signals
t = (0:ts:ttot);
lent = length(t);
sigtxi = cos(2*pi*fcar*t);
sigtxq = cos(2*pi*fcar*t+pi/2);

% generate the pulse mask for 5 pulses
lenx=lent/5;
txmask = zeros(size(t));
len = fix(tau/ts);                  % pulse length in samples
dum = (1:len);
x = ones(size(dum));
for k=0:4
    offset = round(k*lenx);
    txmask(offset+1:offset+len) = x;
end

% mask the in-phase transmit signal to generate five pulses
sigtx = sigtxi.* txmask;

% generate the continuous doppler shifted receive signal
ftxd = (v-vtar)*fcar/(v+vtar);
phase = 2*pi*ftxd*rtar/v;
sigtxd = cos(2*pi*ftxd*t+phase);

% generate the receive mask
rxmask = zeros(size(t));
del1 = fix(2*rtar/(v*ts));          % sample delay to the target
for k=0:4
    offset = round(k*lenx);
    rxmask(offset+del1:offset+del1+len-1) = x;   % mask length equals
                                                 %        pulse length
end

% generate the received signal by multiplying the continuous signal
% by the receive mask and scaling to lower amplitude
sigrx = 0.8*sigtxd.*rxmask;

% plot the transmit and receive signals
plot(t*1000,sigtx,t*1000,sigrx)
grid
title('Transmit Signal and Echo Signal');
xlabel('Time (ms)');
```

```
ylabel('Signal (V)');
legend('transmit signal','receive signal')
pause;

% add noise to the received signal
noise = 0.2*randn(size(sigrx));       % random noise variance = 0.2
sigrx = sigrx+noise;

% plot the transmit and received signal with added noise
plot(t*1000,sigtx,t*1000,sigrx)
grid
title('Transmit Signal and Received Echo with added Noise');
xlabel('Time (ms)');
ylabel('Signal (V)');
pause;

% mix with the transmit reference signals to extract the Doppler
sigdeti = sigrx.*sigtxi;
sigdetq = sigrx.*sigtxq;

% synthesize a Butterworth low-pass filter to remove 2fcar component:
wdop = 2*ts*bdop;
[b,a]=butter(3,wdop);
[h,w]=freqz(b,a,1024);
freq=(0:1023)/(2000*ts*1024);

% plot the filter characteristics
plot(freq,abs(h));
grid
title('Doppler Filter Transfer Function')
xlabel('Frequency (kHz)');
ylabel('Gain')
pause

% filter the received Doppler signal
sigdfili = filter(b,a,sigdeti);
sigdfilq = filter(b,a,sigdetq);

% plot the transmit signal and the received Doppler signal after
  filtering
subplot(211),plot(t*1000,sigtx,'k:',t*1000,sigdfili,'k');
grid
ylabel('Signal (V)');
```

```
subplot(212),plot(t*1000,sigtx,'k:',t*1000,sigdfilq,'k');
grid
xlabel('Time (ms)');
ylabel('Signal (V)');
pause
subplot

% plot the Lissajous figure one sample at a time
% so that the direction of rotation can be seen
for k=1:20:lent
    plot(sigdfili(k),sigdfilq(k),'o')
    axis([-1,1,-1,1]);
    xlabel('I Detected Echo')
    ylabel('Q Detected Echo')
    drawnow
end
```

10.5 DOPPLER SENSORS

Many sensors using ultrasonic, radar, or laser technology make use of the Doppler principle to measure target motion. Such sensors can either operate using continuous wave or pulsed sources. In most cases continuous wave sources have no range discrimination, while pulsed sources can determine both range and velocity. Typical examples of this technology include Doppler ultrasound systems to measure blood flow, microwave burglar alarms, and laser particle velocimeters.

10.5.1 Continuous Wave Doppler Ultrasound

Continuous wave ultrasound systems are used to measure fluid flow and the movement of internal organs (particularly the heart), as shown in Figure 10.15. In this example, the narrow beam pattern and the displacement between the transmit and receive transducers offers some range discrimination. These devices are particularly popular with pediatricians and soon-to-be parents who use them to listen to the fetal heartbeat.

One of the largest applications for acoustic Doppler is in industrial applications where Doppler flow meters are used to measure flow in pipes and canals. In all cases they rely on the acoustic signal reflecting off the moving medium or suspended particulates within it.

Most applications are interested in a single frequency signal which is an indication of the velocity of a single target. However, there can be a large amount of additional information encoded in the Doppler spectrum that can be interpreted to advantage. For example, if a spectrogram is produced, then it is possible to identify specific target characteristics that vary with time. Figure 10.16 shows spectrograms of a walking human subject and a large dog both moving

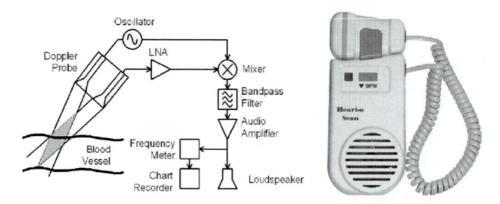

Figure 10.15 Generic continuous wave Doppler instrumentation measuring internal blood flow.

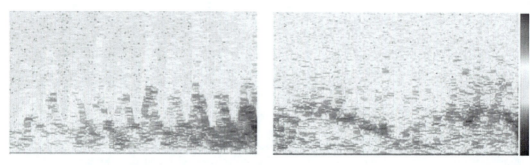

Figure 10.16 Doppler spectrograms of (a) a human subject and (b) a large dog both walking toward the radar (Dropmann 2005).

at the same average speed. Examination of the figure confirms that it is easy to discriminate between the two targets, notwithstanding the fact that the average Doppler is the same.

10.5.2 Continuous Wave Doppler Radar

Continuous wave Doppler radar has a myriad of uses. These include short-range intruder detection, moderate-range sports and police radar for measuring the velocity of moving targets, and long-range aircraft tracking applications.

10.5.2.1 Intruder Detection The main requirement for intruder detection Doppler is low cost. This requires an innovative approach to their manufacture, as shown in Figure 10.17. Instead of building up the circuit with individual components connected by waveguide, everything is housed within a single die-cast module. The oscillator portion of the sensor comprises a cavity containing a Gunn diode, a tuning screw, and sometimes a varactor diode to allow for electronic tuning. This is iris coupled into a second cavity which contains a mixer diode and a second tuning screw. This cavity opens into a horn.

When biased correctly, the Gunn diode becomes a negative resistance device. When inserted into a cavity that is tuned to the correct resonant frequency, it therefore functions as an oscillator.

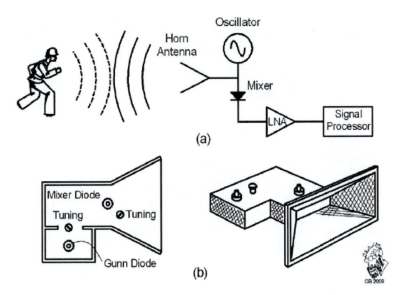

Figure 10.17 Continuous wave Doppler intruder detection: (a) a circuit for measuring Doppler frequency, and (b) Gunn microwave Doppler module.

The tuning screw and varactor diode can be used to alter the cavity resonance and hence change the oscillation frequency slightly. Iris coupling to the second cavity ensures that the coupling is reasonably small, and also that the Q of the oscillator cavity remains sufficiently high to produce stable, narrow-band oscillations.

In the mixer cavity which couples to the outside world through a horn antenna, portions of the transmitted and received signals set up radio frequency (RF) voltages across the mixer diode. Because this device (also sometimes known as a detector) is nonlinear, the RF voltages are mixed (multiplied together), with the result that the sum and difference of the two signal frequencies is produced. A low-pass filter element blocks the sum but allows the difference frequency signal through. If there is a difference between the transmitted and received signals introduced by a moving target-induced Doppler shift, this will be output as the difference frequency.

Specifications of a typical Gunn-based intruder system include:

- Operational frequency typically X-band (8–12 GHz) or K-band (18–27 GHz).
- Output power 1–10 mW.
- Receiver sensitivity −95 dBc (dB with respect to the transmitter power).
- Frequency drift with temperature 500 kHz/°C typical.
- Operational principle, Gunn oscillator iris-coupled to the common antenna port that includes a mixer diode.
- Sometimes a varactor diode is mounted in the Gunn diode cavity to allow for electronic control of the oscillation frequency.
- The antenna port is generally flared into a horn to constrain the beam angle.

Recently, low-cost field effect transistor (FET)-based dielectric resonant oscillators (DROs) have begun to replace the Gunn modules; these are manufactured with either waveguide outputs

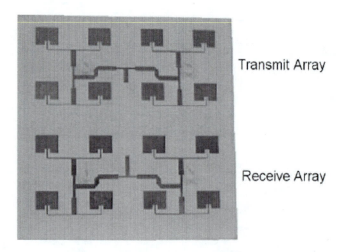

Figure 10.18 Microstrip array antenna for an intruder detection module.

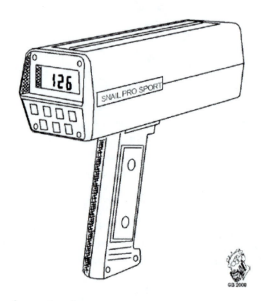

Figure 10.19 Handheld Doppler sports radar.

or as planar devices with a small microstrip antenna array, as shown in Figure 10.18. The operational principle of these is similar to the Gunn devices except that in this case separate transmit and receive antennas are used.

10.5.2.2 Sports Radar   Most sports radar units operate in the K-band because the use of the higher frequency results in a narrower antenna beam from a reasonably sized aperture. The hand-held unit with a pistol grip shown in Figure 10.19 is typical of the genre, although tripod-mounted units are becoming more common (Radar Guns 2007).

Specifications of a typical sports radar include:

- Operational frequency typically K-band (24.15 GHz).
- Output power 40–100 mW.
- Antenna generally a large horn or a horn lens.
- Range depends on antenna size, output power, and target radar cross section (RCS), but it is typically about 50 m for a tennis ball-size object.
- Accuracy ±1 km/hr (typical).

The introduction of high-performance Doppler radars that actually track the ball (see Chapter 13) have introduced a new perspective into a number of sports, including cricket, tennis, and golf, among others. Conventional Doppler radars only measure the radial velocity component of the target and require that the radar be positioned directly in front of the sportsman to obtain an accurate measurement.

By tracking the ball in flight, and therefore obtaining an estimate of its angular velocity, range, and range rate, it is possible to determine its actual velocity from any perspective. In the sports arena, where the initial range to the player is known (e.g., a pitcher on the mound), updates to the range can be obtained by integrating the range rate, and it is not necessary to measure the range independently of the rate.

10.5.2.3 Police Radar Speed Traps A typical police radar includes a pistol-grip gun, a dash-mounted processor, and a display unit. The processor unit has an input for the patrol car's speed which allows these units to be used from a moving vehicle with little loss in accuracy (Radar Guns 2007; Sawicki 2007).

Typical specifications for a range of police radars include

- Operational frequency X-band, K-band, or Ka-band.
- Transmit power 34–130 mW (depending on the manufacturer).
- Antenna beamwidth 13°–26° (depending on the manufacturer).
- Accuracy ±0.2% microwave, ±0.15 mph internal signal processing.
- Patrol car speed range 16–75 mph.
- Target speed range 15–199.
- Target detection distance approximately 1000 m.

10.5.2.4 Worked Example: Police Radar and Detector Comparison This example analyzes the radar and the detector capabilities to confirm that the radar detector can pick up the radiation from the police radar before the latter can get a fix on the vehicle velocity.

Police Radar For a K-band radar with a transmit power of 120 mW and a 3 dB beamwidth of 17.5°, the detection range for a car can de determined using the radar range equation, as follows:
The transmit power, P_t (dBm), is

$$P_t = 10\log_{10} 120 = 20.8 \text{ dBm}.$$

The wavelength is determined from the transmitter frequency and the speed of light to be

$$\lambda = c/f = 0.0124 \text{ m}.$$

The antenna gain assumes that a pencil beam is used, therefore $\theta_{3dB} = \varphi_{3dB} = 17.5°$ and the gain is

$$G = 10\log_{10}\frac{4\pi}{\theta_{3dB}\varphi_{3dB}}$$
$$= 10\log_{10}\frac{4\pi\times57.29^2}{17.5^2} = 21.3 \text{ dB}$$

The system bandwidth is determined by the maximum Doppler shift that is expected by the radar. It should be able to accommodate a patrol car speed of 75 mph plus a maximum target speed of 199 mph ($v_r = 122.5$ m/s). The maximum Doppler shift is therefore

$$f_{dmax} = 2v_r/\lambda$$
$$= (2\times122.5)/0.0124 = 19.75 \text{ kHz}.$$

Assuming a good low-noise amp (LNA) with a noise figure, NF, of 2.5 dB, followed by down-conversion to baseband and a matched filter with a bandwidth, β, of 20 kHz, the receiver noise floor, N (dBm), will be

$$N = 10\log_{10}kT\beta + NF + 30$$
$$= 10\log_{10}(1.38\times10^{-23}\times290\times20\times10^3) + 2.5 + 30 = -128.5 \text{ dBm}.$$

It can be assumed that the RCS of a typical car seen head-on will usually be larger than 5 m². Therefore, assuming that this is the minimum:

$$\sigma_{dB} = 10\log_{10}5 = 7 \text{ dBm}^2.$$

Applying the radar range equation

$$P_r = P_t + 2G + 10\log_{10}\frac{\lambda^2}{(4\pi)^3} + \sigma_{dB} - 40\log_{10}R$$
$$= 20.8 + 2\times21.3\times10\log_{10}\frac{0.0124^2}{(4\pi)^3} + 7 - 40\log_{10}R.$$
$$= -0.7 - 40\log_{10}R$$

The procedure to extract a single target from the received signal generally involves using a comparator with hysteresis to produce a square wave and then counting the number of cycles over a given period. MATLAB code can easily be written to perform this procedure, and the results show that the relationship between the SNR and the percentage error for a nominal count of 50,000 cycles is comparable to that shown in Table 10.1.

Table 10.1 Relationship between SNR and count error for a Doppler radar

SNR (dB)	Error %
10	0
9.2	0.0024
8.5	0.0233
7.96	0.10
7.45	0.29
6.99	0.62
6.57	1.13
6.19	1.87
5.85	2.83
5.53	4.0
5.22	5.36
4.0	15.0

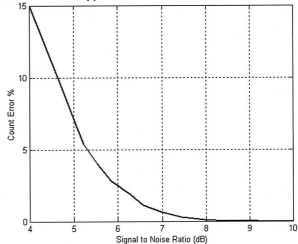

An acceptable error from the radar is stated to be 0.25%, which equates to a SNR of about 7.5 dB. Therefore the minimum detectable signal is

$$S_{min} = N + 7.5$$
$$= -128.5 + 7.5 = -121 \text{ dBm.}$$

Solving for R_{rad} gives

$$-121 = -0.7 - 40\log_{10} R$$

$$R_{rad} = 10^{(121-0.7)/40} = 1017 \text{ m.}$$

Police radars are available in a number of different frequencies driven primarily by the countermeasures race. Soon after the introduction of the first X-band units, enterprising car owners started to use radar warning receivers, and the race began.

Warning Receiver Unfortunately for law enforcement, a receiver can detect the radar emissions at a longer range than the radar can detect the car. A typical radar detector has the following bandwidth specifications:

- X-band: 10.45 GHz ± 25 MHz.
- K-band: 24.15 GHz ± 100 MHz.
- Ka-band: 37.4 GHz ± 1.3 GHz.

It can be assumed that the device is only interested in radars over part of the forward hemisphere, therefore an antenna gain of 6 dB can be assumed. This is followed, for the X-band and K-band channels, by LNAs, a heterodyne down-conversion process, and then detection.

The radar receiver must have a wide enough bandwidth to cover all possible transmit frequencies, which at K-band is specified to be ±100 MHz. The noise floor, N (dBm), including an LNA with a 4 dB noise figure, will be

$$
\begin{aligned}
N &= 10\log_{10} kT\beta + NF + 30 \\
&= 10\log_{10}\left(1.38\times10^{-23}\times290\times200\times10^6\right) + 4 + 30 \\
&= -87 \text{ dBm.}
\end{aligned}
$$

It is shown during the derivation of the radar range equation that the incident power density at the target, S_i (W/m^2), is

$$
S_i = P_t G_t / 4\pi R^2.
$$

The power received by the radar detector is the product of the power density and the effective aperture, A_r (m^2) of the receiver. Given that the relationship between the aperture of an antenna and its gain is

$$
A_r = G_r \lambda^2 / 4\pi,
$$

the received power is

$$
P_r = P_t G_t G_r \lambda^2 / (4\pi)^2 R^2.
$$

In decibel form this is

$$
\begin{aligned}
P_r &= P_t + G_r + G_r + 20\log_{10}\frac{\lambda}{4\pi} - 20\log_{10} R \\
&= 20.8 + 21.3 + 6 - 60 - 20\log_{10} R \\
&= -11.9 - 20\log_{10} R
\end{aligned}
$$

Assuming that the detection and false alarm probabilities are $P_d = 0.9$ and $P_{fa} = 10^{-6}$, then the SNR required is 13.2 dB for a matched filter receiver. The minimum detectable signal level is therefore

$$
S_{min} = -87 + 13.2 = -73.8 \text{ dBm.}
$$

Solving for R_{rec} gives

$$
-73.8 = -11.9 - 20\log_{10} R
$$

$$
R_{rec} = 10^{(73.8-11.9)/20} = 1245 \text{ m.}
$$

This analysis has shown that the radar detector will pick up the transmitted signal at a longer range than the radar can detect the car even though it must operate over a 200 MHz bandwidth.

This margin could be improved considerably if the bandwidth could be narrowed. For example, if the received signal was split into four channels, each with a 50 MHz bandwidth, the noise floor would be reduced by 6 dB and the detection range extended to 2500 m. Another alternative is to sweep the center frequency of an even narrower band receiver to extend the detection range still further.

10.5.2.5 Projectile Tracking Radar   In military applications, it is often sufficient to track an aircraft using the Doppler returns, as these are largely immune to echoes from static sources of clutter (Barton 1976). The schematic of a Doppler tracker and a typical application is shown in Figure 10.20. In this application, the antenna is tracked manually in angles using one of the telescopes, and automatic tracking is carried out only on the velocity. A narrow-band tracking filter in the intermediate frequency (IF) is used to reject noise and crosstalk from the transmitter. This filter is followed by a discriminator that provides an error signal which drives an automatic frequency control (AFC) loop, ensuring that the received frequency ($f_{tx} \pm f_d$) is down-converted to a constant IF, f_o, irrespective of the actual Doppler frequency.

There will be a minimum Doppler frequency below which the transmitter spillover will swamp the received signal. However, at any frequency above this, the Doppler signal will be sufficiently large (as shown in Figure 10.21) to ensure that closed-loop velocity tracking is possible.

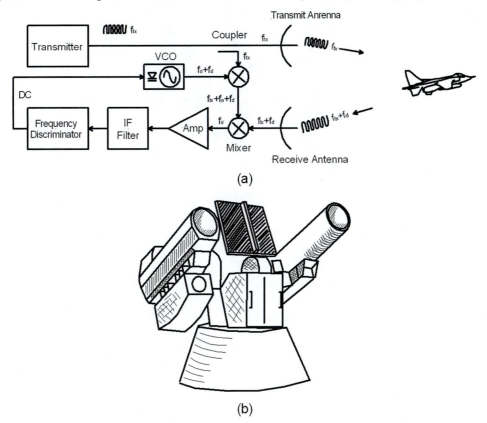

(a)

(b)

Figure 10.20 Continuous wave Doppler aircraft tracking: (a) schematic diagram and (b) application coupled to optical telescopes [adapted from Barton (1976)].

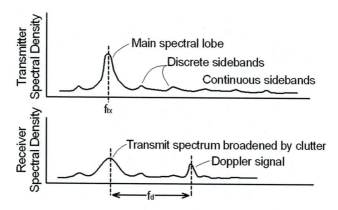

Figure 10.21 Continuous wave Doppler spectra for the tracking radar example [adapted from Barton (1976)].

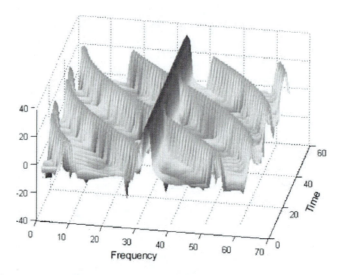

Figure 10.22 Idealized Doppler spectrum of a moving tank.

10.5.2.6 Doppler Target Identification   As shown in Figure 10.16, Doppler signatures of moving targets can be used to differentiate between target types. From a military perspective, this discriminatory capability is extremely useful. For example, a high-cost missile seeker needs to know whether the target it has detected is a high-value tank or helicopter and not a low-value truck. Figure 10.22 shows the Doppler spectrum generated from the tracks of a moving tank, which can easily be distinguished from the single-line return of a moving truck.

Doppler spectra from helicopters are also unique in that they include a body return line surrounded by the hub spectrum and a return from the blades, as shown in Figure 10.23. The blade flash in particular is very wide, but fleeting, so although the instantaneous RCS is high, the total power returned is small.

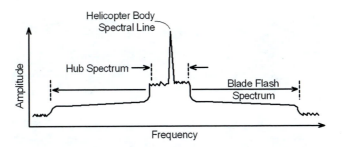

Figure 10.23 Idealized Doppler spectrum of a helicopter [from Currie and Brown (1987)].

10.5.3 Pulsed Doppler Ultrasound

Medical ultrasound imaging is discussed in detail in Chapter 16. At its simplest, it involves the generation of two-dimensional (2D) image slices through the human body using a sector scan probe combined with pulsed time-of-flight range measurement. An intensity image is formed based on the amount of power reflected within each range cell. Because reflections occur with changes in acoustic impedance, these images are representative of the different tissue types encountered.

Range resolution, ΔR (m), is achieved by transmitting a short burst of ultrasound, with the resolution given by

$$\Delta R = c\tau/2, \tag{10.25}$$

where c (m/s) is assumed to be a constant 1540 m/s and τ (sec) is the pulse width, which depends on the required resolution, but is typically about 1 μs long, giving a resolution of 1 mm.

In conventional ultrasound, the received echo is detected and sampled at the appropriate rate and then displayed. However, to extract Doppler information, the echo must be synchronously detected, as illustrated in Figure 10.11, and the Doppler shift in each range cell extracted and displayed in color. Because the direction of movement is important, one of the techniques discussed earlier in this chapter must be used to accomplish this.

False-color Doppler information can be distracting, so most ultrasound instruments have the facility to set up a sampling window that covers a subset of the range and angular coverage provided by the scanner. This is illustrated in Figure 10.24, where blood flow through an umbilical cord is displayed.

As with most sensing applications, a compromise must be reached between the angular resolution, which increases with increasing frequency, and the penetration depth, which increases as the frequency is decreased. As a rule of thumb, the acoustic losses are 0.3 dB/cm/MHz. Typical medical ultrasound can operate over a wide range of frequencies between 1 and 18 MHz.

The image shown in Figure 10.24 is made at a center frequency of 3 MHz, which equates to a wavelength, λ, of 0.5 mm. The Doppler shift corresponding to a velocity of 10 cm/s will be

$$
\begin{aligned}
f_\mathrm{d} &= 2v_\mathrm{r}/\lambda \\
&= \left(2\times10\times10^{-2}\right)/\left(0.5\times10^{-3}\right) = 400 \text{ Hz}.
\end{aligned}
$$

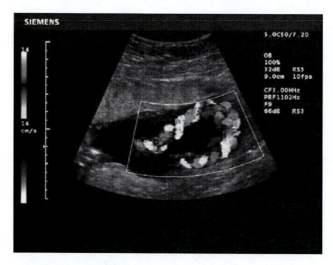

Figure 10.24 Doppler ultrasound image of uterus and umbilical cord.

Obviously such a low frequency cannot be reproduced during a single pulse and the Doppler information must be extracted over a number of pulses, as shown in Figure 10.12c. From the text in Figure 10.24, we know that the image was made at 10 frames per second (fps) with a pulse repetition frequency (PRF) of 1102 Hz, which satisfies the Nyquist sampling criterion by a reasonable margin. The PRF is selected to correspond to the full observable radial speed, from −14 cm/s to +14 cm/s. Using Nyquist again, at a frame rate of 10 fps, the lowest unambiguous frequency is 5 Hz, which corresponds to a velocity of 1.25 mm/s.

10.5.4 Pulsed Doppler Radar

Pulsed Doppler radar systems, such as the NEXRAD weather surveillance radar (National Weather Service 2007) shown in Figure 10.25, operate using the principles discussed earlier in this chapter. The RF return signal in each range gate is down-converted to baseband using the I/Q principle as listed in the specifications.

Transmitter
- Type: S-band, coherent chain (STALO/COHO)
- Frequency: 2700–3000 MHz
- Power: 750 kw peak at klystron output
- Transmitter to antenna loss: site dependent, 2 dB typical
- Average power: 300–1300 W
- Pulse widths: 1.57 and 4.5 μs (−6 dB points)
- PRF short pulse: 318–1304 Hz
- PRF long pulse: 318–452 Hz
- Phase noise (system): −54 dBc required, −60 dBc typical

Figure 10.25 Photograph of a NEXRAD radar in Knoxville, TN.

Antenna/Pedestal
- Type: center-fed paraboloid of revolution 8.5 m in diameter
- Polarization: linear horizontal
- Gain at 2850 MHz: 45.5 dB (including radome loss)
- Beamwidth at 2850 MHz: 0.925°
- First sidelobe: −29 dB (others less than −40 dB beyond 10°)
- Radome: fiberglass foam sandwich, frequency tuned, 12 m sphere
- Radome two-way loss: 0.24 dB at 2850 MHz

Receiver
- Type: coherent (stalo/coho), first down-convert to IF, instantaneous automatic gain control and matched filtering, second convert for synchronous detection (I and Q input to ADC)
- Dynamic range: 93 dB (generally achieves 95 dB)
- Intermediate frequency: 57.55 MHz
- 3 dB bandwidth: 0.630 MHz
- 6 dB bandwidth: 0.798 MHz
- System noise figure: 4.6 dB (540 K)
- Receiver noise: −113 dBm referenced to antenna

Signal Processor
- ADC sample interval: 1.66 μs, 602 kHz (approximately)
- ADC number of bits: 12
- Clutter filter: infinite impulse response (five-pole elliptic)
- Suppression: 30–50 dB, user selectable
- Notch half-widths: 0.5–4 m/s
- Range increment: 250 m
- Azimuth increment: 1°

The specifications for the radar confirm that it uses the conventional architecture with a pulsed transmitter, and in-phase and quadrature receiver outputs. Because the radar operates unambiguously in range and the Doppler frequency is reasonably low, the outputs will vary with time, as shown in Figure 10.12c for $f_d < 1/\tau$). The digitized Doppler spectrum can be extracted from the time waveform of each gate by processing blocks of data through a complex fast Fourier transform (FFT) algorithm. The frequency and phase information output from each range gate is converted to a radial velocity component.

Range and reflectivity or range and velocity maps such as those displayed in Figure 10.26 can then be produced by displaying the data for all range gates over the 360° scan angle of the radar. The reflectivity map gives a good indication of the rainfall rate, while the velocity map can be used to determine the direction in which the storm is traveling and also the wind speed. It must be remembered, however, that only the radial velocity component is measured, so that if the wind direction is at right angles to the beam, velocities will be represented with zero or near-zero values.

10.6 DOPPLER TARGET GENERATOR

Characterizing Doppler sensor performance in the laboratory is not as simple as placing a corner reflector at a known range because the target must be moving. An effective method of generating a moving target is to mount a pair of corner reflectors on a spinning beam, as shown in Figure 10.27. If necessary, the beam can be covered with radar-absorbing material to minimize its contribution to the overall cross section. In addition, the antenna should be aimed so that it illuminates only the one corner reflector facing the sensor.

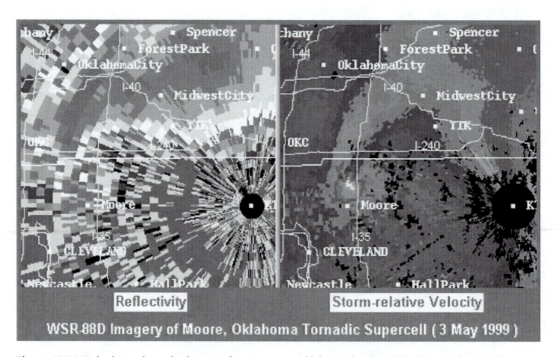

Figure 10.26 Pulsed Doppler radar images of a storm over Oklahoma (National Weather Service 2007).

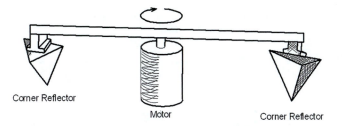

Figure 10.27 Doppler target made up of a pair of spinning corner reflectors.

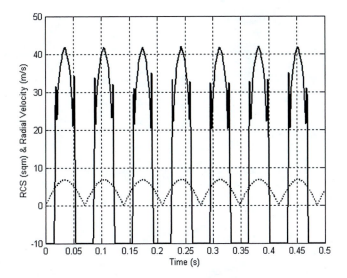

Figure 10.28 Radar cross-section and radial velocity of a single reflector in the Doppler target.

The performance of the Doppler target has been simulated for an arm radius of 0.15 m and a rotation rate of 7 rps. The radial velocity in the direction of the Doppler sensor peaks at 6.6 m/s when the incidence angle is 90°, reducing to zero after the target has rotated by a further 90°. The reflector is modeled from data on trihedral reflectors presented in Chapter 8. It has an appreciable RCS from −45° to +45° with the peak occurring when the radial velocity is a maximum, as shown in Figure 10.28.

For a millimeter wave radar sensor operating at 94 GHz, the amplitude of the return from the reflector is determined by variations in the RCS with the angle. For the specified rotation rate, the flash lasts about 30 ms and produces the spectrum shown in Figure 10.29.

10.7 CASE STUDY: ESTIMATING THE SPEED OF RADIO-CONTROLLED AIRCRAFT

The Doppler shift of the engine frequency can easily be used to determine the speed of a radio-controlled plane. All that is required is a reasonable quality recording of the sound as the aircraft flies past the observer while the aircraft throttle setting is kept constant. For typical aircraft speeds, a 6–10 s recording is sufficient.

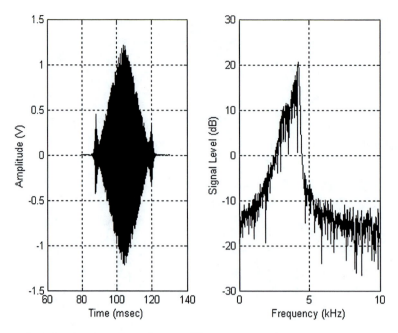

Figure 10.29 Received signal and spectrum from a millimeter wave radar.

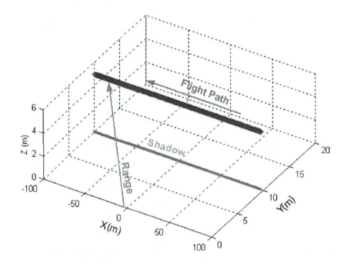

Figure 10.30 Radio-controlled aircraft flight scenario showing a crossing target parallel to the *x*-axis at a height of 5 m and an offset of 10 m.

10.7.1 Background

Consider the scenario shown in Figure 10.30 with an aircraft moving along a route parallel to the *x*-axis at a speed of 33 m/s. At any time during the crossing, the range, *r* (m), to the target can be determined using the Cartesian to polar transform

$$r = \sqrt{x^2 + y^2 + z^2},$$

(10.26)

where the point (x,y,z) (m) is the instantaneous position of the aircraft in Cartesian space. The range rate can easily be obtained by taking the derivative, with respect to time, of the range. These are shown in Figure 10.31.

It is obvious that the steepness of the crossover point is dependent on the minimum range at crossover. In the limit, if the aircraft is flying directly toward the observer, the crossover would be a step.

Because the source is transmitting, the Doppler shift can then be determined from equation (10.7), which has been reproduced here:

$$f_r = \frac{c - v_r}{c - v_s} f_s.$$

Rewriting for $v_r = 0$, because the observer is not moving, gives

$$f_r = \frac{c}{c - v_s} f_s. \tag{10.27}$$

The relationship between the received frequency, the transmitted frequency, and the range rate, v_s, can be obtained. In reality though, the source frequency, f_s, is unknown. In this example it is assumed to be 3.4 kHz, and the received Doppler frequency is then determined from the range rate shown in Figure 10.31. Note that the Doppler frequency, plotted in Figure 10.32, takes the form of the range rate curve with the sign reversed, because an approaching target has a negative range rate but its Doppler shift will be positive.

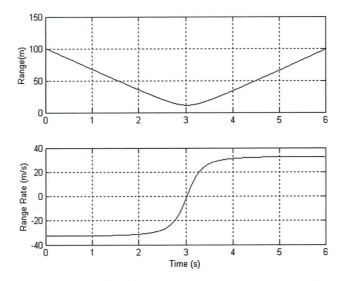

Figure 10.31 Range and range rate of a crossing target.

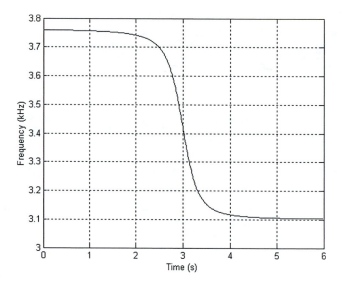

Figure 10.32 Frequency of a crossing target shifted by Doppler effects.

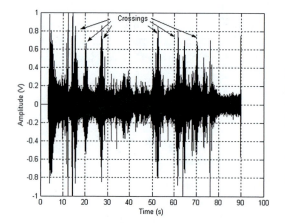

Time (sec)	Manoeuvre
16	Flypast
21	Flypast
26	Flypast
38	Slow flypast
53	Flypast with acceleration
62	Flypast
65	Flypast
71	Flypast with turn
80	Slow flypast

Figure 10.33 Sound file of a ducted fan RC aircraft performing a number of flybys and other maneuvers.

10.7.2 Measured Data

A radio-controlled model plane with an electric motor driving a ducted fan has been recorded during a number of controlled flybys, some circuits, and a sudden acceleration. The sound file is shown in Figure 10.33.

Because of the nature of the motor, the sound spectrum is not a single frequency, but a number of harmonics determined by the number of blades in the fan. This results in a rather complex spectrum with the fundamental at about 600 Hz, with clusters of three harmonics at around 2.5 kHz and 5 kHz, as shown in Figure 10.34.

To make use of the Doppler information from the crossing target, a sequence of closely spaced spectra must be obtained. These produce what is known as a spectrogram, in which the x-axis

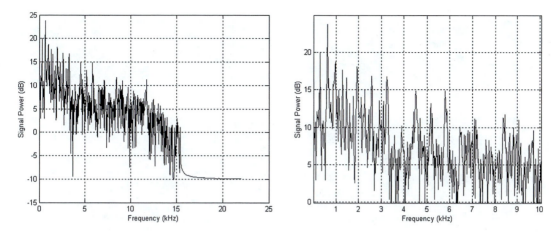

Figure 10.34 Frequency content of engine noise: (a) complete spectrum up to the cutoff frequency of the recorder, and (b) magnified view of the spectrum below 10 kHz showing all of the engine harmonics.

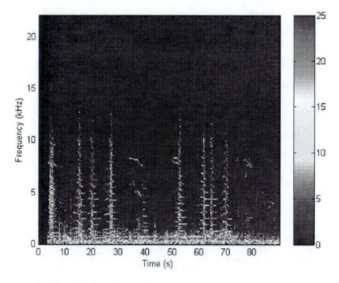

Figure 10.35 Spectrogram for the RC aircraft.

represents time, the y-axis represents the frequency, and color encodes for the magnitude of the frequency component.

In Figure 10.35, each of the "Christmas trees" contains the spectra of a flyby or some other maneuver during which the aircraft approached the sound recorder close enough for the signal to be analyzed. At this resolution, it is just possible to see that the spectra consist of a number of harmonics, but Doppler shifts cannot easily be seen.

In Figure 10.36a and Figure 10.36b, in which a small section of the spectrogram is magnified, more details become visible. In Figure 10.36a, three flybys are visible, with the frequency on the left side of each Christmas tree showing the Doppler-shifted frequency of the approaching

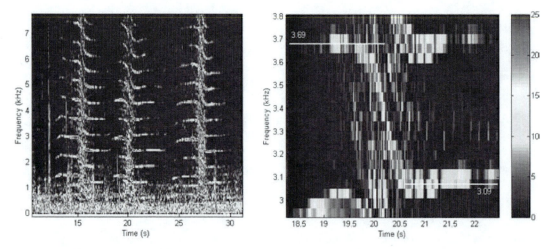

Figure 10.36 Spectrogram for the RC aircraft showing (a) spectra from three flybys, and (b) a single harmonic from a single flyby.

aircraft and that on the right showing the target receding. In each case, more than 10 harmonics are visible. In Figure 10.35b, a single harmonic has been isolated so that the salient features can be measured and the aircraft speed determined. It will be shown that the choice of harmonic is not important so long as the tails are clearly visible. It can be seen that this last figure is very similar to Figure 10.32, and obviously represents a flyby with a minimum range comparable to that used in the simulation.

The easiest features of this figure to measure are the asymptotic frequencies of the approaching and the receding aircraft; these will be referred to as f_1 and f_2, for which the following two equations can be written:

$$f_1 = \frac{c}{c - v_s} f_s,$$ (10.28)

$$f_2 = \frac{c}{c + v_s} f_s.$$ (10.29)

Taking the ratio of $f_1/f_2 = k$ results in an equation that does not include the unshifted frequency, f_s:

$$\frac{f_1}{f_2} = k = \frac{c}{c - v_s} \cdot \frac{c + v_s}{c} = \frac{c + v_s}{c - v_s}.$$ (10.30)

Solving for the target speed, v_s, gives the following:

$$v_s = \frac{c(k - 1)}{k + 1}.$$ (10.31)

From the figure, the asymptotic frequencies are $f_1 = 3.69$ kHz and $f_2 = 3.09$ kHz, making $k = 1.194$. It is important to note that this ratio remains unchanged, irrespective of the harmonic that is chosen.

The speed of sound in air can be estimated from the temperature, T (K) using

$$c \approx 20.05\sqrt{T}. \tag{10.32}$$

Assuming that the air temperature is 10°C, the speed of sound, c, is 337.3 m/s, and the aircraft speed will be

$$v_s = \frac{337.3 \times (1.194 - 1)}{1.194 + 1} = 29.8 \text{ m/s}.$$

10.8 References

Barton, D. (1976). *Radar Systems Analysis*. Norwood, MA: Artech House.

Currie, N. and Brown, C. (1987). *Principles and Applications of Millimeter-Wave Radar*. Norwood, MA: Artech House.

Dropmann, D. (2005). Doppler radar recognition system. University of Sydney, Sydney, NSW, Australia.

Liley, D. (1998). The Doppler principle. Viewed February 2001. Available at http://liley.physics.swin. oz.au.

National Weather Service. (2007). Radar operations center—NEXRAD WSR-88D. Viewed December 2007. Available at http://www.roc.noaa.gov/.

Radar Guns. (2007). Radar guns. Viewed December 2007. Available at http://www.radarguns.com/.

Sawicki, D. (2007). Police traffic radar handbook: a comprehensive guide to speed measuring systems. Viewed December 2007. Available at http://copradar.com/preview/content.html.

Skolnik, M. (1980). *Introduction to Radar Systems*. Tokyo: McGraw-Hill Kogakusha.

11

High Range-Resolution Techniques

11.1 CLASSICAL MODULATION TECHNIQUES

The range resolution of a sensor is defined as the minimum separation (in range) of two targets of equal cross section that can be resolved as separate targets. It is determined by the bandwidth of the transmitted signal, Δf (Hz), which is generated by widening the transmitter bandwidth using one of the following modulation forms:

- Amplitude modulation,
- Frequency modulation,
- Phase modulation.

11.2 AMPLITUDE MODULATION

A special case of amplitude modulation technique is classical pulsed radar, where the amplitude is 100% for a very short period and 0% for the remaining time, as illustrated in Figure 11.1.

11.2.1 Range Resolution

Range resolution is determined from matched filter processing of the rectangular pulse. Consider the case where the transmitted signal consists of a constant frequency signal modulated by a rectangular pulse of width τ (s). The sharp edges of the rectangular time function generate an infinite frequency spectrum, a truncated version of which is shown in Figure 11.2. It can be seen from the frequency response that most energy is concentrated in the frequency band between the zero crossings at $f - 1/\tau$ and $f + 1/\tau$, making the bandwidth Δf_x (Hz),

$$\Delta f_x \approx 2/\tau. \tag{11.1}$$

It is the convention to define the width of a pulse at the points where the power level has dropped to 50% of the peak, the so-called 3 dB points. This occurs over a bandwidth Δf_{3dB} (Hz),

$$\Delta f_{3dB} \approx 1/\tau. \tag{11.2}$$

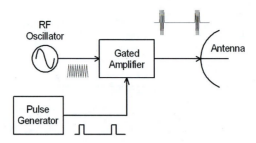

Figure 11.1 On-off amplitude modulation of a sine wave to produce pulses.

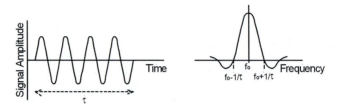

Figure 11.2 The relationship between the waveform and spectrum of a rectangular pulse.

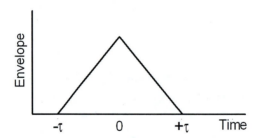

Figure 11.3 Matched filter output for a generic rectangular pulse of duration, τ.

When a rectangular pulse is processed through a perfect matched filter (correlator) it produces a triangular output envelope 2τ wide at the base and with a well-defined central peak, as shown in Figure 11.3. If a second return is displaced from the first by a time delay of τ (s), depending on the relative phases of the two targets, the response envelope can take on any of the shapes shown in Figure 11.4. For all phase differences, the second peak is still identifiable and the two targets are said to be resolved in range.

The range resolution, δR (m), is determined by converting the time delay, τ (s), to the round-trip distance required to achieve that delay:

$$\delta R = c\tau/2. \tag{11.3}$$

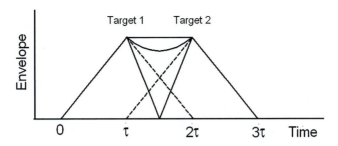

Figure 11.4 Matched filter output of a pair of closely spaced targets showing the limits to the range resolution.

Using the relationship from equation (11.2), the range resolution determined in terms of the pulse width can be rewritten in terms of the effective bandwidth of the signal:

$$\delta R = c/2\Delta f. \tag{11.4}$$

The operational range of a sensor is proportional to the amount of energy transmitted, therefore, to achieve really long range as well as good range resolution, pulsed systems must operate with incredibly high peak powers, sometimes in excess of 10 MW. This requires that special precautions be taken to minimize the problems of ionization and arcing within the waveguide for radar systems or in the air for high-power lasers. Sometimes these requirements are impossible, or at least impractical to implement, and alternatives must be sought.

The first option is to distribute the transmitter spatially by creating a phased array of elements, each of which radiates a moderate amount of power. This solution is discussed in detail in Chapter 12. An alternative is to distribute the transmitted signal temporally by decoupling the range resolution from the duration of the pulse and then lengthening the pulse to decrease the required peak power.

11.3 FREQUENCY AND PHASE MODULATION

If the range resolution of a waveform is considered from a frequency or bandwidth perspective, it is obvious that the poor range resolution of a fixed-frequency continuous wave radar is related to the narrow spectrum of its transmitted waveform. Therefore, if good range resolution is required, the bandwidth of the signal must be widened in some way. Pulsed systems achieve this by reducing the width of the pulse as discussed above, but it is also possible to achieve the same result by applying some form of frequency or phase modulation to a longer pulse. The received signal can then be processed using a matched filter that compresses the long pulse to a duration of $1/\Delta f$ (s).

The process of radiating a long pulse that has been modulated over a bandwidth, Δf (Hz), and then compressing it into a short duration pulse on reception is known as "pulse compression." The performance of these systems is often characterized by the time-bandwidth product, $\Delta f \tau$, of the uncompressed pulse, which is used as a figure of merit.

11.3.1 Matched Filter

A pulse compression receiver is the practical implementation of a matched filter. Figure 11.5 shows that there are a number of different ways of achieving this. In Figure 11.5a, the matched

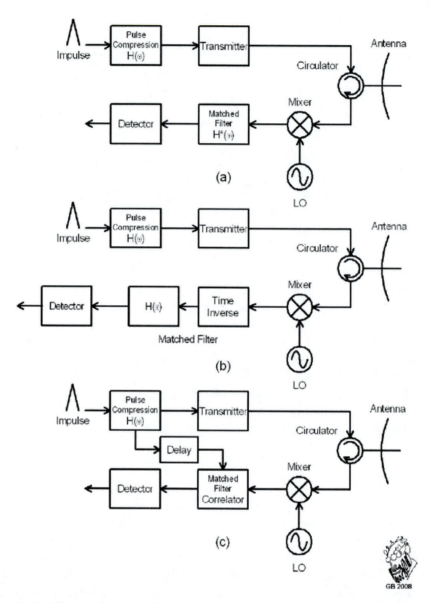

Figure 11.5 Matched filter configurations for pulse compression using (a) conjugate filters, (b) time inverse, and (c) correlation [adapted from Skolnik (1990)].

filter is the complex conjugate, $H^*(\omega)$, of the transfer function, $H(\omega)$, of the pulse expansion process. In Figure 11.5b, the matched filter is implemented by time reversing the received signal before passing it through the same transfer function, $H(\omega)$. Finally, in Figure 11.5c, the process is implemented by correlating the received waveform with the transmitted one.

From a mathematical perspective, a coded signal can be described either by the transfer function, $H(\omega)$, or as an impulse response, $h(t)$, of the coding filter. The received echo is fed into a matched filter whose frequency response is the complex conjugate $H^*(\omega)$ of the coding filter. The output of the matched filter, $y(t)$, is the compressed pulse, which is just the inverse Fourier transform of the product of the signal spectrum and the matched filter response:

$$y(t) = \frac{1}{2\pi} \int_{-\infty}^{\infty} |H(\omega)|^2 \exp(j\omega t)\, d\omega. \tag{11.5}$$

A filter is also matched if the signal is the complex conjugate of the time inverse of the filter's impulse response. This is often achieved by applying the time inverse of the received signal to the pulse compression filter. The output of this matched filter is given by the convolution of the signal, $b(t)$, with the conjugate impulse response $b^*(-t)$ of the matched filter:

$$y(t) = \frac{1}{2\pi} \int_{-\infty}^{\infty} b(\tau) b^*(t - \tau)\, d\tau. \tag{11.6}$$

In essence, the matched filter provides a correlation of the received signal with a delayed version of the transmitted signal, as shown in Figure 11.5c.

The effects of this form of processing on two pulses with the same duration are shown in Figure 11.6. In the continuous frequency (CF) example, the matched filter (correlation) response shows the triangular envelope described earlier. However, in the chirp example, with the same

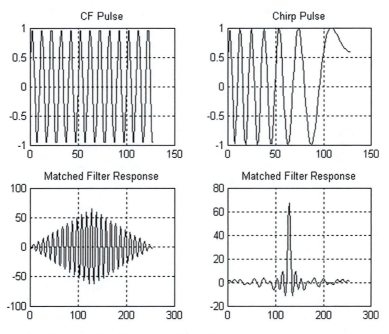

Figure 11.6 Comparison between the ultimate resolution of a rectangular constant-frequency pulse and a chirp pulse of the same duration.

duration, the matched filter generates a sinc function with a much narrower peak because a much wider bandwidth has been transmitted. The introduction of this chirp function serves to decouple the transmitted bandwidth from the pulse duration, as proposed at the start of this chapter.

11.4 PHASE-CODED PULSE COMPRESSION

A simple way to understand pulse compression with a matched filter is to consider the binary phase-shift keying (BPSK) modulation technique. In this modulation, the code is made up of *m* chips which are either in-phase—0° (positive)—or out-of-phase—180° (negative)—with respect to a reference signal, as shown in Figure 11.7.

Demodulation can be achieved by multiplying the incoming radio frequency (RF) signal by a coherent carrier (a carrier that is identical in frequency and phase to the carrier that originally modulated the BPSK signal). This produces the original BPSK signal plus a signal at twice the carrier frequency which can be filtered out. Unfortunately it is quite difficult to reproduce the coherent carrier, and a more common technique that is used widely in radar is shown in Figure 11.8.

The received signals are band-passed by a filter matched to the data rate (not shown); the outputs are then demodulated by I and Q detectors. These detectors compare the phase of the received signal to the phase of the local oscillator (LO), which is also used in the RF modulator. Though the phase of each of the transmitted signals is 0° or 180° with respect to the LO, on receive the phase will be shifted by an amount dependent on the round-trip time and the Doppler velocity. For this reason, two processing channels are generally used, one that recovers the in-phase signal and one that recovers the quadrature signal. These signals are digitized by the analog-to-digital converters (ADCs), correlated with the stored binary sequence, and then combined.

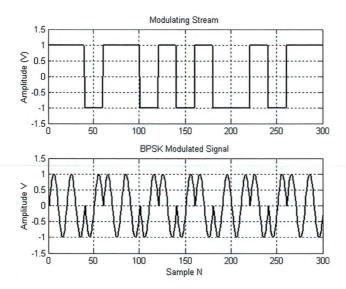

Figure 11.7 Example of BPSK using one cycle per bit.

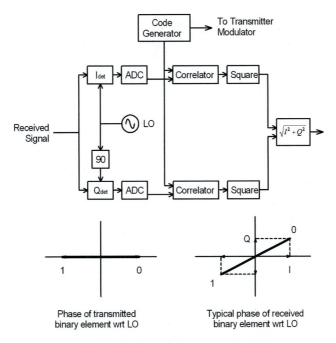

Figure 11.8 BPSK receiver and demodulator [adapted from Skolnik (1990)].

The advantages of this configuration are that they utilize the coherence of the system to produce two quadrature receive channels. If only one channel is implemented, there is a loss in effective signal-to-noise ratio (SNR) of 3 dB and it is possible that the received signal can be sampled at the zero crossing, where no phase information is available (this is akin to the blind speed problem with Doppler radar). The echo is compressed by correlation with the stored reference of the signal that was transmitted in a process that is the discrete equivalent of the matched filter described earlier in this chapter.

11.4.1 Barker Codes

A number of different binary code sequences have been developed. Ideally these should exhibit low range-sidelobe levels to ensure that all of the energy is concentrated in the main lobe. There should also be very low cross correlation with other sequences generated by the same process to minimize the possibility of interference.

Special cases of these binary codes are the Barker codes, where the peak of the autocorrelation function is N (for a code of length N) and the magnitude of the maximum peak sidelobe is 1. All of the known sequences are shown in Table 11.1 because, unfortunately, no Barker codes with lengths greater than 13 have been found. Barker code sequences are called optimum because, for zero Doppler shift, the peak-to-sidelobe ratio is $\pm n$ after matched filtering (where n is the number of bits).

The matched filter for a five-element Barker code, $+++-+$, will have a filter matched to the element length, τ_c (s), with bandwidth $\beta = 1/\tau_c$ (Hz). This is followed by a tapped delay line

Table 11.1 Barker code sequences

Code length	Code elements	Sidelobe level (dB)
2	+– or ++	–6
3	++–	–9.5
4	++–+ or +++–	–12
5	+++–+	–14
7	+++––+–	–16.9
11	+++–––+––+–	–20.8
13	+++++––++–+–+	–22.3

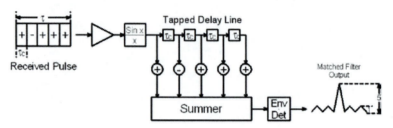

Figure 11.9 Diagram to illustrate the concept of phase-coded pulse compression for a 5-bit Barker code.

having four delays, each τ_c, the outputs of which are weighted by the code +–+++ and summed prior to envelope detection, as shown in Figure 11.9.

In the general case, the output consists of $m - 1$ time sidelobes of unit amplitude G_v and a main lobe with amplitude mG_v each of width τ_c. The ratio of the transmitted pulse width to the output pulse width is $\tau/\tau_c = \beta\tau$, which is the pulse compression ratio. The relative sidelobe power level is $1/m^2$.

The Barker code is the only code that has equal sidelobes at this low level, but this only applies along the zero Doppler axis. If the target that produced the echo pulse is moving toward or away from the radar, then the phase of the echo will change due to the changing range. This will alter the phase relationship between the elements of the expanded pulse and modify the way that they combine on compression. Examination of the ambiguity diagram for a 13-bit Barker code in Figure 11.10 shows that the main lobe decays quickly and sidelobes increase rapidly with increasing Doppler shift (Lewis et al. 1986).

From a resolution perspective, the continuous wave phase-coded techniques offer the same performance as their analog counterparts. However, processing still remains a potential problem as the computationally expensive autocorrelation process is required to extract the range information from the return echo.

11.4.2 Random Codes

Random and pseudo-random codes do not necessarily offer the same low sidelobes as do the Barker codes, but they have many advantages, including ease of generation and no restriction as to length.

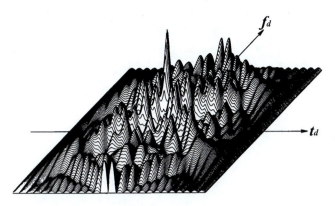

Figure 11.10 Ambiguity diagram for a 13-bit Barker code showing the "thumbtack" main lobe decaying into a sea of increasing delay and Doppler sidelobes (Lewis et al. 1986).

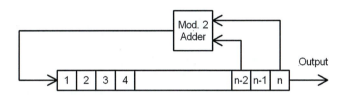

Figure 11.11 Shift register-based pseudo-random code generator.

11.4.2.1 Optimal Binary Sequences The definition of an optimal binary sequence is one whose peak sidelobe of the aperiodic autocorrelation function is the minimum possible for a given code length. Most of the optimal codes are found by computer searches; however, the search time becomes prohibitively long as N increases, and it is often easier to resort to the use of other nonoptimal sequences as long as they posses the desired correlation effects.

Maximal length sequences that are particularly useful are those that can be obtained from linear feedback shift registers. These have a structure similar to random sequences and therefore possess desirable autocorrelation functions. They are often called pseudo-random (PR) or pseudo-noise (PN) sequences.

A typical shift register generator is shown in Figure 11.11. The n stages of the register are preloaded with all 1s or a combination of 1s and 0s (all 0s is not used as it results in an all 0 output). The outputs from specific individual stages of the shift register are summed by modulo-2 addition to form its new input.

Modulo-2 addition depends only on the number of 1s being added. Odd numbers produce the sum of one, and even, the sum of zero. The shift register is clocked and the output at any stage is the binary sequence. When the feedback connections are properly chosen, the output is a sequence of maximal length N, where $N = 2^n - 1$, where n is the number of stages of the shift register. There are a total of M maximal length sequences that can be obtained from a generator with n stages, where M is given by

$$M = \frac{N}{n} \prod \left(1 - \frac{1}{p_i} \right), \tag{11.7}$$

where p_i are the prime factors of N.

The number of different sequences existing for a given n is important, particularly in applications such as collision avoidance, where a number of different radar units will be sharing the same area and the potential for mutual interference exists. A list of optimal binary codes is reproduced in Table 11.2.

Table 11.2 Optimal binary codes (Skolnik 1990)

Length of code (N)	Magnitude of peak sidelobe	Number of codes	Code (octal notation for N > 13)	Octal	Binary
2	1	2	11, 10	0	000
3	1	1	110	1	001
4	1	2	1101, 1110	2	010
5	1	1	11101	3	011
6	2	8	110100	4	100
7	1	1	1110010	5	101
8	2	16	10110001	6	110
9	2	20	110101100	7	111
10	2	10	1110011010		
11	1	1	11100010010		
12	2	32	110100100011		
13	1	1	1111100110101		
14	2	18	36324		
15	2	26	74665165		
16	2	20	141335		
17	2	8	265014		
18	2	4	467412		
19	2	2	1610445		
20	2	6	3731261		
21	2	6	5204154		
22	3	756	11273014		
23	3	1021	32511437		
24	3	1716	44650367		
25	2	2	163402511		
26	3	484	262704136		
27	3	774	624213647		

Table 11.2 *continued*

Length of code (N)	Magnitude of peak sidelobe	Number of codes	Code (octal notation for N > 13)	Octal	Binary
28	2	4	1111240347		
29	3	561	3061240333		
30	3	172	6162500266		
31	3	502	16665201630		
32	3	844	37233244307		
33	3	278	55524037163		
34	3	102	144771604524		
35	3	222	223352204341		
35	3	322	526311337707		
37	3	110	1232767305704		
38	3	34	2251232160063		
39	3	60	4516642774561		
40	3	114	14727057244044		

In this example, a binary sequence is output, with a new value being generated with every clock cycle. However, larger random numbers can be obtained by using a number of sequential bits to form a word.

From a radar perspective, a BPSK sequence of length N will have a time-bandwidth product of N, where the bandwidth of the system is determined by the clock rate. This allows for the generation of large time-bandwidth products (which result in good range resolution) from registers having a reasonably small number of stages. By altering the clock rate and the length and feedback connections on the shift register, it is possible to produce, without additional hardware, waveforms of various pulse lengths, bandwidths, and time-bandwidth products to suit most radar requirements. Table 11.3 lists the length and number of maximal length sequences obtained from shift registers of various lengths along with the feedback connection required to generate one of the sequences.

If the shift register is left in continuous operation, a continuous repeating waveform is generated that can be used for continuous wave operation. Aperiodic waveforms are obtained if the generator output is terminated after one complete sequence, and these are generally used for pulsed radar applications. The autocorrelation functions of the two cases vary in terms of their sidelobe structure.

Maximal length sequences have characteristics which approach the three characteristics ascribed to truly random processes:

- The number of 1s is approximately equal to the number of 0s.
- runs of consecutive 1s and 0s occur, with about half the runs having length 1, a quarter having length 2, an eighth length 3, etc.
- The autocorrelation is thumbtack in nature (peaked in the center and approaching zero elsewhere).

Table 11.3 Maximum length sequences

Number of stages (n)	Length of maximal sequence (N)	Number of maximal sequences (M)	Feedback-stage connections
2	3	1	2,1
3	7	2	3,2
4	15	2	4,3
5	31	6	5,3
6	63	6	6,5
7	127	18	7,6
8	255	16	8,6,5,4
9	511	48	9,5
10	1023	60	10,7
11	2047	176	11,9
12	4095	144	12,11,8,6
13	8191	630	13,12,10,9
14	16383	756	14,13,8,4
15	32767	1800	15,14
16	65535	2048	16,15,13,4
17	131071	7712	17,14
18	262143	7776	18,11
19	524287	27594	19,18,17,14
20	1048575	24000	20,17

Maximal length sequences are an odd length, so to create a power of 2 sequence for processing purposes (as discussed below), a zero is inserted at the start or the end of the sequence. This results in degraded sidelobes.

11.4.3 Correlation

11.4.3.1 Binary Correlation   For binary sequences where the values are restricted to ±1, a comparison counter, as illustrated in Figure 11.12, is used to form the correlation. In this case the transmitted sequence is loaded into the reference register and the input sequence is continuously clocked through the signal shift register. A comparison counter forms a sum of the matches and subtracts the mismatches between corresponding stages of the shift registers on every clock cycle to produce the correlation function. This process can easily be implemented in hardware and is therefore very fast to execute.

11.4.3.2 Circular Correlation   For correlation of two long sequences that are not necessarily binary, the Fourier transform method should be used. This involves taking Fourier transforms of both sequences, followed by the product of the one series with the complex conjugate of the other, and finally, the inverse Fourier transform completes the procedure, as shown in Figure 11.13.

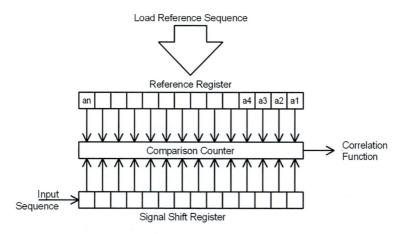

Figure 11.12 Digital correlation (Skolnik 1990).

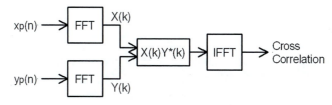

Figure 11.13 Cross correlation using the Fourier transform (Rabiner and Gold 1975).

As an example, a MATLAB simulation has been written using Fourier transform-based circular correlation. This has been used to determine the ability of the technique to extract returns from multiple targets and to compare the effects of processing an even and an odd length sequence.

The results reproduced in Figure 11.14 show that the different target amplitudes are reproduced if the received signals are not binarized. More importantly, if the correct odd length maximal length sequence is generated, the range sidelobes remain very small. In contrast, if an even length sequence is generated, the sidelobe levels fluctuate more, and can reach values up to 100 times as large.

11.5 SURFACE ACOUSTIC WAVE-BASED PULSE COMPRESSION

The original pulse compression radars were based on their pulsed counterparts, as shown in Figure 11.15. A reasonably conventional radar configuration is implemented with two additional components. In the transmit chain, the narrow pulse passes through a surface acoustic wave (SAW) expander prior to transmission, and on the receive side, the echo passes through a SAW compressor prior to the detector.

Considering the transmit side in more detail, when a narrow pulse is generated, it contains a wide range of frequencies as a consequence of its brevity. This pulse passes through a dispersive delay line (SAW expander) in which its components are delayed in proportion to their frequency.

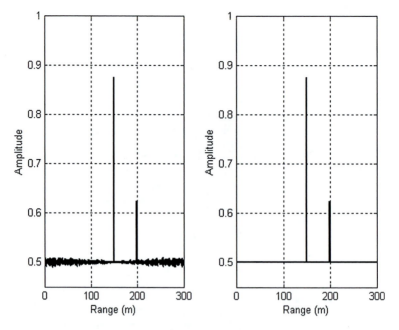

Figure 11.14 Cross-correlation processing showing two targets: (a) BPSK radar noise sequence generated using a 12-bit shift register with 4096 points, and (b) a similar sequence with 4095 points.

In the process, the pulse is stretched to produce the linear chirp seen in Figure 11.16. The dispersion process can be quite considerable, for example, a 10 ns pulse can be lengthened by a factor of 100 to a duration of 1 μs before it is up-converted, amplified, and transmitted.

The echo returns from the target are down-converted and amplified before being fed into a pulse compression network that retards the echo by amounts that vary inversely with frequency to reduce the signal to its original 10 ns length, as shown in Figure 11.17. The compressed echo yields nearly all of the information that would have been available had the unaltered 10 ns pulse been transmitted. A slight sacrifice in range resolution ($\approx$1.3) is the penalty incurred in reducing the range sidelobes from -13.2 dB with no weighting to -43 dB with Hamming weighting.

One of the characteristics of the pulse compression technique is that even if the uncompressed target echoes are overlapping, the matched filter compression process untangles the returns to produce independent targets. This capability is demonstrated in Figure 11.18, in which a pair of overlapping pulses are summed and then processed through a matched filter to produce a pair of target echoes.

The SNR gain achieved by the pulse compression process is approximately equivalent to the pulse time-bandwidth product, $\beta\tau$. Using SAW technology to implement the pulse expansion and compression functions limits the maximum $\beta\tau$ product to about 100, but even so, it is the most common method in use because it is both compact and robust. Modern

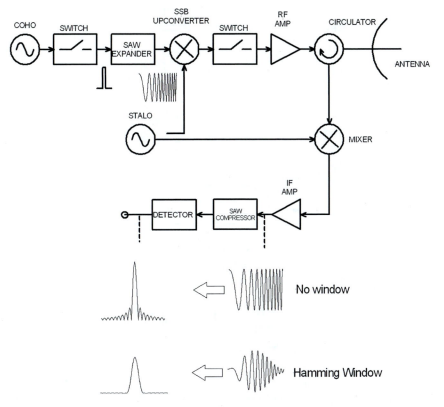

Figure 11.15 Conceptual diagram of a linear chirp pulse compression radar [adapted from (Currie and Brown 1987), reproduced with permission (c) 1987 Artech House].

digital techniques can be used to produce $\beta\tau$ products in the thousands if required (Brookner 1982).

11.6 STEP FREQUENCY

Step-frequency modulation, also known as step-chirp, provides a piecewise approximation of the linear chirp signal. It consists of a sequence of different frequencies spaced Δf (Hz) apart with duration $\tau_f = 1/\Delta f$ (s). The total length, τ_t (s), of the transmission is

$$\tau_t = N/\Delta f = N\tau_f, \tag{11.8}$$

and the transmitted bandwidth, β_t (Hz), is

$$\beta_t = N\Delta f = 1/\tau_c, \tag{11.9}$$

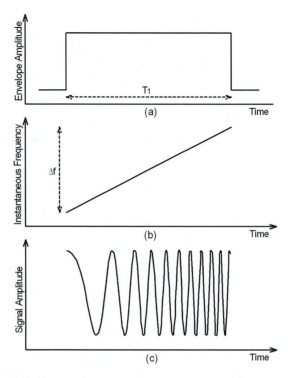

Figure 11.16 Linear chirp pulse (a) transmitter pulse envelope, (b) transmitter pulse frequency, and (c) transmitted pulse RF waveform [adapted from (Wehner 1995)].

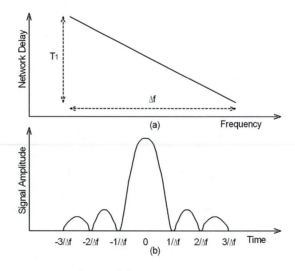

Figure 11.17 Chirp pulse compression characteristics.

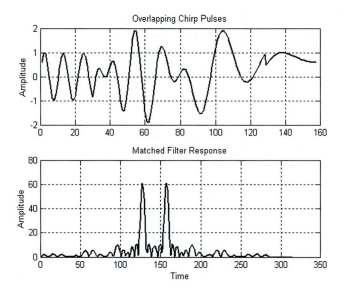

Figure 11.18 Matched filter output for two overlapping chirp pulses.

making the pulse compression ratio

$$\rho = \tau_t / \tau_c = \tau_t \beta_t. \tag{11.10}$$

The step-frequency modulation code is not as Doppler tolerant as the linear frequency modulated (FM) code. Large grating lobes appear in the compressed pulse sidelobes at Doppler shifts that are odd multiples of $\Delta f/2$. Some techniques, including Costas coding, nonlinear FM, and amplitude modulation (AM) have been developed to improve the sidelobe performance.

The step-frequency technique generally relies on a phase-locked oscillator to generate the transmitted signals rather than a free-running voltage controlled oscillator (VCO) used by frequency modulated continuous wave (FMCW). This ensures that although the same homodyne process is applied in the two cases, the magnitude of the close-in phase noise is lower for the step-frequency process. This results in improved performance at short range for low transmit power.

If an interrupted version were implemented, the frequency would have to remain constant for the round-trip time to the target, and hence the synthesis of the required range resolution, that requires N samples, would require N times the period required by FMCW to synthesize the same return.

Step frequency is an ideal method of obtaining extremely high range-resolution measurements if sufficient time is available. The technique is often used in ground penetrating radar and is also the basis for inverse synthetic aperture radar (ISAR) measurements in controlled environments.

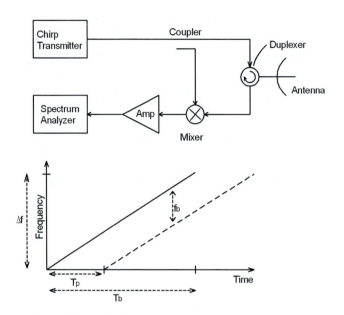

Figure 11.19 Schematic diagram illustrating the FMCW concept.

11.7 FREQUENCY MODULATED CONTINUOUS WAVE RADAR

11.7.1 Operational Principles

A homodyne radar generally refers to a continuous wave radar in which the microwave oscillator serves as both the transmitter and LO. Such a configuration is often used by FMCW radars. As shown in Figure 11.19, the continuous wave signal is, in this case, modulated in frequency to produce a linear chirp that is radiated toward a target through an antenna. The echo received after time T_p (s), by the same antenna, is mixed with a portion of the transmitted signal to produce a beat signal at a frequency f_b (Hz). From the graphical representation of this process, it is clear that the frequency of this signal will be proportional to the round-trip time.

For an analytical explanation, the change in frequency, ω_b (Hz), with time, generally referred to as chirp, can be described by

$$\omega_b = A_b t, \tag{11.11}$$

where A_b is a constant of proportionality and t (s) is the time. Substituting into the standard equation for FM, discussed in Chapter 2, results in

$$v_{fm}(t) = A_c \cos\left[\omega_c t + A_b \int_{-\infty}^{t} t\, dt\right] \tag{11.12}$$

$$v_{fm}(t) = A_c \cos\left[\omega_c t + \frac{A_b}{2} t^2\right]. \tag{11.13}$$

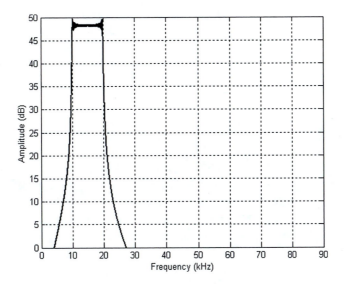

Figure 11.20 Frequency domain representation of a linear FM chirp.

This analysis assumes that the frequency continues to increase indefinitely, but in practice the transmitter has a limited bandwidth and the chirp duration is limited, as shown in Figure 11.19. This results in the spectrum shown in Figure 11.20, which in this example is limited to frequencies from 10 to 20 kHz. Note that the effects of truncation generate ripples across the spectrum, with the largest amplitudes occurring at the two extremities.

In FMCW systems, a portion of the transmitted signal is mixed with the returned echo, by which time the transmit signal will be shifted from that of the received signal because of the time, T_p (s), taken for the signal to propagate to the target and back to the receiver. The transmitted signal can be written as

$$v_{fm}(t - T_p) = A_c \cos\left[\omega_c(t - T_p) + \frac{A_b}{2}(t - T_p)^2\right].$$ (11.14)

Calculating the product of equation (11.13) and equation (11.14), as occurs in the mixer, gives

$$v_{fm}(t - T_p)v_{fm}(t) = A_c^2 \cos\left[\omega_c t + \frac{A_b}{2}t^2\right]\cos\left[\omega_c(t - T_p) + \frac{A_b}{2}(t - T_p)^2\right].$$ (11.15)

Equating using the trigonometric identity for the product of two sines,

$$\cos A \cos B = 0.5[\cos(A + B) + \cos(A - B)],$$ (11.16)

produces

$$v_{out}(t) = \frac{A_c^2}{2} \left[\begin{array}{l} \cos\left\{ (2\omega_c - A_b T_p)t + A_b t^2 + \left(\frac{A_b}{2} T_p^2 - \omega_c T_p \right) \right\} \\ + \cos\left\{ A_b T_p t + \left(\omega_c T_p - \frac{A_b}{2} T_p^2 \right) \right\} \end{array} \right]. \tag{11.17}$$

The first cosine term in equation (11.17) describes a linearly increasing FM signal (chirp) at about twice the carrier frequency with a phase shift that is proportional to the delay time, T_p (s). This term is generally actively filtered out in radar systems because it is beyond the cutoff frequency of the mixer and subsequent receiver components.

The second cosine term describes a beat signal at a fixed frequency, and this can be determined by differentiating, with respect to time, the instantaneous phase term,

$$f_b = \frac{1}{2\pi} \frac{d}{dt} \left[A_b T_p t + \left(\omega_c T_p - \frac{A_b}{2} T_p^2 \right) \right], \tag{11.18}$$

to give

$$f_b = (A_b / 2\pi) T_p. \tag{11.19}$$

It can be seen that the signal frequency is directly proportional to the delay time, T_p (s), and hence is directly proportional to the round-trip distance to the target, as postulated. The spectrum shown in Figure 11.21 includes both the fixed frequency and the chirp terms for illustrative purposes, even though, in general, only the low-frequency component is output.

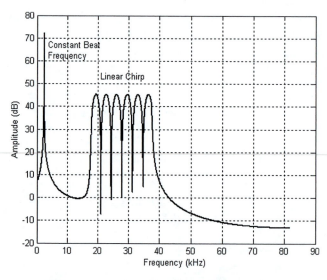

Figure 11.21 Frequency domain representation of the FMCW receiver output, including both the high- and low-frequency components after mixing but before filtering.

In these examples, and for most FMCW implementations, spectral analysis is performed using the standard fast Fourier transform (FFT), as it is robust and reasonably fast. However, there are other potentially more accurate techniques that can be used if the number of spectral components is known. These include autoregressive (AR), autoregressive moving average (ARMA), and minimum-entropy methods, as well as spectral parameter estimation, including the now famous MUSIC algorithm.

11.7.2 Matched Filtering

Linear chirp is the most Doppler-tolerant process that is commonly implemented in analog form today, as it concentrates most of the volume of the ambiguity function into the main lobe and not into the Doppler or delay sidelobes. There is, however, a cross-coupling between Doppler and range, which is discussed at length later in this chapter.

It was shown earlier that a matched filter can be implemented by correlating the received signal with a delayed version of the transmitted signal. In the FMCW case, this function is most often implemented by taking the product of the received signal and the transmitted signal and filtering to obtain a constant frequency beat. The spectrum is determined using the Fourier transform or a similar spectral estimation process. This is not a true matched filter, and losses in SNR occur as a result.

If the chirp duration is T_b (s), the spectrum of the beat signal will be resolvable to an accuracy of $2/T_b$ (Hz), between minima. This assumes that $T_b >> T_p$, so that the signal duration is $\tau \approx T_b$, as shown in Figure 11.19. It is common practice to define the resolution bandwidth of a signal, δf_b (Hz), between its 3 dB (half-power) points, which in this case fall within the $1/T_b$ region centered on f_b, as shown in Figure 11.22.

The rate of change of frequency (chirp slope) in the linear case is constant and equal to the total frequency excursion, Δf (Hz), divided by the chirp time, T_b (s). The beat frequency can then be calculated:

$$f_b = \frac{A_b}{2\pi} T_p = \frac{\Delta f}{T_b} T_p. \tag{11.20}$$

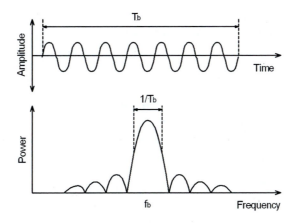

Figure 11.22 Spectrum of the truncated sinusoidal signal output by an FMCW radar.

From the basics of radar, the round-trip time, T_p (s), to the target and back can be written in terms of the range as

$$T_p = 2R/c, \tag{11.21}$$

and substituting into equation (11.20) gives the classical FMCW formula that relates the beat frequency and the target range:

$$f_b = \frac{\Delta f}{T_b} \frac{2R}{c}. \tag{11.22}$$

The range, R (m), to the target can therefore be determined from the beat frequency:

$$R = \frac{T_b c}{2\Delta f} f_b. \tag{11.23}$$

If, instead of writing equation (11.23) in terms of the range and beat frequency, it is written in terms of the range resolution, δR (m), and the frequency resolution, δf (Hz), then the range resolution is

$$\delta R = \frac{T_b c}{2\Delta f} \delta f_b. \tag{11.24}$$

It was shown earlier that $\delta f_b \approx 1/T_b$, which when substituted into equation (11.24) results in the same closed relationship between the total transmitted bandwidth and the range resolution:

$$\delta R = c/2\Delta f. \tag{11.25}$$

This is the same relationship that was derived for equation (11.4) at the start of this chapter, and is a satisfying result as it represents the pulse compression and FMCW equivalents of the classical pulsed radar range-resolution equation, where $\tau = 1/\Delta f$.

11.7.3 The Ambiguity Function

The ambiguity function for linear FM pulse compression, stretch, and for FMCW is derived below. It shows that there is a strong cross-coupling between the Doppler shift and the measured range. For a target with radial velocity, v_r (m/s), the magnitude of the coupling can be determined by replacing the echo signal in equation (11.14) with its Doppler-shifted counterpart:

$$v_{fm}(t - T_p) = A_c \cos\left[\omega_c(t - T_p) + \frac{A_b}{2}(t - T_p)^2 - \frac{2v_r}{c}\omega_c(t - T_p)\right]. \tag{11.26}$$

Processing as before to determine the new beat frequency, f_b, it is just the old beat frequency offset by the Doppler shift

$$f_b = \frac{2v_r}{c} f_c - \frac{A_b}{2\pi} T_p$$
$$= f_d - \frac{A_b}{2\pi} T_p. \tag{11.27}$$

The ambiguity function can be expressed as

$$|\chi_o(t_d, f_d)|^2 = \left[\frac{\sin\left[\pi\left(f_d - \frac{A_b}{2\pi} t_d\right)(T_b - |t_d|)\right]}{\pi\left(f_d - \frac{A_b}{2\pi} t_d\right)(T_b - |t_d|)} \left(1 - \frac{|t_d|}{T_b}\right) \right]^2 \quad \text{for} -T_b \le t_d \le T_b$$
$$= 0 \qquad\qquad\qquad\qquad\qquad \text{for } |t_d| > T_b \tag{11.28}$$

For a typical FMCW radar with a 150 MHz chirp over a 1 ms interval ($A_b \approx 10^{12}$), the beat frequency, in the absence of the Doppler shift, is about 1 MHz at a range of 1 km. The Doppler shift at 94 GHz is 625 Hz/m/s, which equates to just less than the theoretical range resolution of the waveform at 1 m/s. Higher velocities, from a moving vehicle, for example, would introduce significant errors in the measured range which would need to be accounted for.

A cut along the Doppler axis is similar to that of the single pulse because the pulse width is the same, but the modulation is different. A cut along the time delay axis changes considerably, as it is now much narrower and corresponds to the compressed pulse width where $\tau_c = 1/\Delta f$.

In Figure 11.23, an increasing Doppler shift results in a decreasing measure of the range because a rising-frequency chirp is used. This is an indication of the cross-coupling. For a

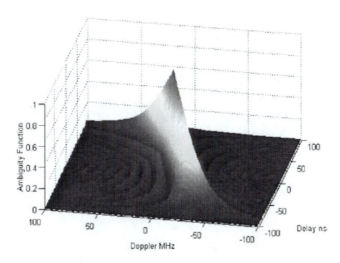

Figure 11.23 Linear FM up-chirp ambiguity diagram for a 100 ns duration signal showing the interaction between delay and Doppler.

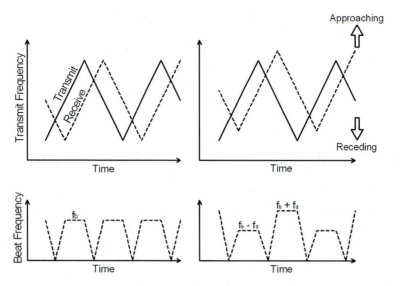

Figure 11.24 Effects of Doppler shift on beat frequency of an FMCW radar.

decreasing-frequency chirp, the sense of the function is reversed to produce a mirror image of this ambiguity diagram. By combining the two slopes using a triangular modulation, it is possible to obtain an unbiased estimate of the target range and the Doppler shift. A moving target will therefore superimpose a Doppler frequency shift on the beat frequency, as shown in Figure 11.24.

For a triangular waveform, one portion of the beat frequency will be increased and the other portion will be decreased. For a target approaching the radar, the received signal frequency is increased (shifted up in the diagram), decreasing the up-sweep beat frequency and increasing the down-sweep beat frequency:

$$f_{b(up)} = f_b - f_d \tag{11.29}$$

$$f_{b(dn)} = f_b + f_d. \tag{11.30}$$

The beat frequency corresponding to range can be obtained by averaging the up and down sections:

$$f_r = \frac{f_{b(up)} + f_{b(dn)}}{2}. \tag{11.31}$$

The Doppler frequency (and hence target velocity) can be obtained by measuring one-half of the difference frequency:

$$f_d = \frac{f_{b(up)} - f_{b(dn)}}{2}. \tag{11.32}$$

The roles are reversed if $f_d > f_b$.

11.7.4 Effect of a Nonlinear Chirp

As shown conceptually in Figure 11.25, if the chirp is not linear, the standard matched filter assumptions for resolution are not satisfied and range resolution will suffer. For illustrative purposes, the transmit frequency consists of a pair of linear sections with different slopes. This results in two distinct beat frequencies, f_{b1} and f_{b2}. However, in general, chirp nonlinearity is a smooth function, and the resultant beat frequency will be smeared.

It can be shown (Brooker 2005) that if the chirp nonlinearity is quadratic in nature, then the range resolution becomes proportional to the slope linearity and the range to the target:

$$\delta R = RLin, \tag{11.33}$$

where the linearity, *Lin*, is defined as the change in chirp slope, $S = df/dt$, normalized by the minimum slope:

$$Lin = (S_{max} - S_{min})/S_{min}. \tag{11.34}$$

This sensitivity to slope linearity is one of the fundamental problems that limits the resolution of real FMCW radar systems. The ultimate range resolution of an FMCW sensor is determined from both the total swept bandwidth and the chirp nonlinearity, which, as they are independent processes, can be combined in quadrature:

$$\delta R_{tot} = \sqrt{\delta R^2 + \delta R_{lin}^2}. \tag{11.35}$$

It should be realized, however, that this is only an approximation, and the true value will depend on the nature of the nonlinearity and the relative contributions of the two components.

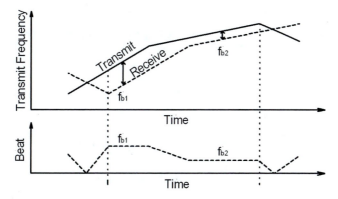

Figure 11.25 Effect of chirp nonlinearity on the beat frequency.

11.7.5 Chirp Linearization

11.7.5.1 Open Loop Techniques A common method to linearize the chirp slope in a VCO-based system uses a programmed correction stored in an erasable programmable read-only memory (EPROM) and clocking this data through a digital-to-analog converter (DAC). Because of the varying characteristics of the VCO with temperature, either the temperature must be controlled, or a number of different curves must be implemented and referenced appropriately. The latter is easily achieved, as shown in Figure 11.26, where the upper bits of the lookup table are driven by a digital representation of the temperature and the lower bits address a particular voltage entry in that table.

Glitches that are generated during some of the DAC transitions generate noise and clock harmonics on the RF signal which are difficult to remove by filtering, and it is only since a new generation of low glitch-power DACs has become available that this technique has become really practical.

An all-analog implementation is shown in Figure 11.27, which can reduce the nonlinearity of a well-behaved VCO by a factor of 10. This circuit uses an analog multiplier chip to produce a quadratic voltage that is added to the linear ramp to perform the correction. A DC offset is often included in the circuit to set the start frequency.

11.7.5.2 Determining the Effectiveness of Linearization Techniques The obvious method of determining the effectiveness of a linearization technique is to examine the beat-frequency spec-

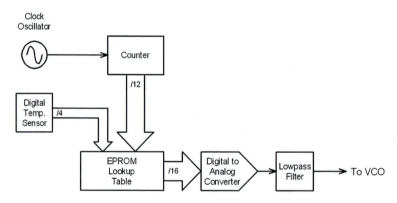

Figure 11.26 Schematic diagram of a linear chirp generator based on a lookup table.

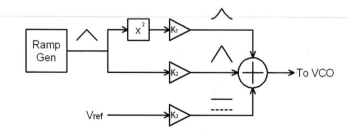

Figure 11.27 Quadratic frequency chirp correction circuit using an analog multiplier chip.

trum for a point target at a reasonable range (>500 m). However, this is often not practical, and an alternative, more compact method is required.

In essence, all an FMCW radar does is mix a portion of the transmitted signal with the received signal to produce a beat signal, the frequency of which is proportional to the range. As the name implies, a delay line discriminator (shown in Figure 11.28) performs the same function using an electrical delay line rather than the genuine round-trip delay to a target and back. The most basic delay line is simply a length of coaxial or fiber-optic cable, but these are usually too bulky for practical applications. It is common to use SAW or bulk acoustic wave (BAW) devices to perform this function.

Disadvantages of SAW delay lines are their high insertion loss (>35 dB), limited bandwidth (<300 MHz), and an operating frequency of less than 1 GHz. For millimeter wave radar applications, the VCO frequency must therefore be down-converted to an appropriate IF to take advantage of commercially available components.

The spectrum of the output of the discriminator is examined to determine the effectiveness of the linearization process. The center frequency, f_c (Hz), defines the chirp slope and the 3 dB bandwidth, β_{3dB} (Hz), of the signal divided by the center frequency, gives the linearity:

$$Lin = \beta_{3dB} / f_c. \tag{11.36}$$

In Figure 11.29, the measured discriminator outputs are shown for a Hughes VCO, both completely unlinearized and after open-loop linearization. In this case, because the nonlinearity is primarily quadratic, the analog linearizer, shown in Figure 11.27, was capable of reducing the width of the signal from 80 kHz to 10 kHz, which implies an improvement in linearity from 0.26 to just over 0.03.

11.7.5.3 Implementation of Closed-Loop Linearization   The delay line discriminator can be used as a feedback element to close the linearization loop using a classical phase-lock loop (PLL), if it is remembered that the loop must maintain a constant rate of change of frequency and not frequency as is more usual.

The delay line discriminator is effectively a differentiator in the frequency domain and produces a constant output frequency if the frequency slope (rate of change of frequency) is constant. Thus to close the loop correctly, an integrator must be included in the feedback path to produce the loop structure shown in Figure 11.30.

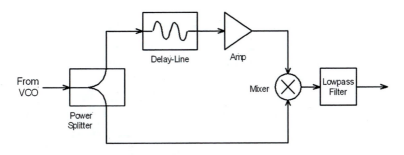

Figure 11.28 Schematic diagram of a delay line discriminator.

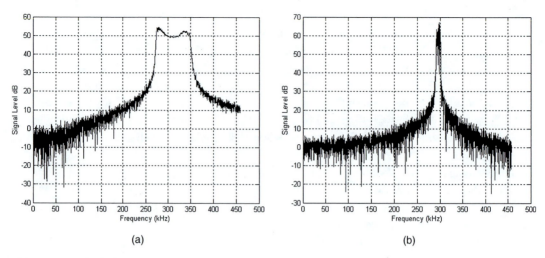

Figure 11.29 Discriminator output spectra for (a) an unlinearized Hughes VCO and (b) after open-loop correction.

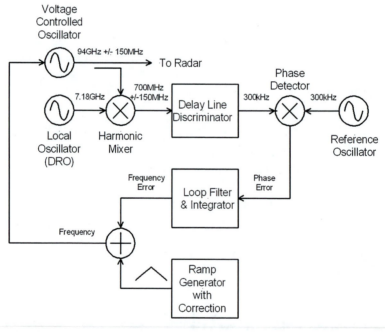

Figure 11.30 Schematic diagram showing the process of chirp linearization based on a combination of open-loop correction and closed-loop delay line discriminator output feedback.

The implementation of a loop filter that exhibits the appropriate locking bandwidth, low phase noise, and good suppression of spurious signals requires careful design and layout. Even so, it is nearly impossible to eliminate all the spurious signals from the receiver spectrum. For example, Figure 11.31 shows the measured discriminator output for open-loop and closed-loop linearization of a 77 GHz radar. In this case, the linearization process is effective, as can be seen from

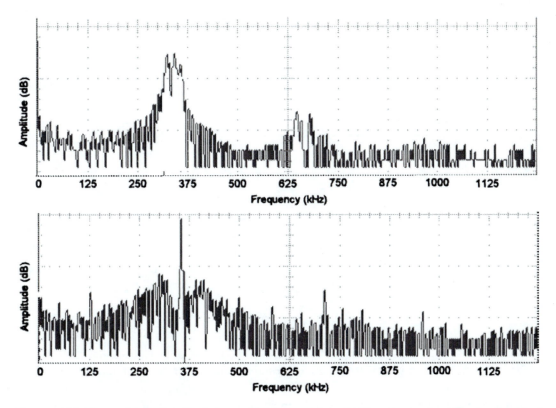

Figure 11.31 Measured delay line discriminator output spectra for a 77 GHz radar showing (above) unlinearized chirp and (below) closed-loop linearized chirp.

the reduction in the width of the peak at 360 kHz. However, a peak at the second harmonic at 720 kHz is also visible, and that can introduce false targets into the receiver.

11.7.5.4 Direct Digital Synthesis   Radar systems are not the only devices that use chirp to obtain good range resolution. Underwater sonar, medical ultrasound, and some terrestrial ultrasound devices also apply this technique. Because the frequency of these applications seldom exceeds a few hundred kilohertz, the chirp signal can easily be generated using direct digital synthesis (DDS).

This method synthesizes sinusoidal signals using the digital process of incrementing the value of a phase accumulator by a specified amount on each clock cycle. This phase accumulator is then used in a sine lookup table to produce the correct value which is passed to a DAC and output as an analog value. The structure of a simple DDS is shown in Figure 11.32.

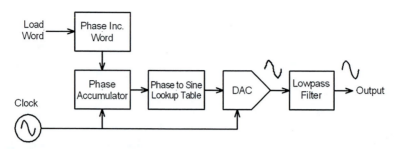

Figure 11.32 Direct digital synthesis.

There are a number of advantages to using DDS for both radar and sonar applications. First, the output frequency is crystal controlled, making it both accurate and repeatable. If a constant phase increment word is used, then the output frequency, f_{out} (Hz), will be

$$f_{out} = f_c \left(\delta\varphi/2\pi \right), \tag{11.37}$$

where $\delta\varphi$ (rad) is the phase increment and f_c (Hz) is the reference clock frequency.

It can be seen that a piecewise linear chirp can be generated if the value of the phase increment word is incremented in a linear manner. Because the phase increment generally occurs many times per output cycle, the process is almost seamless, and after filtering, a clean sinusoidal signal is produced. The DDS output can be amplified and used to drive an acoustic transducer directly, or for higher frequency applications, it can be used as an input to a PLL.

11.7.6 Extraction of Range Information and Range Gating

In some applications, such as the measurement of the liquid level in a tank, only a single target is present and so, if the SNR is sufficiently good, a simple counter can be used to measure the range. For other applications, many targets may be present in the beam and thus a number of beat frequencies will appear in the received signal. To extract them, one of the following forms of spectrum analysis must be performed:

- Bank of band-pass filters,
- Swept band-pass filter and detector (spectrum analyzer),
- Digitization and FFT processing.

11.7.6.1 FFT Processing   As computer hardware has become cheaper and faster, it has become practical to perform the spectrum analysis process using the FFT. This involves digitizing the analog beat frequency at an adequate rate to avoid aliasing and then extracting the spectral components. The power spectrum of a truncated sine wave will have sidelobes only 13.2 dB lower than the main lobe, and this is generally not adequate for high-resolution systems. To reduce these sidelobes so that they do not overwhelm returns from nearby, but smaller targets, the received signal is shaped using one of the weighting or windowing functions described later in this chapter.

If the signal is observed for a time T_d (s), then the width of the FFT frequency bin will be $W = 1/T_d$ (Hz) and main lobe width between nulls is twice that. The 3 dB bandwidth of the filter produced by the FFT process is 0.89 bins for no windowing (rectangle), increasing to 1.3 bins for a Hamming window. In the radar application, these can be considered to be overlapping range bins, as shown graphically in Figure 11.33.

11.7.6.2 Other Range Gating Methods In applications where only a few bins are required, it is often practical to implement a bank of band-pass filters using a SAW device, or even using discrete components as was common in audio test equipment (Randall 1977) Another alternative is to use a swept narrow-band receiver followed by a detector, as implemented in spectrum analyzers. The advantage of this technique is that only a single filter is required, which ensures that the measured response is consistent.

11.7.7 Problems with FMCW

The primary problems with FMCW all relate to transmitting and receiving simultaneously, as the transmitted power can be more than 100 dB higher than the received echo. It is obvious that if even a small fraction of the transmitted power leaks into the receiver it can saturate or even damage the sensitive circuitry. The performance of even well-designed systems is often degraded by 10 to 20 dB compared to their pulsed counterparts (Skolnik 1970). This limitation can be overcome by ensuring that there is good isolation between the receive and transmit antennas by separating them spatially and using antennas with low sidelobe levels. In addition, good performance can be obtained by implementing modern signal processing techniques and hardware to cancel the leakage power in real time.

11.8 STRETCH

In stretch, a linear FM pulse is transmitted and the return echo is demodulated by down-converting using an FM LO signal of identical or slightly different FM slope, as illustrated in Figure 11.34. If the identical slope is used, the echo spectrum corresponds to the range profile. This is a form of pulse compression intermediate between standard pulse compression and FMCW.

If the slope of the LO is different from that of the transmitted chirp, the output of the stretch processor comprises signals with a reduced chirp. These can then be processed using a standard SAW pulse compression system to produce target echoes as described earlier in this chapter.

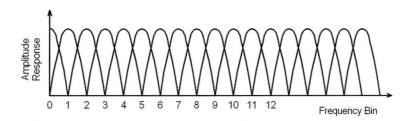

Figure 11.33 Filter bank implemented using the FFT.

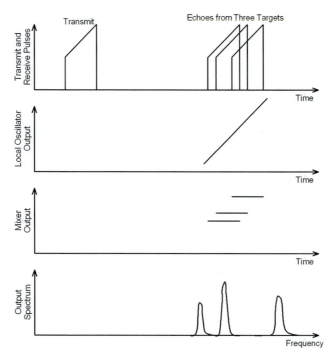

Figure 11.34 Stretch processing of received overlapping chirp echoes [adapted from Wehner (1995)].

11.9 INTERRUPTED FMCW

Known as interrupted FMCW (IFMCW or FMICW), this involves interrupting the FMCW signal to eliminate the requirement for good isolation between transmitter and receiver. It is generally implemented with a transmission time matched to the round-trip time to the target. This is followed by a quiet reception time equal to the transmission time. A duty factor of 0.5 reduces the average transmitted power by 3 dB, but the improved performance due to reduced system noise improves the SNR by more than the 3 dB lost (Currie et al. 1987).

Figure 11.35 shows a simple implementation of the FMICW process using a high-speed PIN switch to interrupt the transmit signal before it is transmitted. It is important that the chirp transmitter continue to run, as this signal is used to down-convert the received echo.

11.9.1 Disadvantages

The major problems are the limited minimum range due to the finite switching time of the transmitter modulator and the need to know the target range to optimize the transmit time. For imaging applications where a whole range of frequencies are received, maintaining a fixed 50% duty cycle is suboptimum except at one range. Probably the major disadvantage is that FFT processing of the interrupted signal results in large numbers of spurious components that can interfere with the identification of the target return, as shown in Figure 11.36 (Brooker 2005).

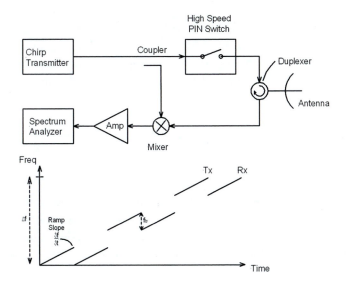

Figure 11.35 FMICW principle of operation.

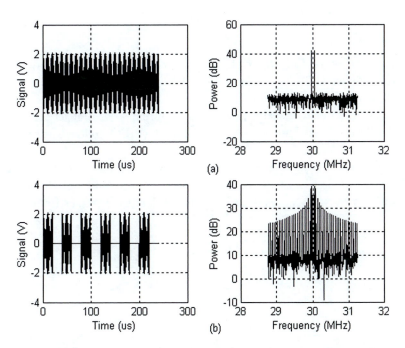

Figure 11.36 Comparison between the received signals and spectra for two closely spaced targets of different amplitudes for (a) an FMCW radar and (b) an FMICW radar with a deterministic interrupt sequence (Brooker 2005).

11.9.2 Optimizing for a Long-Range Imaging Application

Instead of dealing with the comb of spurious signals generated by the interruption sequence, an alternative is to process each of the receive intervals separately, as shown in Figure 11.37. First, the transmit time is matched to the round-trip time for the longest range of interest, where the SNR will be lowest, which in this case is 3 km. This can result in the shorter ranges suffering from the following problems:

- Reduced illumination time → lower SNR.
- Reduced chirp bandwidth → poorer range resolution.
- Suboptimal windowing → higher range sidelobes.

To a large extent these can be overcome on receive by breaking the reception process into blocks according to the range being examined and then processing accordingly. For example, at 1.5 km, only half of the receive window is available, so only that portion of the return is sampled and properly windowed before further processing. This cannot improve the range resolution, but it does reduce the sidelobe levels.

In an imaging application, it is important that the pixel area remain constant irrespective of the range. Constancy can be achieved using synthetic aperture radar (SAR) processing (see Chapter 12), but in this case it is done by compensating for the poorer cross-range resolution (constant beamwidth) as the range increases, with an improvement of the range resolution, to maintain a constant area.

11.9.3 Implementation

The implementation of a millimeter wave FMICW radar incorporates many of the processes discussed earlier. Figure 11.38 shows that the VCO is linearized using the closed-loop technique and this is followed by a PIN switch to isolate the transmit power during receive. In this long-range application, additional transmit power was required, so an impact ionization avalanche transit time (IMPATT) pulsed injection-locked oscillator (PILO) follows the switch to boost the output power to 250 mW.

On the receive side, the system is conventional, with the homodyne mixer followed by a low-noise amplifier (LNA). The LNA also contains an input clamp to minimize switching transients, and this is followed by an analog switch to further isolate the receiver from transients. More

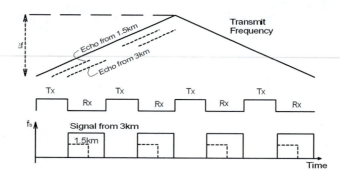

Figure 11.37 FMICW waveform optimized for 3 km.

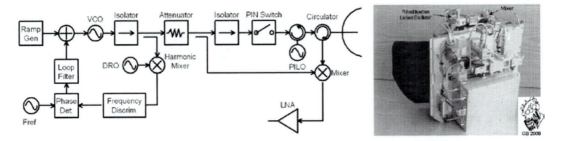

Figure 11.38 FMICW radar front-end block diagram and photograph.

information on the subsequent processing can be found in Brooker (2005) and Brooker et al. (2005).

11.10 SIDELOBES AND WEIGHTING FOR LINEAR FM SYSTEMS

The spectrum of a truncated sine wave output by an FMCW radar for a single target has the characteristic $|\sin(x)/x|$ shape as predicted by Fourier theory. The first range sidelobes in this case are only 13.2 dB lower than the main lobe. This is not satisfactory, as it can result in the occlusion of small nearby targets as well as introducing clutter from the adjacent lobes into the main lobe. To counter this unacceptable characteristic of the "matched filter," the time domain signal is mismatched purposely. This mismatch generally takes the form of amplitude weighting of the received signal.

One method of doing this is to change the FM slope of the chirp pulse near the ends of the transmitted pulse to weight the energy spectrum. This will result in the desired low sidelobe levels after application of the matched filter. It is easy to achieve in SAW-based pulse compression systems, but not really appropriate for conventional FMCW.

It is generally more practical to transmit a constant power signal with linear chirp and then perform the weighting function on the received signal after it has been sampled and digitized. This achieves the same result, and has the added advantage that the window can be adjusted to suit the situation. The actual weighting function applied by a number of different windows is shown in Figure 11.39.

The reduction in sidelobe levels does come at a price though, with the main lobe marginally reduced in amplitude and also widened quite substantially. These effects are summarized in Table 11.4. The rectangular, or uniform, weighting function provides a matched filter operation with no loss in SNR, while the weighting in the other cases introduces a tailored mismatch in the receiver amplitude characteristics, with an associated loss in SNR which can be quite substantial, as can be seen from Table 11.4.

In addition to providing the highest SNR, uniform weighting also provides the best range resolution (narrowest bandwidth), but this characteristic comes with the poorest sidelobe levels. The other weighting functions offer poorer resolution, but improved sidelobe levels, with falloff characteristics that can accommodate almost any requirement, as can be seen in Figure 11.40.

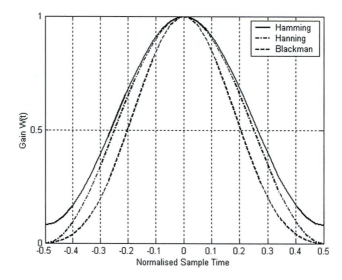

Figure 11.39 Weighting function gains for a number of different windows.

Table 11.4 Properties of some weighting functions

Window	Rectangle	Hamming	Hanning	Blackman
Worst sidelobe (dB)	−13.2	−42.8	−31.4	−58
3 dB beamwidth (bins)	0.88	1.32	1.48	1.68
Scalloping loss (dB)	3.92	1.78	1.36	1.1
SNR loss (dB)	0	1.34	1.76	2.37
Mainlobe width (bins)	2	4	4	6
a_0	1	0.54	0.50	0.42
a_1		0.46	0.50	0.50
a_2				0.08
$W(n) = a_0 - a_1 \cos[2\pi(n-1)/(N-1)] + a_2 \cos[4\pi(n-1)/(n-1)]$				

Of particular interest are the Hamming and Hanning weighting functions, which offer similar loss in SNR and resolutions, but with completely different sidelobe characteristics. In Figure 11.39, Hamming has the form of a cosine-squared-plus-pedestal, while Hanning is just a standard cosine-squared function. In the Hamming case, the close-in sidelobe is suppressed to produce a maximum level of −42.8 dB, but that energy is spread into the remaining sidelobes, resulting in a falloff of only 6 dB/octave, while in the Hanning case, the first sidelobe is higher, −31.4 dB, but with a falloff of 18 dB/octave.

For most FMCW applications, the Hamming window is used, as it provides a good balance between sidelobe levels (−42.8 dB), beamwidth (1.32 bins), and loss in SNR compared to a matched filter (1.34 dB). For imaging applications, where a large dynamic range of target reflectivities is expected, the Hanning window with its superior far-out sidelobe performance is often the function of choice.

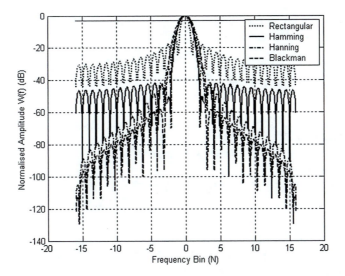

Figure 11.40 Normalized weighting function amplitude spectra for different window functions.

Figure 11.41 Krohne radar is a good example of an industrial FMCW radar.

11.11 HIGH-RESOLUTION RADAR SYSTEMS

11.11.1 Industry

Many short-range industrial sensors operate using FMCW principles. These radars are mostly used for the accurate measurement of liquid level in tanks, but as their sensitivity has improved, applications have extended to include solids. One good example is the Krohne level radar, shown in Figure 11.41, which operates at X-band. The Krohne radar specifications are as follows:

- Frequency 8.5–9.9 GHz
- Range 0.5–100 m (longer if required)
- Swept bandwidth 1 GHz
- Closed-loop linearization
- Linearity correction 98%
- Accuracy better than ±0.5% for a still target
- Fourier processing
- Rate of change of level less than 10 m/min.

11.11.2 Automotive Radar

A large variety of automotive radar systems using different modulation techniques are under development or being tested (Lowbridge 1995; Meinel 1996; Wenger 1998). Of these, most are based on FMCW principles (Eriksson and Broden 1996; Honma and Uehara 2001; Langheim et al. 1999; Li et al. 1999; Neininger 1977; Robinson et al. 1998; Rohling 1998), with a few using pulsed Doppler (Gresham et al. 2000), pulsed FM (Ishikawa et al. 1997), conventional pulsed AM techniques, or even noise radar (Lukin et al. 2001). At present, most of the radar systems operate in the 77 GHz band, which has been reserved for these applications, or the 24 GHz ISM (industrial, science, and medical) band.

The major players in the automotive radar sensor business have focused on pulsed Doppler or FMCW (Clarke 1998; Nelson 2001, 2004). The reasons that FMCW systems are so popular are that the required transmitter power is low and they are capable of resolving both position and velocity (see Table 11.5) if a triangular waveform is generated. Because radar performance is determined to a large extent by the average radiated power, pulsed and pulsed Doppler systems (which operate with lower duty cycles) compensate by transmitting higher peak powers, require correspondingly more complex modulation schemes, and are more expensive (Roe 1991) than the simple uninterrupted homodyne process used by FMCW radars.

In addition, the slow FM sweep of FMCW systems confines most of the transmitted energy within the sweep bandwidth (Stove 1991). In contrast, pulsed systems (FM or AM) with the same

Table 11.5 Typical automotive radar characteristics (Yamano et al. 2004)

Manufacturer	Fujitsu Ten	ADC	Autocruise	Delphi	Bosch	Denso	Hitachi
Image							
Size	89×107×86	136×133×68	98×98×63	137×67×100	91×124×79	77×107×53	80×108×64
Detection Range	4 – 120m	1 – 150m	1 – 200m	1 – 150m	2 – 120m	2 – 150m	1 – 150m
Azimuth Scan Angle	16°	10°	11°	15°	8°	20°	16°
Scan method	Mechanical	Beam Conversion	Beam Conversion	Mechanical	Beam Conversion	Phased Array	Monopulse
Modulation	FMCW	Pulsed FM	Step Frequency	FMCW	FMCW	FMCW	2 frequency CW
Transmitter	MMIC	Gunn	MMIC	Gunn	Gunn	MMIC	MMIC

mean spectral width generate sidelobes in the frequency domain due to the fast rise and fall time of the pulse. Pulsed Doppler units generally have longer pulse widths and poorer range resolutions than their pulsed cousins, but they do provide radial velocity measurements directly, while conventional pulsed units must use successive range measurements to infer the relative velocity.

From Figure 11.42, which shows a number of different automotive radars, it can be seen that the aperture is quite small, typically less than 100 mm across, and therefore the antenna gain is quite low. Angular target positions are either obtained by physically scanning the antenna in azimuth or, in some automotive applications, by processing the returns from a number of overlapping beams.

Restrictions imposed by the Australian Communications Authority (ACA), and similar authorities from other countries, limit the width of the spectrum that can be used in the 77 GHz automotive and 94 GHz experimental bands. At 77 GHz, the limit is only 2 GHz (76–78 GHz), and at 94 GHz the limit is 4.9 GHz (94.1–100 GHz) in two bands.

If even a moderate number of radars with bandwidths of 200 MHz (Eriksson et al. 1996), transmitter power of 10 mW, and antenna gain greater than 34 dB operate in overlapping bands, mutual interference between two or more units is bound to occur at some stage (Brooker 2007). This issue has been appreciated for some time (Eriksson and As 1997) and a concerted effort has

Figure 11.42 A number of different automotive radar system configurations from different manufacturers.

been made to minimize the effects of the interference by either transmitting a number of different modulations (Jian-Hui et al. 2001; Rohling and Meinecke 2001; Sanmartin-Jara et al. 1999), by notching out the offending sequence (Tullsson 1997), or by instituting sophisticated post-processing (Moon-Sik and Yong-Hoon 2001).

11.11.3 Research Radars

Universities and other research institutes all over the world develop high range-resolution radars for experimental applications. These include landmine detection, autonomous navigation, and volcano imaging, among a host of others.

The AVTIS radar shown in Figure 11.43, developed at the University of St. Andrews, is typical of the genre (Macfarlane and Robertson 2004; Robertson and Macfarlane 2004; Wadge et al. 2006). It is a 94 GHz long-range imaging radar that is capable of generating three-dimensional (3D) images of volcanoes at a range of up to 7 km. It uses a yttrium iron garnet (YIG) oscillator operating at around 7 GHz (open loop) to generate a linear FM chirp, followed by an active IMPATT multiplier to reach 94 GHz, where the signal is amplified to about 200 mW by an IMPATT injection locked oscillator (ILO) before transmission. With such a high transmitted power, a reflected power canceller would generally be included. However, in this case it has proven to be unnecessary because the Cassegrain antenna used has a very low voltage standing wave ratio (VSWR) and the circulator leakage is low.

At the Australian Center for Field Robotics, we have built a number of imaging radars to enhance situational awareness and for navigation, as shown in Figure 11.44. These include radars that operate at 77 GHz and 94 GHz. They have been mounted on unmanned aerial vehicles (UAVs) and autonomous ground vehicles, as well as on a number of large mining machines (Brooker et al. 2005, 2006, 2007; Goktogan et al. 2003; Widzyk-Capehart et al. 2006).

11.12 WORKED EXAMPLE: BRIMSTONE ANTITANK MISSILE

Possibly one of the most successful applications for high range-resolution FMCW is the millimeter wave radar seeker in the Brimstone missile. This is one of three missile variants developed

Figure 11.43 AVTIS radar for imaging volcanoes, showing inserts of a photograph and a radar image of a volcano. (Courtesy of the University of St. Andrews).

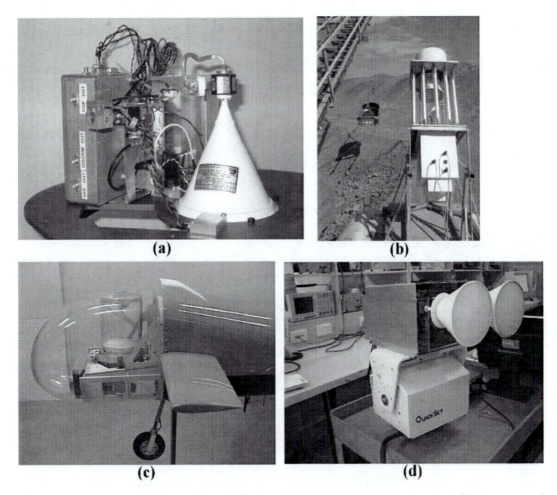

Figure 11.44 Some high-resolution radars developed at the ACFR: (a) 77 GHz radar front end, (b) radar mounted on a 3D mirror scanner for imaging, (c) miniature 77 GHz radar with mirror scanner and gimbals mounted on a UAV, and (d) wide-band high-power 94 GHz noise radar.

for the Longbow Apache AH-64D attack helicopter, and subsequently operated from various other airborne and ground-based platforms. It is shown in Figure 11.45 with the radome removed to expose the FMCW seeker.

11.12.1 System Specifications

- Length: 1.8 m
- Diameter: 178 mm
- Mass: 50 kg
- Operation: 24 hr, day/night, all weather
- Mode: totally autonomous, fire-and-forget, lock-on after launch (LOAL)
- Resistant to camouflage, smoke, flares, chaff, decoys, jamming
- Operational range: 8 km

- Designation: Accepts any or no target information
- Motor: boost/coast, burns for 2.75 s with a thrust of 7.5 kN
- Guidance: digital autopilot, 2 gyros (25°/hr drift), 3 accelerometers

11.12.2 Seeker Specifications (known)

- 94 GHz active radar
- Low power, narrow beam
- Dual polar, dual look
- Fast 96002 processor
- Detection/classification software

The processor for a typical radar-based seeker consists of a number of circular PC boards connected to a common bus that can be mounted rigidly within the missile frame. Boards developed for a similar application are shown in Figure 11.46; these include an RF processor, a high-speed ADC and DAC card, as well as the digital signal processor and memory. Note that the monkey

Figure 11.45 The Brimstone missile with radome removed showing the FMCW seeker.

Figure 11.46 Processor boards for an FMCW seeker and a monkey.

is not actually part of the autopilot, but was the team mascot at AMS in South Africa where these boards were developed.

11.12.3 Operational Procedure: Lock-On After Launch

The operational procedure for lock-on after launch is as follows:

- Rough target designations including range, bearing, and estimated target speed are down-loaded to missile autopilot.
- The missile is fired in the general direction of the target.
- It updates its designation from initial target position and speed.
- Flies up to 7 km toward the target using inertial navigation system (INS) guidance only.
- In the last 1 km it activates the radar seeker and searches for a target.
- A push-broom search scans the search box in 200 ms, as shown in Figure 11.47.
- Acquisition algorithms map all targets in the search area and exclude trucks.
- This track-while-scan process enables optimum decisions on target priority.
- The algorithm tags air defense units (ADUs) and main battle tanks (MBTs) as targets.
- Moving armor is given the highest priority.
- The missile autopilot selects the target with the highest score and locks on.

11.12.4 System Performance (Speculated)[1]

11.12.4.1 Target Detection and Identification   Target identification is based on a combination of the high range-resolution and polarization characteristics of the radar echo. The system transmits horizontal polarization (H) and receives vertical (V) and horizontal (H) returns. The range gate size is matched to the radar bandwidth for high resolution (≈ 0.5 m), which puts between 6

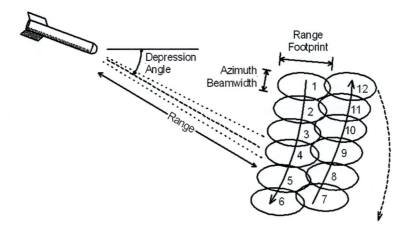

Figure 11.47 Push-broom search relies on the forward motion of the missile and antenna scan in azimuth to cover a swathe of ground.

---

1 Because of the limited digitization and processing power available when the Brimstone was developed, it actually uses fewer gates for search and detection before performing higher resolution processing for the target identification phase.

and 10 range cells on a typical 3 m × 5 m MBT. Doppler processing based on differences between the returns during the up and down chirp are used to distinguish moving targets from their static counterparts.

11.12.4.2 Radar Front End To give the radar a low probability of intercept (LPI), the transmit power will be low and spread spectrum, which almost certainly implies using the FMCW technique. FMCW operation through a single antenna generally limits the transmit power to less than 50 mW. However, with good matching and active leakage compensation, transmit powers can be as high as 1 W. It is believed that the Brimstone transmit power, P_{tx}, is approximately 100 mW (20 dBm), as it includes an injection-locked amplifier stage.

The transmitter swept bandwidth, Δf, is 300 MHz, to meet the 0.5 m range resolution requirement:

$$\delta R_{chirp} = \frac{c}{2\Delta f} = \frac{3 \times 10^8}{2 \times 3 \times 10^6} = 0.5\,\text{m}.$$

To allow for Doppler processing a triangular waveform will be used as shown in the figure.

For an operational range of 1 km with a 0.5 m bin size, 2000 gates are required. It is speculated that a 4096-point FFT is implemented to produce 2048 range gates for both the copolar and cross-polar receive channels.

Because the time available to perform a search is limited, the data rate should be as high as possible. However, there is a limit to the speed at which the loop linearization and the ADC can operate. It is assumed that the total sweep time is 1 ms, 500 μs for each of the up and down sweeps as shown.

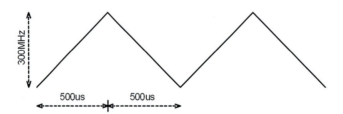

The beat frequency for a FMCW radar is given by

$$\begin{aligned} f_b &= \frac{\Delta f}{T} \cdot \frac{2R}{c} \\ &= \frac{300 \times 10^6}{500 \times 10^{-6}} \frac{2 \times 1000}{3 \times 10^8} = 4\,\text{MHz} \end{aligned}$$

Using the Nyquist criterion, the minimum sample rate required to digitize a signal with a 4 MHz bandwidth is 8 MHz. Because of non-brick-wall characteristics of the anti-aliasing filter, the sample rate is generally 2.5 times the maximum beat frequency, making the sample rate 10 MHz.

Because it is not practical to use an automatic gain control (AGC) with an FMCW radar to ensure sufficient dynamic range, an ADC with at least 12 bits of resolution is required. This results in an instantaneous dynamic range, on the input, of about 70 dB.

A total of 5000 samples can be taken over each up and down sweep. This is just about perfect for the 4096-point FFT because the sweep linearity is generally poor at the start and end of each ramp. A schematic block diagram of the seeker processing is shown in Figure 11.48. It can be seen that a closed-loop lineariser generates the linear ramp and that two homodyne stages produce the H and V polarised outputs on receive. These are amplified, filtered and digitised for further processing.

11.12.4.3 Antenna and Scanner For a missile diameter of 178 mm, the antenna cannot be much more than 160 mm across. For $\lambda = 3.2$ mm at 94 GHz, the 3 dB beamwidth will be

$$\theta_{3dB} = 70\lambda/D = (70 \times 3.2)/160 = 1.4°.$$

The pencil beam antenna uses an interesting Cassegrain configuration with a scanned parabolic mirror, shown schematically in Figure 11.49. The critical aspect of this antenna is the subreflector beam shaping that allows a limited scan using the parabolic prime reflector without generating large sidelobes. The gain of the pencil beam antenna will be approximately

$$G = \frac{4\pi\eta A}{\lambda^2} = \frac{4\pi \times 0.6 \times \pi \times 0.08^2}{0.00319^2} = 14897\,(41.7\,\text{dB})$$

From the beamwidth, at a range of 1 km, the width of the footprint will be 24.5 m and the length of the footprint will be a function of the operational height, as determined in Table 11.6.

To limit the amount of potential shadowing of the target area due to trees and undulating terrain (as illustrated in Figure 11.50), while still maintaining a reasonable size footprint on the

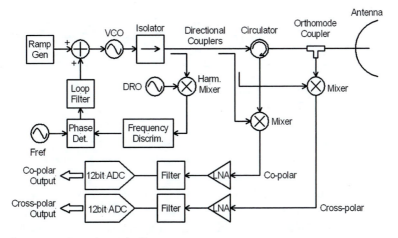

Figure 11.48 Brimstone seeker schematic diagram.

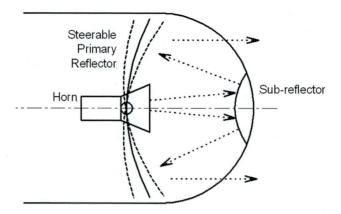

Figure 11.49 Schematic of the Cassegrain antenna used by Brimstone.

Table 11.6 Relationship between radar height and beam footprint length

Height (m)	Angle$_1$ (deg)	Angle$_2$ (deg)	R_2 (m)	Footprint (m)
10	0.57	1.97	290.29	709.71
20	1.15	2.55	449.83	550.17
30	1.72	3.12	550.67	449.33
40	2.29	3.69	620.13	379.87
50	2.86	4.26	670.87	329.13
100	5.71	7.11	801.64	198.36

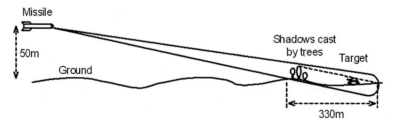

Figure 11.50 Shadow effects due to low grazing angle.

ground, an operational height of 50 m would be reasonable. This results in a footprint length of 330 m. It can be assumed that a single mechanical scan takes place in the 200 ms search time.

Because the missile is coasting, it will have limited lateral acceleration capability, and so it is pointless to search beyond the boundaries that the missile can reach. It is therefore reasonable to assume that a square search area of 330 m × 330 m will be covered. At a range of 1000 m, this equates to an angular scan of about 18° if the antenna beamwidth is considered. To scan 18° in 200 ms requires an angular rate of 90°/s, which is easily achieved.

11.12.4.4 Signal Processing The time-on-target for a beamwidth of 1.4° and an angular rate of 90°/s is 15.5 ms. For a total sweep time of 1 ms, a total of nearly 16 hits per scan occurs. This allows for 16 pulse integrations to improve the SNR if it is required; it also gives the processor more information to identify the target type. Each target can be identified using the following information:

- 5–10 gates that span it in range.
- 16 time slices.
- 2 orthogonal polarizations.

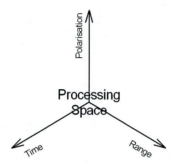

This is probably sufficient information to discriminate between a truck and an MBT, even if the vehicles are not moving.

11.12.4.5 Signal-to-Clutter Ratio: Clutter Levels The single-look signal-to-clutter ratio (SCR) is determined from the target radar cross section (RCS), the clutter reflectivity, $\sigma°$, and the area of the range gate. Figure 11.51 shows the measured clutter reflectivity data at 94 GHz for grass and crops (Currie et al. 1992).

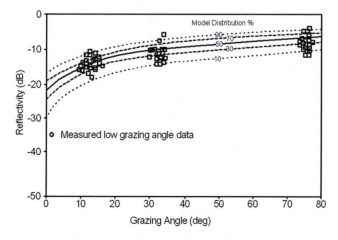

Figure 11.51 Measured and modeled clutter reflectivity for grass and crops at 94 GHz [adapted from Currie et al. (1992)].

At a grazing angle of between $3°$ and $4°$, the mean reflectivity of grass will be about -20 dBm²/m² (which reduces to decibels). The clutter cross section is the product of the clutter reflectivity, $\sigma°$, and the area of the gate footprint, $R_{gate}R\theta_{3dB}$ (m²), on the ground for flat terrain (the beamwidth must be in radians), as described in Chapter 8:

$$\begin{aligned}\sigma_{clut} &= \sigma° + 10\log_{10}\left(R_{gate}R\theta_{3dB}\right)\\ &= -20 + 10\log_{10}(0.5 \times 1000 \times 1.4 \times \pi/180)\\ &= -9\,\text{dBm}^2.\end{aligned}$$

Because tank commanders are aware that they are vulnerable when out in the open, they tend to make use of available local cover and will position themselves on tree lines or within forests. Figure 11.52 shows the measured reflectivity of an unbroken canopy of deciduous trees as seen from above. However, the reflectivity of lines of trees observed broadside is much higher than that of the canopy, as can be seen in Figure 11.53, which shows rows of pine trees between orchards and a double line of eucalyptus straddling a railway line. Measurements confirm that the mean reflectivity of individual deciduous trees seen side-on is typically -10 dBm²/m².

The clutter RCS in this case is the product of the area of trees illuminated by the radar and the reflectivity. Because of the narrow gate, there will be areas where the tree reflectivity is very strong and areas where it is very low depending on the actual reflecting area, as can be seen in Figure 11.54. There will also be areas where the tank is sticking out from under the tree, in which case the clutter level is determined by the ground clutter only. If a 4 m hedge of trees the width of the range gate is illuminated, then the RCS will be

$$\sigma_{clut} = \sigma°hR\theta_{3dB} = -10 + 10\log_{10}(4 \times 1000 \times 1.4 \times \pi/180) = 10 \text{ dBm}^2.$$

In general, however, a much smaller section of the tree will be illuminated within a single gate. For a tree 4 m tall and 3 m wide, roughly elliptical in shape, a maximum area of 8 m² will be illuminated within each gate. This reduces the clutter level to

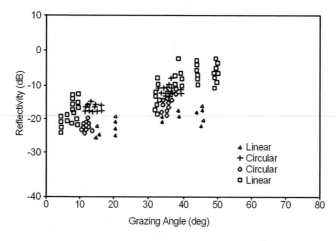

Figure 11.52 Clutter reflectivity for a deciduous tree canopy at 94 GHz [adapted from Currie et al. (1992)].

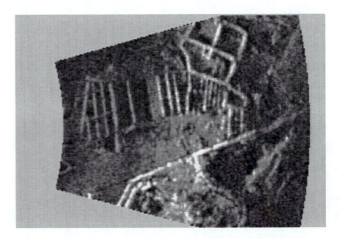

Figure 11.53 A 94 GHz radar image of trees and scrub gives an indication of the difficulties inherent in detecting small targets in ground clutter.

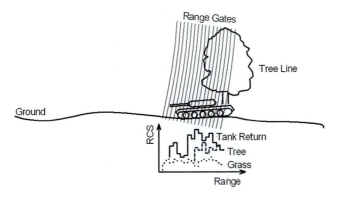

Figure 11.54 RCS profile of a tank under a tree.

$$\sigma_{\text{clut}} = \sigma^\circ A = -10 + 10\log_{10}(8) = -1 \text{ dBm}^2.$$

11.12.4.6 Target Levels The RCS of a tank depends on the observation angle, as shown in Figure 11.55. The maximum RCS can reach 40 dBm^2 and the minimum seldom falls below 10 dBm^2. Hence, to ensure that the vehicle is always detected irrespective of the angle, the 10 dBm^2 threshold must be selected.

11.12.4.7 Signal-to-Clutter Ratio In open ground, the SCR is the ratio of the target level to that of the clutter return from grassland:

$$\text{SCR} = \sigma_{\text{tan}} - \sigma_{\text{clut}} = 10 - (-10) = 20 \text{ dB.}$$

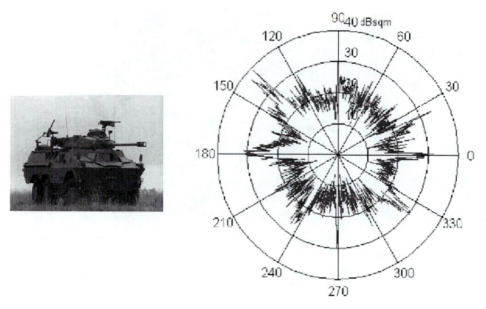

Figure 11.55 Radar cross-section of a Ratel armored personnel carrier.

For the tank under the tree, the worst case will be

$$\text{SCR} = \sigma_{\text{tan}} - \sigma_{\text{clut}} = 10 - 10 = 0\,\text{dB}.$$

However, the typical SCR will be more reasonable:

$$\text{SCR} = \sigma_{\text{tan}} - \sigma_{\text{clut}} = 10 - (-1) = 11\,\text{dB}.$$

Without resorting to the statistics of the variation in tank RCS and that of trees, it can be seen that if the range bin is sufficiently narrow, parts of the tank will be visible if it is parked on the border of a row of trees. When the radar is looking for a moving target, the clutter signals will be reduced because they are either completely static or only moving very slowly (swaying in the wind).

11.12.4.8 Signal-to-Noise Ratio   The SNR is determined using the characteristics of the radar and the target as they are related in the radar range equation. The total noise at the output of the receiver, N (dB), can be considered to be equal to the noise power output from an ideal receiver multiplied by a factor called the noise figure, NF, where $NF_{\text{dB}} \approx 15$ dB for an FMCW radar.

In this case, the matched filter bandwidth, β (Hz), is the bandwidth of a single bin output by the FFT and widened by the window function 1.3×5 MHz/2048 ≈ 3 kHz. Therefore the noise floor is

$$N_{\text{dB}} = 10\log_{10} P_N NF = 10\log_{10} kT_{\text{sys}}\beta + NF_{\text{dB}} = -154\,\text{dBW}.$$

Because the transmitter power is in milliwatts, this value is generally converted from dBW to dBm by adding 30 dB:

$$N_{dB} = -154 + 30 = -124\,dBm.$$

Writing the range equation for a monostatic radar system in decibels gives

$$10\log_{10}(P_r) = 10\log_{10}(P_t) + 10\log_{10}\left(\frac{\lambda^3}{(4\pi)^3}\right) + 20\log_{10}(G) + 10\log_{10}(\sigma)$$
$$-10\log_{10}(L) - 40\log_{10}(R) - 2\alpha R_{km}$$

This is best tackled in MATLAB as shown in Figure 11.56, as the attenuation, α, is a function of the weather conditions. The SNR is sufficient for detection up to a rain rate of about 10 mm/hr.

11.12.4.9 Target Identification: Doppler Processing The bandwidth of each bin output by the FFT is about 3 kHz, which is equivalent to a Doppler velocity of

$$v_r = f_d\lambda/2 = \left(3\times10^3 \times 0.00319\right)/2 = 4.8\,m/s.$$

Because a Doppler shift causes an upward shift for half the sweep and a downward shift for the other half, as illustrated in Figure 11.24, the range profiles generated by the up and down sweeps will diverge. For a target with a radial velocity of 4.8 m/s, this will be 2 bins, and will increase to 6 bins at a speed of 50 km/h, which is reasonable for a tank on the move.

A simple form of moving target discrimination can be obtained by taking the difference between the up-sweep and the down-sweep range profiles. Static targets will cancel if the correct shift, to compensate for the missile velocity, is applied, but moving targets will appear as two large peaks, as shown in Figure 11.57.

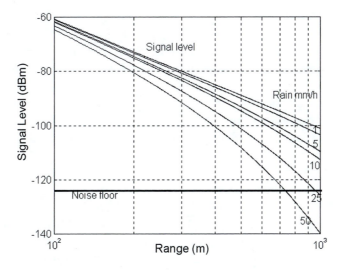

Figure 11.56 Brimstone performance in adverse weather.

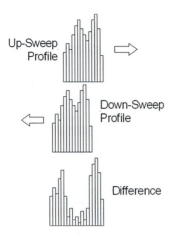

Figure 11.57 Moving target detection.

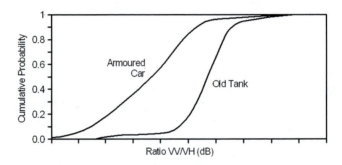

Figure 11.58 Polarization ratio used to identify vehicles.

11.12.4.10 Target Identification: Other Techniques   Different target types are identified by the differences in their copolar and cross-polar signatures. For example, targets with lots of corners and attachments tend to reflect signals back to the radar after more than one bounce, and that rotates the polarization. Because there are lots of scatterers, each rotating the polarization by a different amount, the overall return will have a random polarization that is uniformly spread. The signal is said to be depolarized. On the other hand, smooth targets reflect with a single bounce, so the polarization is not rotated. This discrimination process is shown in Figure 11.58, where VV and VH ratios of two different military vehicles are compared.

11.12.5 Tracking and Guidance

Once the correct target has been selected and its offset angle noted, the antenna slews back to that designation and the target is reacquired. Range and angle tracking loops are closed to maintain lock using techniques discussed in Chapter 13 and Chapter 14, and the seeker errors drive the control surfaces, shown in Figure 11.59, to guide the missile toward the target.

   A number of different guidance laws can be applied, of which the simplest is command to line-of-sight (CLOS). This algorithm is appropriate if the missile velocity is much higher than

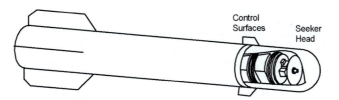

Figure 11.59 Missile schematic showing control surfaces.

Figure 11.60 Antitank missile scoring a direct hit (Courtesy White Sands Missile Range Museum).

that of the target, as it continually directs the missile towards the target's current position. If, however, the target velocity is high, proportional navigation (prop-nav) is the technique of choice, as it maintains a constant line-of-sight angle between the missile and the target. This directs the missile toward the point where it will intercept the target trajectory if each continues to travel in a straight line. The results are as shown in Figure 11.60.

11.13 References

Brooker, G. (2005). Long-range imaging radar for autonomous navigation. Ph.D. dissertation. University of Sydney, Sydney, NSW, Australia.

Brooker, G. (2007). Mutual interference of millimeter-wave radar systems. *IEEE Transactions on Electromagnetic Compatibility* 49(1):170–181.

Brooker, G., Birch, D., and Solms, J. (2005). W-band interrupted FMCW imaging radar. *IEEE Transactions on Aerospace and Electronic Systems* 41(3):955–972.

Brooker, G., Hennessey, R., Bishop, M., Lobsey, C., and Durrant-Whyte, H. (2006). High-resolution millimeter-wave radar systems for visualization of unstructured outdoor environments. *Journal of Field Robotics* 23(10):891–912.

Brooker, G., Scheding, S., Bishop, M., and Hennessy, R. (2005). Development and application of millimeter wave radar sensors for underground mining. *IEEE Sensors Journal* 5(6):1270–1280.

Brooker, G., Widzyk-Capehart, E., Hennessey, R., Bishop, M., and Lobsey, C. (2007). Seeing through dust and water vapor: millimeter wave radar sensors for mining application. *Journal of Field Robotics* 24(7):527–557.

Brookner, E., ed. (1982). *Radar Technology*. Norwood, MA: Artech House.

Clarke, P. (1998). Adaptive cruise control takes to the highway. *EEtimes*, available at http://www.eetimes.com/story/OEG19981020S0007.

Currie, N. and Brown, C. (1987). *Principles and Applications of Millimeter-Wave Radar*. Norwood, MA: Artech House.

Currie, N., Hayes, R., and Trebits, R. (1992). *Millimeter-Wave Radar Clutter*. Norwood, MA: Artech House.

Eriksson, L. and Broden, S. (1996). High performance automotive radar. *Microwave Journal* October: 24–38.

Eriksson, L.H. and As, B. (1997). Automotive radar for adaptive cruise control and collision warning/avoidance. IEE International Conference on Radar, Edinburgh, UK.

Goktogan, A., Brooker, G., and Sukkarieh, S. (2003). A compact millimeter wave radar sensor for unmanned air vehicles. 4th International Conference on Field and Service Robotics, Yamanashi, Japan, July 14–16.

Gresham, I., Jain, N., Budka, T., Alexanian, A., Kinayman, N., Ziegner, B., Brown, S., and Staecker, P. (2000). A 76–77 GHz pulsed-Doppler radar module for autonomous cruise control applications. *IEEE International Microwave Symposium* 3:1551–1554.

Honma, S. and Uehara, N. (2001). Millimeter-wave radar technology for automotive applications. Tokyo: Mitsubishi Electric.

Ishikawa, Y., Tanizaki, T., Nishida, H., and Taguchi, Y. (1997). 60 GHz band FM-pulse automotive radar front end using new type NRD guide and dielectric lens antenna. 1997 Topical Symposium on Millimeter Waves, Kanagawa Japan, July 7–8.

Jian-Hui, Z., Guo-Sui, L., Hong, G., and Wei-Min, S. (2001). A novel transmit signal based on high range-resolution concept for FLAR or AICC system applications. CIE International Conference on Radar, Beijing, China, October 15–18.

Langheim, J., Henrio, J.-F., and Liabeuf, B. (1999). Autocruise ACC radar system with 77 GHz MMIC radar. Autocruise, Florence, Italy.

Lewis, B.L., Kretschmer, F.F., and Shelton, W.W. (1986). *Aspects of Radar Signal Processing*. Norwood, MA: Artech House.

Li, D.D., Luo, S.C., Pero, C., Wu, X., and Knox, R. (1999). Millimeter-wave FMCW/monopulse radar front-end for automotive applications. *IEEE MTT-S International Microwave Symposium Digest* 1:277–280.

Lowbridge, P. (1995). Low-cost mm-wave radar for cruise control. *GEC Review* 10(2):47–102.

Lukin, K.A., Mogyla, A.A., Alexandrov, Y.A., Zemlyaniy, O.V., Lukina, T., and Shiyan, Y. (2001). W-band noise radar sensor for car collision warning systems. *Fourth International Kharkov Symposium on Physics and Engineering of Millimeter and Sub-Millimeter Waves* 2:870–872.

Macfarlane, D. and Robertson, D. (2004). A 94 GHz dual-mode active/passive imager for remote sensing. SPIE Passive Millimeter-Wave and Terahertz Imaging and Technology, London, October 27–28.

Meinel, H. (1996). Millimeterwaves for automotive applications. *26th European Microwave Conference* 2:830–835.

Moon-Sik, L. and Yong-Hoon, K. (2001). New data association method for automotive radar tracking. *IEE Proceedings—Radar, Sonar, and Navigation* 148(5):297–301.

Neininger, G. (1977). A FM/CW radar with high resolution in range and Doppler application for anti-collision radar for vehicles. *IEEE Radar* 77, London, pp. 526–530.

Nelson, R. (2004). Radar gives cars eyes. *Test & Measurement World*, May 1, 2002; available at http://www.tmworld.com/article/CA214128.html.

Nelson, S. (2001). Measurement and calculation of powdered mixture permittivities. *IEEE Transactions on Instrumentation and Measurement* 50(5):1066–1070.

Rabiner, L. and Gold, B. (1975). *Theory and Application of Digital Signal Processing*. Upper Saddle River, NJ: Prentice Hall.

Randall, R. (1977). *Application of B & K Equipment to Frequency Analysis*, 2nd ed. Nærum, Denmark: Bruel & Kjaer.

Robertson, D. and Macfarlane, D. (2004). AVTIS: all-weather volcano topography imaging sensor. 29th International Conference on Infrared and Millimeter Waves, Karlsruhe, Germany, Sept. 27–Oct. 1.

Robinson, J., Paul, D., Bird, J., Dawson, D., Brown, T., Spencer, D., and Prime, B. (1998). A millimetric car radar front end for automotive cruise control. IEE Colloquium on Automotive Radar and Navigation Techniques, Feb. 9, 1998.

Roe, H. (1991). The use of microwaves in Europe to direct, classify and communicate with vehicles. IEEE MTT-S International, Boston, MA, June 10–14.

Rohling, H. (1998). A 77 GHz automotive radar system for AICC applications. International Conference on Microwaves and Radar, Workshop, Krakow, Poland, May 20–22.

Rohling, H. and Meinecke, M.-M. (2001). Waveform design principles for automotive radar systems. CIE International Conference on Radar, Beijing, China, October 15–18.

Sanmartin-Jara, J., Burgos-Garcia, M., and Retamose-Sanchez, J. (1999). SS-FH signals used for very low interference in vehicular cruising control systems. IEEE VTS 50th Vehicular Technology Conference, Amsterdam, Netherlands, September 19–22.

Skolnik, M. (1970). *Radar Handbook*, New York: McGraw-Hill.

Stove, A.G. (1991). 80 GHz automotive radar. Eighth International Conference on Automotive Electronics, London, UK, October 28–31.

Tullsson, B.-E. (1997). Topics in FMCW radar disturbance suppression. *Radar 97* (Conference Publication No. 449), Edinburgh, UK, October 14–16.

Wadge, G., Macfarlane, D., James, M., Odbert, H., Applegarth, L., Pinkerton, H., Robertson, D.A., Loughlin, S.C., Strutt, M., Ryan, G., Dunkley, P.N. (2006). Imaging a growing lava dome with a portable radar. *EOS, Transactions American Geophysical Union* 87(23):226–227.

Wehner, D. (1995). *High Resolution Radar*, 2nd ed. Norwood, MA: Artech House.

Wenger, J. (1998). Automotive mm-radar: status and trends in system design and technology. Colloquium on Automotive Radar and Navigation Techniques, London, UK, Feb. 9.

Widzyk-Capehart, E., Brooker, G., Scheding, S., Hennessy, R., Maclean, A., Lobsey, C. (2006). Application of millimeter wave radar sensors to environment mapping in surface mining. 9th International Conference on Control, Automation, Robotics and Vision (ICARCV 2006), Singapore, December 5–8.

Yamano, S., Higashida, H., Shono, M., Matsui, S., Tamaki, T., Yagi, H., and Asanuma, H. (2004). 76 GHz millimeter wave automobile radar using single chip MMIC. *Fujitsu Ten Technical Journal* 22(1):12–19.

12

High Angular-Resolution Techniques

12.1 INTRODUCTION

For imaging systems (not null steering trackers), the angular resolution is limited by the beam divergence. This is generally defined as the half-power, or 3 dB beamwidth, θ_{3dB}, and is a function of the wavelength and the aperture size:

$$\theta_{3dB} = k\lambda/d, \tag{12.1}$$

where

θ_{3dB} = 3 dB beamwidth,
   k = constant (typically 70 for degrees and 1.22 for radians),
   λ = wavelength (m),
   d = aperture diameter (m).

The cross-range resolution for real aperture systems, δx_r (m), is the product of the beamwidth, θ_{3dB} (rad), and range, R (m):

$$\delta x_r = R\theta_{3dB} = 1.22R\lambda/d. \tag{12.2}$$

As the operational wavelength is often fixed by the atmospheric window or other propagation effects, and it is difficult to make the antenna diameter, d (m), arbitrarily large, for structural reasons or other manufacturing limitations, the resolution of a single antenna is limited.

One solution to this problem is to use two or more antennas in an array to synthesize an effective linear aperture equal to the array baseline. Another is to use the forward motion of the antenna to synthesize a larger aperture (in one dimension only).

This chapter examines the synthesis and performance of one-dimensional (1D) and two-dimensional (2D) phased arrays and their application in sonar and radar before considering the elegant algorithms for Doppler beam sharpening and synthetic aperture radar (SAR) that can be used to produce a "beam" with no divergence at all.

Figure 12.1 Phased array antennas: (a) millimeter wave radio telescope, (b) Jindalee over-the-horizon radar, (c) Patriot missile system, and (d) acoustic array for atmosphere profiling.

12.2 PHASED ARRAYS

It is shown later in this chapter that a cluster of individual, but synchronized, antenna elements can replace a physical aperture to produce a collimated beam with similar characteristics. This technique is used by phased array radars and sonar systems as well as long-baseline radio telescopes. Phased arrays come in a myriad of forms, as can be seen in Figure 12.1, from the more conventional 2D grid and linear structures to the almost ad hoc configurations used in radio astronomy.

12.2.1 Advantages of Using Phased Arrays

The main advantages of using arrays include the following:

- Inertia-less rapid beam steering.
- Multiple, independent beams.
- Control of radiation pattern.
- Electronic beam stabilization.
- Potential for large peak and average powers.
- Graceful degradation.
- Convenient aperture shape.

The beam generated by a single aperture has a number of limitations due to the manner in which it is synthesized. To perform a search or to produce an image, the antenna must be physically rotated or the beam deflected by using a mirror as discussed in Chapter 6. As will be made clear later in this chapter, the beam from a phased array can be made to point in any direction by adjusting the relative phase excitations of the elements. Because there is no physical movement involved, this process can be almost instantaneous and is extremely reliable.

An extension to this basic premise is to use the individual elements, or groups of elements (known as subarrays), to generate a number of independent beams that can be used simultaneously to track a number of targets or continue to search for new targets while simultaneously tracking others. In a similar manner, it is possible to generate a null in place of a beam in a specific direction if, for example, a jammer or other noise source is situated there.

Because the number of elements in an array can be large (thousands in some cases), it is possible to produce high output powers using individual elements that radiate only moderate amounts of power. Not only does this allow extremely long range operation, but even if some of the elements fail the system continues to operate effectively. This is known as graceful degradation and is an essential aspect of most military systems.

A really exciting aspect of phased array technology is the fact that the antenna can be configured to conform to the available physical configuration of the carrier vehicle. Ultimately this should allow such systems to be built as part of the skin of an aircraft or a submarine without encountering the problems associated with radome bubbles and the like.

12.2.2 Array Synthesis

As an introduction to the idea of beam steering, consider the array of elements shown in Figure 12.2. In this case, each of the elements is radiating the same frequency signal, with the same phase in Figure 12.2a, but with a different phase—one that increases linearly across the array—in Figure 12.2b. If a line of constant phase is drawn for each of the elements, a linear phase front is produced parallel to the array in the first case and at an angle to the array in the second. This phase front defines the direction in which the beam is synthesized, or the direction from which a signal will be received.

In a conventional array, the power received by each element in the array is the sum of the received powers scattered by target, P, from all the transmit elements. The voltage outputs of all N elements are summed via lines of equal length, without delays or phase shifts, to give E_a (V), as shown in Figure 12.3. Since each element observes the same phase,

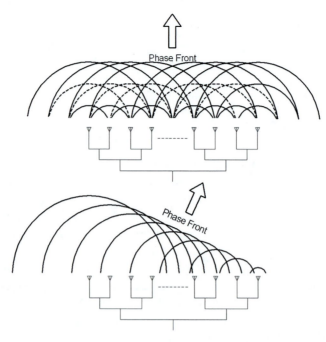

Figure 12.2 Radiation from individual elements with (a) equal phase and (b) a linearly increasing phase shift combine to form a linear phase front in a specific direction determined by the phase difference between successive elements.

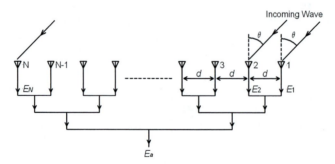

Figure 12.3 The array beam-forming process is achieved by summing the signals from all the array elements in phase.

φ_k (rad), the output after summing the signals from all N elements directly, with no phasing, is given by

$$E_a = \sum_{k=1}^{N} \sin(\omega t + \varphi_k). \tag{12.3}$$

For the example of a uniform phase front approaching the array at an angle θ (rad), the relative phase shifts are 0, Ψ, 2 Ψ, 3 Ψ, ..., $N\Psi$, as shown in Figure 12.4.

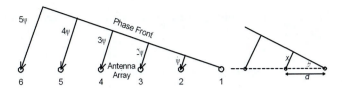

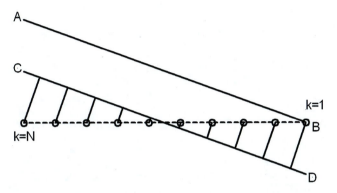

Figure 12.4 Schematic diagram showing (a) the relative phase differences for a plane wave approaching a phased array and (b) construction used to determine the magnitude of the phase difference.

Figure 12.5 Diagram showing the geometric effect of measuring the phase front relative to the geometric center of the phased array rather than to element one.

It is easy to see that the displacement, x (m), corresponds to the phase shift between sequential elements and that it can be determined from the array spacing, d (m), and the angle of arrival of the plane wave, θ (rad):

$$x = d \sin \theta. \tag{12.4}$$

The phase shift, Ψ, that corresponds to the distance x is then

$$\Psi = \frac{2\pi}{\lambda} x. \tag{12.5}$$

By substituting for x into equation (12.5),

$$\Psi = \frac{2\pi d}{\lambda} \sin \theta. \tag{12.6}$$

In general, it is more usual to measure the phase shifts relative to the geometric center of the array, and not element one. The effect of this is to displace the phase front, represented by the line AB, down to CD, which passes through the geometric center, as illustrated in Figure 12.5.

For the construction shown, for the phase front AB, the relative phase shift of the elements starting at point B ($k = 1$) is summed as

$$E_a = \sum_{k=1}^{N} \sin[\omega t + (k-1)\Psi].$$

(12.7)

The phase front CD that passes through the geometric center of the array is displaced by $(N-1)\Psi/2$, allowing the output to be rewritten as

$$\begin{aligned} E_a &= \sum_{k=1}^{N} \sin\left[\omega t + (k-1)\Psi - \frac{N-1}{2}\Psi\right] \\ &= \sum_{k=1}^{N} \sin\left[\omega t + k\Psi - \frac{N+1}{2}\Psi\right] \end{aligned}.$$

(12.8)

For the six-element array, $N = 6$, the phase shifts are shown in Figure 12.6.

12.2.3 Two-Point Array

Examination of the elements on either side of the array center shown in Figure 12.6 shows that the phase shifts are $-\Psi/2$ and $+\Psi/2$, and the voltage sum will be

$$E_a = \sin(\omega t + \Psi/2) + \sin(\omega t - \Psi/2).$$

(12.9)

Using the trigonometric identity $2\sin A \cos B = \sin(A + B) + \sin(A - B)$, equation (12.9) can be rewritten as

$$E_a = \sin(\omega t)2\cos(\Psi/2).$$

(12.10)

The trigonometric identity $\sin 2A = 2\sin A \cos A$ can also be written as $2\cos A = \sin 2A/\sin A$, and this can be used to further simplify equation (12.10) to

$$E_a = \sin(\omega t)\frac{\sin(2\Psi/2)}{\sin(\Psi/2)}.$$

(12.11)

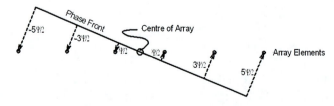

Figure 12.6 Displaced phase shift across the array.

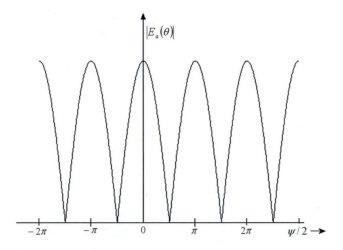

Figure 12.7 Field intensity pattern generated by two isotropic radiators.

Plotting the absolute value of equation (12.11), $|E_a(\theta)|$, as a function of $\Psi/2$ produces the lobed field intensity pattern shown in Figure 12.7.

12.2.4 Four-Point Array

Extending to a four-point array, and once again examining the elements on either side of the array center in Figure 12.6, the sum will be

$$E_a = \sin(\omega t + 3\Psi/2) + \sin(\omega t + \Psi/2) + \sin(\omega t - \Psi/2) + \sin(\omega t - 3\Psi/2) \qquad (12.12)$$

$$(1) \qquad\qquad (2) \qquad\qquad (3) \qquad\qquad (4)$$

Grouping terms (1) and (3), and (2) and (4), and applying the trigonometric identity $2\sin A \cos B = \sin(A + B) + \sin(A - B)$ gives

$$E_a = 2\sin(\omega t)\cos(3\Psi/2) + 2\sin(\omega t)\cos(\Psi/2). \qquad (12.13)$$

Now, let $A + B = 3\Psi/2$ and $A - B = \Psi/2$ and reapply the same trigonometric identity results in

$$E_a = \sin(\omega t)4\cos(2\Psi/2)\cos(\Psi/2). \qquad (12.14)$$

Multiplying by $\sin(2\Psi/2)/\sin(2\Psi/2)$, simplifying, and applying the trigonometric identity $\sin 2A = 2\sin A \cos A$ gives

$$E_a = \sin(\omega t) \frac{2\sin(4\Psi/2)\cos(\Psi/2)}{\sin(2\Psi/2)}. \tag{12.15}$$

Applying the trigonometric identity again, and simplifying further, the canonical form for the sum of a four-point array is finally obtained:

$$E_a = \sin(\omega t) \frac{\sin(4\Psi/2)}{\sin(\Psi/2)}. \tag{12.16}$$

12.2.5 The General Case

It can be seen that the only difference between the two-element and four-element cases is the size of the numerator. In the general case, for N elements, the numerator can therefore be rewritten as $\sin(N\Psi/2)$ and the equation for the sum becomes

$$E_a = \sum_{k=1}^{N} \sin\left[\omega t + k\psi - \frac{N+1}{2}\psi\right] = \sin(\omega t) \frac{\sin(N\psi/2)}{\sin(\psi/2)}. \tag{12.17}$$

Substituting for Ψ from equation (12.6) results in the classical equation that defines the voltage output of the 1D linear phased array:

$$E_a = \sin(\omega t) \cdot \frac{\sin\left\{\frac{N\pi d}{\lambda}\sin\theta\right\}}{\sin\left\{\frac{\pi d}{\lambda}\sin\theta\right\}}. \tag{12.18}$$

The second term of the equation is known as the amplitude factor of the array. The field intensity pattern, $E_a(\theta)$, for the antenna is the magnitude of this amplitude factor. The general expression for the field intensity pattern is therefore

$$|E_a(\theta)| = \left|\frac{\sin\left\{\frac{N\pi d}{\lambda}\sin\theta\right\}}{\sin\left\{\frac{\pi d}{\lambda}\sin\theta\right\}}\right|. \tag{12.19}$$

If this equation is considered in detail, it is obvious that there are nulls where the numerator is zero, where $(N\pi d/\lambda)\sin\theta = 0$, $\pm\pi$, $\pm2\pi$, etc. The denominator is also zero at $(\pi d/\lambda)\sin\theta = 0$, $\pm\pi$, $\pm2\pi$, etc.

Applying L'Hopital's rule where $E_a = 0/0$, it is found that E_a is a maximum where $\sin\theta = \pm n\lambda/d$, and these maxima all have the value N. There is a special case maximum where $\sin\theta = 0$, which is called the main lobe, and all the other lobes on either side of it are called grating lobes, as can be seen in Figure 12.8.

A special case occurs when $d/\lambda = 0.5$. In this instance the grating lobe does not appear for $n = \pm1$ in real space because $\sin\theta > 1$, which is not possible. If $d/\lambda = 1$, the grating lobes appear

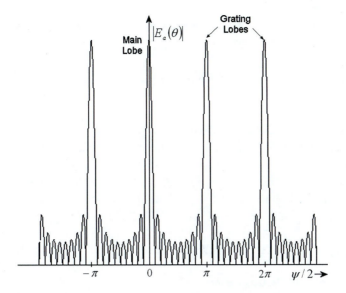

Figure 12.8 Ten-element synthesis showing the position of the main lobe and grating lobes.

at ±90°, and as most real radiating elements do not radiate much at $\theta = 90°$, the grating lobes will therefore be suppressed.

It follows that, to minimize the number of elements required to create a given aperture for a nonscanning array, the best element spacing is $d = \lambda$, while for a scanned array the best spacing is d $\approx \lambda/2$.

12.3 THE RADIATION PATTERN

12.3.1 Linear Array

The radiation pattern is defined as the normalized square of the amplitude factor:

$$G_a(\theta) = \frac{|E_a|^2}{N^2} = \frac{\sin^2\left\{\dfrac{N\pi d}{\lambda}\sin\theta\right\}}{N^2\sin^2\left\{\dfrac{\pi d}{\lambda}\sin\theta\right\}}. \tag{12.20}$$

This can be simplified by substituting $Nd = L$ (the length of the antenna) and using the small angle approximation $\sin\alpha \approx \alpha$:

$$G_a(\theta) \approx \frac{\sin^2\left\{\dfrac{\pi L}{\lambda}\sin\theta\right\}}{\left\{\dfrac{\pi L}{\lambda}\sin\theta\right\}^2}. \tag{12.21}$$

It can be shown that for $d = \lambda/2$, with N as the number of elements in the array, the half-power beamwidth, θ_{3dB} (rad), is

$$\theta_{3dB} = 1.73/N. \qquad (12.22)$$

If N is sufficiently large, the antenna pattern will be equivalent to a uniformly illuminated aperture, and the first sidelobe will be 13.2 dB down.

For directive elements, the overall antenna pattern is the product of the patterns of the individual elements, known as the element factor, $G_e(\theta)$, and the pattern generated by an array of isotropic elements, the array factor, $G_a(\theta)$, as shown in Figure 12.9. It can be seen from this figure that even though the array generates large sidelobes, and a pair of grating lobes, these are suppressed by the narrower beam pattern of the individual elements:

$$G(\theta) = G_e(\theta)G_a(\theta). \qquad (12.23)$$

12.3.2 Radiation Pattern: 2D Rectangular Array

The radiation pattern for a three-dimensional (3D) array may be approximated as the product of the patterns of the two planes that contain the principle axes of the antenna:

$$G(\theta, \varphi) = G(\theta)G(\varphi). \qquad (12.24)$$

Substituting for the radiation patterns defined in equation (12.20) with arrays of length M and length N:

$$G(\theta, \varphi) = \frac{\sin^2\left\{\dfrac{N\pi d}{\lambda}\sin\theta\right\}\sin^2\left\{\dfrac{M\pi d}{\lambda}\sin\varphi\right\}}{\sin^2\left\{\dfrac{\pi d}{\lambda}\sin\theta\right\}\sin^2\left\{\dfrac{\pi d}{\lambda}\sin\varphi\right\}}. \qquad (12.25)$$

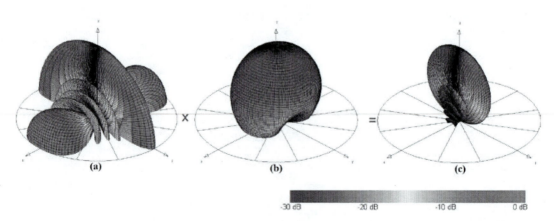

Figure 12.9 Effect of directive elements on the antenna pattern: (a) the array factor, (b) the element factor, (c) the final antenna pattern.

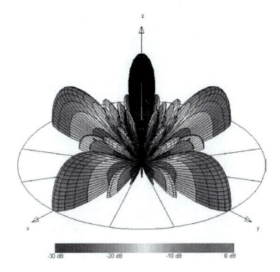

Figure 12.10 3D polar pattern for a 5 × 5 element phased array.

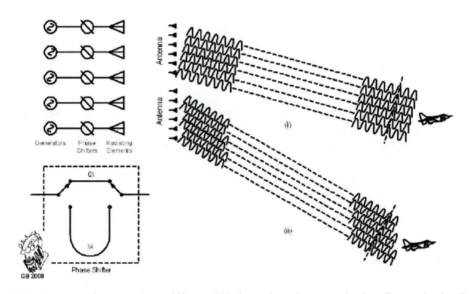

Figure 12.11 Beam steering uses phase shifts to shift the region of constructive interference in the direction required. Note from (b, ii) that the phase shift need never exceed one wavelength.

The 3D antenna pattern for a 5 × 5 array of isotropic elements is shown in Figure 12.10.

12.4 BEAM STEERING

As mentioned earlier in this chapter, the direction of the beam is determined by the relative phase shifts of the elements across the array. As shown in Figure 12.11a, this direction can be achieved by switching phase shift elements into the path between a bank of synchronized oscillators and the radiating elements. If there is an element spacing of less than $\lambda/2$, Figure 12.11b shows that any squint angle between $-90°$ and $+90°$ can be obtained for phase shifts of less than one wavelength.

From an analytical perspective, if the same phase is applied to all the elements of the array, the main beam will be broadside to the array and $\theta = 0$. The direction of the main beam will be θ_o (rad) if the relative phase difference, φ (rad), is

$$\varphi = (2\pi d/\lambda)\sin\theta_o. \tag{12.26}$$

The phase of each element will be $\varphi_c + m\varphi$, where $m = 1, 2, 3, \ldots, N-1$ and φ_c is a constant phase applied to all of the elements. In this case the radiation pattern is

$$G_a(\theta) = \frac{|E_a|^2}{N^2} = \frac{\sin^2\left\{\dfrac{N\pi d}{\lambda}(\sin\theta - \sin\theta_o)\right\}}{N^2\sin^2\left\{\dfrac{\pi d}{\lambda}(\sin\theta - \sin\theta_o)\right\}}, \tag{12.27}$$

and grating lobes will occur at

$$\frac{\pi d}{\lambda}\left(\sin\theta_g - \sin\theta_o\right) = \pm n\pi. \tag{12.28}$$

If the total scan requirement is the full forward sector from $-90°$ to $+90°$, the element spacing should be $d = \lambda/2$ (m). However, this is impractical, as the aperture size decreases to zero as the scan angle approaches $90°$. For a practical array that can scan from $-60°$ to $+60°$, the spacing can be increased to $d > 0.54\lambda$. If the element spacing is not reduced to $d = \lambda/2$ (m), then as soon as the beam scans away from $\theta = 0$, the grating lobes will appear as shown in Figure 12.12.

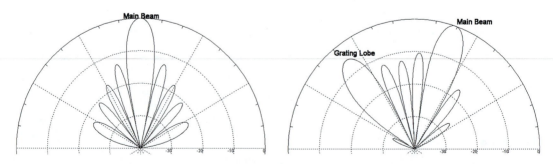

Figure 12.12 Beam scanning for an element spacing $d = \lambda$ showing the appearance of a grating lobe as soon as the beam is scanned away from $\theta = 0$.

12.4.1 Active and Passive Arrays

With the advent of low-cost monolithic microwave integrated circuits (MMICs) it is now practical to manufacture individual transceiver modules to produce active arrays. Such arrays have been manufactured at X-band for less than $15 per module (Brookner 2006). A comparison between the architecture of passive and active arrays is shown in Figure 12.13.

12.4.2 Corrections to Improve Range Resolution

Individual elements have to be fitted with phase delay circuitry to steer the beam. However, as Figure 12.14a shows, if the squint angle is great, the pulse gets spread with a resultant

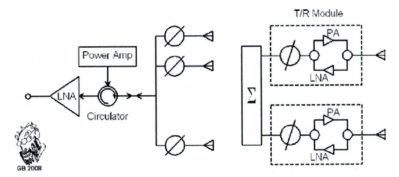

Figure 12.13 Schematic diagram showing the configuration for (a) passive and (b) active arrays.

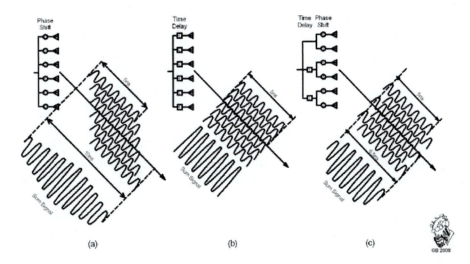

Figure 12.14 Difference between phase delay and time delay. In (a), which uses only phase delays, the 5 ns pulse is stretched out to 10 ns. In (b), only time delays are implemented to maintain the 5 ns pulse. In (c), a compromise is implemented to minimize the expense of using individual time delay circuits for each element [adapted from Brookner (1985)].

degradation in the range resolution. To counter this, it is possible to use adjustable time delay circuits as shown in Figure 12.14b. Because time delay elements are expensive, it is practical to time delay complete subarrays using dedicated phase shifters before individual elements are steered.

12.5 ARRAY CHARACTERISTICS

12.5.1 Antenna Gain and Beamwidth

For large arrays, the nonscanned antenna gain can be approximated by the gain of a uniformly illuminated aperture:

$$G_o = 4\pi A / \lambda^2. \tag{12.29}$$

However, as the array is scanned, the gain is reduced by the scan angle, θ_o (rad), to accommodate the reduction in size of the projected aperture:

$$G(\theta_o) = \frac{4\pi A \cos \theta_o}{\lambda^2} \tag{12.30}$$

For the same reasons that the antenna gain decreases as the angle is scanned off-axis, the half-power beamwidth, θ_{3dB} (rad), is increased and, for angles not close to endfire, can be approximated by

$$\theta_{3dB} \approx 0.886\lambda / Nd \cos \theta_o. \tag{12.31}$$

Generally some form of taper is used to reduce the sidelobe levels. Using cosine on a pedestal, as discussed earlier, where $A_n = a_0 + 2a_1 \cos(2\pi n/N)$, where $0 < 2a_1 < a_0$, the beamwidth will be widened:

$$\theta_{3dB} \approx \frac{0.886\lambda}{Nd \cos \theta_o} \left\{ 1 + 0.636 \left[\frac{2a_1}{a_0} \right]^2 \right\}. \tag{12.32}$$

12.5.2 Matching and Mutual Coupling

The impedance of the array elements varies with the scan angle, and spurious lobes may appear due to the mismatch. This is a difficult problem to solve analytically, and is often determined experimentally by exciting a single element and terminating all of the surrounding ones. Coupling is proportional to $1/d$ for $d = \lambda/2$, so the pattern and impedance are drastically altered by surrounding elements. Generally the surrounding 5×5 or even 9×9 elements must be considered to obtain an accurate model.

12.5.3 Thinned Arrays

Because phased arrays can have thousands of elements, it is very expensive and not particularly productive to populate each space with a fully functioning transceiver element. Reducing the

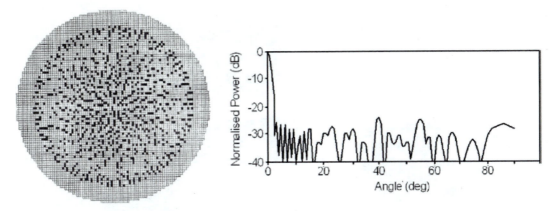

Figure 12.15 Characteristics of a thinned array [adapted from Skolnik (1970)].

number of elements has hardly any effect on the main-lobe pattern, but degrades the sidelobe levels slightly. In the example, shown in Figure 12.15, the 4000-element array is thinned to 900 elements, which results in the mean sidelobe level increasing to −31.5 dB, which is still acceptable. The peak power in both the sidelobes and the main lobe reduces in proportion to the number of elements. If the removed elements are replaced by matched dummy elements, the pattern remains unchanged and only the gain is decreased.

12.5.4 Conformal Arrays

It is theoretically possible to place array elements on any arbitrary surface and by applying the correct phase, amplitude, and polarization, radiate a beam in the required direction. However, in practice this is difficult to achieve for the following reasons:

- It is difficult to control the beam shape and to get low sidelobes.
- The feeding mechanism is difficult to manufacture.
- Each element in the array is different (unless there is symmetry) so impedance and coupling will be different.
- They are computationally difficult to optimize, as there are no element or array factors.

To date, most of the work has been done on truncated cones, cylinders, or ogives, but with the availability of low-cost transmit/receive (TR) modules, as illustrated in Figure 12.13, and the increased processing power available, it is becoming feasible to populate more complex shapes. The ultimate aim is to have an "active skin" on aircraft that will replace all of the antennas that are now mounted on the surface.

12.6 APPLICATIONS

Acoustics and radar provide the majority of applications of phased array technology. These extend from commercial medical ultrasound equipment, through sidescan and multibeam underwater sonar, to the radar systems, both large and small, developed for the military. In the following section, a few research and commercial phased arrays are considered.

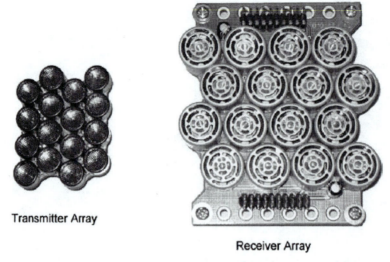

Figure 12.16 Acoustic array hardware for 3D imaging (Thompson 2003).

12.6.1 Acoustic Array

The imaging acoustic array shown in Figure 12.16 was developed as part of an honors thesis at the Australian Center for Field Robotics (ACFR) (Thompson 2003). Both of the arrays are made from commercial 40 kHz piezoelectric transducers packed as closely as possible. The configurations of the two modules have been designed as subarrays so that larger apertures can be constructed as required.

In this application, the transmitter elements are phased in a random fashion to produce floodlight insonification of a region of space approximately 0.6 m × 0.6 m × 0.6 m with a single pulse. The receiver array then samples the resulting echo over the round-trip time, from which a complete 3D image can be reconstructed. The spatial resolution of the 4 × 4 receive array is 14° and temporal resolution is 0.28 ms, giving an overall voxol size of 50 mm × 60 mm × 60 mm at a range of 250 mm, with the result that high-resolution images can be produced, as shown in Figure 12.17.

Large-scale research arrays operating using similar principles have been developed for underwater applications. These are particularly useful for the close scrutiny of sea mines in turbid waters. An example of one system developed by Thomson Marconi Sonar (TMS) in Sydney, NSW, Australia, is shown in Chapter 16.

12.6.2 New Generation MMIC Phased Arrays

The advent of low-cost MMIC technology has unleashed a generation of high-performance phased array radars for military applications. Some of these are illustrated in the composite in Figure 12.18 and Figure 12.19 (Brookner 2006).

12.6.3 Early Warning Phased Array Radar

The cold war saw the development and application of a number of really large phased array radar systems for early warning and attack characterization of ballistic missiles. One of the more

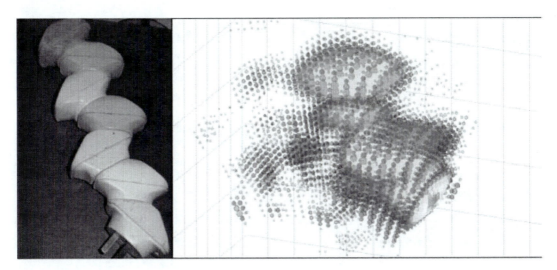

Figure 12.17 Imaging capability of acoustic phased array: (a) photograph and (b) acoustic image (Thompson 2003).

Figure 12.18 New generation MMIC-based arrays for military aircraft (Brookner 2006).

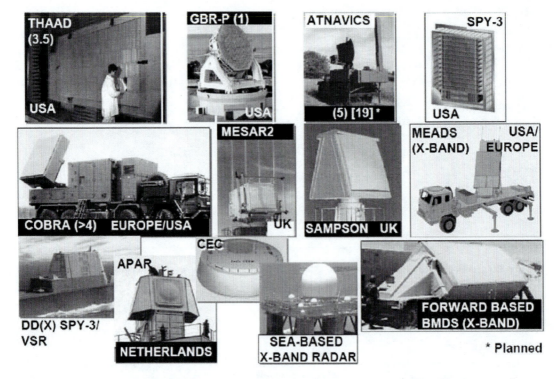

Figure 12.19 New-generation MMIC-based arrays for ground and sea-based applications (Brookner 2006).

successful, AN/FPS-115 (PAVE PAWS), is shown in Figure 12.20. These are long-range early warning radar systems positioned at Cape Cod Air Force Station (Massachusetts), Beale Air Force Base (California), and Clear Air Force Station (Alaska). Each face of each radar contains 1792 active elements, each radiating 322 W peak power, and 885 dummy elements on a 30 m wide area. This allows them to detect targets with a 10 m^2 cross section at a range of about 4800 km (Brookner 1985).

Each of the modules has the following performance specifications (Brookner 1982):

- Frequency 433 ± 13 MHz
- Radio frequency (RF) peak power 284–440 W
- Pulse width 0.25 μs–16 ms
- Duty cycle up to 25%
- Efficiency 42% (typical)
- Phase track error 14° rms
- Harmonics −90 dBc
- Gain 27 ± 1 dB
- Noise figure less than 2.9 dB
- Limiter 440 W, 16 ms, 25% DF
- Phase tracking 10° rms

- Phase shifter bits 4
- Phase shifter error 4.6° rms

Note that to achieve the long range, a low frequency (in the ultra-high-frequency range [UHF]) is used to minimize atmospheric attenuation, and the peak power radiated by each face is high, up to 585 kW.

Figure 12.20 Photograph of a Pave Paws early warning radar and the coverage diagram for the three radar systems.

Figure 12.21 Photograph of the SBX midcourse defense radar mounted on an oil rig base. The insert shows the gimbals and the phased array (covered). Courtesy of Raytheon.

Figure 12.21 shows one of the new-generation radars. This is the sea-based X-band (SBX) radar which has been developed for ground-based midcourse defense. Its specifications are as follows:

- Built on a semisubmersible oil-drilling platform 65 m to the top of the radome
- Physical aperture 384 m^2
- Active aperture 248 m^2
- Number of gallium arsenide (GaAs) transmit/receive modules more than 45,000
- Phase and mechanical steering
- Azimuth coverage ±270°
- Elevation coverage 2°–90°
- Rotating weight 2400 tons

12.7 SIDESCAN SONAR

12.7.1 Operational Principles

Sidescan sonar operates using similar principles to side-looking airborne radar (SLAR). The sonar antenna is a short ($\approx 50\lambda$) linear transducer array made of a piezoelectric material that is towed behind a boat in a straight line at a fixed depth, as illustrated in Figure 12.22.

In a pulsed system, the transducer is excited by a short ($\tau \approx 3\ \mu$s) high-voltage sinusoidal stimulus at a frequency close to the resonant frequency of the array which the array converts to vibrations and radiates into the water. Because of its shape, the array produces a fan-beam pattern with a narrow azimuth beamwidth (typically 0.75° to 1.5°) determined by the length of the array and a wide elevation beamwidth (typically 35° to 65°) determined by the vertical aperture of each element, as shown in Figure 12.23.

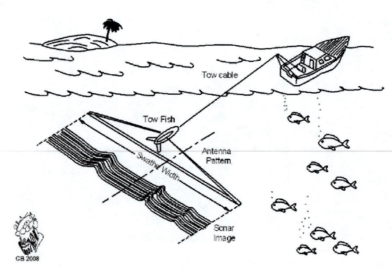

Figure 12.22 Sidescan sonar deployment.

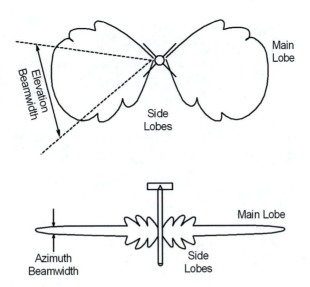

Figure 12.23 Array azimuth and elevation beam patterns [adapted from Fish and Carr (2001)].

The beam propagates through the water until it hits the bottom. The roughness of the floor and any projections reflect sound waves back in the direction of the sonar, where they are detected by the same array and converted back to an electrical signal and processed.

The frequency of choice is determined by the maximum operational range and resolution requirements, and is generally between 50 kHz and 500 kHz, with some short-range units operating up to 1 MHz. The operational range at 50 kHz will be about 500 m, decreasing to 50 m at 1 MHz due to increased attenuation through water (see Chapter 7).

The array is generally mounted within a streamlined body called a "towfish" that looks like a torpedo. Arrays are placed on either side of the towfish and angled slightly downward to produce symmetrical beams, as shown in Figure 12.23. To minimize the geometric distortion introduced by the differences between the slant range distance measured by the sonar and its projection on the sea bottom, operation is generally restricted to shallow water, typically 20% of the maximum operating range of the sonar.

Forward velocity is limited by the width of the cross-range footprint and the round-trip time to the furthest target of interest. For example, sound takes about 650 ms to travel out to a range of 500 m and back, and this limits the boat speed to 7.5 m/s if the resolution required is 5 m.

12.7.2 Hardware

A towfish is a torpedo-shaped structure about 1.5 m long and 100 mm in diameter with a hydrodynamic nose and large fins, as can be seen from the photograph in Figure 12.24. Sidescan arrays are mounted on the two sides and are typically about 600 mm long and about 20 mm high.

The configuration of the towfish and the rigging ensure that it remains horizontal as it glides through the water. Cables from the boat include the high-tensile tow cable as well as power for

Figure 12.24 Towfish showing one transducer (Fish et al. 2001).

Table 12.1 Specifications of ITC-5202 sidescan array

Size of unit	68.5 cm × 4 cm × 3.8 cm
Array dimensions	1.27 cm × 53 cm shaded active array
Resonance frequency	117 kHz
Usable frequency range	111–126 kHz
Beam pattern 53 cm line	≈ 1.5° at 117 kHz
1.27 cm line	≈ 60° at 117 kHz
Efficiency	>40%
Input power <5% duty cycle	1500 W
Operating depth	Unlimited
Weight	4.3 kg
Housing	Aluminum

the high-voltage transducers and signal lines to communicate echo information back to the boat for display and logging. A typical moderate-range transducer is the ITC-5202 array (ITC 2008), with specifications as listed in Table 12.1 and in Figure 12.25.

12.7.3 Operation and Image Interpretation

The array is towed through the water at a speed between 1 kt and 15 kt, depending on the maximum operational range and the azimuth beamwidth, and the returned echo intensity is plotted as a function of round-trip time on the range axis and synchronized to the towboat velocity on the cross-range axis.

An examination of Figure 12.26 shows a typical result for a sidescan image of a shipwreck. The black line running vertically through the center of the image is the bang pulse, and it gives an idea of the range resolution and the size of the image pixels from which the image is made.

The Model ITC-5202 is a full ocean rated side scan transducer assembly. This unit has a very narrow beam and is shaded with low side lobes. It is suitable for most side scan or other high directivity needs.

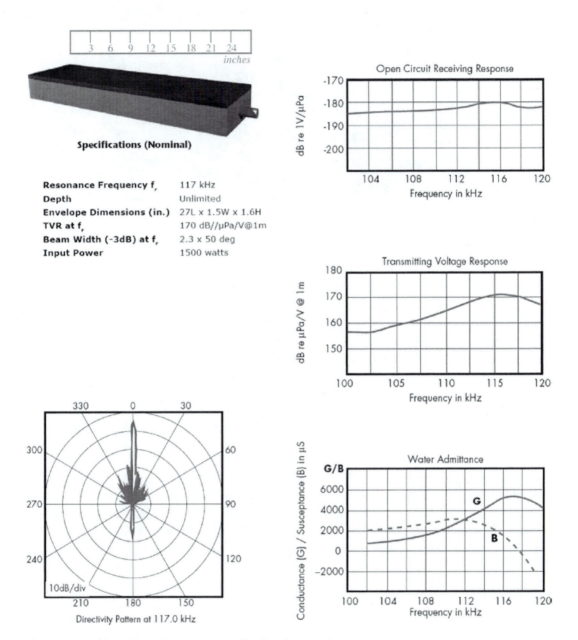

Specifications (Nominal)

Resonance Frequency f$_r$	117 kHz
Depth	Unlimited
Envelope Dimensions (in.)	27L x 1.5W x 1.6H
TVR at f$_r$	170 dB//µPa/V@1m
Beam Width (-3dB) at f$_r$	2.3 x 50 deg
Input Power	1500 watts

Figure 12.25 Sidescan transducer array specifications (ITC 2008).

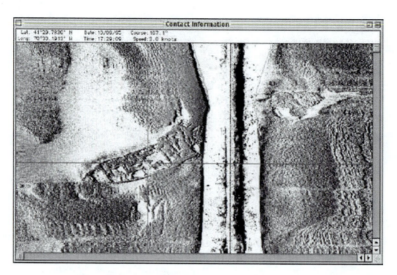

Figure 12.26 Sidescan sonar image of the wreck of the *Port Hunter*.

The reasonably bare white region on either side of this line is the slant range from the array to the intersection of the lower edge of the beam with the sea bottom. Dark patches within this region are caused by reverberation and echoes from clutter, such as fish, krill, etc. The remainder of the image is determined by the reflectivity properties of the sea bottom and any objects lying on it.

It should be remembered that the images of the wreck and the sea bottom are distorted, with the worst effects occurring at short range, just after the white regions, and decreasing progressively as the range increases. This form of distortion is discussed in more detail in the section on SAR later in this chapter.

Although the image of the *Port Hunter* appears to be flat when compared to the actual ship shown in Figure 12.27, because of the low grazing angles and the exposed superstructure, shadowing can be pronounced for other targets. In Figure 12.28, an image produced by the Advanced Unmanned Search System (AUSS) (Fish et al. 2001) takes advantage of this feature to obtain far more information from the image than would otherwise be available. In this image, a complete profile of the ship, including deck structures and the mast, are clearly visible.

12.7.4 Signal Processing

The standard matched-filtering principles discussed in Chapter 5 are applied to sonar systems to ensure that the maximum SNR is achieved. From an imaging perspective, most sidescan systems are real beam in that their cross-range resolution, δx_r (m), is a function of range, R (m), and the azimuth beamwidth, θ_{az} (rad):

$$\delta x_r = R\theta_{az}. \tag{12.33}$$

Figure 12.27 Photograph of the *Port Hunter* (Courtesy Stan Lair).

Figure 12.28 Shadowing may be useful to identify the height of the target, as can be seen in this sidescan image of a wreck (Fish et al. 2001).

However, because the propagation speeds are slow, digital techniques can be applied to correct for phase front curvature to some extent. This is known as focusing, and it can be used to achieve a fairly constant linear beamwidth with range. This process is not the same as that used by SAR, although the results are similar.

That limitation notwithstanding, exceptionally high-resolution images can be achieved with the appropriate hardware and good signal processing techniques, as demonstrated time and again

Figure 12.29 High-resolution sidescan sonar image of the *Fritzen* (Hultqvist 2007).

by Sture Hultqvist (Hultqvist 2007). Figure 12.29 shows the image of the 124 m long German coal carrier *Fritzen* lying at a depth of 75 m made by Sture using a 500 kHz sidescan system. Sture's untimely death in 2007 was a serious loss to the sonar community.

Simultaneous multifrequency operation (e.g., 100 kHz and 600 kHz) is possible for simultaneous high-resolution, short-range operation and lower resolution long-range operation. Note, for example, the fine range and angular resolutions that are achieved in Figure 12.29, if the frequency is high and operational range short.

12.8 WORKED EXAMPLE: PERFORMANCE OF THE ICT-5202 TRANSDUCER

Determine the performance of a sidescan sonar based on the ITC-5202 transducer at a range of 500 m. From Figure 12.25, it can be seen that the transducer operates at a resonant frequency, f_r, of 117 kHz, which equates to a wavelength, λ, of 13 mm if the velocity of sound is taken to be $c = 1522$ m/s. The usable frequency range extends from 111 to 126 kHz, making the quality factor, Q, of the transducer (Knutstal and Bunker 1992)

$$Q = f_r/(f_u - f_l) = 117/(126 - 111) = 7.8. \tag{12.34}$$

As an approximation, the rise time of any pulse generated by the transducer is related to its resonant frequency and quality factor:

$$\tau_{rise} = Q/f_r = 7.8/(117 \times 10^3) = 66.7\,\mu s. \tag{12.35}$$

The minimum pulse width must be at least twice the rise time if the pulse is to reach its peak value. As a rule of thumb, for a "rectangular" pulse, the total pulse width is $5\,\tau_{rise} = 333\,\mu s$. This equates to a pulse width and an effective range resolution of

$$\delta R = c\tau/2 = (1522 \times 333 \times 10^{-6})/2 = 0.25\,m. \tag{12.36}$$

Note that this is a slight excursion from the standard matched filter approximation which states that the pulse width, τ (s), can be approximated by the reciprocal of the bandwidth, β (Hz). This is because a longer pulse is used to improve the range performance at the expense of the poorer range resolution.

If, instead of relying on the pulse width to determine the range resolution, a chirp-based pulse compression system is implemented, the range resolution, δR_{pc} (m), will be determined by the chirp bandwidth of $\Delta f = f_u - f_l = 126 - 111 = 15$ kHz to be

$$\delta R_{pc} = \frac{c}{2\Delta f} = \frac{1522}{2 \times 15 \times 10^3} = 0.05 \, \text{m}. \tag{12.37}$$

The cross-range resolution, δx_r (m), at 500 m for the given beamwidth of 1.5° is the product of the beamwidth, θ_{az} (rad), and the range, R (m), if no focusing occurs:

$$\delta x_r = R\theta_{az} = 500(1.5/57.3) = 13.1 \, \text{m}. \tag{12.38}$$

To operate out to a maximum unambiguous range of 500 m, the maximum pulse repetition frequency, f_{pr} (Hz), is

$$f_{pr} = c/2R_{max} = 1522/(2 \times 500) = 1.52 \, \text{Hz}. \tag{12.39}$$

The transmitter power is limited to a maximum of 1500 W for a duty cycle, χ, of less than 5%. The duty cycle in this case is

$$\chi = 100 f_{pr}\tau = 100 \times 1.52 \times 333 \times 10^{-6} = 0.05\%, \tag{12.40}$$

which is much smaller than the limit, so the maximum power can be applied to the transmitter.

For the chirp-based system, the limiting pulse length is determined by the 5% duty cycle, therefore

$$\tau_{pc} = \chi/100 f_{pr} = 5/(100 \times 1.52) = 32.9 \times 10^{-3} \, \text{s}. \tag{12.41}$$

This is 100 times longer than the 333 μs length of an uncompressed pulse, and therefore, given that the peak power is unchanged at 1500 W, the average power and therefore the detection range will be increased proportionally.

If the transducer is omnidirectional, the power density at a range of 1 m is the product of the electrical power, P_{elec}, and the conversion efficiency, η, divided by the surface area of a sphere with a radius of 1 m:

$$I_{iso} = P_{elec}\eta/4\pi = (1500 \times 0.4)/4\pi = 47.7 \, \text{W/m}^2. \tag{12.42}$$

The antenna gain, known as the directivity index, *DI*, which is defined in terms of the power with respect to an isotropic radiator, can be calculated from the elevation and azimuth beam-widths (see Chapter 5):

$$G = \frac{4\pi}{\theta_{3dB}\varphi_{3dB}} = \frac{4\pi \times 57.3^2}{1.5 \times 60} = 458.4. \tag{12.43}$$

The actual power density in the direction of the peak gain is the product of the gain and the isotropic value:

$$I = I_{iso}G = 47.7 \times 458.4 = 21{,}864 \, \text{W/m}^2. \tag{12.44}$$

The sound pressure level (SPL), *S*, is generally given in decibels relative to 1 μPa at a range of 1 m. This can be calculated from the power density and the acoustic impedance, *Z*, of the water.

The acoustic impedance of water is the product of the density and the velocity:

$$Z = \rho_o c = 1026.4 \times 1522 = 1.56 \times 10^6 \, \text{kg/m}^2\text{s}. \tag{12.45}$$

The relationship between the acoustic pressure, *P* (Pa), the power density, *I* (W/m^2), and the impedance, *Z*, is

$$P^2 = IZ. \tag{12.46}$$

This can be rewritten for the acoustic pressure in microPascals for the sound pressure level, *S*:

$$S = \left(10^6 P\right)^2 = 10^{12} IZ. \tag{12.47}$$

This is generally written in decibel form:

$$10\log_{10} S = 20\log_{10}\left(10^6 P\right) = 10\log_{10} 10^{12} + 10\log_{10}(IZ)$$
$$S_{dB} = 120 + 10\log_{10}\left(21864 \times 1.56 \times 10^6\right) = 120 + 105.3 = 225.3 \, \text{dB}. \tag{12.48}$$

If the transmitting voltage response (TVR), shown in Figure 12.25, is examined, it can be seen that at 117 kHz, the response is 170 dB relative to 1 μPa/V at 1 m.

The electrical power input, P_{elec}, is related to the root mean square (RMS) voltage, *V*, and the transducer conductance, *G*:

$$P_{elec} = V^2 G. \tag{12.49}$$

For a conductance $G = 5.5\text{k}\ \mu\text{mho}$ from Figure 12.25 and a power $P_{elec} = 1500\ \text{W}$:

$$V = \sqrt{\frac{P_{elec}}{G}} = \sqrt{\frac{1500}{5.5\times10^{-3}}} = 522\,\text{V (rms)}. \qquad (12.50)$$

The sound pressure level, S_{dB} (dB), for an RMS voltage of 522 V applied to the transducer is

$$S_{dB} = 170 + 20\log_{10}(V) = 170 + 20\log_{10}(522) = 224.3\,\text{dB}. \qquad (12.51)$$

This is as expected, because it should equal the sound pressure level calculated from the maximum input power. The 1 dB difference is due to the difficulty of reading the graph accurately.

As a signal propagates through water, the sound pressure level reduces because the wave is expanding on a spherical wavefront and due to attenuation. The transmission loss in decibels is H and is determined as

$$H = 20\log_{10}(R_2/R_1) + \alpha_{dB}(R_2 - R_1). \qquad (12.52)$$

Because the sound pressure is determined relative to the level existing at 1 m from the effective center of the sound source, the equation can be rewritten for this reference distance if $R \gg 1$:

$$H = 20\log_{10} R + \alpha_{dB} R. \qquad (12.53)$$

The attenuation, α (dB/m), at 5°C as a function of frequency, f (kHz), is determined from equation (7.44), and reproduced here:

$$\begin{aligned}
\alpha_{dB} &= \frac{0.036 f^2}{f^2 + 3600} + 3.2\times10^{-7} f^2 \\
&= \frac{0.036\times117^2}{117^2 + 3600} + 3.2\times10^{-7}\times117^2. \\
&= 0.0329\,\text{dB/m}
\end{aligned}$$

The target strength, TS (dB), is defined by the ratio of the reflected sound pressure scattered by the target at a distance of 1 m from the effective center of the scattered sound to the incident sound pressure on the target:

$$TS = 20\log_{10}(P_r/P_i). \qquad (12.54)$$

This target strength is determined by its size, shape, and the fraction of sound that is reradiated. If the scattering cross section is σ (m^2), then TS (dB) is

$$TS = 10\log_{10}(\sigma/4\pi). \qquad (12.55)$$

As with the radar case, a sphere with a radius a (m), much larger than the wavelength, will have a cross section equal to the projected area, and TS can be rewritten to take that into account:

$$TS = 10\log_{10}\frac{\pi a^2}{4\pi} = 10\log_{10}\left(\frac{a}{2}\right)^2 = 20\log_{10}\frac{a}{2}. \tag{12.56}$$

The target strength of World War II submarines varied between about 10 dB at the bow to 25 dB broadside and 15 dB at the stern. Target strengths of various other shapes are listed in Chapter 8 (Waite 2002).

For a spherical target, the echo sound pressure level, E_{dB}, relative to 1 μPa at a range of 1 m from the receiver is

$$E_{dB} = S_{dB} - 2H + TS. \tag{12.57}$$

Substituting for H and TS from equation (12.53) and equation (12.56) gives

$$E_{dB} = S_{dB} - 40\log_{10} R - 2\alpha_{dB} R + 20\log_{10}(a/2). \tag{12.58}$$

For a sphere with a diameter of 1 m, this reduces to

$$E_{dB} = 225.3 - 40\log_{10} R - 0.0656R - 12.$$

If the target is the sea floor, the cross section will be the product of the range resolution and the cross-range resolution modified by a scaling factor to take into account the scattering strength, S_b (dB), of the bottom:

$$TS = 10\log_{10}\frac{\sigma^O A}{4\pi} = 10\log_{10}\frac{\delta R.\delta x_r}{4\pi} + S_b = 10\log_{10}\frac{\delta R.R.\alpha_{az}}{4\pi} + S_b, \tag{12.59}$$

and the echo sound pressure level will be

$$E_{db} = S_{dB} - 40\log_{10} R - 2\alpha_{dB} R + 10\log_{10}\frac{\delta R.\theta_{az}}{4\pi} R + S_b. \tag{12.60}$$

Substituting for
  $S_{dB} = 225.3$ dB,
  $\alpha_{dB} = 0.0328$ dB/m,
  $S_b = -20$ dB (pebbles and rock at f $\approx$ 100 kHz),
  $\theta_{az} = 1.5°$ (0.0262 rad),
  $\delta R = 0.25$ m (pulsed),
  $\delta R_{pc} = 0.05$ m (pulse compression),

$$E_{dB} = 225.3 - 40\log_{10} R - 0.0656R - 52.8 + 10\log_{10} R.$$

This reduces by $10\log_{10}(\delta R/\delta R_{pc}) = 7$ dB in the pulse compression case because the area of the individual gates is reduced by a factor of five due to the improved range resolution.

As with the radar range equation, detection can only take place if the signal exceeds the noise level by a specific margin. It will be assumed that the noise is white and the signal is sinusoidal so that the same graphs that are used for radar can be used.

Noise level at sea is mostly generated by wind and wave action on the surface. It is proportional to sea state and inversely proportional to frequency, as is clear in Table 12.2. Many of the sources of noise are shown in Chapter 9.

The noise pressure density generated from the sea surface decreases from the level at 1 kHz, specified in Table 12.2, up to the transducer frequency, as explained in Chapter 9. This relationship can be described by

$$N_f = N_1 - 17\log_{10} f_{kHz}. \tag{12.61}$$

For a sea state of 3 and a transducer frequency of 117 kHz,

$$N_f = 65 - 17\log_{10} 117 = 30\,dB.$$

The total noise pressure level, L_N (dB), relative to 1 μPa must take into account the bandwidth, β (Hz), of the matched filter and its directivity or gain, G, because the sensor is only looking at a small part of the hemisphere and hence only picks up a small portion of the total noise:

$$L_N = N_f + 10\log_{10}\beta - 10\log_{10} G. \tag{12.64}$$

In the case of the pulsed option, the receiver bandwidth can be approximated by $\beta = 1/\tau = 3$ kHz:

$$L_N = 30 + 10\log_{10}(3\times10^3) - 10\log_{10}(458.4) = 38\,dB,$$

Table 12.2 Noise pressure density as a function of sea state

Sea state	Wind speed (knots)	Noise pressure density N_1 (dB rel 1 μPa) Isotropic, 1 Hz bandwidth at 1 kHz
6	35	70
3	15	65
2	10	60
1	5	55
0.5	2	50
0	0	45

while, in the case of the pulse compression system, the receiver bandwidth can be approximated by the reciprocal of the uncompressed pulse width (or signal observation time), making $\beta_{pc} = 1/\tau_{pc} = 30$ Hz, and the noise pressure level, L_{Npc}, is reduced by 20 dB:

$$L_{Npc} = 30 + 10\log_{10}(30) - 10\log_{10}(458.4) = 18\,dB.$$

A second source of noise called volume reverberation noise, which is discussed in Chapter 8, is not included in this analysis as it is typically much smaller than the sea noise.

Assuming that the targets are nonfluctuating, and that $P_d = 0.9$ and $P_{fa} = 10^{-6}$ are required, the detection threshold needs to be 13.2 dB above the noise floor. The echo level for two perfectly reflecting spheres of different sizes, the ground, and the noise pressure are plotted in Figure 12.30, from which it can be seen at a glance that attenuation through the water plays an important role in limiting the maximum range.

Because the noise floor has been offset by 13.2 dB, the intersection of this sound pressure with the returns from the targets marks the detection range in each case. It is obvious that the detection range for the pulse compression sonar is greater than that for the conventional pulsed unit in all cases, with the smaller sphere being detectable at 750 m and 550 m, respectively.

Figure 12.30 also shows that even though the return from the sea floor is higher for the pulsed sonar due to the larger footprint, the longer observation time with the resultant decrease in the

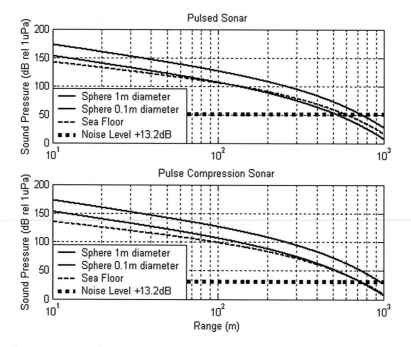

Figure 12.30 Sidescan sonar performance.

noise floor of the pulse compression unit more than makes up for it, and the detection range in the two cases is 750 m and 600 m, respectively. Finally, in the case of the larger sphere, the detection ranges are 950 m and 730 m.

The actual voltage output by the transducer is determined from the transducer specifications. The open circuit receive response at 117 kHz is −180 dB relative to 1 V/μPa. For a signal pressure of 50 dB, which corresponds to the minimum detectable signal level in the pulsed case, the output is

$$20\log_{10}(V) = 50 - 180 = -130\,\text{dB}. \tag{12.65}$$

This can be converted to a voltage

$$V = 10^{-130/20} = 320\,\text{nV (rms)}, \tag{12.66}$$

which is sufficiently small that the noise figure of the receiver should be included when the detection characteristics of the system are being considered.

The noise floor of 38 dB for the pulsed option corresponds to a voltage output of 79 nV rms, and the noise specification of a good low-noise op amp is $1.2\,\text{nV}/\sqrt{\text{Hz}}$ at 1 kHz and above. Therefore, in the pulsed case, at a frequency of 117 kHz with a bandwidth of 3 kHz, the total noise contribution of the op amp is

$$\begin{aligned} V_{noise} &= 1.2 \times 10^{-9} \times \sqrt{3 \times 10^3} \\ &= 66\,\text{nV (rms)} \end{aligned}$$

This increases the noise floor to 102 nV and the effective SNR from 13.2 dB to 9.9 dB at the specified detection range.

12.9 DOPPLER BEAM SHARPENING

As an introduction to synthetic aperture processing, it is useful to consider a similar, but more basic signal processing technique called Doppler beam sharpening. It uses the different Doppler shifts across the antenna beam footprint to narrow the effective beamwidth.

The resolution of a real aperture imaging radar is determined by the range resolution and the azimuth beamwidth, as shown in Figure 12.31. If the radar is moving at a constant velocity, the dotted lines in the figure show lines of constant Doppler shift (isodop lines).

If the beam is offset from the direction of travel, then the isodop lines get closer together and it becomes easier to isolate smaller sections of the beam by their relative Doppler shifts. From a different perspective, if the beam is considered in terms of lines of constant velocity (isovel lines) separated from each other by ΔV (m/s), the angular spacing, θ_N (rad), is

$$\theta_N = \arccos\left(1 - \frac{N\Delta V}{V_t}\right). \tag{12.67}$$

For V_t = 250 m/s and ΔV = 1.25 m/s, the isovel lines will be at the angles shown in Figure 12.31, which correspond to those in Table 12.3.

The Doppler shift of a target viewed from an aircraft traveling at V_t (m/s), at an angle θ (rad) to the direction of travel is

$$f_d = (2V_t/\lambda)\cos\theta. \qquad (12.68)$$

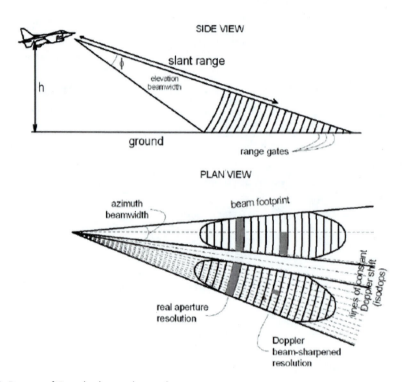

Figure 12.31 Process of Doppler beam sharpening.

Table 12.3 Isovel contours

Velocity (m/s)	Azimuth angle (deg)
1.25	5.73
2.5	8.11
3.75	9.93
5	11.48
6.25	12.84
7.5	14.07

If the radar operates at a frequency of 10 GHz, these isovel lines become isodop lines with a separation of 83 Hz.

One of the limitations to the amount of sharpening that can occur is the observation time. Obtaining a resolution of 83 Hz in the received Doppler requires that the target be observed for $\tau = 1/\beta = 1/83 = 12$ ms. However, at a speed of 250 m/s, the target will have moved 3 m during this period, and that distance may be bigger than the range resolution. This is referred to as range walk, and is accommodated in both unfocused and focused SAR, but not in Doppler beam sharpening. For a fixed transmission frequency, there is therefore a trade-off that must be made between the range resolution and the amount of sharpening that occurs.

An alternative is to increase the transmitted frequency, as this will increase the Doppler shift. For example, if the radar operates at 94 GHz, the separation in the isodop lines increases to 800 Hz and the observation time is reduced to 1.2 ms, during which time the aircraft only moves 0.3 m.

A reflectivity image can be constructed by physically scanning the antenna beam to the one side (or both) of the direction of travel, with the pixel size determined by the range resolution and the sharpened angular resolution. Figure 12.32 shows a comparison between a real beam

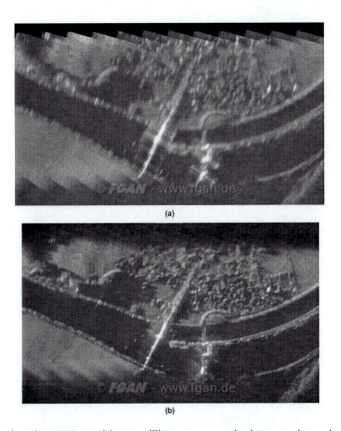

(a)

(b)

Figure 12.32 Comparison between a real beam millimeter wave radar image and one that has been Doppler beam sharpened (Hagelen 2005). Courtesy FGAN.

millimeter wave image and one that has been Doppler beam sharpened. An examination of point targets in the two images shows that even though the cross-range resolution has only been improved by a factor of two, the image quality is much improved.

12.10 OPERATIONAL PRINCIPLES OF SYNTHETIC APERTURE

The term "synthetic aperture" refers to the distance that the sensor travels during the time that the reflectivity data are collected from a single point. Energy from each scatterer is made to arrive in phase at the output of the processor for all of the samples to realize the narrow beamwidth. As with Doppler beam sharpening, the process of synthesis relies on the fact that the wide antenna beamwidth can be subdivided into narrow slivers using the differences in the Doppler shifts across the beam, as shown in Figure 12.33.

As shown in Figure 12.34, the process to determine the radiation pattern for unfocused SAR is similar to that used for a fixed array of the same length, the primary difference being that the

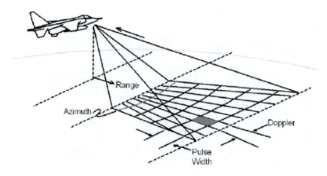

Figure 12.33 Synthetic aperture generation.

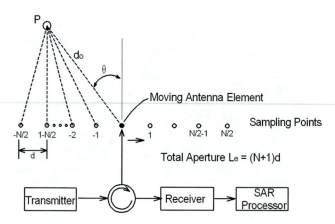

Figure 12.34 Generation of a synthetic array.

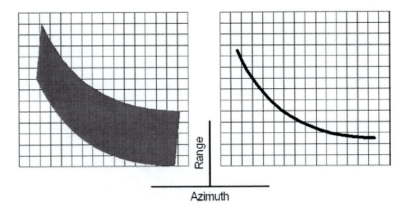

Figure 12.35 Range-azimuth response to a point scatterer shows range walk and range curvature (a) before pulse compression and (b) after pulse compression [adapted from Mensa (1981)].

signal received by each element is due only to the received power scattered by target P from one transmitter element. This results in a slightly different radiation pattern for SAR. The beamwidth is narrower by a factor of two, but the sidelobes are higher than that for the equivalent phased array (Mensa 1981).

For L_e, the synthetic array length, the normalized gain is

$$G_{SAR}(\theta) \approx \frac{\sin^2\left\{\dfrac{2\pi L}{\lambda}\sin\theta\right\}}{\left\{\dfrac{2\pi L}{\lambda}\sin\theta\right\}^2}. \tag{12.69}$$

12.11 RANGE AND CROSS-RANGE RESOLUTION

In most SAR applications, good range resolution is achieved by pulse compression and cross-range resolution obtained by using synthetic aperture processing. It is useful to plot the radar response to a point scatterer on the range-azimuth plane, as this allows the various processes which determine the overall resolution to be illustrated graphically. Figure 12.35 shows the raw data output by the radar before and after pulse compression.

12.11.1 Unfocused SAR

In unfocused SAR, all aircraft motion that deviates from a straight line is compensated for. In addition, for a forward squinted radar, the reduction in range induced by the radar velocity, called range walk, is also removed prior to processing. This step is shown graphically in Figure 12.36.

However, unfocused SAR makes no attempt to correct for the relative phase shift to a point target across the aperture. This results in a phenomenon known as range curvature. Because the

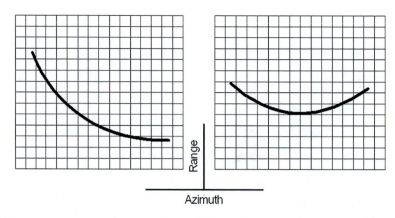

Figure 12.36 Effect of range walk removal on a point scatterer signature shows (a) the signature before range walk removal and (b) the signature after range walk correction [adapted from Mensa (1981)].

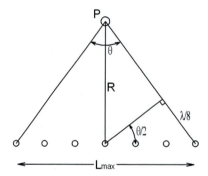

Figure 12.37 Largest possible synthetic aperture for an unfocused SAR.

SAR process involves taking the sum of all of the returns, the limiting condition is the point at which the round-trip phase error reaches $\lambda/4$,[1] as determined using the construction shown in Figure 12.37.

$$\frac{\lambda}{8} = \frac{L_{max}}{2}\sin\left(\theta/2\right) \tag{12.70}$$

1 Different authors use different values for the maximum phase error allowed and so obtain slightly different resolutions.

for $\sin(\theta/2) \approx \dfrac{L_{\max}}{2R}$,

$$\frac{\lambda}{8} \approx \frac{L_{\max}^2}{4R} \tag{12.71}$$

$$L_{\max} = \sqrt{\frac{R\lambda}{2}}. \tag{12.72}$$

A second limiting condition is that the true beamwidth, prior to processing, is sufficiently wide to illuminate the target at point P during the whole time that the aperture is being synthesized. Therefore $L_{\max} \lesssim R\theta_{3dB}$.

The final beamwidth, θ_{SAR} (rad), after processing is obtained by determining half-power points for equation (12.69), $G_{\text{SAR}}(\theta) = 0.5$, and solving graphically or using an iterative technique:

$$\frac{2\pi L_e}{\lambda} \sin \theta_{SAR} \approx \frac{0.886\pi}{2} = 1.39 \tag{12.73}$$

$$\sin \theta_{SAR} = \frac{0.886\pi\lambda}{4\pi L_e}. \tag{12.74}$$

The cross-range resolution $\delta x_r = R\theta_{\text{SAR}} = R\sin\theta$ for small angles is

$$\delta x_r = R 0.886\pi\lambda / 4\pi L_e. \tag{12.75}$$

Substituting $L_e = L_{\max} = \sqrt{\dfrac{R\lambda}{2}}$ and simplifying,

$$\delta x_r = 0.313\sqrt{R\lambda}. \tag{12.76}$$

This is a significant improvement on the real aperture case where $\delta x_r \propto R$.

12.11.2 Focused SAR

In the focused SAR case, both range walk and range curvature are removed so that all of the returns from a point scatterer are concentrated at a single range, as shown in Figure 12.38. This allows the signal to be correlated across the azimuth plane to produce the required cross-range resolution.

A complete derivation of the focused SAR equations is beyond the scope of this book, but if the process is examined from a Doppler perspective one gains insight about how a good resolution can be achieved.

Consider a point scatterer that enters the forward edge of the beam, as shown in Figure 12.39. It will have Doppler frequency of

$$f_d = \frac{2v_r}{\lambda} = \frac{2v}{\lambda} \cos \frac{\theta_{3dB}}{2}. \tag{12.77}$$

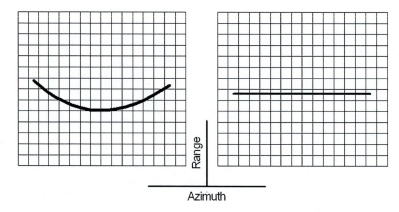

Figure 12.38 Effect of range curvature removal on a range compressed point scatterer signature shows (a) the signature before range curvature removal and (b) the signature after range curvature correction [adapted from Mensa (1981)].

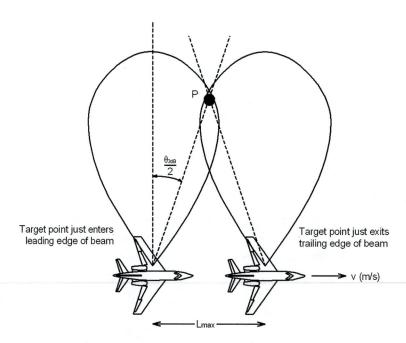

Figure 12.39 Antenna beamwidth limitations to maximum aperture.

For reasonably small beamwidths, the Doppler frequency will decrease almost linearly to zero and then increase again. The angle to the target, as a function of time, can be approximated by $\theta \approx vt/R$. The Doppler frequency can also be written as a function of time:

$$f_d(t) = \frac{2v_r}{\lambda} = \frac{2v}{\lambda}\cos\frac{vt}{R}. \tag{12.78}$$

Take the derivative to obtain the rate of change of Doppler frequency, or the Doppler slope:

$$\frac{df_d}{dt} = \frac{2v}{\lambda}\cdot\frac{v}{R}\sin\frac{vt}{R}, \tag{12.79}$$

equating at $t = 0$,

$$\frac{df_d}{dt} = \frac{2v^2}{R\lambda}. \tag{12.80}$$

If T_d (s) is the total time that the point P remains within the beam from $\theta = -\theta_{3dB}/2$ to $+\theta_{3dB}/2$, the total change in frequency, Δf_d (Hz), is

$$\Delta f_d = \frac{2v^2}{R\lambda}T_d. \tag{12.81}$$

By analogy to the linear FM range resolution, the signal can be passed through a matched filter to give a spectral resolution of

$$\delta f = 1/T_d. \tag{12.82}$$

The cross-range resolution after SAR processing, δx_r (m), is the cross-range resolution for the real beam case, δ_b (m), scaled by the ratio of the spectral resolution, δf (Hz), to the total frequency change (or bandwidth), Δf_d (Hz), of the received Doppler signal:

$$\delta x_r = \delta_b \frac{\delta f}{\Delta f_d}. \tag{12.83}$$

The cross-range resolution for the real beam $\delta_b = L_{\max} = L_e$, therefore

$$\delta x_r = L_e \frac{\delta f}{\Delta f_d}. \tag{12.84}$$

Substituting equation (12.81) and equation (12.82) into equation (12.84),

$$\delta x_r = L_e \frac{R\lambda}{2v^2 T_d}\frac{1}{T_d} = L_e \frac{R\lambda}{2v^2 T_d^2}. \tag{12.85}$$

The aperture is the forward velocity of the aircraft multiplied by the total observation time. It can also be written in terms of the range and the real beamwidth, which can, in turn, be written in terms of the real aperture of the antenna, d (m), and the wavelength, λ (m):

$$L_e = vT_d = R\theta_{3dB} = R\frac{\lambda}{d}. \tag{12.86}$$

Substituting for those various approximations into equation (12.85) and then simplifying gives

$$\delta_{cr} = \frac{R\lambda}{2L_e} = \frac{R\lambda}{2} \cdot \frac{d}{R\lambda} = \frac{d}{2}. \tag{12.87}$$

The cross-range resolution for focused SAR is independent of the range R and only dependent on the size of the antenna.

12.11.3 Resolution Comparison

For a radar with a real aperture of 120 mm and a frequency of 94 GHz, Figure 12.40 shows the relationship between the cross-range resolution of a real aperture antenna, unfocused SAR, and focused SAR.

12.12 WORKED EXAMPLE: SYNTHETIC APERTURE SONAR

Three targets are placed with cross-range separations of 20 mm and 35 mm at a range of 2 m with coordinates [20,2000], [40,2000], [75,2000] mm. The antenna aperture $d = 10$ mm produces a 3 dB beamwidth of 49°.

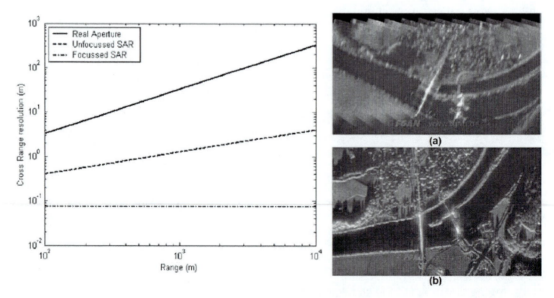

Figure 12.40 Cross-range resolution comparison for a radar with a 120 mm antenna operating at 94 GHz and a comparison between (a) a real aperture image and (b) a SAR image also made at 94 GHz. Courtesy FGAN.

As the sonar moves past the targets the received phase of the echoes follows a parabolic trajectory, as shown in Figure 12.41a. The focused returns are generated by subtracting a reference phase profile generated by a fixed target at [0,2000], making the phase shift with radar position now almost linear, as can be seen in Figure 12.41b.

The received signal is reconstructed by taking the cosine of the individual phases and summing the results to produce the signal shown in Figure 12.42.

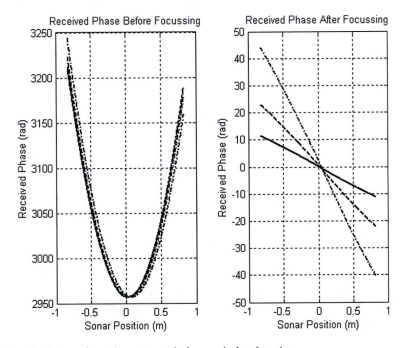

Figure 12.41 Received phase from three targets before and after focusing.

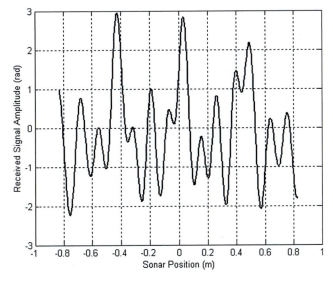

Figure 12.42 Received signal from the three targets.

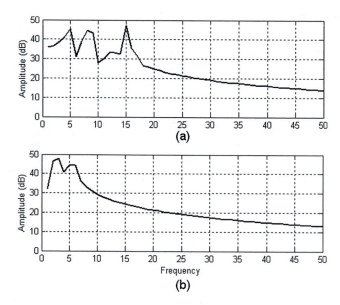

Figure 12.43 Resolved target cross-range positions for an antenna aperture (a) where $d = 10$ mm and (b) where $d = 30$ mm.

The individual targets are resolved in cross range by examining the spectrum of the received signal, as shown in Figure 12.43. If the antenna aperture size is $D = 10$ mm, then all three targets are resolved, but if the aperture is increased to 30 mm, the targets that are only 20 mm apart merge into a single return.

```
% Synthetic Aperture Sonar Model
% sar01.m


% variables
degrad = pi/180;            % degrees to radians
c = 340;                    % speed of sound (m/s)
f = 40e+03;                 % operational frequency (Hz)
d = 10.0e-03;               % antenna aperture (m)
lam = c/f;                  % wavelength (m)


% generate three targets at the same range separated by
% a fraction of the beamwidth
x0 = 0;
y0 = 2.0;                             % reference target to determine phase shift for focus
```

```
x1 = 20.0e-03;
y1 = 2.0;
x2 = 40.0e-03;
y2 = 2.0;
x3 = 75.0e-03;
y3 = 2.0;

% assume a rectangular aperture and determine the 3dB beamwidth
thet3 = asin(0.887*lam/d);

% generate a moving sonar system along the x axis
lmax=2*y1*sin(thet3/2);
dx = lmax/511;
xrad = -lmax/2:dx:lmax/2;
yrad = zeros(size(xrad));

% calculate the range to each of the targets
dx1 = xrad-x1;
dy1 = yrad-y1;
dx2 = xrad-x2;
dy2 = yrad-y2;
dx3 = xrad-x3;
dy3 = yrad-y3;

[az1,r1]=cart2pol(dx1,dy1);
[az2,r2]=cart2pol(dx2,dy2);
[az3,r3]=cart2pol(dx3,dy3);

% calculate the range to the reference target for focusing
dx0 = xrad-x0;
dy0 = yrad-y0;
[az0,r0]=cart2pol(dx0,dy0);

% calculate the phase due to the round trip distance to each target

ph1 = 4*pi*f.*r1./c;
ph2 = 4*pi*f.*r2./c;
ph3 = 4*pi*f.*r3./c;

ph0 = 4*pi*f.*r0./c;                    % reference phase
```

```
subplot (121), plot(xrad,ph1,'k',xrad,ph2,'k-',xrad,ph3,'k-.')
grid
title('Received Phase Before Focusing')
xlabel('Sonar Position (m)')
ylabel('Received Phase (rad)')


% subtract the reference phase to focus

dph1 = ph1-ph0;
dph2 = ph2-ph0;
dph3 = ph3-ph0;

subplot(122), plot(xrad,dph1,'k',xrad,dph2,'k-',xrad,dph3,'k-.');
grid
title('Received Phase After Focusing')
xlabel('Sonar Position (m)')
ylabel('Received Phase (rad)')


pause
subplot

% generate the time domain signal from the received phases
sig1 = cos(dph1);
sig2 = cos(dph2);
sig3 = cos(dph3);
sig = sig1+sig2+sig3;

plot(xrad,sig,'k');
grid
title('Received Combined Signal')
xlabel('Sonar Position (m)')
ylabel('Received Signal Amplitude (rad)')
pause

% extract the target returns by looking at the received spectrum
sigf = fft(sig);
db = 20*log10(abs(sigf(1:255)));
subplot(211), plot(db,'k');
grid
title('Received Signal Spectrum')
xlabel('Frequency')
ylabel('Amplitude (dB)')
axis([0,50,0,50])
```

12.13 RADAR IMAGE QUALITY ISSUES

12.13.1 Perspective of a Radar Image

The perspective of terrain as seen by a side-looking radar is different from that seen by an aerial photograph because, in radar, two objects coincide if they are at the same range, whereas in an aerial photograph they appear to coincide if they are at the same angle. These characteristics are illustrated in Figure 12.44. This issue with perspective results in the subtle forms of distortion discussed in this section.

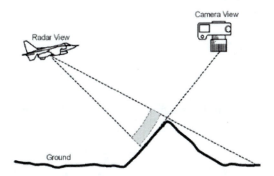

Figure 12.44 Difference in perspective for a radar view and that from an aerial photograph.

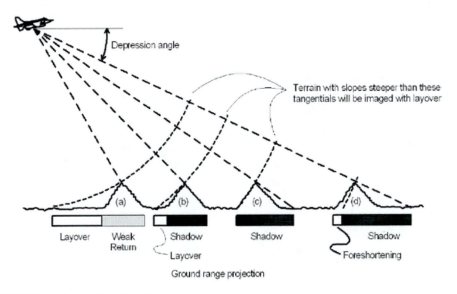

Figure 12.45 Some image distortion effects.

12.13.2 Image Distortion

The most common forms of distortion that are suffered by radar images made from the air are summarized in Figure 12.45, and include the following:

- Layover, when the range to the top of an object is smaller than the distance to its base.
- Foreshortening, when the near side of elevated objects appears steeper than it actually is.
- Stretching, due to the differences between the ground and slant ranges.
- Shadowing, when a tall opaque object blocks the signal path behind it and no returns are received.

12.13.2.1 Stretching   As shown in Figure 12.46, the mapping from the ground range to the slant range results in an image that is progressively more and more compressed as the depression angle increases.

12.13.2.2 Shadowing   As illustrated in Figure 12.47, shadowing occurs mostly at low grazing angles when an object blocks the signal path to cast shadows which cannot be imaged by the radar. This is a common effect when rough terrain (such as mountains) is being imaged from a reasonably shallow grazing angle or if man-made objects are in the scene. As discussed in the side-scan section, shadowing can be extremely useful as it allows the observer to determine the height of objects in an image which would otherwise be unknown.

12.13.3 Speckle

Because the measurement process is coherent, in some of the image pixels the scatterers will combine constructively to produce high-intensity returns, and in some the summation will be destructive to produce dark returns, as shown in Figure 12.48. Various filtering and averaging techniques are used to reduce these effects and to make the images easier to interpret (Henderson and Lewis 1998).

12.14 SAR ON UNMANNED AERIAL VEHICLES

12.14.1 TESAR

Probably the best known of all the SAR systems designed for use in unmanned aerial vehicles (UAVs) is the Tactical Endurance Synthetic Aperture Radar (TESAR), also called the MAE UAV SAR, installed in the General Atomics' Predator. An overview of the specifications of this radar is listed in Figure 12.49.

The TESAR is a strip mapping SAR that provides continuous 0.3 m (1 ft) imagery. The focused imagery is formed onboard the Predator aircraft, compressed, and sent to the ground control station over a Ku-band data link. The imagery is reformed and displayed in a scrolling manner on the SAR workstation displays. As the imagery is scrolling by, the operator has the ability to select 1000 × 1000 pixel image patches (approximately 800 m × 800 m box at a height of 15,000 ft) for exploitation.

There are two modes of operation. Mode 1 provides a noncentered strip map; that is, the map center moves with respect to the aircraft motion. Mode 2 is the classic strip map mode in which mapping occurs over a predetermined scene centerline, independent of the aircraft motion.

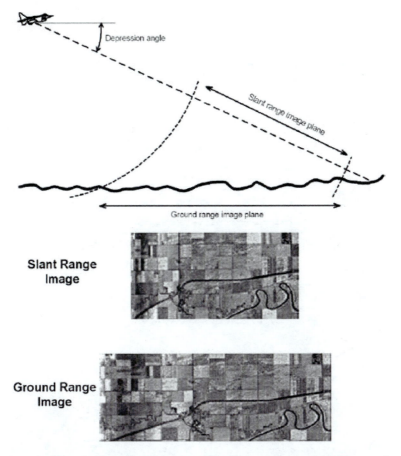

Figure 12.46 The radar image is distorted and compressed by the nonlinear mapping between the ground range and the slant range.

Figure 12.47 For low grazing angle applications, buildings and mountains cast shadows that are not imaged by the radar. These appear as black regions in SAR images.

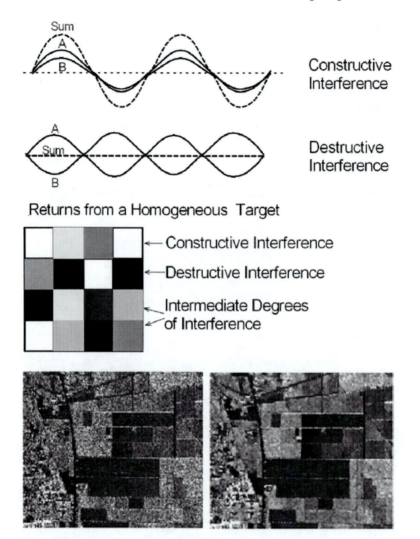

Figure 12.48 SAR Speckle explained and an example of multilook averaging used to reduce the effect.

The radar is designed to map while squinting up to ±45° off the velocity vector. At ground speeds from 25 to 35 m/s, the swathe width is 800 m, while at speeds beyond 35 m/s, the swathe width decreases proportionally with the increase in ground speed (TESAR 2007).

12.14.2 MiniSAR

By the end of 2006 Sandia National Laboratories was flying the smallest SAR ever to be used for reconnaissance on near-model-airplane-sized UAVs. Weighing less than 10 kg, the miniSAR

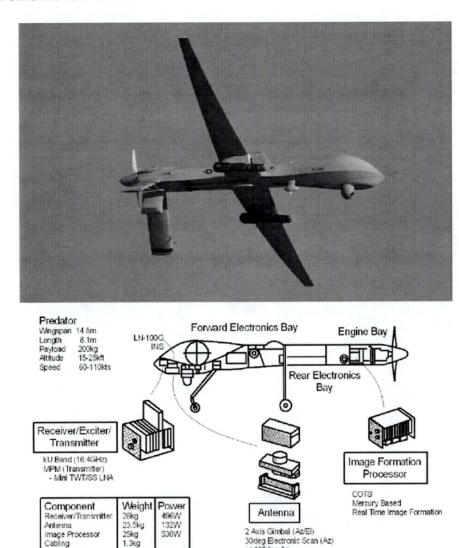

Figure 12.49 TESAR in the Predator UAV [adapted from GlobalSecurity.org (2007)].

is one-fourth the weight and one-tenth the volume of its predecessors currently flying on larger UAVs such as the Predator. It is the latest design produced by Sandia based on more than 20 years of related research and development. The new miniSAR, shown in Figure 12.50, will be able to take high-resolution (4-inch) images through bad weather, at night, and in dust storms out to a range of about 15 km (Sandia Corp. 2007).

12.15 AIRBORNE SAR CAPABILITY

Almost all modern radar systems fitted to defense aircraft include both strip map and spotlight SAR capabilities. The capability of airborne SAR is extraordinary, with submeter resolutions made at long standoff ranges, as illustrated so effectively in Figure 12.51. The image of the airport is typical of a multilook SAR image that has been processed to reduce speckle. The sensitivity of the system is sufficiently good that backscatter from the smooth runway even reveals some

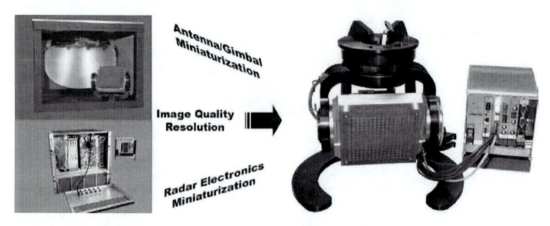

Figure 12.50 Sandia Laboratories miniSAR developed for UAVs (Sandia Corp. 2007).

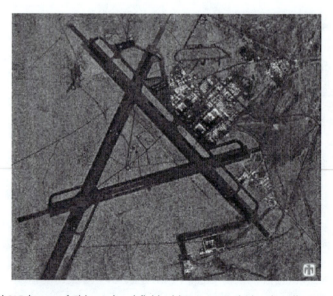

Figure 12.51 X-band SAR image of China Lake airfield with a 3 m resolution (Sandia Corp. 2004).

texture. Returns from the surrounding grass adjacent to the runways and from the rougher terrain behind the building complex reveal some of the strengths of SAR in imaging natural or agricultural terrain. The airport complex and adjacent village are also indicative of the excellent processing in which gate leakage has been all but eliminated.

In the case of the tank formation shown in Figure 12.52, the low grazing angle makes the tracks through the terrain visible. The resolution in this image is so good that the vehicles are identifiable as tanks and experts at image interpretation could probably hazard a guess as to the type. In this image, the large RCS return of the corner reflectors does result in some leakage.

12.16 SPACE-BASED SAR

To achieve good angular resolutions from real-aperture space-borne radars is impossible at lower frequencies because the size of the antenna becomes prohibitively large. As derived previously, the large synthetic aperture results in a cross-range resolution independent of range, $\delta_{cr} = d/2$, where d is the antenna aperture, and good range resolution is achieved by transmitting a wide bandwidth chirp, $\delta_r = c/2\Delta f$.

The primary advantage of space-borne SAR is that the trajectory of the satellite or shuttle is precisely known and stable. A typical trajectory, antenna footprint, and image swathe are shown in Figure 12.53. This means that motion compensation is not required and exceptionally high-quality images can be produced.

Figure 12.52 SAR image (a) of tanks and a ring of corner reflectors and (b) a photograph of the same scene taken from the ground.

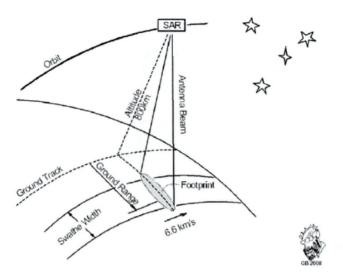

Figure 12.53 Trajectory and footprint of a spaceborne SAR.

Spaceborne Imaging Radar-C and X-Band Synthetic Aperture Radar (SIR-C/X-SAR) is part of NASA's Mission to Planet Earth. The radar uses three microwave wavelengths: L-band (24 cm), C-band (6 cm), and X-band (3 cm). These multifrequency data are available to the international research community and are being used to better understand the global environment.

SIR-C was developed by NASA's Jet Propulsion Laboratory and X-SAR was developed by the Dornier and Alenia Spazio companies for the German space agency, Deutsche Agentur für Raumfahrtangelegenheiten (DARA), and the Italian space agency, Agenzia Spaziale Italiana (ASI), with the Deutsche Forschungsanstalt für Luft und Raumfahrt e.v.(DLR), the major partner in science, operations, and data processing of X-SAR.

Figure 12.54 shows an image of the Mississippi River delta where the river enters into the Gulf of Mexico. This image was acquired by SIR-C/X-SAR aboard the space shuttle Endeavour on October 2, 1995. The image is centered on latitude 29.3°N and 89.28°W. The area shown is approximately 63 km × 43 km.

The main shipping channel of the Mississippi River is the broad stripe running northwest to southeast down the left side of the image. The bright spots within the channel are ships. Visible in the Web version of the book, the colors in the image are assigned to different frequencies and polarizations of the radar as follows: red is L-band vertically transmitted, vertically received; green is C-band vertically transmitted, vertically received; blue is X-band vertically transmitted, vertically received (Deutsches Zentrum für Luft- und Raumfahrt 2005).

Because of its ability to penetrate cloud and rain, as well as the fact that it can operate day and night, SAR has become the primary sensor for remote sensing of the earth. To maximize their effectiveness, SAR systems are generally multiband and polarization diverse, as this gives a large number of degrees of freedom from which to isolate a myriad of surface effects.

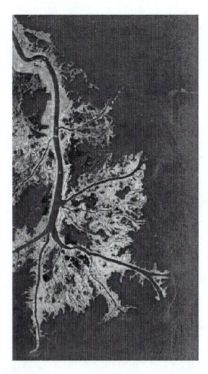

Figure 12.54 Space-based SAR image of the Mississippi Delta (Deutsches Zentrum für Luft- und Raumfahrt 2005). Courtesy NASA/JPL-Caltech.

The use of SAR data for agricultural applications is very promising. For example, phase information has been used to identify the differences between plant types, and even whether they have been harvested or not. Given the world's concern with deforestation, global warming, and carbon credits, the ability of SAR systems to estimate forestry biomass and even to distinguish quickly between species over large areas will become increasingly important in identifying illegal logging and policing compliance. Geological applications for SAR are also common, and it has been reasonably successful in identifying lava flows and other geological features such as alluvial fans, sand dunes, and moraines. Finally, SAR is also used extensively in hydrology for soil moisture estimation, flood mapping, and the measurement of snow and ice cover (Henderson et al. 1998).

12.16.1 Interferometry

Because SAR is concerned with the phase relationships between scatterers on the ground, if two similar images are produced using offset antennas or on subsequent passes over the same area, the interference patterns can be used to determine the true height of the objects on the ground.

In addition to being useful for mapping ground features, this technology has a number of other uses:

- Observing local deformation of the earth's crust as an early warning of earthquakes or volcanoes.
- Ground subsidence due to mining activities or excessive use of groundwater.

12.17 MAGELLAN MISSION TO VENUS

The radar sensor, Magellan's sole scientific instrument, produced high-resolution SAR images of more than 98% of the surface of Venus with a resolution of better than 150 m between August 1990 and September 1992. The high-gain antenna, clearly visible in Figure 12.55, was also used to communicate with the earth (NASA 2008).

The key radar characteristics are as follows:

- Synthetic aperture radar (SAR)
- Frequency 2.385 GHz
- Peak power 325 W
- Pulse length 26.5 μs
- PRF 4400–5800 Hz
- Swathe width 25 km (variable)
- Data acquisition rate 806 kbps
- Downlink quantization 2 bits
- Operates in SAR, altimeter, and radiometer modes
- SAR resolution 150 m range/150 m azimuth
- Altimeter resolution 30 m
- Radiometer accuracy 2°C

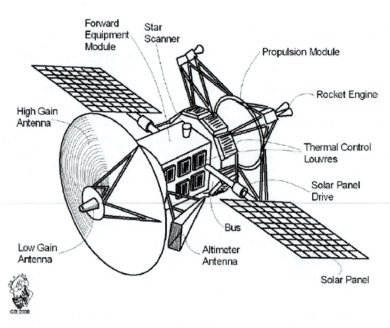

Figure 12.55 Magellan showing the high-gain antenna used to SAR map the surface of Venus.

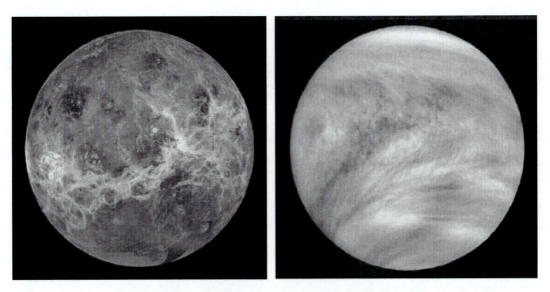

Figure 12.56 Magellan SAR map of Venus and visible image (NASA 2008). Courtesy NASA/JPL-Caltech.

Figure 12.57 Sif Mons 2 km height and 300 km diameter, 3D image produced by combining SAR and altimetry data (NASA 2008). Courtesy NASA/JPL-Caltech.

Figure 12.56 shows the map of Venus generated by the Magellan SAR, and Figure 12.57 shows a computer generated 3D image of Sif Mons made using both the SAR and altimeter data.

12.18 References

Brookner, E., ed. (1982). *Radar Technology*. Norwood, MA: Artech House.

Brookner, E. (1985). Phased-array radars. *Scientific American* 252(2):94–102.

Brookner, E. (2006). Modern Phased Array Radar. Radar and Industry Workshop, Adelaide, South Australia, May 10.

Deutsches Zentrum für Luft- und Raumfahrt. (2005). Mississippi Delta. Viewed January 2008. Available at http://www.op.dlr.de/ne-hf/SRL-2/p45759_mississipi.html.

Fish, J. and Carr, A. (2001). Acoustics and sonar (AUSS Ltd). Viewed March 2001. Available at http://www.marine-group.com/acoustic.htm.

GlobalSecurity.org. (2007). TESAR (tactical endurance synthetic aperture radar). Viewed January 2008. Available at http://www.globalsecurity.org/intell/systems/tesar.htm.

Hagelen, M. (2005). Doppler beam sharpening im Millimeterwellenbereich. Viewed January 2008. Available at http://www.fhr.fgan.de/fhr/fhr_c358_f3_de.html.

Henderson, F. and Lewis, J., eds. (1998). *Principles and Applications of Imaging Radar. Manual of Remote Sensing*. New York: John Wiley & Sons.

Hultqvist, S. (2007). Swedish east coast wrecks. Viewed January 2008. Available at http://www.abc.se/~m10354/uwa/wreck-se.htm.

ITC. (2008). Sidescan transducer model ITC-5202. Viewed January 2008. Available at http://www.itc-transducers.com/html/underwater_transducers.html.

Knutstal, E. and Bunker, W. (1992). Guidelines for specifying underwater electroacoustic transducers. Paper presented at UDT 1992 Conference, London.

Mensa, D. (1981). *High Resolution Radar Imaging*. Norwood, MA: Artech House.

NASA. (2008). Magellan mission to Venus. Viewed January 2008. Available at http://www2.jpl.nasa.gov/magellan/.

Sandia Corp. (2004). X-band synthetic aperture radar imagery. Viewed November 2004. Available at http://www.sandia.gov/RADAR/imageryx.html.

Sandia Corp. (2007). Sandia miniSAR—miniaturized synthetic aperture radar. Viewed January 2008. Available at http://www.sandia.gov/RADAR/minisar.html.

Skolnik, M. (1970). *Radar Handbook*. New York: McGraw-Hill.

Thompson, P. (2003). Design and construction of an ultrasound imaging system using phased arrays. University of Sydney, Sydney, NSW, Australia.

Waite, A. (2002). *Sonar for Practicing Engineers*. New York: John Wiley & Sons.

13

Range and Angle Estimation and Tracking

13.1 INTRODUCTION

The analysis in this book, so far, has concentrated on using sensors to detect the presence of targets in noise and clutter. However, sensors are capable of much more than that. Consecutive measurements of a target's position (and sometimes velocity) can be used to calculate the target state (position, velocity, and acceleration), from which a good estimate of the future position can be made. This process of estimation is generally referred to as "tracking." For most time-of-flight sensors that operate in polar space, this involves following the target independently in both range and angles to obtain good estimates of its position in three dimensions. This chapter is concerned with the mechanics of the tracking process.

13.2 RANGE ESTIMATION AND TRACKING

Range tracking is an extension to the independent measurement of range on a pulse-by-pulse or block-by-block basis. It involves using the previous estimate of the range and range rate (and sometimes acceleration) to predict the position of the target in time for the next range measurement. The most common configuration uses a type 2 loop (two cascaded integrators or their equivalent) to provide a zero lag estimate for constant range rate targets.

In the radar context, standard estimation techniques were first introduced to automate the process of plot extraction for individual aircraft targets observed by surveillance radars. This is a process called "track while scan," which is discussed in Chapter 14.

13.2.1 Range Gating

To determine the position of a target accurately in range, the detected video signal is sampled at a specified time after a pulse has been transmitted. This involves measuring the amplitude of this signal over a short period using some form of electronic switch. In most cases the switch closes for a short period during which time a small capacitor is allowed to charge. An alternative is to use a ready-built sample-and-hold (S&H) integrated circuit (IC) which performs a similar function.

The sample period should be about equal to or shorter than the length of the transmitted pulse so that the maximum amount of pulse energy and the minimum amount of noise is incorporated into each sample. An example of a S&H circuit and its application are shown in Figure 13.1.

13.3 PRINCIPLES OF A SPLIT-GATE TRACKER

Sampling the received pulse only once during every transmit cycle is not sufficient to extract any information apart from the presence or absence of an echo. To obtain information about the relative position of the pulse with respect to the sample time requires that a pair of closely spaced samples be made. Together, these make up what is known as a split-gate tracker (Hughes 1988). It consists of two S&H circuits known as the early and late gates, separated in time (range) by about one pulse width. In early analog systems, these were in fact field effect transistor (FET) switches that allowed charge to flow into an integrator for the gate period τ_e and τ_l, as shown in Figure 13.2.

13.3.1 Range Transfer Function

The relationship between the error voltage, ε_r (V), and the relative position of the sample, ΔT (s), with respect to the center of the echo pulse is known as the range transfer function. Consider that at the time corresponding to the estimated target range, these S&Hs are triggered, the first

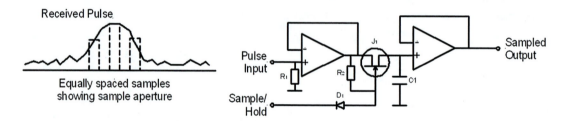

Figure 13.1 Sampling an echo pulse is known as range gating.

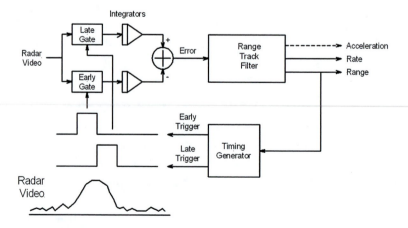

Figure 13.2 Analog split-gate tracker.

at one-half pulse width prior to the estimated range delay, and the second the same period afterwards. If the range estimate is accurate, the two circuits sample equal amplitudes of the target echo pulse on either side of the peak, as shown in Figure 13.2, and the difference between the two range gate voltages, $V_L - V_E = 0$. However, if there is a small range error, the contents of the one gate will be larger than the other and the difference will not be zero.

To measure the transfer function, it is possible to keep the gates still and move the target through the gates, or vice versa. Figure 13.3 shows the measured range transfer function for an ultrasonic radar simulation. Note that the sign and magnitude of the error will drive the tracking error toward zero only over a limited range (in this case, about 20 samples) and the "linear" region over which the magnitude of the range error can be used to estimate the actual error is more limited (about 8 samples).

A closed-loop range tracker uses this error to drive the smoothed estimate of the target range until the amplitudes of the early and late gate samples are equal, and the output is zero. It is for this reason that this is sometimes called a "null tracker." If, however, the error falls outside the bounds shown by the transfer function, then the error voltage falls to zero (in a noise-free environment), there is no driving function, and tracking cannot occur.

13.3.2 Noise on Split-Gate Trackers

The best thermal noise performance is achieved when the matched filter is matched to the pulse spectrum ($1 < \beta\tau < 2$) and then the range gate width is matched to the pulse width ($\tau < \tau_g < 2\tau$) (Barton 1976). This leads to an error slope, $k_r = 2.5$, and a normalized root mean square (RMS) measurement error, σ_r:

$$\sigma_r = \frac{\tau}{2.5\sqrt{2SNR}}. \tag{13.1}$$

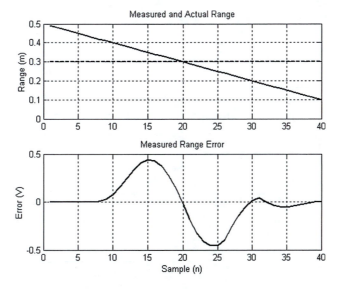

Figure 13.3 Measured transfer function for an ultrasonic range tracker.

The measurement error takes on the dimensions of τ, so it can be a range error in meters or a delay error in seconds. Note that the error is ultimately determined by the signal-to-noise ratio (SNR) and therefore, as this value increases, the RMS error decreases proportionally.

13.4 RANGE TRACKING LOOP IMPLEMENTATION

The range tracking process applies simple estimation theory to predict where the target will be, and placing the gates at the correct range prior to the measurement taking place improves the measurement accuracy. Placing the range gates as accurately as possible has two primary functions. In the first instance it ensures that any residual errors are within the linear region of the range transfer function and are therefore close to the true error. Second, sampling the echo signal close to its peak ensures that the measurement SNR is as high as possible.

Typically a range tracker will include three gates, the early and late gates discussed earlier, and a sum gate which straddles the other two gates and samples the whole received pulse. This gate can be a physical realization, as shown in Figure 13.4, or it can be the sum of the voltage signals from the early and late gates.

To maintain a constant loop bandwidth, the error function $V_L - V_E$ must be normalized, otherwise targets with a large radar cross section (RCS) or closer to the radar will produce larger errors than those with a small RCS or further away, even if the true range error is the same. Normalization is achieved using an automatic gain control (ACG) driven by the sum channel range gated video signal.

The normalized range error drives a second-order tracker, represented in the figure by a pair of cascaded integrators. In most modern trackers digital implementations of the tracking loop are used, as they are easy to implement and can be optimized to suit specific conditions. Problems with pure digital implementations such as the one shown in Figure 13.5 are because the S&H and analog-to-digital converter (ADC) must have sufficient dynamic range to digitize the full range of target amplitude variations. For radar systems, target RCS variations and the operational

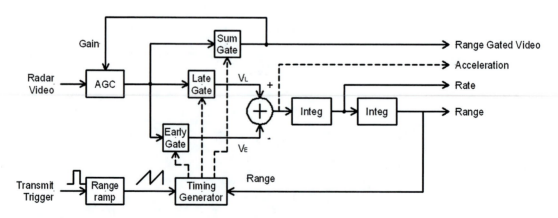

Figure 13.4 Analog tracking loop implementation using cascaded integrators.

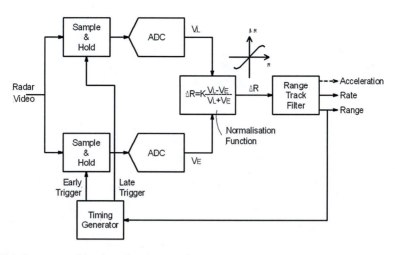

Figure 13.5 Digital range tracking loop implementation.

range can result in variations in the received signal level in excess of 100 dB, which is impractical to digitize directly and therefore some form of analog limiting or an AGC must be implemented.

The primary objectives of range tracking filters are to ensure that the tracking gates remain centered on the target echo, and to provide a good estimate of the target position and velocity by minimizing the noise variance under specific conditions. Unfortunately these two objectives are often mutually incompatible because keeping the tracking gates centered on the return of an accelerating target requires a wide bandwidth loop, whereas producing the best estimate of the target state is usually required when the target is moving at a constant speed and requires a narrow bandwidth loop. Kalman filters and α-β trackers are widely used for track-while-scan and single-target tracking applications to try to achieve these goals.

13.4.1 The α-β Filter

The α-β tracker is a fixed-gain formulation of the Kalman filter and is still widely used because it is easy to implement and performs well under most circumstances (Blackman 1986). The implementation of most tracking filters follows a two-stage process, first of smoothing and then of prediction. The equations defining these two stages are shown below for the α-β tracker implementation:

- Smoothing

$$\hat{R}_n = \hat{R}_{pn} + \alpha \left(R_n - \hat{R}_{pn} \right) \tag{13.2}$$

$$\hat{V}_n = \hat{V}_{pn} + \frac{\beta}{T_s} \left(R_n - \hat{R}_{pn} \right) \tag{13.3}$$

• Prediction

$$\hat{R}_{p(n+1)} = \hat{R}_n + \hat{V}_n \cdot T_s \qquad (13.4)$$

$$\hat{V}_{p(n+1)} = \hat{V}_n \qquad (13.5)$$

where
    $\hat{R}_n$ = smoothed estimate of range,
    $\hat{V}_n$ = smoothed estimate of range rate,
    R_n = measured range,
    $\hat{R}_{p(n+1)}$ = predicted range after T_s seconds (one sample ahead),
    $\hat{V}_{p(n+1)}$ = predicted range rate after T_s seconds (one sample ahead),
    $\hat{R}_{pn}$ = predicted range at the measurement time,
    $\hat{V}_{pn}$ = predicted velocity at the measurement time,
    T_s = sample time,
    α, β = smoothing constants.

The equations shown above are only one of the ways that the tracking filter can be described. Other ways include the block diagram representation (Mahafza 2000), state-space representations, or rewriting the equations as a position transfer function from the difference equations:

$$R_p(z) = R(z)z^{-1} + T_s \dot{R}(z)z^{-1} \qquad (13.6)$$

$$R(z) = R_p(z) + \alpha \left[R_m(z) - R_p(z) \right] \qquad (13.7)$$

$$\dot{R}(z) = \dot{R}(z)z^{-1} + \frac{\beta}{T_s} \left[R_m(z) - R_p(z) \right] \qquad (13.8)$$

where $R_p(z)$ is the predicted position, $R(z)$ is the estimated position, $\dot{R}(z)$ is the rate estimate, and $R_m(z)$ is the measured position.

The position transfer function will therefore be

$$H_1(z) = R(z)/R_p(z), \qquad (13.9)$$

and it can be obtained by manipulation of the equation (13.6) to equation (13.8):

$$H_1(z) = \frac{\alpha + (\beta - \alpha)z^{-1}}{1 + (\beta + \alpha - 2)z^{-1} + (1 - \alpha)z^{-2}} \qquad (13.10)$$

This is a convenient form of the filter to be used in MATLAB, where the filter coefficients, as described in Chapter 2, are

$$B = [\alpha, (\beta - \alpha)]$$
$$A = [1, (\beta + \alpha - 2), (1 - \alpha)].$$

One possible filter optimization (Benedict and Bordner 1962) is to minimize the output noise variance at steady state and the transient response to a maneuvering target as modeled by a ramp function. This results in the following relationship between the two gain coefficients:

$$\beta = \alpha^2/2 - \alpha. \tag{13.11}$$

Other criteria can also be used. For example, it has been suggested that the filter should have the fastest possible step response (critical damping), in which case the coefficients are related as

$$\alpha = 2\sqrt{\beta} - \beta. \tag{13.12}$$

The actual values of the gains depend on the sample period, the predicted target dynamics, and the required loop bandwidth.

One of the main disadvantages of the fixed gain α-β tracker is that it estimates the position of an accelerating target with a constant lag. However, because the magnitude of this lag can easily be estimated, the filter can use adaptive gains to improve the RMS tracking accuracy under these circumstances. The cost function that is minimized in this case is $[\text{lag}^2 + \sigma_r^2]$.

For estimates that include accelerations, the α-β-γ tracker can be substituted. However, its performance if the target is moving with constant velocity is poorer than that for the simple α-β tracker because of the additional noise introduced by the acceleration estimate.

13.4.2 The Kalman Filter

A simplified view of the Kalman filter when compared to one of the suboptimal filters is that it uses optimum weighting coefficients (somewhat analogous to α and β) that are dynamically computed each update cycle. In addition, a more precise model of the target dynamics can be used. The Kalman filter will minimize the mean squared error so long as the target dynamics and the measurement noise are accurately modeled (Blackman 1986; Mahafza 2000).

The benefits of the additional complexity include:

- Improvement in tracking accuracy.
- A running measure of the accuracy.
- A method of handling measurements of variable accuracy, nonuniform sample rate, or missing samples.
- Higher order systems are easier to handle.

13.4.3 Other Tracking Filters

Because of the difficulties involved in accurately modeling target dynamics and measurement noise, many radar engineers are skeptical of the advantages offered by using the Kalman filter, with the result that a number of interesting alternatives have been developed. One of these is the Centroid-β filter, which produces a good estimate of the target position from a moving average and then uses a recursive filter to estimate the velocity (Brooker 1983; Harris and Clarke 1981). The rationale behind this structure is that the long moving average (typically 1 s) produces a good estimate even if the noise is correlated.

- Smoothing

$$\hat{V}_n = \hat{V}_{n-1} + \frac{\beta}{T_s}\left(R_n - \hat{R}_{pn}\right) \tag{13.13}$$

$$\hat{R}_n = \frac{1}{N}\sum_{j=k-N+1}^{k} R_j + \frac{N-1}{2}T_s\hat{V}_n \tag{13.14}$$

- Prediction

$$\hat{R}_{p(n+1)} = \hat{R}_n + \hat{V}_n \cdot T_s \tag{13.15}$$

where
$\quad \hat{R}_n$ = smoothed estimate of range,
$\quad \hat{V}_n$ = smoothed estimate of range rate,
$\quad R_n$ = measured range,
$\quad R_j = j$th sample of the measured range for the moving average,
$\quad \hat{R}_{pn}$ = predicted range at the measurement time,
$\quad \hat{R}_{p(n+1)}$ = predicted range after T_s seconds (one sample ahead),
$\quad T_s$ = sample time.
$\quad \beta$ = smoothing constant for range rate estimate,
$\quad N$ = length of moving average.

13.5 ULTRASONIC RANGE TRACKER EXAMPLE

An ultrasonic range tracker based on the matched filter simulation described in Chapter 5 has been developed to evaluate the effects of various gate and tracking filter parameters. In Figure 13.6, an approaching target is shown starting at a range of 1.5 m ($\approx$ 9 ms) and moving toward the sensor. The tracking gates are triggered after a delay of 6 ms.

To perform the split-gate tracking function, two gates are shown, the early and the late gates (*,+). Due to receiver noise, the tracking gates will move in a random fashion until they are reached by the target echo, at which time they will lock onto it, as seen in Figure 13.7, and track its motion. Note that the early and late gates are sitting astride the target echo peak, which would produce a small tracking error voltage that drives the tracking loop to maintain its position straddling the target echo perfectly, even though the latter is moving.

13.6 TRACKING NOISE AFTER FILTERING

An α-β filter (with Benedict-Bordner gains) takes the error signal produced by the split gate tracker and produces estimates of the position and the rate. The one-sample-ahead position estimate is fed back into the range gate timing circuitry to trigger the early and late gate samples. The results of this process are shown in Figure 13.8.

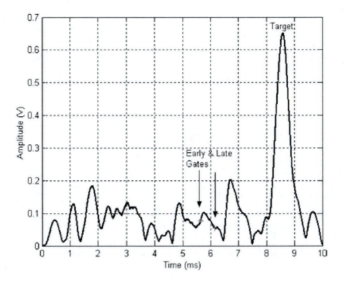

Figure 13.6 The target echo and the split-gate position at the start of the simulation.

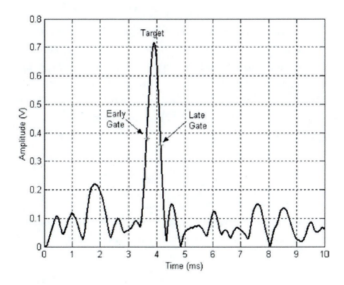

Figure 13.7 Target echo and split-gate position during tracking.

Note that for large α, the measured noise has a larger effect on the gate position than that for small α, but that the filter settles onto the target much more quickly once the two coincide. In general this settling period can be reduced by injecting coarse estimates of the target range rate into the tracking filter, or by starting out with a filter with a wider bandwidth and progressively reducing it as the tracking error decreases.

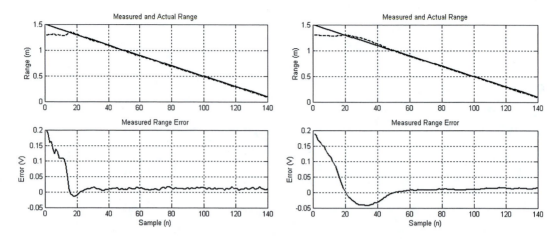

Figure 13.8 Target range, measured range, and track error with time for (a) $\alpha = 0.3$, and (b) $\alpha = 0.1$.

The tracking noise, being the difference between the measured and the true target position (for a constant velocity target) is inversely proportional to the filter bandwidth, as can be seen in Figure 13.8. The filter performance is often characterized by the ratio in the variances of the output and input noise. This is known as the variance reduction ratio (VRR). In an averaging filter, measurements made over n samples combine in the filter to provide an output whose RMS noise is reduced by $1/\sqrt{n}$ compared to a single pulse measurement.

In terms of the equivalent noise bandwidth, β_n (Hz), of the range filter,

$$n = f_r/2\beta_n = 1/VRR, \tag{13.16}$$

where VRR is the variance reduction ratio, f_r (Hz) is the pulse repetition frequency, and β_n (Hz) is the bandwidth of the tracking filter. For the split-gate tracker, the filtered thermal noise output will therefore be

$$\sigma_r = \frac{\tau}{2.5\sqrt{(SNR)(f_r/\beta_n)}}. \tag{13.17}$$

Figure 13.9 shows the filter transfer function for the position estimate of the tracking filter using various values of α and β (Benedict-Bordner relationship). This is obtained using equation (13.10) and the freqz function in MATLAB. Even though the transfer function does not conform to the classical Butterworth shape discussed in Chapter 2, the half-power bandwidth is still measured at $|H(z)| = 0.707$.

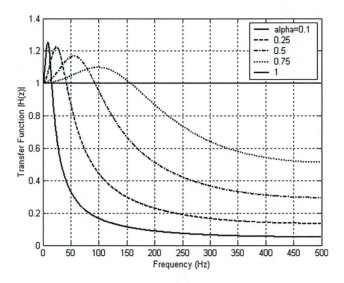

Figure 13.9 The α-β filter transfer function for a sample rate of 1 kHz.

```
% plot the alpha beta filter position transfer function
% ab_trans.m
%
% Variables
fs = 1000;                        % Sample frequency (Hz)
alpha = 0.1                       % Position gain
beta = alpha.^2/(2-alpha);        % Velocity gain (Benedict Bordner
                                    Relationship


% Calculate the filter coefficients for the position transfer function
B=([alpha,beta-alpha])
A=([1,(beta+alpha-2),(1-alpha)]);
[H1,F]=freqz(B,A,1024,fs);
% Plot the magnitude of the transfer function
plot(F,abs(H1),'k')
grid
xlabel('Frequency (Hz)')
ylabel('Transfer Function |H(z)|')
```

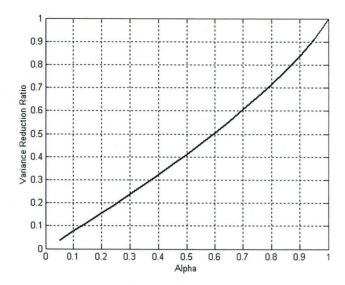

Figure 13.10 The α-β filter variance reduction ratio as a function of α.

Figure 13.9 shows that the filter bandwidth increases as α increases, but that the relationship is not completely linear. As stated Chapter 2, a filter can be characterized completely by its impulse response. This allows the VRR to be determined as a ratio of the power in the impulse response to the power in the impulse. This is valid because the frequency spectrum of an impulse is flat, whereas the response will be low-passed. Alternative methods of calculating the VRR are either to inject white noise of known variance into the filter and to measure the variance on the output, or to integrate the area under the $|H(z)|^2$ curve in the frequency domain. In general, the impulse response method converges very quickly and so has been used to determine the VRR relationship with α that is plotted in Figure 13.10. Using the VRR formulation, the filtered tracking noise can be rewritten as

$$\sigma_r = \frac{\tau}{2.5\sqrt{2SNR}}\sqrt{VRR}. \tag{13.18}$$

13.7 TRACKING LAG FOR AN ACCELERATING TARGET

It was stated earlier that the α-β filter produces a position estimate with zero lag for a constant velocity target and that the lag in the position estimate of an accelerating target is a function of the sample time T_s (s), the filter gains, and the magnitude of the acceleration, $\ddot{r}$ (m/s^2):

$$L = \ddot{r}\frac{T_s^2}{\beta}(1-\alpha), \tag{13.19}$$

where

L = lag (m),
$\ddot{r}$ = range acceleration (m/s^2),
T_s = sample interval (s),
α,β = filter gains.

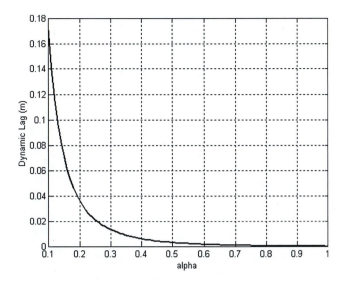

Figure 13.11 Benedict-Bordner α-β filter dynamic lag for a sample time frequency of 1 kHz and an acceleration of 10 m/s^2 as a function of α.

For a sample rate of 1 kHz ($T_s = 1$ ms) and an acceleration of 1 m/s^2, Figure 13.11 shows the lag as a function of α for a Benedict-Bordner α-β filter. Because of the fast sample rate, this lag is insignificant unless the target acceleration is very high.

13.8 WORKED EXAMPLE: RANGE TRACKER BANDWIDTH OPTIMIZATION

Determine the optimum bandwidth of an α-β tracker optimized to track an accelerating missile. The following parameters are assumed:

- Target acceleration 1 g.
- Split gate tracker with a gate size of 3 m.
- SNR is assumed to be 10 dB (thermal noise only).
- Sample rate 50 Hz.

As discussed earlier in this chapter, as the tracking filter bandwidth increases, the dynamic lag decreases. To minimize the mean squared tracking error, the cost function that is minimized is $[\text{lag}^2 + \sigma_r^2]$.

Using the equations developed to determine the RMS noise and the dynamic lag, develop the MATLAB code to calculate the RMS sum of the two parameters.

```
% determine the optimum tracking gains for an alpha beta filter tracking a
missile
% ablag.m

% Variables
acc = 10;                    % target acceleration (m/s.s)
fs = 50;                     % sample frequency (Hz)
snrdb = 10;                  % SNR (dB)
tau = 3;                     % Range gate size(m)


snr = 10.^(snrdb/10);

% Calculate the RMS range tracking noise out of the split gate
% from the pulse-width and the signal-to-noise ratio
sigma = tau/(2.5*sqrt(2*snr));

k = 0
ts = 1/fs
sigmaout = [];
alp = [];
lag = [];

for alpha=0.1:0.01:1,
    k=k+1;
    beta = alpha.^2/(2-alpha);

% position transfer function for the alpha beta filter
    b=([alpha, beta-alpha]);
    a=([1,beta+alpha-2,1-alpha]);

% measure the power in the impulse response to determine the VRR

    sig=[zeros(1,500),ones(1),zeros(1,500)];
    out=filter(b,a,sig);
    alp(k)=alpha;
    vrr=sum(out.^2);

% calculate the noise after the track filter
    sigmaout(k) = sigma*sqrt(vrr);

% calculate the dynamic lag from the formula
    lag(k) = acc*ts.^2*(1-alpha)/beta;
end
```

```
comb = sqrt(lag.^2+sigmaout.^2);
plot(alp,lag,'k-',alp,sigmaout,'k-.',alp,comb,'k');
grid
xlabel('alpha')
ylabel('Dynamic lag and RMS noise (m)')
legend('Lag','Noise','Combined')
```

From the graph in Figure 13.12, the optimum position gain is $\alpha = 0.29$, and using the Benedict-Bordner relationship, the velocity gain, β, is 0.049. The filter bandwidth can then be obtained by interpolation and scaling from Figure 13.9 or by running the ab_trans.m MATLAB script listed earlier. It comes out to be about 3.7 Hz.

13.9 RANGE TRACKING SYSTEMS

Most range tracking systems are associated with angle trackers to produce a full three-dimensional (3D) tracking capability, as discussed in Chapter 14. However, there are a number of applications in which the angular pointing function is either nonexistent or performed manually. Good examples of the latter are combined optical and ranging systems like close-in weapon systems or even some radar or lidar speed traps.

13.9.1 Lidar Speed Trap

As discussed in Chapter 6, laser range finders provide the most cost-effective method of measuring long range in benign environments because a very narrow beamwidth can be achieved using low-cost optics. This allows high angular resolution measurements to be made.

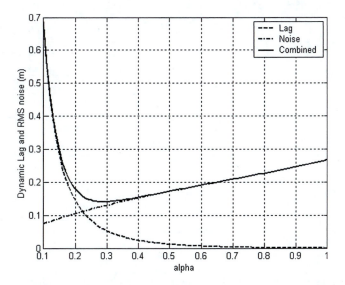

Figure 13.12 RMS noise and dynamic lag and the RMS sum determine the optimum gain.

In the typical configuration shown in Figure 13.13, the standard pulsed time-of-flight measurement technique is followed by a second microcontroller that processes the measured data to estimate the position and speed of the target. Estimates of velocity can be made using a tracking filter of the kind discussed above or by measuring the change in range as determined by the varying time-of-flight between successive pulses. A comparison between the performance of the two methods is shown in Figure 13.14, where measurement data are corrupted by noise. It can be seen that once the tracking filter has settled, the speed estimates are smoother than those produced by the difference method. The amount of smoothing is determined by the value of the gains and the sample rate, and in this case they are $\alpha = 0.5$, $\beta = 0.167$, and $f_s = 20$ Hz. In practical applications, the filter is seeded with a reasonable estimate of the rate based on the position difference over a few samples, and this speeds convergence considerably.

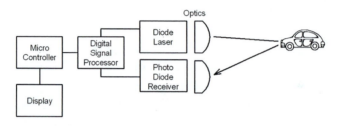

Figure 13.13 Laser radar operational principle.

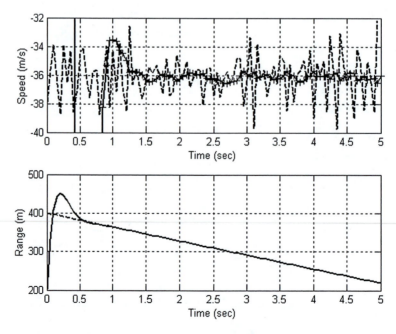

Figure 13.14 Tracker performance comparison between an α-β filter estimate of the speed and a simple difference estimator shows the superiority of filtering.

In the interests of simplicity and low cost, lidar speed scopes, such as the one shown in Figure 13.15, do not implement tracking filters, but use changes in the measured range to estimate the velocity directly. The Riegl FG21-P laser traffic speed meter has the following specifications:

- Operating principle: pulsed time-of-flight
- Beamwidth 2.5 mrad
- Measuring time 0.4 to 1 s
- Speed 0–250 km/hr
- Accuracy ±3 km/h to 100 km/h ±3% of reading above 100 km/h
- Range 30–1000 m
- Accuracy ±10 cm typical
- Target marker, circular reticle matched to beam diameter

Figure 13.16 shows the boresight video of a lidar speed scope showing the spot size of the laser beam. It is clear that the small footprint compared to Doppler radar based speed traps makes accurate measurements on congested roads easy to achive.

13.10 SEDUCTION JAMMING

One of the common electronic countermeasures is seduction jamming, also known as range gate pull-off (RGPO). This technique is mostly used to try to deceive range tracking radar systems as to the true position and velocity of the target being tracked. At its simplest, such a jammer performs the following functions:

- Phase I: When the electronic warfare (EW) system of an aircraft detects that it is being tracked by a missile or a fire control radar (FCR), it engages the first phase of the seduction process. This involves storing and retransmitting every received pulse, while gradually increasing the power level. As far as the radar is concerned, all that is happening is that the RCS of the target is increasing, and so its AGC decreases the receiver gain. This continues

Figure 13.15 Laser traffic speed meter Riegl FG21-P (Riegl 2006).

Figure 13.16 Effectiveness of a narrow beam for measuring speed on congested roads.

for some time until the rebroadcast signal is at least 30 dB higher than the unenhanced echo.

- Phase II: Instead of retransmitting the echo immediately, the jammer starts to delay the rebroadcast by increasingly longer periods. This process used to be achieved by a memory loop that could store the complete analog signal, and hence was able to reproduce even complex chirp waveforms. However, as the sophistication of EW systems has grown, modern systems down-convert and sample the waveform for storage in digital memory. As the delay increases, it appears to the radar that the target velocity changes slightly. At the end of this phase, the delay can reach 20 μs, with the result that the radar range gates are no longer straddling the real echo, but one at a longer range—3000 m longer in this case.

In digital systems, if any pulse repetition frequency (PRF) stagger is known, it is possible to preempt the receipt of a pulse and rebroadcast the stored signal from the previous cycle to pull the tracking gate toward the radar and not away from it This is known as range gate pull-in (RGPI). Either way, the range measured by the radar is incorrect.

Modern fire control radars include a velocity tracking loop based on the received Doppler shift. This Doppler velocity is compared to the measured range rate output by the range tracking loop, and if there is a significant difference, then the radar recognizes that the range gate pull-off (RGPO) process is in progress and the appropriate action is taken (Schleher 1986).

If the fire control radar (FCR) is directing a gun or a command guidance missile, then the incorrect range will affect the calculations of the lead angle, and the projectile may pass harmlessly to one side of the aircraft. However, in the case of a sophisticated missile, RGPO alone is seldom sufficient to save the aircraft, and additional measures are required.

- Phase III: The jammer now starts to on-off modulate the delayed pulse at a frequency designed to upset the angle tracking loops. If tracking uses the conical scan technique, discussed later in this chapter, the modulation frequency is selected to be at the nutation rate, with almost guaranteed effects. If, however, monopulse is used, then the modulation attempts to upset the AGC and consequently to introduce instabilities and resonances into the angle tracking loops. A successful break-lock is far less likely in the latter case.

13.11 ANGLE MEASUREMENT

13.11.1 Amplitude Thresholding

At its most simple, a received echo amplitude threshold can be used to determine that a target is within the beam, and this gives a very rough measure of the target direction. It can be seen from Figure 13.17, that because the antenna gain drops off sharply, the angular position uncertainty is usually constrained to within the 10 dB (one-way) beamwidth of the antenna.

The operational principle is to transmit a signal via a directional antenna, lens, or acoustic transducer and then monitor the receiver for an echo that exceeds the detection threshold. Both pulsed and continuous wave techniques can be used over a broad range of frequencies. In most cases the signal is modulated with some unique waveform for the following reasons:

- Discrimination against ambient solar radiation.
- Elimination of interference from fluorescent lights.
- Reduction in the probability of interference from other sensors.

Although this method works well in the infrared (IR) and visible band, where the collimated beamwidth can be very narrow, at lower frequencies, the wider antenna beamwidth results in progressively worse cross-range resolution with increasing range, as shown in Figure 13.18.

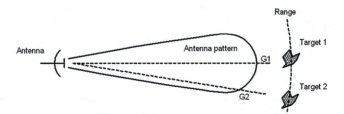

Figure 13.17 Antenna gain as a function of angle.

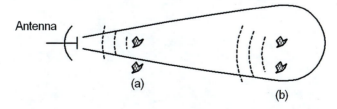

Figure 13.18 The cross-range resolution degrades with range as shown. At position (a) the targets can be resolved, but at position (b) the targets, with the same spacing, are not resolved.

The main applications of this sensor modality are limited to continuous wave or modulated IR proximity detectors for industrial and robotic applications and some acoustic proximity detection devices.

13.11.2 Proximity Detector Example

Infrared proximity detectors are commonly used for short-range robotic and control applications. These sensors are not normally used to measure range, just the presence of a target within the beam. They operate in the near-IR range (at a wavelength slightly longer than visible light, typically 940 nm) and are visible to charge-coupled devices (CCDs), so can be observed using a video camera.

A typical receiver is shown Figure 13.19, along with a near-IR light-emitting diode (LED) transmitter. This receiver package includes a photodiode, amplifiers, filters, and a limiter. It responds to a burst-modulated 38 kHz signal with an on-off period of 600 + 600 μs. The digital output goes low if a reflected signal is detected.

13.12 ANGLE TRACKING PRINCIPLES

For many applications, the angular accuracy provided by the sensor beamwidth is sufficient, but in others, it is important to know the bearing to a target to within a fraction of a degree. The following section discusses some of the techniques that have been developed to achieve this performance.

13.12.1 Scanning Across the Target

Using the antenna beam pattern, it is possible to get a more accurate bearing on the target by sweeping the beam across it and noting changes in the signal amplitude with angle, as illustrated in Figure 13.20. This process can increase the angular resolution significantly with only a marginal effect on the range accuracy. The technique performs best if the SNR is good and similar amplitude levels are reached to the left and right of the target. The bearing to the target can be estimated to be midway between the two levels if the beam pattern is symmetrical.

Figure 13.19 Generic IR proximity detector module and IR LED.

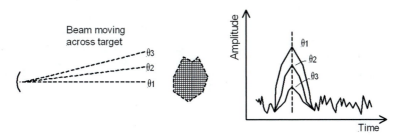

Figure 13.20 Using the beam pattern to estimate angle.

13.12.2 Null Steering

Null steering involves the subtraction of the returns from a pair of overlapping beams to produce a "null" when the target is aligned with the beam axis of symmetry. Extremely accurate angle measurements can be achieved. As with the split-gate range measurement case, for a point target, the RMS angle measurement accuracy, σ_t, in thermal noise is limited only by the SNR:

$$\sigma_t = \frac{\theta_{3dB}}{k\sqrt{2SNR}},\qquad(13.20)$$

where θ_{3dB} is the antenna beamwidth and k is a constant dependent on the tracking type. The RMS tracking accuracy takes on the units in which the beamwidth is measured. A typical value of k for conscan is 1.4, while for monopulse it is generally between 1.5 and 2.3.

The following are the most common null steering techniques:

- Lobe switching.
- Conical scan.
- Monopulse.

13.13 LOBE SWITCHING (SEQUENTIAL LOBING)

The technique of sequential lobing or lobe switching involves sequential transmission from two or more antennas with overlapping, but offset beam patterns (Barton 1976; Skolnik 1980). As shown in Figure 13.21, sequential returns will be amplitude modulated if the target is not on boresight. This amplitude modulated signal can be used to generate an angle error estimate or used to drive the antenna pair mechanically to null the error in one axis. Two additional switching positions are needed to obtain the angular error in the orthogonal axis, so four pulses are required to control the antenna in two dimensions.

This technique can be used for electromagnetic or acoustic tracking systems as it is simple to implement. It was probably the first method used to automate radar tracking systems developed in World War II. Today most nonscanning collision avoidance radars use this technique, with either two or three lobes to determine the angular offset of cars. In addition, a beacon/receiver version of lobe switching is often used for direction-finding applications, as discussed in an example at the end of this chapter.

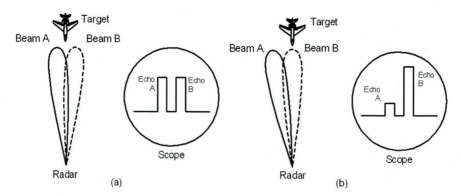

Figure 13.21 Principle of sequential lobing to determine angle error [adapted from Skolnik (1970)].

13.13.1 Main Disadvantages of Lobe Switching

The primary shortcomings of this method of angle estimation are due to the fact that at least four pulses are required to resolve the target direction in two dimensions. This results in a reduction in the available bandwidth for tracking. If the target is moving (which it probably would be), then fluctuations in the signal level due to variations in the echo strength on a pulse-by-pulse basis reduce the tracking accuracy. In addition, it is susceptible to modulation by the target, either natural (propellers, wing beats, etc.) or as part of the electronic countermeasures.

It can be seen in Figure 13.21a that when the target is on boresight, the antenna gains in the two antennas is lower than the peak. This difference is typically a few decibels and consequently has a significant effect on the SNR, with a resultant reduction on the maximum operational range.

13.14 CONICAL SCAN

Conical scan is an evolution of the sequential lobing process in which a single beam displaced in angle by less than the antenna beamwidth is nutated on its axis.[1] Rather than tilting the whole antenna off boresight and then nutating the whole structure, this beam squint is most often achieved by incorporating a rotating subreflector or feed at the focal point of a parabolic dish. As shown in Figure 13.22, the nutation rate is usually much lower than the PRF, with the result that a sequence of echo pulses are received which are amplitude modulated in a roughly sinusoidal fashion.

The scan rate is generally limited to between 5 and 25 pulses per revolution for long-range operation, but at short range, where the PRF is higher, many more pulses may be received. Amplitude modulation (AM) of target returns will be a function of the position of the boresight with respect to the target. The modulation depth is proportional to the absolute offset (within the linear region) and the relative phase of the signal gives the direction. To extract this information in order to drive the two orthogonal axes, a pair of phase detectors, also known as angle-error

---

1 Nutated means spinning without rotating the polarization.

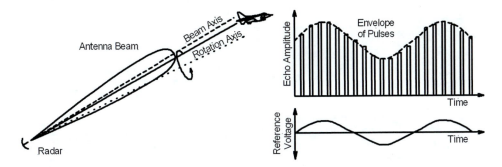

Figure 13.22 Principle of conical scanning to determine angle error [adapted from Skolnik (1970)].

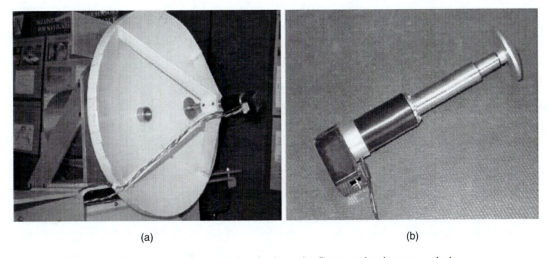

(a)   (b)

Figure 13.23 Cassegrain antenna with a canted, spinning subreflector to implement conical scan.

detectors, using quadrature phase references, demodulate the received AM signal out of the range tracking gate to generate the angle tracking error.

A typical conical scan implementation is shown in Figure 13.23a. A Cassegrain antenna configuration that consists of a feed horn that illuminates a hyperbolic subreflector, which in turn illuminates the parabolic primary reflector, produces a collimated beam. The subreflector, shown enlarged in Figure 13.23b, is canted slightly and attached to the shaft of a motor so that when it spins, the illumination on the primary reflector is offset and produces a squinted beam that nutates around the boresight.

Attached to the back of the motor shaft is an encoder that is used to produce the quadrature reference signals for demodulation. Depending on the subsequent processing of the received echo stream, there are a number of different methods that can be used to perform this demodulation function, as shown in Figure 13.24. An implementation of the digital solution in software is discussed in detail in the following section.

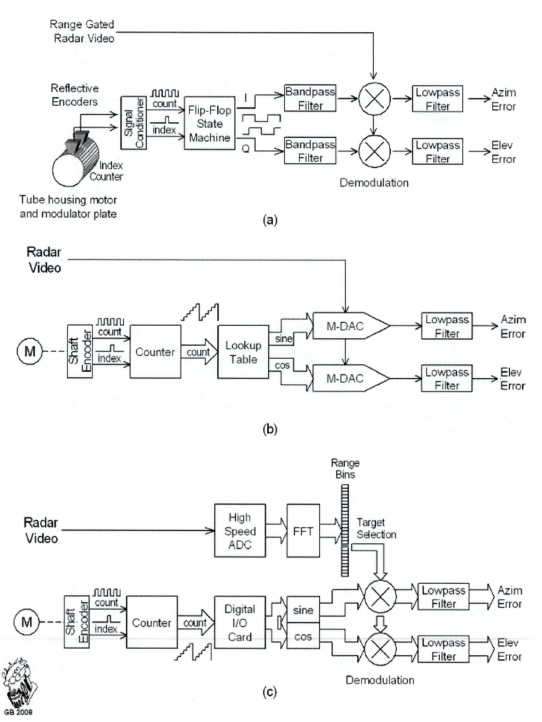

Figure 13.24 Some conscan angle error demodulation techniques: (a) analog, (b) digital in hardware, and (c) digital in software.

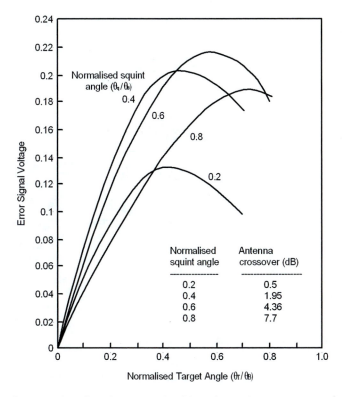

Figure 13.25 Error voltage as a function of target angle with squint angle as a parameter [adapted from Skolnik (1980)].

13.14.1 The Squint Angle Optimization Process

The performance of a null tracking system is determined to a large extent by the slope of the error function at the crossover point and the SNR. It can be seen in Figure 13.25 that the error voltage is a function of the squint angle and the target angular error normalized with respect to the 3 dB beamwidth of the antenna.

The greater the slope of the error signal, the more accurate the angle tracking, and the maximum occurs where the normalized squint angle, θ_q/θ_B, is just greater than 0.4. However, at this squint angle, the gain on boresight is about 2 dB down on the peak. The range tracking accuracy is determined by the SNR, and so requires the maximum on boresight gain, which occurs when the normalized squint angle $\theta_q/\theta_B = 0$, which is not feasible. For active sensors, a compromise is used with the normalized squint angle $\theta_q/\theta_B = 0.28$. This corresponds to a one-way antenna gain about 1 dB below the peak (Barton 1976).

13.14.2 Measuring the Conscan Antenna Transfer Function

In the same way that a split-gate range tracker can be characterized by its range transfer function, so too can an angle tracking system be characterized by its angle error transfer function. To perform this measurement, a conscan antenna mounted on a pan/tilt unit, as illustrated in Figure 13.26, is swept past a point source of radiation (in the far field of the antenna). The received

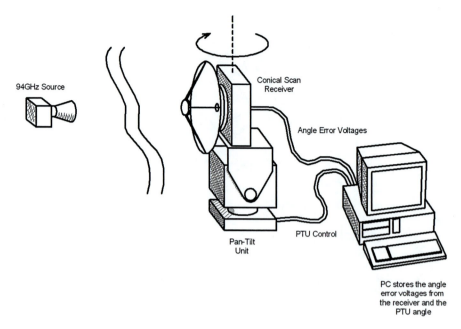

94GHz Source

Conical Scan
Receiver

Angle Error Voltages

PTU Control

Pan-Tilt
Unit

PC stores the angle
error voltages from
the receiver and the
PTU angle

Figure 13.26 Mechanism for measuring the conscan error transfer function.

signal level at the output of the antenna is logged as a function of angle (or time if the angular rate is constant).

The signal from the conscan receiver as a function of time is shown in Figure 13.27a, where it has the following characteristics:

- The alternating current (AC) component of the signal level starts out low, when the offset is large.
- It then increases to a peak when the beam squint angle equals the target offset on the one side.
- On axis, the beam offset is symmetrical so there is no modulation, and so the AC component reduces to zero.
- At the cross-over point, the phase of the modulation is reversed but the shape of the modulation is the mirror image due to the symmetry of the process.

In Figure 13.27b, both the modulation signal and the reference signal are shown. Note the 180° phase shift at crossover, which is an indication that the sign of the error is reversed. In Figure 13.27c, the synchronously demodulated signal is shown. Note that there is a DC component and an AC component at twice the modulation frequency. Finally, Figure 13.27d shows the filtered demodulated signal showing only the DC component, which is the conscan angular transfer function for the Cassegrain antenna shown in Figure 13.23.

13.14.3 Application

Conscan systems are typically used in tracking radars, as shown in Figure 13.28. They are simple to implement, using a single beam and a single receiver and transmitter, and beacon tracking can be implemented without the transmitter and without the range gating circuitry. This makes them ideal for tracking cooperative targets like missiles and cannon shells in ballistic test ranges.

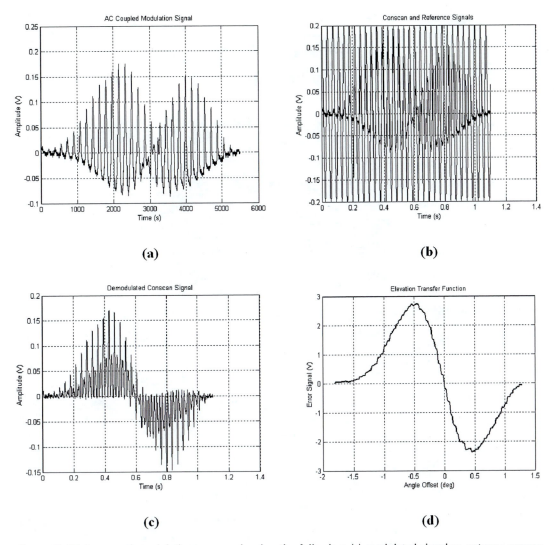

Figure 13.27 Conscan demodulation process showing the following: (a) modulated signal as antenna sweeps past source, (b) modulated signal and reference signal showing phase reversal, (c) the demodulation product of the signal and the reference, and (d) filtered demodulated error signal mapped to the angle.

The complete schematic diagram of a conscan tracker is shown in Figure 13.29. It can be seen that a standard radar transceiver is implemented, with the signal transmitted through a duplexer and the received echo passed into a receiver with an AGC. It is important that the bandwidth of the AGC be much lower than the conscan frequency, otherwise it would remove the conscan modulation.

A range tracker (not shown) maintains lock and keeps a range gate centered on the target echo. The output of this gate is amplitude modulated in a way that is related to the position of the target relative to the antenna boresight, as discussed previously, and so is demodulated using the

Figure 13.28 SCR-584 Conscan radar tracker on a pedestal (Courtesy Antiaircraft Command).

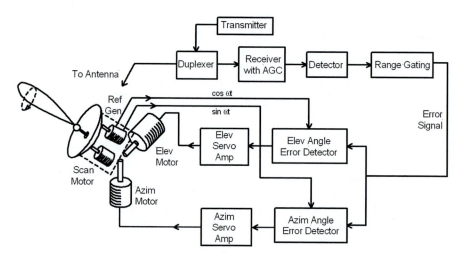

Figure 13.29 Block diagram of a conical scan radar [adapted from Skolnik (1980)].

azimuth and elevation reference signals to produce the two angle error signals. These angle errors drive the angle servos, which in turn control the position of the antenna and drive it to minimize the error (a null tracker).

13.14.4 Main Disadvantages

The main disadvantages of the conscan system are similar to those described for sequential lobing. The tracking bandwidth is reduced because of the number of pulses required to produce each error estimate and fluctuations in the echo signal amplitude induce tracking errors, making these systems sensitive to target modulation, and the on-boresight antenna gain is reduced. In addition, the tracking range is limited because the beam direction cannot be allowed to move too far between transmission and reception.

The problem with target modulation and the ease with which conscan systems can be deceived have been known since World War II, when the technique was invented, and so a number of methods have been developed to reduce the effects. A novel Russian design used on the Flapwheel FCR (Blake 1995–1996) uses a dual-beam conscan method that cancels out deliberately induced jamming modulations. Another more common technique is to use conscan on receive only (COSRO) so that the conscan frequency is not advertised to potential targets. It also reduces the on-boresight loss in gain.

13.14.5 Other Considerations

Automatic gain control, required to normalize the pulse amplitude for range tracking, must be carefully designed not to interfere with the conscan modulation. However, because the amplitude of the modulation is proportional to the RCS of the target, the AGC must perform a normalization function to maintain a constant error slope.

13.15 INFRARED TARGET TRACKERS

An interesting extension of the basic conscan mechanism is used by IR missiles such as the Sidewinder to obtain angle error and target discrimination information. The hardware consists of a single thermal IR detector element of the type discussed in Chapter 3, which observes the scene through a spinning disk called a reticle. The reticle is mostly transparent with an opaque pattern, shown in Figure 13.30, which introduces a modulation onto the detector in the presence of a bright IR source.

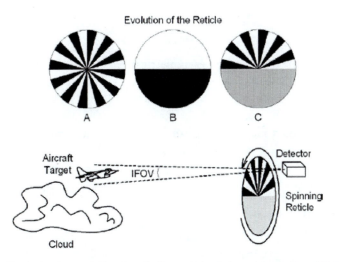

Figure 13.30 Target discriminating and direction finding reticles [adapted from Kopp (2005)].

Reticle A in the diagram is used for background suppression, particularly in regard to differentiation between a point target (such as an aircraft engine) and a distributed target such as a sunlit cloud. The chopping action on a point target produces a narrow-band signal, while the cloud will subtend a larger area and produces a wider bandwidth, lower frequency output. Reticle B determines the direction to the target by producing a square wave for a point target source, with the direction determined by the signal phase. Reticle C combines the two functions, with the top half taking on the grid pattern and the bottom half being semitransparent with the same average transmittance as the top. When the segmented half passes over the target source, a series of pulses are output with some superimposed modulation from the background, and when the semitransparent half passes over the target region, the output corresponds to the average brightness of the scene. The final output therefore resembles B, but with a burst of pulses rather than an individual pulse making up one-half cycle of the square wave.

Filtering separates the point target information from the clutter, the phase of the block of bursts determines the direction of arrival, and the magnitude of the error is obtained by measuring the amplitude of the pulses. This is possible because a circular spot from the point target is projected onto the reticle, so as the radial distance from the center of the disk increases, the gap between wedges increases and a larger portion of the spot energy is passed through to the detector. As with conscan, the signal amplitude and phase is demodulated to produce an angle error and magnitude that is used to control the missile flight path.

13.16 AMPLITUDE COMPARISON MONOPULSE

Amplitude comparison monopulse is the angle estimation method of choice for high-performance tracking radar systems for a number of reasons that will be made clear in the following section. In essence, this technique uses two overlapping antenna beams for each of the two orthogonal axes that are generated from a single reflector illuminated by four adjacent feed horns. The difference in the amplitude (or phase) of the signals output by these beams is used to derive the angle errors in both elevation and azimuth simultaneously from a single pulse—hence the name (Sherman 1984).

13.16.1 Antenna Patterns

The sum pattern, Σ, of the four horns is a symmetrical pencil beam, as illustrated in Figure 13.31, and is used on transmit and for range measurement on receive. Angle error measurement on receive is accomplished by the synthesis of four individual squinted beams, shown in Figure 13.32, one from each of the four horns.

When viewed from the front, the four beam patterns and the sum channel pattern are as seen in the cross section shown in Figure 13.33. It can be seen that the difference channel beam patterns, labeled A through D, are squinted with respect to the sum channel pattern, but that they overlap each other and on boresight.

13.16.2 Generation of Error Signals

The difference channel patterns ΔAz and ΔEl are produced on receive using a microwave hybrid circuit called a monopulse comparator. The hybrids perform phasor additions and subtractions of the RF signals to produce the output signals shown in Figure 13.34.

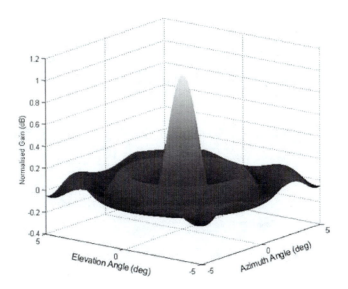

Figure 13.31 Monopulse sum channel beam pattern.

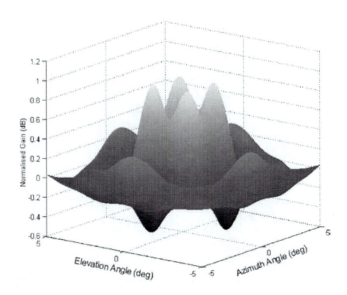

Figure 13.32 Monopulse difference channel beam patterns.

If a target is on boresight, then the amplitudes of the signals received in the four channels (A, B, C, D) will be equal, and so the azimuth and elevation difference signals will be zero. However, as the target moves off boresight, the gain in one or two of the beams will increase and that in the others will decrease, so the amplitudes of the signals received will differ. The difference in signal output by the hybrid will take on the sign and magnitude proportional to the error.

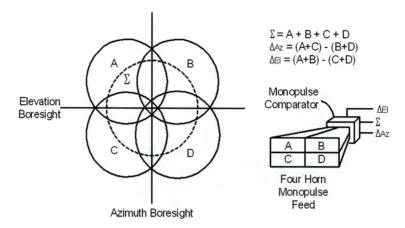

$\Sigma = A + B + C + D$
$\Delta_{Az} = (A+C) - (B+D)$
$\Delta_{El} = (A+B) - (C+D)$

Figure 13.33 Combining beams using a microwave hybrid circuit (monopulse comparator) to produce sum and difference channel signals.

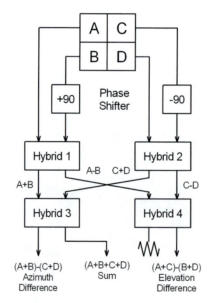

Figure 13.34 Hybrid configuration to produce sum and difference channel signals at the transmit frequency [adapted from Skolnik (1970)].

As with the conscan mechanism, the amplitude of the difference channel signals is dependent on the RCS of the target and the range. These signals must therefore be normalized with respect to the sum channel amplitude to produce an error signal that is independent of the echo amplitude. This ratio can be obtained using an AGC circuit that operates on the two difference channels and is driven by the detected sum channel output of the tracking gate, or by division in a digital tracker. The resultant angle error transfer function is shown in Figure 13.35.

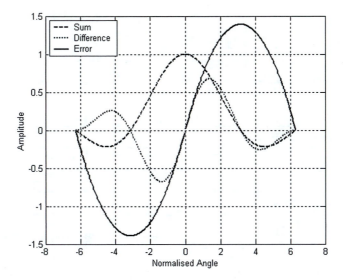

Figure 13.35 Normalized sum and difference channel gains as a function of angle off boresight.

As shown in Figure 13.36, phase-sensitive detectors demodulate the azimuth and elevation error signals using the sum channel intermediate frequency (IF) signal as a reference to produce the two error voltages. These outputs must also be range gated so that their magnitudes represent the error signals from the correct target.

The sum and difference channel voltage signals can be modeled quite accurately in the main lobe of the sum channel using (Barton 1988)

$$E_{\text{sum}} = \cos^2(1.144\Delta)$$
$$E_{\text{dif}} = 0.7.7 \sin(2.28\Delta), \tag{13.21}$$

where E_{sum} (V) is the normalized sum channel output voltage, E_{dif} (V) is the difference channel output voltage normalized with respect to the sum channel, and Δ is the angle offset from the beam axis normalized with respect to the half-power sum channel beamwidth.

13.17 COMPARISON BETWEEN CONSCAN AND MONOPULSE

The monopulse option gives a slightly larger SNR for the same size target due to the higher on-boresight antenna gain. This is because the sum channel gain is a maximum on boresight and only the gains of the difference channel signals are reduced by the beam squint angle. Figure 13.37 shows the normalized error slope at crossover as a function of the normalized squint angle. It can be seen that in the monopulse case, the slope peaks for a smaller squint angle and with a larger value than the conscan case. The steeper error slope near the origin and the lower on-boresight loss (labeled antenna crossover in the figure) results in superior tracking accuracy.

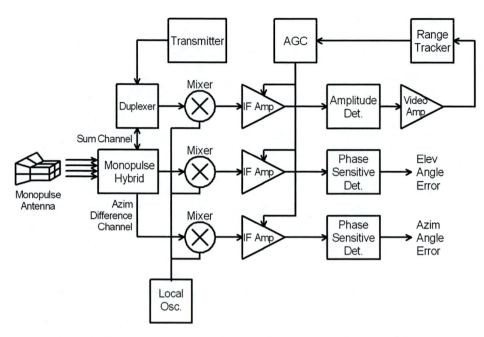

Figure 13.36 Schematic diagram of a monopulse front end [adapted from Skolnik (1980)].

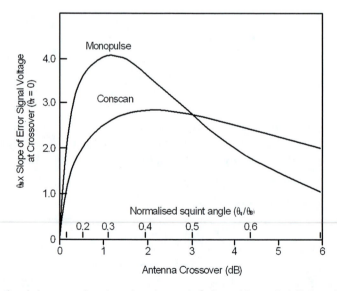

Figure 13.37 Error signal slope as a function of squint angle [adapted from Skolnik (1980)].

In addition, because new tracking information is generated with each new pulse, tracking is not degraded by fluctuations in echo amplitude. From an FCR perspective, this insensitivity to amplitude fluctuations, and hence immunity to jamming, is probably the most important advantage of the monopulse technique.

13.18 ANGLE TRACKING LOOPS

Unlike the range tracking loop in which the gate position is controlled electronically, in an angle tracker, the angle of the beam must be physically displaced to minimize the azimuth and elevation tracking errors. This physical displacement is achieved by supplying the angle servos with the error signals from the conscan or monopulse modules. These servos power the motors which rotate the antenna in the correct direction, as illustrated in Figure 13.38.

The RMS tracking error due to thermal noise will be the measurement noise described in equation (13.20) modified by the bandwidth of the angle servo loop relative to the pulse repetition frequency:

$$\sigma_t = \frac{\theta_{3dB}}{k\sqrt{2SNR f_r / \beta_n}},$$
(13.22)

where

θ_{3dB} = antenna beamwidth (deg),
k = constant dependant on the tracking type (1.4 for conscan and $1.5 < k < 2.3$ for monopulse),
SNR = signal-to-noise ratio,
f_r = radar pulse repetition frequency (Hz),
β_n = angle servo bandwidth (Hz).

As before, the units used are determined by those used in defining the antenna beamwidth.

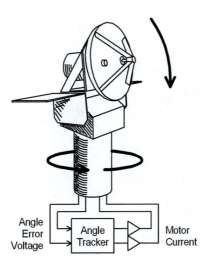

Figure 13.38 Angle tracking loop rotates the antenna to minimize tracking error.

13.19 ANGLE ESTIMATION AND TRACKING APPLICATIONS

In combination with range trackers described earlier in this chapter, monopulse techniques are generally used for modern tracking radar systems. The technique can be used for direction-finding systems or beacon tracking in passive receivers. The principle can be used by electromagnetic or acoustic systems. Lidar and passive optical trackers use quadrant detectors that apply the same basic principles to obtain angle error information.

13.19.1 Instrument Landing System

The instrument landing system (ILS) facility is a highly accurate and dependable means of navigating an aircraft to the runway in low visibility. It operates using an extension of the standard direction-finding principle by generating a pair of squinted beams with different modulations that can be detected by a receiver on the approaching aircraft, as illustrated in Figure 13.39. It consists of the following hardware:

- A localizer transmitter.
- A glide path transmitter.
- An outer marker (can be replaced by a nondirectional beacon or other fix).
- The approach lighting system.

A category I ILS provides guidance information down to a decision height (DH) of 200 ft, and with good equipment, ILS can even be used for category II approaches of 100 ft on the radar altimeter. The ILS provides both the lateral and vertical guidance necessary to fly a precision approach if glide slope information is provided.

13.19.1.1 Localizer Transmitter   The transmitter provides lateral guidance. It operates at very high frequency (VHF) in the range 108.1 to 111.95 MHz. The transmitter and antenna are situated on the centerline at the opposite end of the runway from the approach threshold. The antenna radiates two vertical fan-shaped beams that overlap at the extended centerline of the runway. To differentiate between the two overlapping beams that are radiating at the same frequency, the right-hand side of the beam (as seen by an approaching aircraft) is modulated at 150 Hz and the left-hand beam is modulated at 90 Hz. The total width of the beam pair can be

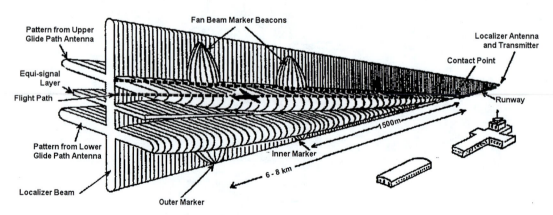

Figure 13.39 ILS signal patterns [adapted from Jacobowitz (1965)].

varied between 3° and 6°. It is adjusted to provide a track signal 700 ft wide at the runway threshold, increasing to 1 nm wide at a range of 10 nm.

13.19.1.2 Localizer Receiver The localizer receiver activates the needle of one of the cockpit instruments. For an aircraft to the right of the beam, in the 150 Hz region only, the needle will be deflected to the left, and visa versa in the 90 Hz region. In the overlap region, both signals apply a deflection force to the needle, causing a deflection in the direction of the strongest signal so that when the aircraft is precisely aligned, there will be zero net deflection, and the needle will point straight down.

13.19.1.3 Glide Slope Equipment The glide slope equipment consists of a transmitter and antenna operating in the ultra high frequency (UHF) range at a frequency between 329.30 and 335.00 MHz. It is situated 750 to 1250 ft down the runway from the threshold, offset 400 to 600 ft to one side of the centerline, and is monitored to a tolerance of ±0.5°. It consists of two overlapping horizontal fan beams modulated at 90 Hz and 150 Hz, respectively, with the thickness of the overlap being 1.4°, or 0.7° above and below the optimum glide slope. The glide slope may be adjusted between 2° and 4.5° above the horizontal plane, depending on any obstructions along the approach path. Because of the antenna construction, no false signals can be obtained at angles below the selected glide slope, but are generated above at multiples of the glide slope angle. The first is at about 6°. It can be identified because the instrument response is the reciprocal of the correct response.

The glide slope signal activates the glide slope needle in a manner analogous to that of the localizer. If there is sufficient signal, the needle will show full deflection until the aircraft reaches the point of signal overlap. At this time the needle will show partial deflection in the direction of the strongest signal. When both signals are equal, the needle is horizontal, indicating that the aircraft is precisely on the glide path. With 1.4° of overlap, the glide slope area is approximately 1500 ft thick at 10 nm, reducing to less than 1 ft at touchdown. A single instrument provides indications for both vertical and lateral guidance.

13.20 WORKED EXAMPLE: COMBINED ACOUSTIC AND IR TRACKER FOR A SMALL AUTONOMOUS VEHICLE

This worked example outlines the development of a low-cost tracker that combines the wide field of view offered by an IR system with the more accurate, but narrower field of view obtained using acoustic monopulse (Brooker et al. 2007). Range measurement can be undertaken using standard pulsed ultrasound or, more effectively, using a beacon that transmits an electromagnetic and an acoustic signal simultaneously, then using a receiver to measure the difference in the arrival times, a technique known as time difference of arrival (TDOA).

The main advantage of this method is that it allows the fast electromagnetic signal to carry a unique identification (ID) code which would identify a specific beacon to a specific receiver. TDOA systems have been developed for indoor localization over the past decade and include the well-known MIT cricket (Priyantha 2005), as well as various other pulsed systems (Girod and Estrin 2001; Ward et al. 1997). Issues with multipath and acoustic interference have resulted in the use of spread-spectrum techniques (Bortolotto et al. 2001; Palmer 2002). Some researchers have opted to use a sequential approach to overcome the interference problems (Fukuju et al. 2003).

13.20.1 Operational Principles of a Prototype

Initial experiments were conducted using RF modules designed for keyless entry. However, it was found that those tested have a substantial (>100 ms) and variable period between the transmission of the encrypted signal and its acknowledgement, making them unsuitable for the measurement of arrival time differences of less than 15 ms which are required for a 5 m maximum range. An alternative technique using standard IR remote control hardware, shown conceptually in Figure 13.40, was implemented to produce a range measurement system that works as follows:

- The beacon transmits a 600 μs burst of IR light with a wavelength of 960 nm on-off modulated at 38 kHz.
- It then transmits a 1 ms burst of ultrasound at a frequency of 40 kHz.
- The receiver detects the IR pulse, which, to all intents is instantaneous, followed shortly afterwards by the slower acoustic pulse.
- Because the speed of sound is reasonably constant (about 340 m/s) and slow, it is easy to measure the time difference, ΔT, from which the range can be calculated.

The relative angle of the beacon with respect to the receiver is determined using a variation of the conventional amplitude comparison monopulse technique (Leonov and Fomichev 1986; Skolnik 1990). Although this process is usually applied using electromagnetic signals, a number of acoustic applications have been made (Aguilar and Meijer 2002; Kuc 2002); its implementation as an aid to the blind has been quite successful.

One of the novel aspects of this sensor is that both IR and acoustic implementations of the angle measurement are made. In the IR implementation, two receivers are placed between an opaque septum, as shown in Figure 13.41, so that as the angle of arrival varies: first one receiver is illuminated, then both, and finally the other. This produces four states: no signal (if the beacon is outside the angle limits of both receivers), angle right (if one receiver is illuminated), angle straight on (if both are illuminated), and angle left (if the other receiver is illuminated). This can be used to drive a bang-bang controller for coarse tracking of the beacon. The size of the septum that divides to two receivers can be adjusted to control the angle over which both receivers are illuminated, and so defines the "dead" band of the controller.

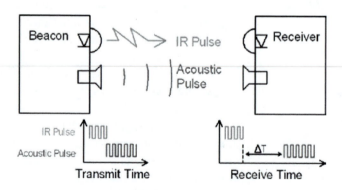

Figure 13.40 Conceptual diagram of the system showing the range measurement technique.

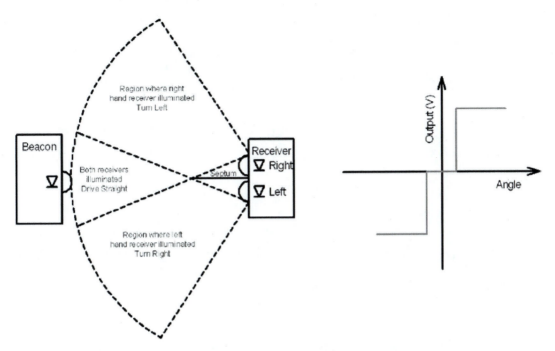

Figure 13.41 Conceptual diagram showing digital angle measurement and transfer function (Brooker et al. 2007).

Because of the finite size of the receive aperture lens and diffraction around the septum, there will be a fuzzy region of transition between the receiver seeing the beacon and not. This will depend on the intensity of the beacon and the ambient light level. It has been estimated that a dead band of much less than 10° could not be reliably obtained.

In the case of the acoustic system, because the amplitude of the received signal is determined by the receiver beam pattern, a far more subtle measure of the angle is possible. Two receiver transducers are mounted with a slight outward squint so that if the acoustic signal arrives at an angle, the amplitude received by one of the receivers will be higher than that received by the other, as illustrated in Figure 13.42. When the illumination is symmetrical, the two signal amplitudes are equal and there is no dead band region because the sensor becomes a null tracker.

To visualize the effect of adjusting some of the parameters of these transducers, a simulation model was built using the ultrasonic transducer normalized gain pattern supplied by the manufacturer and reproduced as a polar plot in Figure 13.43.

The difference between the signal voltages received from the left and the right transducer can be used to produce the receive antenna transfer function, shown in Figure 13.44a. It can be seen that the peak magnitude of the error signal is dependent on the squint angle and that there is a region of about ±20° over which the relationship between the voltage and the beacon angle is almost linear. The error slope at the origin is a good indication of the tracking accuracy, and it can be seen that this increases as the squint angle is increased. The downside is that as the squint

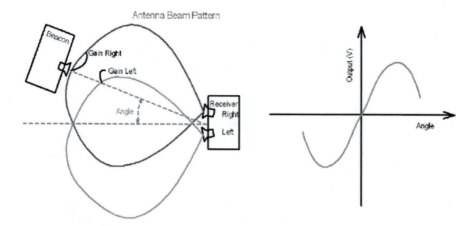

Figure 13.42 Conceptual diagram showing analog angle measurement and the transfer function (Brooker et al. 2007).

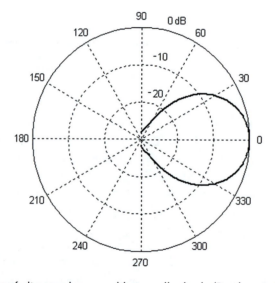

Figure 13.43 Polar pattern of ultrasound sensor with normalized gain (Brooker et al. 2007).

angle is increased, the on-boresight signal amplitude decreases, as shown in Figure 13.44b, and at a half squint of more than 15° it becomes too low and the SNR is compromised.

In addition, the amplitude of each of the received signals decreases considerably as the beacon angle increases off the beam axis in the vertical direction. For a symmetrical beam pattern, at an elevation offset of 30°, the gain is reduced by 10 dB, and in the end a threshold will be reached below which the SNR will be too low for reliable measurements to be made. Because the signal level is also dependent on the range to the beacon, the angle limit threshold will also be range

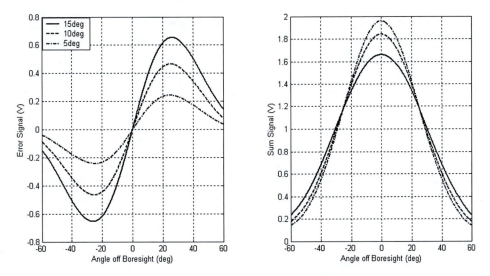

Figure 13.44 Antenna transfer function (a) error voltage signal and (b) sum voltage signal (Brooker et al. 2007).

dependent. These factors result in a complex operational volume envelope around the receiver.

13.20.2 Theoretical Performance

As discussed in Chapter 7, in acoustic systems, the acoustic power density and sound pressure levels (SPLs) in air are determined at a reference distance of 30 cm from the transmitter. Therefore, for an acoustic transmitter radiating a power, P_t, equally in all directions (isotropically), the power density, I_{iso} (W/m^2), at a range of 30 cm will be the total radiated power divided by the surface area of a sphere with a radius of 30 cm:

$$I_{iso} = P_t/4\pi 0.3^2. \tag{13.23}$$

The antenna gain is defined as the ratio of the solid angle subtended by the surface of the sphere (4π) to the area of the beam:

$$G \approx 4\pi/\theta\varphi, \tag{13.24}$$

where θ, φ (rad) are the two orthogonal 6 dB beamwidths.

The power density in the direction of the beam is the product of this gain and the isotropic power density:

$$I = P_t G/4\pi 0.3^2. \tag{13.25}$$

Unlike electromagnetic radar, the performance of acoustic systems is seldom determined in terms of the power density, but rather in terms of the SPL. Atmospheric pressure is measured in Pascals (N/m^2), where the air pressure at sea level is defined to be 101,325 Pa. Sound pressure is the difference between the instantaneous pressure generated by the sound and the air pressure. These differences are very small, with an unbearably loud noise having a sound pressure of 20 Pa and one that is at the threshold of hearing having a pressure of 20 μPa.

The SPL, L_p (dB), is defined as the square of the ratio of the sound pressure, P (Pa), to the P_{ref} = 20 μPa threshold level. It is typically represented in decibel form as

$$L_p = 10\log_{10}(P/P_{ref})^2. \tag{13.26}$$

The relationship between the sound pressure, P (Pa), and the transmitted power density, I (W/m^2) is

$$P^2 = IZ_{air}, \tag{13.27}$$

where Z_{air} is the acoustic impedance of air. This can be determined by taking the product of the air density, ρ_{air} (kg/m^2), and the speed of sound, c (m/s):

$$Z_{air} = \rho_{air}c = 1.229 \times 340 = 418\,\Omega. \tag{13.28}$$

Rewriting equation (13.26) in terms of equation (13.27) and equation (13.28) gives the SPL, L_p (dB), in terms of the acoustic power density, I (W/m^2), at a given range:

$$L_p = 10\log_{10}\left(IZ_{air}/P_{ref}^2\right)$$
$$= 10\log_{10} I + 10\log_{10} Z_{air} - 20\log_{10} P_{ref}$$
$$= 10\log_{10} I + 26.21 + 93.9$$
$$L_p = 10\log_{10} I + 120. \tag{13.29}$$

As the sound propagates out to a target, the SPL drops with the sum of the square of the range plus an attenuation component. This transmission loss, H (dB), is given by

$$H = 20\log_{10}(R/R_{ref}) + \alpha R, \tag{13.30}$$

where R is the range to the target (m), R_{ref} is the range at which the SPL is defined (0.3 m in this case), and α is the attenuation (dB/m). The SPL at the receiver, L_r (dB), can be calculated from these two equations:

$$L_r = L_p - H. \tag{13.31}$$

In this application the ultrasonic beacon and receiver are muRata piezoelectric transducers with the following specifications (muRata 2008):

- MA40S4S: SPL 120 dB at 40 kHz; 0 dB relative to 0.0002 μbar per 10 Vpp at a range of 30 cm.

- MA40B8R: sensitivity −63 dB at 40 kHz; 0 dB relative to 1 Vrms/μbar.
- Beamwidth 55° (6 dB).

The acoustic power density 30 cm from the transmitter driven by 20 Vpp can be determined easily if it is realized that 1 μbar = 0.1 Pa, making 0.0002 μbar = 20 μPa, which is the standard reference level for SPL.

The SPL increases with the square of the applied voltage, therefore

$$L_p = 120 + 20\log_{10}(20/10) = 126 \text{ dB}.$$

The transmission loss at a range of 3 m is then calculated using the attenuation, α, for sound through air, which is approximately 1.2 dB/m:

$$H = 20\log_{10}(3/0.3) + 1.2 \times 3 = 23.6 \text{ dB}.$$

The received SPL is

$$L_r = L_p - H = 126 - 23.6 = 102.4 \text{ dB}.$$

This excites the receive transducer to produce an electrical signal. For the specified sensitivity of −63 dB, relative to 1 V/μbar the output voltage is

$$20\log_{10} V_{out} = 102.4 - 63 + 20\log_{10} 0.0002.$$
$$= 102.4 - 63 - 74$$

Therefore

$$V_{out} = 10^{(102.4-63-74)/20} = 18.6 \text{ mV}.$$

This voltage is more than adequate to ensure a good SNR when followed by a high-gain preamplifier.

13.20.3 Tracker Implementation

13.20.3.1 Beacon The beacon is made as light as possible so that it can easily be attached to a person's belt. From Figure 13.45a it can be seen that it consists of a PIC microcontroller (PIC16F62X) which drives two IR LEDs with a peak current of 80 mA for a 600 μs pulse with a modulation frequency of 38 kHz. The pulse repetition interval is 50 ms, making the duty cycle $0.012 \times 0.5 = 0.6\%$ and reducing the average current consumption to 0.48 mA per LED.

The microcontroller also drives two ultrasonic transducers at 40 kHz through switching transistors and step-up transformers to provide each with 20 Vpp, which is their maximum rated drive level. Two transducers are used to widen the azimuth coverage so that the beacon will be seen by the receiver even if it is at an angle.

Figure 13.45b shows the layout for the beacon transducers. It is important that the IR LEDs and the ultrasonic transducers be squinted in the horizontal plane to improve angular coverage. It is also important that the two ultrasonic transducers be mounted one above the other, as close

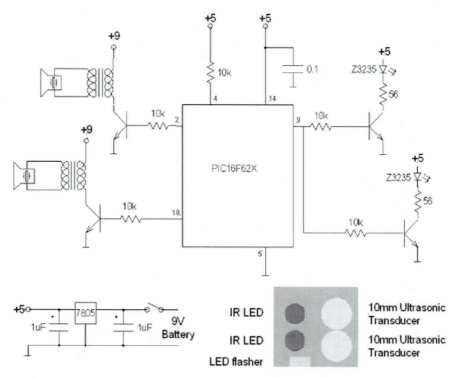

Figure 13.45 Details of the beacon showing (a) the schematic diagram and (b) the transducer layout on the face (Brooker et al. 2007).

together as possible, to reduce the magnitude of the elevation lobing. A visible LED, not shown in the schematic, can be included if necessary and strobed for a brief period once every second to indicate that the beacon is switched on and working.

In the prototype, the 9 V supplied by an alkaline battery was regulated down to 5 V using a linear regulator. This is inefficient and it is proposed that a small DC/DC converter should be used, which would double the battery life to about 12 hr for the alkaline battery. A higher energy density rechargeable battery would improve the time still further.

13.20.3.2 Receiver The receiver comprises a pair of muRata piezoelectric ultrasonic transducers amplified by a conventional stereo preamplifier, as shown in Figure 13.46. These transducers and the amplifiers should be matched to ensure that the output voltage is balanced or a measurement bias will result.

The piezoelectric transducers receive the 1 ms long ultrasound pulse and convert it to an electrical signal. Because of the narrow bandwidth of both the transmit and receive transducers, the pulse shape is not rectangular, but has a finite rise and fall time, as can be clearly seen in Figure 13.47. These received signals are amplified are then detected and filtered to produce a rectified envelope of the pulse, as shown.

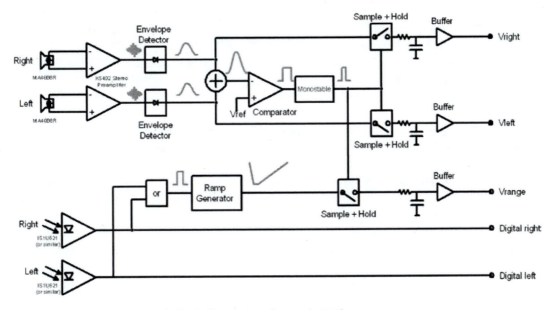

Figure 13.46 Receiver schematic block diagram (Brooker et al. 2007).

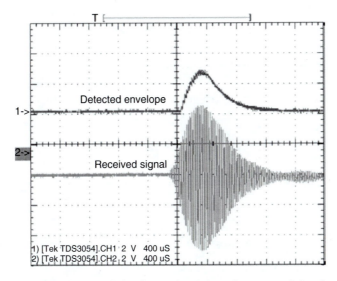

Figure 13.47 Received acoustic signal and the detected envelope from one of the channels (Brooker et al. 2007).

From the schematic diagram it can be seen that the sum of the envelope voltages from the left and right channels is then converted to a digital pulse using a comparator configured with a Schmidt trigger input and with an adjustable threshold level. The leading edge of this pulse then triggers a monostable circuit with a duration of 0.5 ms. The reason for this is to ensure that the envelopes are sampled at their peak, and that occurs after about 0.5 ms, as can be seen in the measured data.

The comparator output is a transistor-transistor logic (TTL) signal which remains high for the period that the pulse envelope has crossed the detection threshold. Its duration is about 1 ms, but because the pulse is not rectangular, will depend on the size of the detected pulse and the threshold setting.

Because the comparator output can vary, it is used only to trigger the monostable circuit, which is then used to perform a number of functions. In the first instance it officially marks the receipt of the acoustic pulse for the range measurement circuitry, which is explained later. Second, it triggers an S&H which samples the detected envelope for both the receivers. These values are then held for the 50 ms interpulse period.

The S&H circuitry consists of an analog switch followed by a resistor and a capacitor, the output of which is buffered by a noninverting follower. The resistor-capacitor (RC) time constant of this circuit is designed to smooth the sampled voltage slightly (otherwise only a capacitor could be used). During the 0.5 ms that the analog switch is closed, the capacitor charges and then tracks the envelope voltage. This is known as the sample period or sample aperture. When the switch is opened, the high input impedance of the buffer and the high off resistance of the analog switch ensure that the charge remains on the capacitor for the complete interpulse period with minimal droop. This is known as the hold period.

The IR circuitry consists of a pair of buffered IR receivers that produce a TTL low on receipt of an IR signal modulated at 38 kHz. These signals are read into a PIC microcontroller (not shown in the schematic) for further processing. The pulsed inputs are latched into the PIC on every cycle and are used to determine in which quadrant the beacon is located, that is, whether the left, the right, or both receivers have been triggered. These outputs can be used to control a search algorithm in the absence of an acoustic signal.

As shown in Figure 13.48, an output from the PIC, which is the logical OR of the two IR receiver inputs, serves as a low-going trigger pulse for the analog range voltage generator. The

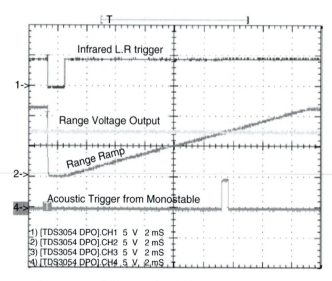

Figure 13.48 Analog range signals showing the IR trigger that starts the timing ramp and the acoustic trigger that is used to sample the ramp voltage (Brooker et al. 2007).

range circuit comprises a constant current generator that charges a capacitor to produce a linear voltage ramp with a slope of about 0.75 V/ms. This ramp capacitor is discharged through an analog switch when the trigger from the IR inputs is received. The duration of this pulse is determined by the beacon and is about 600 μs, which is sufficient time to completely discharge the capacitor. When the trigger pulse ends, the capacitor is again free to start charging up from 0 V for the remainder of the 50 ms cycle, as can be clearly seen in Figure 13.48. After about 18 ms it reaches the supply voltage (12 V) and can increase no further. However, this is not important, as the maximum range of interest is reached after 15 ms.

A similar S&H circuit to those wired for the angle signals is also triggered by the output of the monostable circuit to sample the ramp voltage when the acoustic pulse is received. This results in a range voltage that remains constant for a complete 50 ms cycle until a new sample is taken. This range voltage is directly proportional to the time between the receipt of the IR pulse and the receipt of the acoustic pulse, and hence is representative of the range between the beacon and the receiver. At a speed of 340 m/s, the conversion factor is 2.94 V/m.

13.20.4 Construction

Construction of the prototype was straightforward. The ultrasonic transducers were soldered, at the correct squint angle, to the input posts on the stereo preamplifier. This board was mounted in a die-cast box with the transducers protruding through large holes drilled in the one side, as can be seen in Figure 13.49. The remainder of the electronics were assembled onto a printed circuit board (PCB) and mounted above the preamplifier using standoffs. Additional holes were drilled in the wall of the box to allow the IR transducers an unrestricted view. Finally, a cardboard septum was constructed to restrict the view from the IR transducers, as can be seen in Figure 13.49, so that only about 10° of overlap occurred.

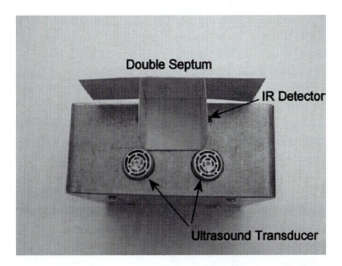

Figure 13.49 Photographs of receiver box showing both ultrasound and IR receivers and a construction to adjust the viewing angle of the IR receivers (Brooker et al. 2007).

13.20.5 Control Algorithms

The tracker was mounted on a small radio control (RC) vehicle modified for the purpose. An Atmel Atmega-128 8-bit microcontroller runs the software which performs the tracking, throttle, and steering functions. In order for the vehicle to properly track the target, the presence of the acoustic signal and at least one IR signal is required. An additional output from the sensor indicates this presence and notifies the controller that the incoming sensor data are valid. With these valid data, the vehicle can track the target using proportional control. The microcontroller digitizes the range and the two angle voltages, and samples the digital outputs of the sensor.

The digitized range voltage, V_{range}, is subtracted from a set point, V_{set}, to produce the difference which controls the speed of the drive wheels, $\dot{R}$:

$$\dot{R} = K_1(V_{range} - V_{set}). \tag{13.32}$$

To control the vehicle steering, a simple linear tracking algorithm was implemented by estimating the offset error, ΔAz, using the left and right receiver voltages (V_{left} and V_{right}):

$$\Delta Az = K_2 \frac{V_{left} - V_{right}}{V_{left} + V_{right}}. \tag{13.33}$$

The value is normalized to full scale and used to set the steering angle. Normalization of the error signal is necessary to provide accurate steering control independent of signal amplitude. Unfortunately, at low SNRs, this normalization process cannot be used, as amplification of the noise introduces unwanted inputs into the control algorithm. For this reason, the amplitude of the received acoustic signal is monitored and if it drops below a threshold, the acoustic data are ignored. The algorithm also includes a small dead band to prevent oscillation around the center point.

Because the IR detection range is longer than that of the acoustic system, it is desirable that the vehicle track the beacon even when the acoustic data are not present. In this mode, the vehicle turns until the IR signal illuminates both optical sensors. If the acoustic signal is out of range, the vehicle moves forward using a bang-bang controller based on the IR tracking signal only, until the acoustic signal is acquired. It must also be noted that due to the mechanical configuration of the vehicle, it can only turn toward the beacon when it is also moving forward. Proportional control was implemented for both the steering and range parameters, which was sufficient, as the slow response of the mechanical components added considerable damping to system.

The beacon-following capability of the system was good if the beacon remained within the receiver beam. The vehicle was capable of following a person carrying the beacon at a fast walk and was capable of performing a 180° turn with a radius of less than 3 m. Figure 13.50 shows a plot of the beacon and the vehicle that was extracted from a video taken of one of the indoor trials.

13.21 ANGLE TRACK JAMMING

The section on seduction jamming mentioned that the synthetic target echo could be on-off modulated to introduce instabilities in the angle tracking loops of conscan and monopulse

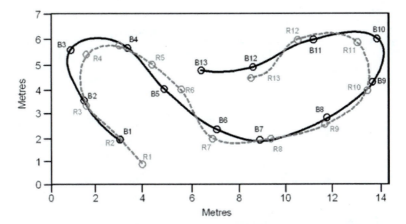

Figure 13.50 Plot of the vehicle receiver R_x and the beacon B_x positions (Brooker et al. 2007).

systems. In the conscan case, this effect is straightforward because the angle error is determined directly from the amplitude and phase of the received signal. The jammer monitors the radar transmission and rebroadcasts its on-off modulation in antiphase. This ensures that the sense of the errors generated by the radar are incorrect, with the result that instead of turning the antenna toward the target, it pushes it in the opposite direction. Break-lock is almost instantaneous. In the event that a conical scan receive only (COSRO) unit is used and the jammer is unable to detect the conscan frequency, it will transmit a modulation that varies in frequency over the typical range of conscan frequencies until a match is found and the lock is broken.

Because the monopulse technique is not sensitive to fluctuations in the amplitude of the detected signal, the on-off modulation method cannot induce the loss of lock. However, because the angle tracking loops often rely on an AGC to normalize the angle errors, modulation can introduce some instabilities into the angle tracking accuracy. When the jammer is turned off and the radar loses lock it may need to reacquire it in angle as well as in range, and this takes time.

13.22 TRIANGULATION

Although triangulation is not really a range or angle estimation technique, it is probably the most common method in use today for measuring the position of a target in space. The basic technique is thousands of years old and can be applied using

- Bearing only, also known as direction finding.
- Range only, as used in the Global Positioning System (GPS).
- Hybrid range and bearing, often used in surveys with theodolites containing electronic distance measurement (EDM) modules.

Two of the common triangulation methods are illustrated in Figure 13.51, in which three range or bearing measurements have been made on a two-dimensional surface. In each case, some errors have been included in the measurements which translate into an uncertainty shown by the shaded area at the intersection of the measurements.

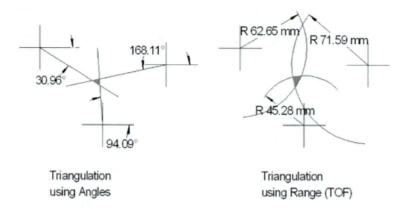

Triangulation
using Angles

Triangulation
using Range (TOF)

Figure 13.51 Position estimation by triangulation.

Applications of triangulation include terrestrial and airborne navigation using GPS, Omega, and Loran-C, among others, which use time-of-arrival methods from beacons at known positions. Angle-only methods are probably the oldest, dating back thousands of years, and include the classical survey to trigonometric beacons using theodolites. More recently this technique has been used for radio direction finding as used by Chain Home (discussed in Chapter 1).

13.22.1 Loran-C

Loran is an acronym for long-range navigation. It is a highly accurate (though not as accurate as GPS), highly available, all-weather system for navigation in the coastal waters around the United States and many other countries. An absolute accuracy of better than 0.25 nm is specified within the region of coverage. And as with GPS, it is also used as a precise time reference.

13.22.1.1 Summary of Operation   A chain of three or more land-based transmitting stations each separated by a couple of hundred kilometers is used instead of a constellation of satellites. Within the chain, one station is designated as the master (M) and the other transmitters as secondary stations, conventionally designated W, X, Y, and Z.

The master station and the secondaries transmit bursts of radio pulses at a carrier frequency of 100 kHz at precisely timed intervals. A Loran-C receiver onboard a ship or aircraft measures the slight differences in the time of arrival of the pulses. The difference in the time of arrival for a given master-secondary pair observed at a point in the coverage area is a measure of the difference in distance from the vessel to the two stations.

The locus of points having the same time delay from the pair is a curved line of position (LOP). These curved lines are hyperbolas, or more correctly spheroidal hyperbolas, on the curved surface of the earth, and the intersection of two or more LOPs from different master-secondary pairs determines the position of the user, as shown in Figure 13.52.

13.22.1.2 Measurement Process   The master station (M) transmits a burst of eight pulses, each separated by 1 ms from the previous one, followed by a ninth identification pulse spaced 2 ms later. After a fixed delay, the secondary station (W) transmits its burst of eight pulses spaced 1 ms

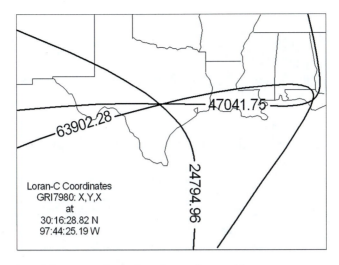

Figure 13.52 Intersection of three LOPs determines the receiver position.

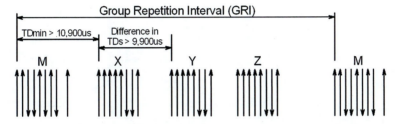

Figure 13.53 Pulse configuration for Loran-C with the direction of the arrows indicating the relative phase of each pulse.

apart, and after another fixed delay, secondary station (X) does the same. This continues for up to five slave stations before the cycle repeats itself, as illustrated in Figure 13.53. The period of this cycle is known as the group repetition interval (GRI), and it is the characteristic that identifies a particular chain. This value is quoted in tens of microseconds, as seen in Figure 13.52, which identifies the chain as GRI17980.

To ensure that the time interval measurement is as accurate as possible, not only are the bursts phase encoded (binary phase-shift keying [BPSK]) to allow for correlation, but each pulse is also carefully constructed, as can be seen in Figure 13.54. Coarse timing information can be obtained using the relative delays of the pulse envelopes because the time from the start of the pulse to the peak is fixed at 65 μs. However, accurate time delay measurements use the zero crossing after the third full cycle, that is, 30 μs from the start, or 35 μs from the peak of the envelope. The pulse is intentionally damped so that the trailing skywave does not contaminate the next burst of

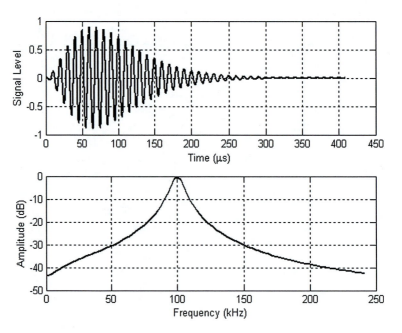

Figure 13.54 Construction of a Loran-C pulse.

pulses. It also limits the bandwidth of the signal to 20 kHz to minimize spectral usage (U.S. Department of Transportation 1994).

13.22.1.3 Advantages of Loran-C   Why would anyone want Loran-C when GPS is so good?

- It is inexpensive to operate, at $17 million per year, and user equipment is low cost and of proven capability. In contrast, to maintain the GPS infrastructure requires in excess of $500 million per year.
- Loran's signal format has an integrity check built in, GPS does not. Loran is easy to service—just drive to the transmitting station—while GPS replacement requires a new satellite launch schedule.
- On-air time for Loran-C is about 99.99%, while GPS is only about 99.6%, and even lower in some regions. These features have convinced more than 25,000 users to sign petitions to keep Loran-C online to 2015.
- Other nations have purchased new solid-state Loran-C transmitters to form Loran-C chains in Europe, China, Japan, and Russia, as shown on the map reproduced in Figure 13.55. This land-based navigation system is viewed as a desirable, stable complement to GPS.
- Finally, Loran is totally unclassified and is operated by host country authorities and not the U.S. military (Brusin 2008).

Why does the military, who designed the GPS system, not rely on it totally?

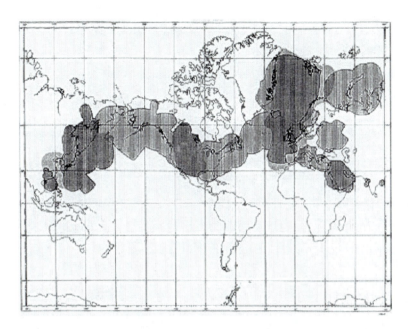

Figure 13.55 Loran-C coverage (Courtesy Rosetta Project).

- The ease with which GPS can be jammed either on purpose or by unexpected interferences is certainly a major reason. The deliberate jamming is well documented by the U.S. military, the International Association of Lighthouses, and the Civil Aviation Authority (UK).
- A 1 W jammer, costing about US$100, with dimensions of 5 cm × 5 cm × 10 cm, and a 10 cm antenna can block out GPS over a 60 km diameter circle.
- If you want a jammer for GPS and Glonass (the Russian equivalent of the GPS), such units are offered for sale by the Aviaconversia Company, Russia, which displayed them at a recent Moscow Air Show. Their jamming range was said to be 200 km. What was the reaction of the Federal Aviation Administration (FAA)? "Nothing new," because there are "hundreds of these devices on the market" (Brusin 2008).

13.23 References

Aguilar, R. and Meijer, G. (2002). Low-cost ultrasonic fusion sensor for angular position. *Proceedings of SeSense 2002*, pp. 594–597.

Barton, D. (1976). *Radar Systems Analysis*. Norwood, MA: Artech House.

Barton, D. (1988). *Modern Radar Systems Analysis*. Norwood, MA; Artech House.

Benedict, T. and Bordner, G. (1962). Synthesis of an optimal set of radar track while scan smoothing equations. *IRE Transactions on Automatic Control* AC-7:27–32.

Blackman, S. (1986). *Multiple-Target Tracking with Radar Application*. Norwood, MA: Artech House.

Blake, B., ed. (1995–1996). *Jane's Radar and Electronic Warfare Systems*, 7th ed. Alexandria, VA: ITP.

Bortolotto, G., Masson, F., and Bernal, S. (2001). USRPS—ultrasonic short range positioning system. IX Reunion de Trabajo en Prosesamiento de la Informacion y Control (RPIC 2001), Santa Fe, Argentina, September 12–14.

Brooker, G. (1983). Design of the tracking complex for a tracking radar system. University of the Witwatersrand, Johannesburg, South Africa.

Brooker, G., Lobsey, C., and McWilliams, K. (2007). Combined infrared and acoustic beacon tracker implementation on an autonomous vehicle. 2nd International Conference on Sensing Technology, ICST'07, Palmerston North, New Zealand, November 26–28.

Brusin, E. (2008). Hangar talk: options; Loran C: an expanding utility for worldwide use. Available at http://www.landings.com/_landings/reviews-opinions/loran-c.html.

Fukuju, Y., Minami, M., Morikawa, H., and Aoyama, T. (2003). DOLPHIN: an autonomous indoor positioning system in ubiquitous computing environment. IEEE Workshop on Software Technologies for Future Embedded Systems (WSTFES'03).

Girod, L. and Estrin, D. (2001). Robust range estimation using acoustic and multimodal sensing. Intelligent Robots and Systems (IROS 2001), Maui, Hawaii, Oct. 29–Nov. 3.

Harris, C. and Clarke, A. (1981). A comparison of alpha-beta and centroid beta variances effect on future position and resulting range of beta available. WES-0064, Marconi.

Hughes, R. (1988). *Analog Automatic Control Loops in Radar and EW*. Norwood, MA: Artech House.

Jacobowitz, H. (1965). *Electronics Made Simple*. New York: Doubleday.

Kopp, C. (2005). Heat-seeking missile guidance. Viewed January 2008. Available at http://www.ausairpower.net/TE-IR-Guidance.html.

Kuc, R. (2002). Binaural sonar electronic travel aid provides vibrotactile cues for landmark, reflector motion and surface texture classification. *IEEE Transactions on Biomedical Engineering* 49(10):1173–1180.

Leonov, A. and Fomichev, K. (1986). *Monopulse Radar*. Norwood, MA: Artech House.

Mahafza, B. (2000). *Radar Systems Analysis and Design Using MATLAB*. Boca Raton, FL: CRC Press.

muRata. (2008). Piezoelectric ceramic sensors (PIEZOTITE). Viewed January 2008. Available at http://www.murata.com/catalog/p19e.pdf.

Palmer, R. (2002). A spread spectrum acoustic ranging system: an overview. IEEE Canadian Conference on Electrical & Computer Engineering.

Priyantha, N. (2005). The Cricket indoor location system. PhD dissertation, MIT.

Riegl. (2006). Eyesafe laser rangefinder FG21. Viewed March 2008. Available at http://www.riegl.com/.

Schleher, D. (1986). *Introduction to Electronic Warfare*. Norwood, MA: Artech House.

Sherman, S. (1984). *Monopulse Principles and Techniques*. Norwood, MA: Artech House.

Skolnik, M. (1970). *Radar Handbook*. New York: McGraw-Hill.

Skolnik, M. (1980). *Introduction to Radar Systems*. Tokyo: McGraw-Hill Kogakusha.

Skolnik, M. (1990). *Radar Handbook*. New York: McGraw-Hill.

U.S. Department of Transportation. (1994). *Specification of the Transmitted Loran-C Signal*. Washington, DC: U.S. Department of Transportation, U.S. Coast Guard.

Ward, A., Jones, A., and Hopper, A. (1997). A new location technique for the active office. *IEEE Personal Communications* 4(5):42–47.

14

Tracking Moving Targets

14.1 TRACK-WHILE-SCAN

The track-while-scan (TWS) concept originated with some of the earliest radar systems developed during World War II. In principle, the position of a moving target is collapsed onto a two-dimensional (2D) surface where its position is updated at regular intervals. At its most simple, this process can be performed by a human operator by marking the positions of all of the targets and associating sequences of returns with individual tracks.

The introduction of the surveillance radar with outputs to a circular cathode ray tube display called a plan-position indicator (PPI), shown in Figure 14.1, allowed this process to be automated. The PPI display positions the radar at the center and displays a 2D representation of the surroundings in polar space, with the range to any targets proportional to the radial distance from the center of the display, and their bearing relative to true north typically reproduced as angles with respect to the top of the display.

As the radar rotates, any received echoes are reproduced on the display at the correct range and bearing by exciting a long-persistence phosphor on the screen. The rotation rate is dependent on the maximum range of the radar and can vary from a maximum of about 1 rps for short-range surveillance down to about 6 rpm for really long-range early warning radars.

The PPI will display both static and moving targets unless the former are filtered out. This may be necessary because the returns from the terrain surrounding the radar are often sufficiently large to swamp the smaller returns from any moving targets, as is illustrated in Figure 14.2b.

Modern PPIs are generally reconstructions on a computer display, such as the one shown in Figure 14.2a. This has the advantage that graphical overlays and additional information can easily be included with the "raw" echo information.

Automation of the tracking process proceeds as illustrated in Figure 14.3 for a single target, as it is assumed that static target returns have been suppressed using Doppler or moving target indicator (MTI) processing.

1. A moving target is detected when the received echo exceeds a threshold. Initially there is no information about its velocity so the software constrains the uncertainty to a reasonable value for an aircraft target (large circle in 1).

Figure 14.1 Examples of plan position indicator display hardware (Skolnik 1970).

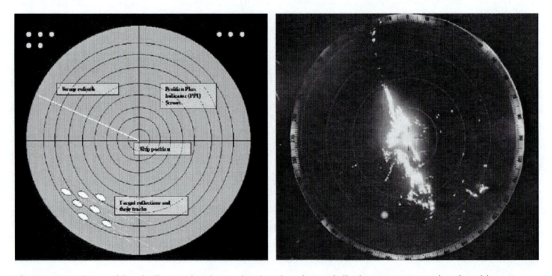

Figure 14.2 Plan position indicator showing a simulated and a real display. Courtesy National Archives.

2. A target is detected again, displaced in range and angles, but within the uncertainty boundary. The software sets up a possible track and makes a crude velocity estimate from two points. It then predicts the target's next position. However, the uncertainty is still large as the position and velocity estimates are not very good. A tracking filter is initialized with the position and velocity estimates.

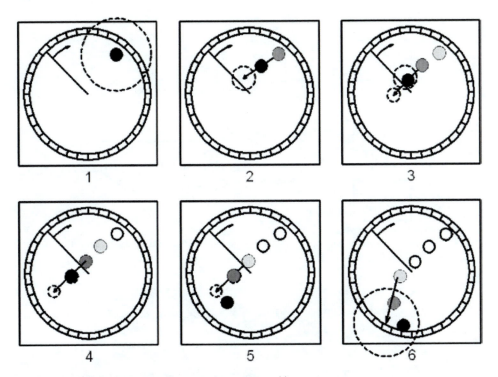

Figure 14.3 PPI display sequence to illustrate the target tracking process.

3. The target appears within this uncertainty boundary. The tracking filter estimates of position and velocity improve, and the next sample prediction shows a smaller position uncertainty.
4. As with (3).
5. The actual target position falls outside the position uncertainty boundary because it has accelerated and the prediction algorithm used only position and velocity. Track is lost.
6. A new target is detected with unknown velocity and the cycle repeats.

One method of performing the filtering and prediction function is to use tracking filters on the measured polar coordinates (R, θ). However, a target moving in a straight line appears to accelerate in polar space unless the trajectory is radial. This results in lag errors on the estimated positions. A better alternative is to convert the measured positions from polar to Cartesian space (x, y) before filtering, as shown in the block diagram in Figure 14.4.

14.2 THE COHERENT PULSED TRACKING RADAR

Pulsed tracking radar systems, such as the BAE Systems unit shown in Figure 14.5, are mostly used for military applications such as fire control. These fire control radars (FCRs) track fast-moving aircraft or missiles with high accuracy and then use the estimates of the target position

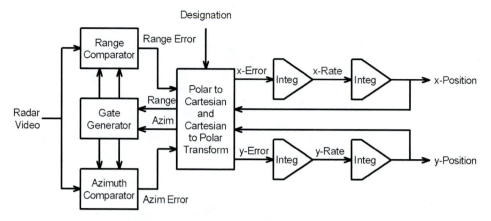

Figure 14.4 Track-while-scan processing of a single target [adapted from (Barton 1976), reproduced with permission © 1976 Artech House].

Figure 14.5 Cluster of fire control tracking radars.

and velocity to direct missiles or antiaircraft guns. The system shown consists of a pair of X-band prime focus antennas mounted one below the other, which could be used as an interferometer to improve low-angle tracking. The larger antenna in the center uses the twist Cassegrain configuration and is the primary sensor. On the left is a cluster of optical units which would include low-light TV and infrared cameras. These are all mounted on the elevation positioner, which is supported on a yoke above the azimuth positioner. Slip-ring clusters typically feed signals and power to and from the radar hardware mounted below decks.

Most FCRs are low or medium pulse repetition frequency (PRF) pulsed Doppler systems. Such coherent radar systems extract both amplitude and phase information from the signal reflected by a target. Because the length of a single pulse is too short to resolve typical target Doppler frequencies, as shown in Figure 14.6c, to extract Doppler shift, the returns from many pulses over an observation time T must be analyzed so that the spectrum can be resolved down to a bandwidth $\beta \approx 1/T$. For this process to work, a deterministic phase relationship must be maintained over the observation time T.

14.2.1 Single Channel Detection

The block diagram of a single-channel coherent radar is shown in Figure 14.7. This transmitter configuration is known as a master oscillator power amplifier (MOPA) and it generates the carrier by mixing a radio frequency (RF) signal from a stable local oscillator (STALO) with a frequency f_{rf} with a coherent oscillator (also very stable) with a frequency f_{if}. The resulting frequency (after filtering) is f_o and is pulse modulated, amplified, and transmitted.

The received signal is down-converted to intermediate frequency (IF) (typically 30–60 MHz) by mixing with the STALO signal. After amplification and filtering it is down-converted further to baseband (video) by mixing with the coherent local oscillator (COHO) signal. A consequence of this single-channel down-conversion is that there is no direction information in any Doppler modulation since a target receding would produce exactly the same Doppler signature as one approaching

$$V_{out}(t) = k \sin\left(2\pi f_d t + \varphi_o\right), \tag{14.1}$$

where

$V_{out}(t)$ = video output voltage,

k = amplitude of the video signal,

f_d = Doppler frequency (Hz),

φ_o = phase shift (rad).

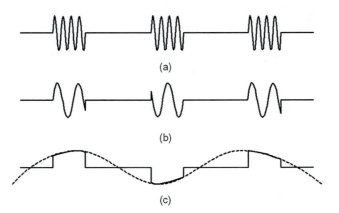

(a)

(b)

(c)

Figure 14.6 Doppler modulation of a pulse train [adapted from Skolnik (1980)].

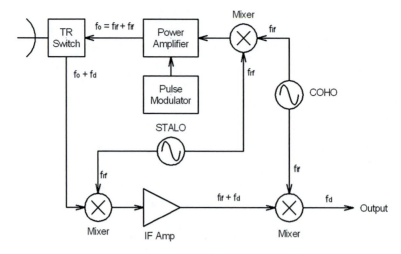

Figure 14.7 Single-channel coherent pulsed Doppler radar [adapted from Morris (1988)].

14.2.2 I/Q Detection

In an I/Q detector, the IF signal is split into two channels with the quadrature (Q) channel being phase shifted by 90° with respect to the in-phase (I) channel, as can be seen in Figure 14.8. In this case, although the two Doppler outputs will have identical frequency whether the target is approaching or receding, their phase relationship with each other will reverse, and so direction information can be obtained as discussed in Chapter 10.

By sampling the I and Q outputs and using a complex fast Fourier transform (FFT), the magnitude and phase of the combined Doppler spectrum can be obtained. Another benefit of this dual-channel detection method is that a 3 dB gain in signal-to-noise ratio (SNR) is obtained after processing.

14.2.3 Moving Target Indicator

If the actual Doppler frequency is not important, but only the fact that the target is moving is, a process called moving target indication can be used. If the video output of the I or Q channel is examined for a moving target, the amplitude will vary on a pulse-to-pulse basis due to the changing phase between the transmitted and received signals, as shown Figure 14.6. In contrast, for a static target the phase will remain unchanged and the amplitude will remain constant.

A simple MTI based on a delay line canceller operates by taking the difference of the amplitudes of successive pulses, as shown in Figure 14.9. This is, in effect, a finite impulse response (FIR) filter with a high-pass characteristic that rejects signals that are unchanging or that are changing very slowly. It is also known as a nonrecursive filter. Because it is a sampled data system, as discussed in Chapter 2, the frequency response is repeated with a period of $1/T$, as shown in Figure 14.10.

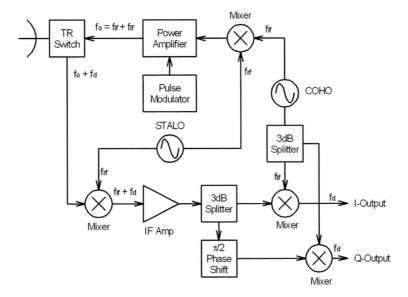

Figure 14.8 Coherent pulsed radar with a synchronous (I-Q) detector [adapted from Morris (1988)].

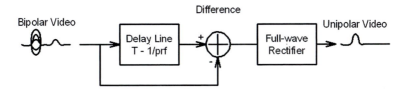

Figure 14.9 Delay line canceller used to suppress the returns from static targets.

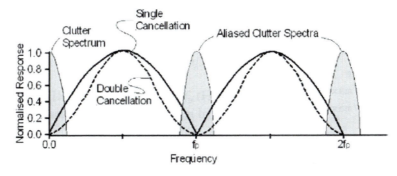

Figure 14.10 Transfer function of single and double delay line cancellers for MTI [adapted from Skolnik (1980)].

high PRF. However, a high PRF becomes ambiguous at a short range, which is also not ideal for surveillance or long-range tracking radar applications.

14.2.3.2 Staggered PRF and Blind Speed   The effect of blind speeds can be reduced by operating at more than one PRF. This has the effect of overlapping the different filter characteristics and filling in the zeros at multiples of the PRF, as can be seen in Figure 14.13.

As the ratio of the PRI, T_1/T_2, approaches unity, the greater will be the value of the first blind speed. However, the first null also gets deeper, and so the rejection of slowly moving clutter will be compromised.

To cater for the moving clutter problem, a bank of Doppler filters can be implemented instead of a delay line canceller. In modern radars, this process is generally implemented digitally using the complex FFT. The outputs (excluding sidelobes) of such a filter implementation are shown Figure 14.14. To obtain sufficient rejection of unwanted signals in adjacent bins the filter side-

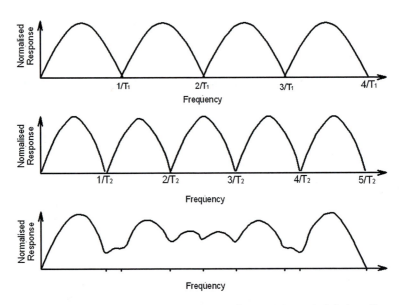

Figure 14.13 Effect of staggered PRF on MTI transfer function [adapted from Skolnik (1980)].

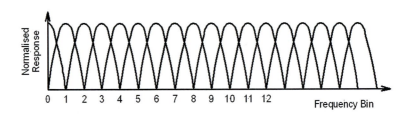

Figure 14.14 Filter bank implemented using the FFT.

Table 14.1 Properties of weighting functions

Weighting function	Peak sidelobe level (dB)	Main-lobe width
Rectangular	−13.26	0.886
Hanning	−31.5	1.42
Hamming	−42.5	1.32
Taylor $n = 5$	−34	1.19
Taylor $n = 6$	−40	1.25
Dolph Chebyshev	−40	1.2

lobes must be made as low as possible. A trade-off exists between the width of the main lobe and the sidelobe level with different windowing (weighting) functions, as discussed in Chapter 11 and summarized in Table 14.1.

14.3 LIMITATIONS TO MTI PERFORMANCE

Subclutter visibility is the ratio by which the moving target power may be lower than the clutter in the same range bin and still be detected with a specified P_d and P_{fa}. It is usually limited by internal instabilities in the amplitude and phase of the various waveforms generated by the radar, by the motion of a scanning antenna, and the finite time on target, which have the effect of widening the clutter spectrum. It is also affected by the bandwidth characteristics of real clutter, for example, rain is blown by the wind, the sea moves, as do leaves and grass.

14.4 RANGE GATED PULSED DOPPLER TRACKING

The Doppler signal extraction and the range tracking loop are interrelated, as can be seen in the block diagram in Figure 14.15. Doppler extraction operates on video sampled at the range indicated by the range tracker, while range tracking is usually accomplished by means of the split-gate (early/late) technique. However, before the signals in the early and late gates are compared to derive the range tracking error, each is passed through a bandpass filter (usually implemented by an FFT) to reject returns from stationary objects at the same range.

The range tracking loop can be updated at the PRF, f_p (Hz), or at a slower rate if the process to extract Doppler requires more than one return pulse. For example, if an FFT processes N_p pulses to perform the Doppler analysis, the update frequency is reduced to f_p/N_p (Hz).

Initiation of tracking for an FCR requires the near-simultaneous initialization of all four tracking coordinates: range, two angles, and Doppler frequency. This is achieved as follows (Morris 1988):

- The operator, who identifies a target during the search phase, designates the selected target on the display using a cursor.
- On the following scan, if the target is still present, the range, azimuth, elevation angles, and Doppler frequency are recorded.

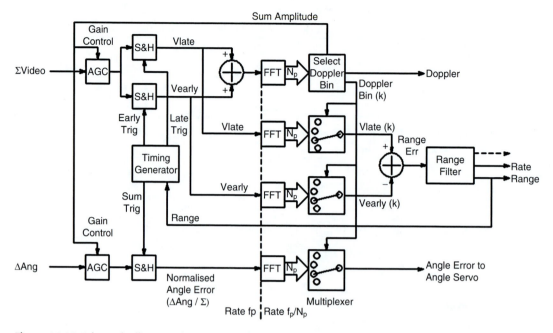

Figure 14.15 Schematic diagram of a range-gated Doppler tracker.

- The scan sequence is interrupted and the radar antenna returns automatically to the designated area.
- The range gate pair is moved to the designated range where they begin a small search to compensate for uncertainties in the designation accuracy.
- The sum channel video received by the combined early and late gates, or by a third target gate that straddles the two, is processed by a Doppler analyzer (e.g., FFT).
- If a moving target is found at the designated range and sufficiently close to the Doppler of the detection that triggered the acquisition sequence, then the FFT bin containing the target return is identified and stored.
- Video signals in both the early and late gates are processed through FFTs and the outputs of the same bins are selected.
- The difference between the selected bins in the late and early gates produces a range error that feeds the range filter.
- The range tracking loop is closed by feeding the estimated range back to control a timing generator that positions the early and late gates over the target echo.
- The video signals in both the azimuth and elevation monopulse angle error channels are sampled by a range gate slaved to the target gate of the range tracker.
- Angle errors are generated by normalizing the error video using an automatic gain control (AGC) controlled by the sum channel signal. These data are sampled by the timing generator sum signal which straddles both the early and late gate periods.

- After a brief period to allow the range tracking loop to settle, the angle tracking loops are closed by feeding the angle errors to their servo amplifiers which control motors that drive the antenna to point at the target. This completes the transition sequence.

The transition from search to track is not a trivial problem, and it is made even more difficult in a military scenario where targets are not passive and will either start to maneuver or try to disrupt the process by deploying chaff or some other electronic means such as noise jamming (see Chapter 9) or seduction jamming (see Chapter 13).

14.5 COORDINATE FRAMES

In many applications, more than one sensor is used to obtain information about moving targets. These sensors can be collocated, as seen in the FCR shown in Figure 14.5, or they can be separated by varying distances which may or may not remain constant. An example of the latter is a convoy of ships, each of which has its own surveillance and tracking radars.

14.5.1 Measurement Frame

Measurements made by radars and other active sensors are made in polar space (R, θ, φ), as they can only measure the target position in range, elevation, and bearing (azimuth).

14.5.2 Tracking and Estimation Frame

The equations of motion that govern the profile of a target operate in Cartesian space, (x, y, z), so it is advantageous to transform the coordinate system from polar to Cartesian space for filtering and then back to polar space to direct the sensor.

The transform from polar to Cartesian space, as defined in Figure 14.16, is

$$x = r \cos \theta \cos \varphi$$
$$y = r \cos \theta \sin \varphi$$
$$z = r \sin \theta, \tag{14.4}$$

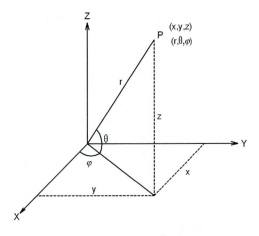

Figure 14.16 Relationship between Cartesian and polar coordinate systems.

and the transform back from Cartesian to polar space is

$$r = \sqrt{x^2 + y^2 + z^2}$$

$$\varphi = \tan^{-1}\frac{y}{x}$$

$$\theta = \sin^{-1}\frac{z}{\sqrt{x^2 + y^2 + z^2}} \qquad (14.5)$$

Generally this frame of reference will remain centered at the radar, however, some fire control systems translate and rotate the frame to make it target centered, as the aircraft dynamics can be better modeled in this frame. If more than one sensor is involved in the tracking function, and these sensors are not collocated (they may be on different platforms that move relative to each other), then an earth centered Cartesian frame is generally used.

Translation of coordinate frames is a straightforward operation involving the addition of a displacement vector to the position of the target measured in Cartesian space. However, rotation is more complicated and relies on the Euler angles. To give an object a specific orientation it is subjected to a sequence of three rotations described by the Euler angles. Unfortunately the order in which the rotations occur is critical, and this has never been agreed on, so it needs to be specified when making any change in the frame of reference. Euler angles have two disadvantages. First, the equations contain many trigonometric functions which can be slow to execute. However, more importantly, singularities exist at angles around 90° which can present numerical problems. A solution to these problems is to replace the three Euler angles with four quaternions.

14.6 ANTENNA MOUNTS AND SERVO SYSTEMS

The pencil beam of a mechanically directed tracking radar must be pointed at the target for tracking to occur. This is quite a challenge, as a typical tracking radar has a 3 dB beamwidth between 1° and 2°, or smaller in the case of the radar for a close-in weapon system onboard ship. As can be seen in Figure 14.17, the radar system provides three primary outputs, range from the range tracker and two angle errors from the phase sensitive detectors described in Chapter 13. These angle errors can be used as inputs to control the current (and hence torque) of servo amplifiers which, in turn, power the motors to direct the antenna.

The positioner (see Figure 14.18), controlled by a servo system, is used to drive the antenna in the direction that minimizes the tracking errors. Most servo systems are type 2, or zero velocity error systems, since in theory, no steady-state error exists for a constant velocity (angular rate) input. With type 2 systems, dynamic lags proportional to the magnitude of the target acceleration do occur. To accommodate this, the tracking bandwidth can be adjusted to minimize the tracking error which is due to a combination of measurement noise and dynamic lag.

At long range, where the angular motion of the target is small, a very small tracking bandwidth can be tolerated. However, at short range, where target angular rates and accelerations are large, a wider bandwidth becomes acceptable. Secant correction increases the azimuth error signal gain as a function of the elevation angle.

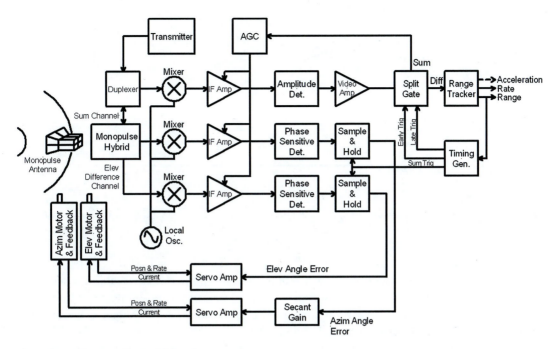

Figure 14.17 Monopulse tracking radar.

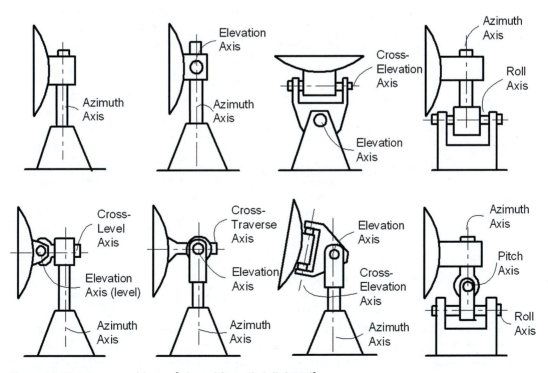

Figure 14.18 Antenna positioners [adapted from Skolnik (1980)].

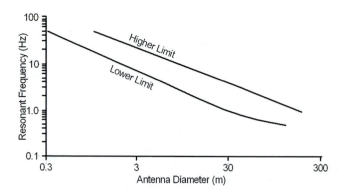

Figure 14.19 Lowest resonant frequency as a function of antenna diameter [adapted from Skolnik (1980)].

Another restriction on tracking bandwidth is that it must be small (10%) compared to the lowest natural resonant frequency of the positioner to reduce the risk of instabilities occurring. Figure 14.19 gives an indication of the resonant frequency of typical positioners as a function of the size of the antenna they are carrying. A typical FCR antenna has a diameter of about 1.5 m, resulting in a resonance between 10 and 30 Hz.

The dynamic requirements of the positioner must be capable of accommodating both target motion and its own base motion. This requirement can be quite onerous in ship- or ground vehicle-mounted sensors, as the roll rate and bandwidth can be quite high. Another issue of base motion is that it can affect the polarization angle relative to the ground. In a two-axis positioner, for example, though the antenna can be made to point anywhere over the complete hemisphere, the optimum polarization for ground clutter suppression may not be maintained. This leads to the requirement for additional axes.

14.7 ON-AXIS TRACKING

The best tracking occurs using null steering when the antenna is pointed toward the target with an accuracy of only a few milliradians. This is known as on-axis tracking. It maximizes the power incident on the target as well as reducing cross coupling between the axes by minimizing cross-polar levels. It also reduces the effects of system nonlinearities, with the result that the tracking stability is improved.

It requires the following:

- Removal by prior calibration of biases.
- A filter than can perform one-sample-ahead prediction.
- The selection of the appropriate coordinate system for tracking.

14.7.1 Crossing Targets and Apparent Acceleration

Target dynamics dictate the real and apparent accelerations of a crossing target, shown in Figure 14.20, when measured in a polar coordinate system. The tracking loop bandwidth then determines the tracking accuracy, which is a combination of the acceleration-induced dynamic lag and the noise.

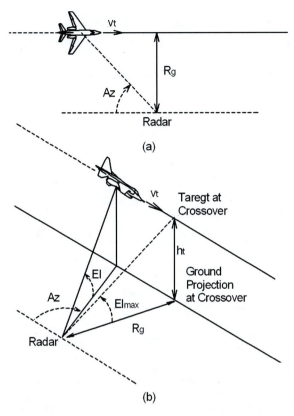

Figure 14.20 Geometry of a crossing target (a) in plan and (b) in perspective [adapted from (Barton 1976), reproduced with permission © 1976 Artech House].

When viewed in radar polar coordinates, the maximum target range and angular rates are

$$\dot{R}_{\max} = v_t,$$

$$\dot{Az}_{\max} = \frac{v_t}{R},$$

$$\dot{El}_{\max} = \frac{v_t}{R}. \qquad (14.6)$$

The real accelerations are dependent on the actual target acceleration, a_t (m/s^2), in Cartesian space:

$$\ddot{R}_{\max} = a_t,$$

$$\ddot{Az}_{\max} = \frac{a_t}{R},$$

$$\ddot{El}_{\max} = \frac{a_t}{R}. \qquad (14.7)$$

The polar coordinates of a crossing target are shown in Figure 14.21. These graphs were produced using the MATLAB code below. In the example shown, the target velocity is 100 m/s with a minimum crossing range of 25 m and a height of 25 m. The graph shows the range decreasing to a minimum of 35 m at $t = 0$ and then increasing again. The azimuth angle passes through zero at $t = 0$, while the elevation angle peaks at 0.78 rad (45°) at the same time.

Figure 14.22 shows the range derivatives for the crossing target. On the incoming leg, the range rate starts out at about −100 m/s, crosses through zero at $t = 0$, before approaching

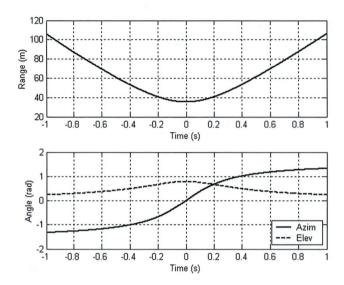

Figure 14.21 Polar coordinates of a crossing target.

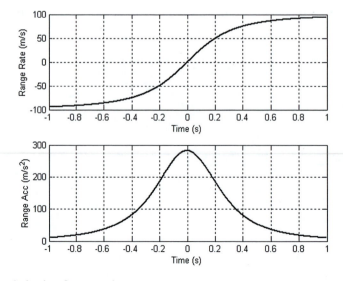

Figure 14.22 Range derivatives for a crossing target.

+100 m/s asymptotically on the outbound leg. The range acceleration peaks at nearly 300 m/s^2 at $t = 0$. This is close to 30 g and is much larger than any true acceleration that an aircraft piloted by a human being could reach.

Figure 14.23 shows the azimuth and elevation angle derivatives for the crossing target. The azimuth rate peaks at -4 rad/s at $t = 0$, while the elevation rate peaks at ± 1 rad/s at $t = \pm 0.2$ s. The accelerations peak at ± 10 rad/s^2 and -8 rad/s^2, respectively.

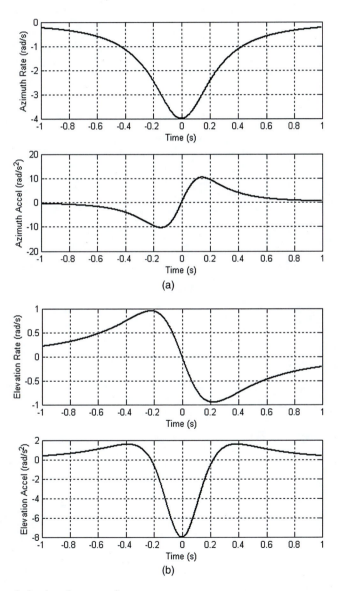

Figure 14.23 Angular derivatives for a crossing target.

```
% determines the rate and accelerations for a crossing target
% cross.m

% Variables
xa0=0;                          % Initial acceleration vector
ya0=0;
za0=0;

xv0=0;                          % Initial velocity vector
yv0=100;
zv0=0;

x0=25;                          % Initial position
y0=-100;
z0=25;

dt = 0.001;                 % Time increment
t=(-1:dt:1);                % Time

% Develop the target vectors in Cartesian space by integrating
% acceleration and then velocity to obtain position
xa = xa0*ones(size(t));
ya = ya0*ones(size(t));
za = za0*ones(size(t));

xv = xv0 + cumsum(xa).*dt;
yv = yv0 + cumsum(ya).*dt;
zv = zv0 + cumsum(za).*dt;

x = x0 + cumsum(xv).*dt;
y = y0 + cumsum(yv).*dt;
z = z0 + cumsum(zv).*dt;

% Convert to polar space
[th,ph,r] = cart2sph(x,y,z);

subplot(211),plot(t,r,'k')
grid
xlabel('Time (s)')
ylabel('Range (m)')
subplot(212), plot(t,th,'k',t,ph,'k-')
grid
xlabel('Time (s)')
```

```
ylabel('Angle (rad)')
legend('Azim','Elev',4)
pause

% Obtain range rate and range acceleration by differentiating
rv = diff(r)./dt;
ra = diff(rv)./dt;

subplot(211),plot(t(1:length(rv)),rv,'k');
grid
xlabel('Time (s)');
ylabel('Range Rate (m/s)')
subplot(212),plot(t(1:length(ra)),ra,'k');
grid
xlabel('Time (s)');
ylabel('Range Acc (m/s^2)')
pause

% Obtain azimuth rate and acceleration by differentiating
thv = diff(th)./dt;
tha = diff(thv)./dt;

subplot(211), plot(t(1:length(thv)),thv,'k')
grid
xlabel('Time (s)');
ylabel('Azimuth Rate (rad/s)')
subplot(212), plot(t(1:length(tha)),tha,'k')
grid
xlabel('Time (s)');
ylabel('Azimuth Accel (rad/s^2)')
pause

% Obtain elevation rate and acceleration by differentiating
phv = diff(ph)./dt;
pha = diff(phv)./dt;

subplot(211), plot(t(1:length(phv)),phv,'k')
grid
xlabel('Time (s)');
ylabel('Elevation Rate (rad/s)')
subplot(212), plot(t(1:length(pha)),pha,'k')
grid
xlabel('Time (s)');
ylabel('Elevation Accel (rad/s^2)')
```

Conventional servomechanism theory can be used to determine the lag errors if conventional loops are implemented. Using this method, the lag error can be determined from the angular rate of the target, ω_t (rad/s), the angular acceleration, $\dot{\omega}_t$ (rad/s^2), and the next derivative, often called jerk, $\ddot{\omega}_t$ (rad/s^3), etc.:

$$\delta_a = \frac{1}{K_o} + \frac{\omega_t}{K_v} + \frac{\dot{\omega}_t}{K_a} + \frac{\ddot{\omega}_t}{K_3} + \dots, \tag{14.8}$$

where the coefficients K_o, K_v, K_a, and K_3 are the servo error coefficients, the values of which increase with increasing loop gain and bandwidth.

Servos are classified according to the first coefficient that is finite in the loop design. A type 1 servo has a finite K_v (but infinite position error constant K_o), a type 2 has finite K_a but infinite K_v, etc.

The acceleration error coefficient is intimately connected to the closed-loop bandwidth of the servo. If the bandwidth is expressed in terms of the equivalent noise bandwidth, β_n (Hz), then

$$K_a = 2.5\beta_n^2 = \frac{0.63}{t_0^2}, \tag{14.9}$$

where t_o (s) is the equivalent averaging time for the tracking loop. The dynamic lag, ε_t (rad), will be

$$\varepsilon_t = \frac{\dot{\omega}_t}{K_a} = \frac{\dot{\omega}_t}{2.5\beta_n^2} = 1.6\dot{\omega}_t t_o^2. \tag{14.10}$$

This formula is applicable to azimuth, elevation, and range tracking lags with their corresponding acceleration components.

If thermal noise and dynamic lag are the primary sources of error and a real target acceleration component $\dot{\omega}_t = a_t/R$ exists, then using the relationship from equation (14.9), the total mean squared error can be written as

$$\sigma_\theta^2 = \frac{\theta_{3dB}^2 \beta_n}{2k_m^2 f_r \beta\tau SNR} + \frac{a_t^2}{6.25R^2\beta_n^4}. \tag{14.11}$$

Differentiating with respect to β_n and setting the result to zero makes it possible to find a closed-form equation for the optimum bandwidth, β_o (Hz), which minimizes σ_θ^2:

$$\beta_o = \left[\frac{1.28a_t^2 k_m^2 f_r \beta\tau SNR}{\theta_{3dB}^2 R^2}\right]^{1/5}, \tag{14.12}$$

where
     a_t = acceleration (real or geometric) (m/s^2),
     k_m = monopulse gain constant (typically 1.6),
     f_r = pulse repetition frequency (Hz),
     $\beta\tau$ = IF bandwidth and pulse width (see matched filter),
    SNR = single-pulse signal-to-noise ratio,
    θ_{3dB} = antenna 3 dB beamwidth (rad),
     R = crossing range (m).

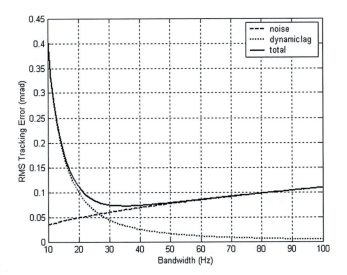

Figure 14.24 Typical noise and lag optimization.

This is taken from Barton (Barton 1988), but the formula that he derives appears to be incorrect by a factor of 2.

As an example of the optimization, consider a monopulse radar with an antenna beamwidth $\theta_{3dB} = 17.4$ mrad, $\beta\tau = 1$ and PRF, $f_r = 50$ kHz, tracking a target at a range of 3000 m to give a SNR = 10. For an aircraft accelerating at 30 m/s^2, the noise variance and lag contributions are shown in Figure 14.24 along with the total of the two contributions. For the parameters selected, the bandwidth that results in the lowest root mean square (RMS) tracking error is about 36 Hz (35.12 Hz from equation (14.10)). It can be seen that the form of the graph is very similar to that obtained for the α-β filter bandwidth optimization in Chapter 13.

```
% Tracking error as a function of servo bandwidth
% track_opt.m

% Variables
degrad = pi/180;
bw = 1;                        % beamwidth (deg)
km = 1.6;                      % for monopulse
betatau = 1;                   % matched filter
snr = 10;                      % signal-to-noise ratio
fr = 50e03;                    % prf
a = 30;                        % acceleration (m/s.s)
r = 300;                       % range (m)
band = (10:100);               %servo bandwidth (Hz)

nvar = (bw*degrad).^2*band./(2*km*km*fr*betatau*snr);
nlag = a*a./(6.25*r*r*band.^4);
```

```
plot(band,1000*sqrt(nvar),'k-',band,1000*sqrt(nlag),'k:',band,1000*sqrt(nvar+nlag),'k');
legend('noise','dynanic lag','total');
grid
%title('RMS Angle Tracking Error')
xlabel('Bandwidth (Hz)')
ylabel('RMS Tracking Error (mrad)')


num = 1.28*a.^2*km.^2*fr*betatau*snr;
den = (bw*degrad).^2*r.^2;
opt = (num/den).^0.2
```

In the crossing target case, the real acceleration term, a_t/R, can be replaced by the geometric acceleration term, $\dot{\omega}_t$, determined in Figure 14.23. Because this value will vary throughout the trajectory, the optimum bandwidth will also vary from a minimum at long range, peaking at the crossing point, and then decreasing again as the target recedes.

14.7.2 Millimeter Wave Tracking Radar

Tracking slowly moving targets in ground clutter is one of the more difficult tasks that must be accomplished by tracking radar systems. This is because the target radial velocity is often too low to discriminate using the Doppler shift, particularly if there is a breeze to stir up the vegetation and the clutter spectrum is nonzero.

This section discusses the development of an experimental millimeter wave tracking radar and some of the results that were obtained while tracking a small armored personnel carrier called a Buffel. The trial was conducted from a spur overlooking a valley with the vehicle driving along a dirt road through vynbos (scrub), shown in the photograph in Figure 14.25.

The radar transmits a 100 ns pulse with a peak power of 5 W at 94 GHz. A 100 mm aperture horn lens antenna containing a conscan mechanism provides an angle reference. A successive detection log amplifier (SDLA) followed by a conventional split-gate sample-and-hold circuit

Figure 14.25 Ground target tracking environment with the target Buffel APC as an inset.

provides sum and difference voltage outputs. The sum channel signal provides an indication of target amplitude and is demodulated to provide angle errors. The difference channel signal provides an estimate of the range tracking errors directly without AGC because the signals have already been compressed in the SDLA. The measured transfer functions showed a linear region of ±0.5° for the angle error and ±2 m for the range error. These signals are used as inputs into the conventional analog loops shown in Figure 14.26 to control the range and angles.

Typical gains and time constants for the azimuth positioner with an inertia of approximately $I_{az} = 1.5$ kgm² are listed. In the elevation case where the inertia, the servo amp characteristics, and the motor torque constants are different, a different set of filter parameters are calculated.

A portable measurement facility was constructed by mounting the radar on a positioner which was in turn mounted onto the roof of a custom modified Combi camper, as shown in Figure 14.27.

In this example, manual designation occurs in angles using visual feedback from the boresight TV camera, while range acquisition can be automatic, using a sawtooth input to the range filter, or manual using a joystick. As soon as the sum channel signal exceeds the constant false alarm rate (CFAR)-controlled detection threshold, the difference channel output of the split-gate circuit switches into the input of the range filter. This closes the range loop feedback path and range tracking commences. About 100 ms later the azimuth and elevation error signals output by the conscan demodulator are switched into the input of the angle filters to close the angle tracking loops, completing the transition from acquisition to track.

The range filter output voltage provides a measure of the target range, while the angle potentiometers, attached to the positioner, provide voltage outputs proportional to the azimuth and

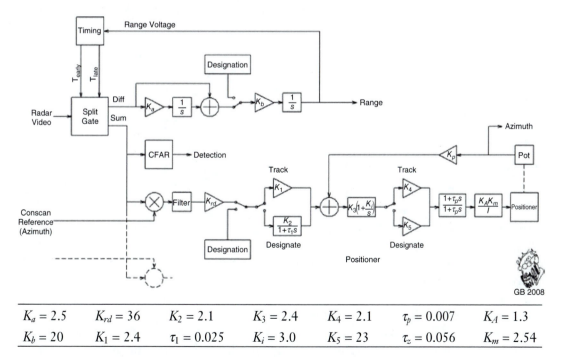

| $K_a = 2.5$ | $K_{rd} = 36$ | $K_2 = 2.1$ | $K_3 = 2.4$ | $K_4 = 2.1$ | $\tau_p = 0.007$ | $K_A = 1.3$ |
| $K_b = 20$ | $K_1 = 2.4$ | $\tau_1 = 0.025$ | $K_i = 3.0$ | $K_5 = 23$ | $\tau_z = 0.056$ | $K_m = 2.54$ |

Figure 14.26 Range and angle loop schematic diagram.

Figure 14.27 Millimeter wave tracking radar (a) mounted on the data acquisition vehicle, and (b) close-up showing radar and TV camera mounted on positioner.

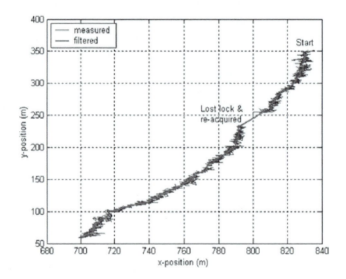

Figure 14.28 Measured and filtered position estimates of the vehicle.

elevation angles. During this tracking test, which lasted about 140 s, the data was sampled at 20 Hz and stored for later processing.

Figure 14.28 shows the measured position of the Buffel, converted into Cartesian space, as it drives slowly ($\approx$ 10 km/hr) along the dirt road. It can be seen that the measured data are extremely noisy and that the target was lost as it disappeared behind a spur, but was reacquired about 10 s later and tracked continuously for the remainder of the run.

Because the vehicle is traveling very slowly, and the range and angle servo bandwidths are very wide, it can be assumed that there will be no dynamic lag and the tracking noise is zero mean. Relatively noise-free position estimates are generated from the data by a pair of α-β filters, with gains of 0.05 and 0.0013, respectively. These produce a filter bandwidth of 0.24 Hz when the sample rate is 20 Hz, and so all but the lowest frequency noise is rejected. This position estimate can be subtracted from the measured position to obtain the glint noise for the target shown in Figure 14.29. The variances of the noise signals are $\sigma_x^2 = 1.9 \text{ m}^2$ and $\sigma_y^2 = 3.15 \text{ m}^2$, with the difference being due to the differences in the range and angular measurement noise and the relative dimensions of the target.

From the figure, it is obvious that this glint noise does not look much like ordinary Gaussian white noise—it is too spiky—and as such has much in common with the measurement noise out of many other sensor types. The probability density function (PDF) for these errors, in Figure 14.30, shows a distribution that has extremely long tails, extending out to 9 m (5 standard deviations) on either side. As discussed in Chapter 13, the tails of this glint error exceed the dimensions of the target by a large margin, although in this case some of the error may be contributions from adjacent sources of clutter.

14.8 TRACKING IN CARTESIAN SPACE

One method of maintaining the noise performance of the system while minimizing the dynamic lag is to operate with a wide angle-servo bandwidth and perform the tracking and smoothing in Cartesian space. As there are no geometric accelerations in Cartesian space, it is possible to reduce the filter bandwidth to less than 1 Hz. This determines the noise performance of the tracker. A simplified block diagram of this concept is shown in Figure 14.31.

As discussed in Chapter 13, even when tracking in Cartesian space, the optimum bandwidth is a function of the true target accelerations. It is therefore common practice to use adaptive tracking filters followed by fixed bandwidth angle and range servos.

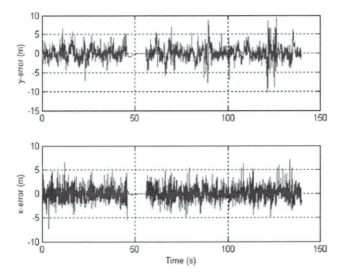

Figure 14.29 Measured glint noise in the *x* and *y* directions.

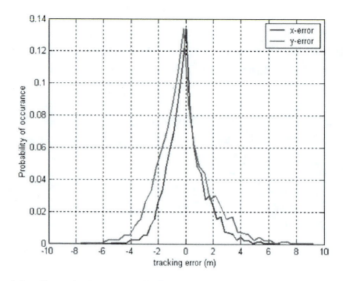

Figure 14.30 PDFs of the glint noise in the *x* and *y* directions.

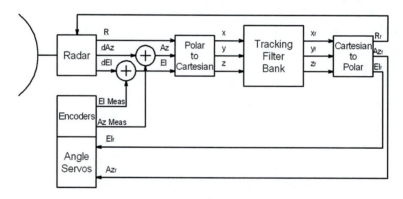

Figure 14.31 Tracking in Cartesian space.

Because the overall loop bandwidth is governed by the tracking filters operating in Cartesian space, the angle servo bandwidth can then be made as wide as necessary (typically 10 to 100 Hz for a real system) to minimize the lags due to geometric accelerations.

14.9 WORKED EXAMPLE: FIRE CONTROL RADAR

A conventional mechanically directed ship-borne FCR is to be designed to meet the following requirements:

14.9.1 Requirements

- Ships motion
  Radar on the deck of a ship 8 m above the water
  Roll ±25°, pitch ±10°, yaw ±5°
  Period about 5 s
- Designation
  From a surveillance radar (once per second)
  Elevation accuracy ±5°, azimuth accuracy ±2°
  Range accuracy ±25 m
  No velocity information
- Environment
  Up to sea state 5 (very rough, wave height 2.4 to 3.6 m)
  Rainfall up to 25 mm/hr
- Target types
  Fixed-wing aircraft (frontal radar cross section [RCS] 1 m^2 independent of frequency)
  Aircraft height >60 m above the sea
  Sea skimming missiles (frontal RCS 0.1 m^2 independent of frequency)
  Sea skimmer height 3 m above the sea
- Detection performance
  Probability of detection, $P_d = 0.95$
  Probability of false alarm, $P_{fa} = 10^{-6}$
  Detection time from receipt of designation (excluding slew) 0.5 s
  Detection range shown in Table 14.2
- Tracking performance
  Minimum tracking range, $R_{min} = 50$ m
  Range tracking accuracy (<1 m RMS)
  Angle tracking accuracy (<1 mrad RMS)
  Tracking aircraft directly overhead at $h > 100$ m
  Aircraft and missile velocity <280 m/s
- Safety constraints
  Average transmit power <100 W

14.9.2 Selection of Polarization

For low-flying aircraft over the sea we want to minimize the sea clutter. In calm seas at low grazing angles, the difference between the reflectivity for horizontal polarization and vertical

Table 14.2 Detection and tracking range requirement

Weather	Aircraft	Sea skimmer
Clear air	15 km	6 km
Rain 12.5 mm/hr	10 km	4 km
Rain 25 mm/hr	5 km	2 km

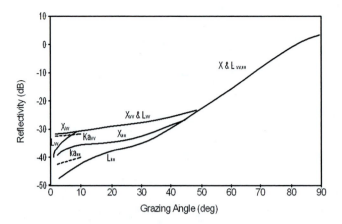

Figure 14.32 Sea clutter reflectivity as a function of grazing angle.

polarization exceeds 12 dB as can be seen in Figure 14.32. As the roughness increases the difference decreases. The radar will operate using horizontal polarization, and it will be assumed that the surface reflectivity, σ°, is -35 dB.

14.9.3 Positioner Specifications

The maximum combined roll and pitch angle is

$$\varphi_{\max} < \sqrt{25^2 + 5^2}.$$

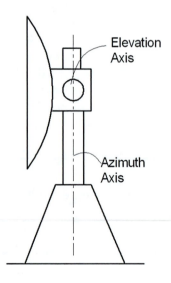

This does not exceed 30°. The antenna can be stabilized with respect to this tilt without resorting to a third axis; however, it can result in a significant rotation of the polarization.

The positioner/pedestal will be of the type elevation, θ, over azimuth, φ. θ is defined as positive up from the horizontal and φ is defined as positive counterclockwise from the x-axis.

For a mounting height of approximately 8 m above the sea, with the minimum tracking angle for a sea skimmer at $R = 50$ m ($\theta = -6°$), and a combined roll and pitch angle of 30°, a minimum angle of $-36°$ ($\theta_{\min} = -40°$) is required. To allow the antenna to track over the vertical, and to have time to slew around in azimuth without losing lock at the maximum combined roll and pitch angle requires $\varphi_{\max} = 90 + 30 = 120°$ (use 125°).

14.9.4 Radar Horizon

For $b_r = 8$ m and $b_t = 60$ m, the radar horizon is given by

$$d = 130\left(\sqrt{b_r(km)} + \sqrt{b_t(km)}\right) = 43\,km.$$

The radar horizon is not a consideration for this design.

14.9.5 Selection of Frequency

A reasonable maximum diameter for the antenna on a ship-borne radar is 1.5 m. This makes the beamwidth a function of frequency as tabulated in Table 14.3. A narrow beam decreases the effect of multipath and limits the area of clutter within the tracking gate, while attenuation increases with frequency (particularly in the rain). The minimum elevation angle when tracking a target at $h = 60$ m and $R = 15$ km is $\theta = 0.19°$ and the minimum tracking angle when tracking a sea skimmer at $h = 3$ m and $R = 50$ m is $\theta = -6°$.

Though it may be possible to minimize the effects of multipath, it is not possible to eliminate them when the radar is looking down at the target, as illustrated in Figure 14.33. Multipath has a major effect on tracking accuracy, as can be seen in Figure 14.34, which compares the measured data for two systems with different antenna beamwidths tracking the same target.

The best compromise is to use the narrowest beamwidth possible and to use multipath reduction techniques such as off-angle tracking. Application of these techniques can maintain an RMS elevation tracking accuracy of between 0.05 and 0.1 beamwidths (Currie et al. 1987). Assuming that an improvement down to 0.1 beamwidths can be achieved, the tracking accuracy is given in Table 14.4. This excludes the X-band option, as the tracking accuracy does not meet the accuracy criteria defined earlier.

14.9.6 Adverse Weather Effects

Typical attenuation as a function of frequency under different weather conditions is shown in Table 14.5.

Table 14.3 Antenna beamwidth for different frequencies

Frequency (GHz)	Band	Beamwidth (deg)
10	X	1.4
35	Ka	0.4
94	W	0.13

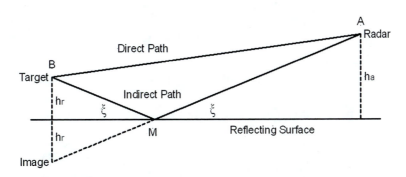

Figure 14.33 Multipath propagation for the radar looking down at a target.

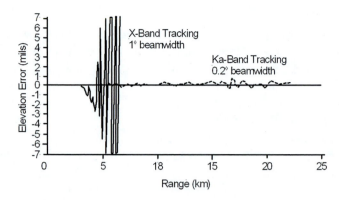

Figure 14.34 Effect of beamwidth on low-angle tracking accuracy [adapted from Currie and Brown (1987)].

Table 14.4 Tracking accuracy with multipath mitigation techniques

Band	Beamwidth (deg)	RMS tracking (mrad)
X	1.4	2.4
Ka	0.4	0.69
W	0.13	0.22

Table 14.5 Atmospheric attenuation at different frequencies and in different weather conditions

Band	Clear (dB/km)	Rain 12.5 mm/hr (dB/km)	Rain 25 mm/hr (dB/km)
X	0.02	0.25	1
Ka	0.15	3	7
W	0.3	7	12

14.9.7 Required Single-Pulse SNR

The required SNR to achieve the specified P_d and P_{fa} is a function of the target distribution and its fluctuation characteristics. For a nonfluctuating target this is 13.6 dB, as determined using the curves in Figure 14.35.

Swerling 1: Many independent scatters if similar RCS. This results in slow fluctuations with time.

Swerling 2: One major scatterer and many smaller scatterers. This is typical of an aircraft nose on. This results in fast fluctuations with time.

The additional SNR required to achieve these probabilities of the target fluctuating is determined from Figure 14.36. It will be 10.4 dB. The single-pulse SNR required is thus $13.6 + 10.4 = 24.0$ dB.

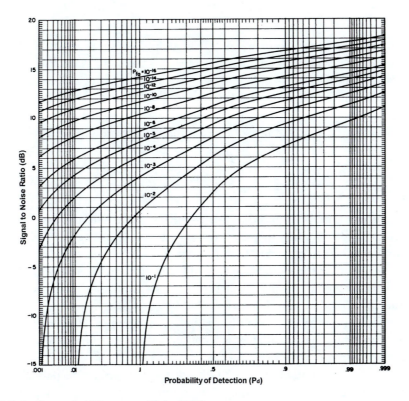

Figure 14.35 Detection probability curves (Blake 1986).

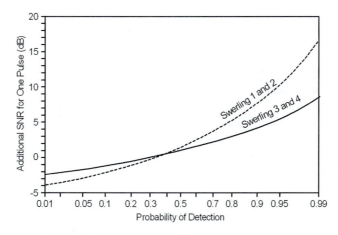

Figure 14.36 Fluctuating target effects [adapted from Skolnik (1980)].

14.9.8 Tracking Gate Size

Ideally the tracking gate size is matched to the target length to maximize the amount of power returned to the radar. A typical jet fighter is 15 m long, so a gate size between 15 and 20 m would be ideal. In this example, a gate size of 20 m has been selected as appropriate. If the gate size is made larger, then the amount of sea clutter would be increased without increasing the target return. However, it does complicate the target acquisition process, as the designation accuracy is only ±25 m, so at least three gates would be required to span the uncertainty.

14.9.9 Signal-to-Clutter Ratio

To make the analysis as simple as possible, it is assumed that the reflectivity of the sea is the same in the X-, Ka-, and W-band. It is also safe to assume that the same signal-to-clutter ratio (SCR) is required for detection, as in the SNR.

For a gate size of 20 m, the illuminated area will be a function of the antenna beamwidth and the range. Assuming that the antennas for all three options are the same size, then the beamwidth, and hence the illuminated area, will be as listed in Table 14.6.

For an average reflectivity $\sigma^{o} = -35$ dB in all cases, the clutter RCS, which is a product of the reflectivity and the illuminated area, is listed in Table 14.7. For an aircraft RCS of 1 m^2 (0 dBm^2) or a sea skimmer with an RCS of 0.1 m^2 (−10 dBm^2), an extra 20 to 30 dB of signal is needed to achieve the required SCR of 24 dB (for the required single-pulse detection probability). It should be remembered that integration cannot be used to improve the SCR because, unlike the white thermal noise used for target detection, clutter is correlated, so integration will not be as effective. The primary difference between the targets and the clutter is that in most cases the target has a significant radial velocity while the clutter is static (or slow moving in high seas).

Table 14.6 Illuminated sea area as a function of frequency

Band	Beamwidth (deg)	Area (6 km) (m^2)	Area (15 km) (m^2)
X	1.4	2928	7320
Ka	0.4	840	2100
W	0.13	264	660

Table 14.7 Clutter RCS as a function of frequency

Band	RCS (6 km) (dBm^2)	RCS (15 km) (dBm^2)
X	−0.33	3.6
Ka	−5.8	−1.8
W	−10.8	−6.8

14.9.10 Moving Target Indicator

Some form of MTI will have to be implemented to achieve the subclutter visibilities listed in Table 14.8 and Table 14.9. This can be achieved using a delay line canceller; however, because the ship will be moving, the clutter will also have an effective velocity, so this is not a good technique to use. A better alternative is to take more samples, to window them, and to use an FFT to isolate moving from "static" targets.

Ideally a block of I and Q outputs of the phase sensitive detector at each range will be processed through a complex FFT to produce a spectrum that will separate the target Doppler from that of the clutter. With a Hamming window, a static return rejection of 42 dB can be obtained. However, because the PRF must be sufficiently slow to ensure that measurements are unambiguous in range, it is unlikely that they will also be unambiguous in velocity.

14.9.11 Pulse Repetition Frequency

To be unambiguous in range out to 15 km, the maximum allowed PRF, f_p (Hz), is

$$f_p = c/2R_{max}$$
$$= 10\,kHz.$$

Using the Nyquist criterion, this is unambiguous in velocity up to a Doppler frequency of $f_d = 5$ kHz. Therefore the maximum unambiguous radial velocity is given by

$$v_r = f_d\lambda/2 = f_p\lambda/4,$$

and listed in Table 14.10 for the three frequency options.

If the output is sampled at a rate slower than the Doppler frequency, the signal will be aliased, or folded down into the unambiguous velocity range, and it is possible that the folded target

Table 14.8 Subclutter visibility requirements for sea skimmer

Band	Clutter RCS at 6 km (dBm^2)	Sea skimmer RCS (dBm^2)	Subclutter visibility for SCR = 24 dB (dB)
X	−0.33	−10	33.67
Ka	−5.8	−10	28.2
W	−10.8	−10	23.2

Table 14.9 Subclutter visibility requirements aircraft

Band	Clutter RCS at 15 km (dBm^2)	Aircraft RCS (dBm^2)	Subclutter visibility for SCR = 24 dB
X	3.6	0	27.6
Ka	−1.8	0	22.2
W	−6.8	0	17.2

Table 14.10 Unambiguous velocity as a function of frequency

Band	Unambiguous velocity, v_r (m/s)
X	75
Ka	21.5
W	8

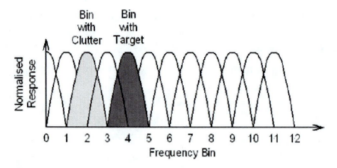

Figure 14.37 Separating target and clutter returns in a Doppler filter bank.

velocity will fall directly into the bin (or bins) containing the clutter. To account for this possibility, the PRF can be changed on a block-by-block basis. This will shift the relative positions of the clutter and the folded target Doppler frequencies until a PRF is found in which the target is not in the same bin as the clutter. It is also quite likely that the clutter velocity spread will exceed 8 m/s, so there may not be an acceptable PRF at W-band.

Because both the clutter and the ship may be moving, the actual gate in which the clutter will be found will not be around DC, as can be seen in Figure 14.37, and it will need to be tracked so that it is not mistaken for a real target.

14.9.12 Search Requirement

The radar needs to search a volume $10° \times 4° \times 50$ m in less than 0.5 s to meet the specification. Typical search patterns include vertical and horizontal raster scans and spirals, shown in Figure 14.38.

Option (a) requires that the positioner change direction more often than option (b), and that takes time as well as using more energy, and there is more of an overlap between scans with option (c), which also wastes time, so option (b) is probably the best choice. In reality, the probability of the target appearing at a specific elevation angle is not uniform, and this distribution should be considered when formulating a search strategy.

For a 50% beam overlap between scans, the total distance that must be traveled to search the area is determined from the number of vertical scans required to cover the 4° in azimuth. The results of this comparison are tabulated in Table 14.11. Because this must occur in 0.5 s or less

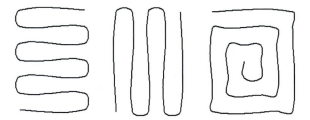

Figure 14.38 Search pattern options.

Table 14.11 Total search scan distance as a function of frequency

Band	Vertical scans (*n*)	Total distance (deg)	Scan speed (deg/s)
X	$4/1.4 \times 1.5 = 4.28$ [5]	50	100
Ka	$4/0.4 \times 1.5 = 15$ [15]	150	300
W	$4/0.13 \times 1.5 = 46$ [46]	460	920

Table 14.12 Hits per scan for the different frequency options

Band	Hits per scan (*N*)
X	140
Ka	13.3
W	1.4

to meet specifications, the scan speed can be calculated. This assumes that changing direction is instantaneous.

Hits per scan (the number of pulses that illuminate the target as the beam passes over it) can then be calculated from the required scan rate and the antenna beamwidth. These numbers are listed in Table 14.12.

From the table, it is obvious that there are not sufficient hits at either Ka-band or W-band to generate an FFT with sufficient bins, therefore the overlap must be decreased to meet the specification. The revised figures for scan speed and hits per scan are listed in Table 14.13, if the overlap is decreased to 0% on the assumption that there will still be enough gain in the overlap to ensure detection.

There is the added burden that the target, moving at 280 m/s, may have moved up to 140 m during the search. This requires that a large bank of gates spanning the original 50 m uncertainty plus 140 m on each side (330 m) must be examined. At least 17 gates will be required to span this range.

In terms of the detection of moving targets, it is now feasible to use a 16-point FFT at Ka-band or a 128-point FFT at X-band on each of the 17 range gates required during the search phase.

Table 14.13 Scan distance and hits per scan for 0.5 s detection time

Band	Vertical scans (n)	Total distance (deg)	Scan speed (deg/s)	Hits per scan (N)
X	3	30	60	233.0
Ka	10	100	200	20
W	31	310	620	2.1

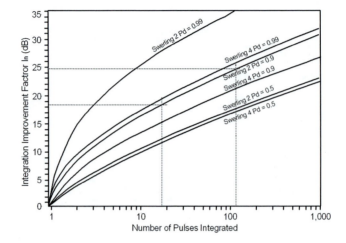

Figure 14.39 Integration improvement factor.

14.9.13 Integration Gain

The FFT process is the equivalent of a coherent integrator, which will produce gains of $10\log_{10}(N)$. However, because the target is fluctuating, this effect will vary with the observation time compared to the fluctuation rate and the probability of detection.

The graph shown in Figure 14.39 is used even though it is not quite correct, as it assumes postdetection integration. For a $P_d \approx 0.95$ (use 0.9) and a Swerling 2 target, the integration gain is about 18.5 dB for the 16-point FFT at Ka-band, and for the 128-point FFT at X-band, the integration gain is about 25 dB. Although these are higher than the theoretical maxima of 12 and 21 dB, respectively, the extra is taken care of in the 10.4 dB increase in SNR to accommodate fluctuation loss required for detection described earlier.

14.9.14 Matched Filter

It is assumed that the pulse is rectangular and that the filter is made up of five cascaded tuned band-pass sections. For a pulse width of 20 m (133 ns), the bandwidth, β, is 5 MHz using the optimum relationship presented in Table 14.14.

14.9.15 Transmitter Power

The average power allowed is 100 W, the pulse width is 133 ns, and the PRF is 10 kHz, making the duty cycle 0.133%. This makes the peak transmitted power, P_t, 75 kW. This can be achieved using a magnetron or traveling wave tube (TWT) at Ka- or X-band.

14.9.16 System Configuration

A magnetron can be used in a (pseudo) coherent radar configuration if the COHO is primed by the random start phase of the magnetron on every pulse, as illustrated in the system block diagram shown in Figure 14.40.

Table 14.14 Matched filter characteristics

Input signal	Filter	Optimum $\beta\tau$	Loss in SNR compared to matched filter (dB)
Rectangular pulse	Single tuned circuit	0.4	0.88
Rectangular pulse	Two cascaded tuned circuits	0.613	0.56
Rectangular pulse	Five cascaded tuned circuits	0.672	0.5

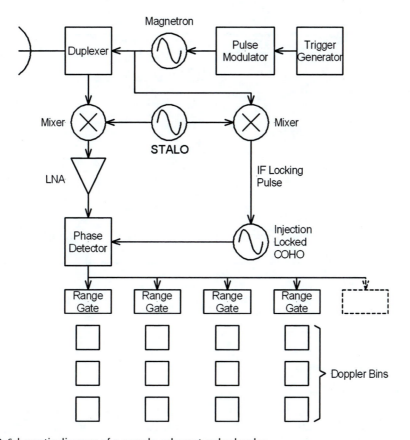

Figure 14.40 Schematic diagram of a pseudo-coherent pulsed radar.

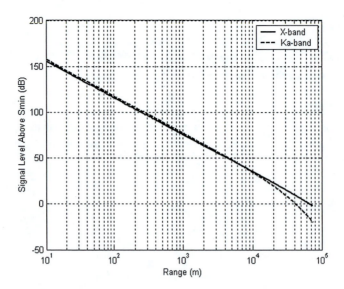

Figure 14.41 Radar performance, aircraft target, and clear air.

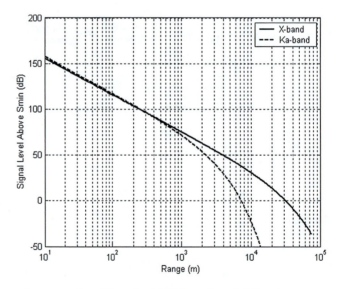

Figure 14.42 Radar performance, aircraft target, and 12.5 mm/hr rainfall.

14.9.17 Free-Space Detection Range

Applying the radar range equation when there is significant attenuation is best achieved using MATLAB. The graphs shown in Figure 14.41 and Figure 14.42 can be produced by adjusting the values for atmospheric attenuation (shown in bold) and switching between the RCS values for the aircraft and the sea skimmer.

```
% Fire Control Radar

% Variables
r=(10:10:75000);      % Range (m)
sigmat = 1;           % Aircraft (sqm)
sigmas = 0.1;             % Sea skimmer (sqm)
dant = 1.5;           % Antenna Diameter (m)
pt = 75e03;           % Transmit power (W)
lamx = 30e-03;            % Wavelength (m)
lamk = 8.6e-03;
nfxdb = 3;            % Noise fig (dB)
nfkdb = 4.5;
tau = 133e-09;            % Pulse-width (s)
k = 1.38e-23;             % Boltzmann (Js)
t = 290;
nintxdb = 25;             % Integ improve(dB)
nintkdb = 18.5;
ltxx = 3;             % Tx plumbing loss (dB)
ltxk = 4;
lscanx = 3.2;             % 2D Scan loss (dB)
lscank = 3.2;
lmiscx = 3;           % Misc loss (dB)
lmisck = 3;
alphax = 1;           % Atmos attn (dB/km) atten = 0.02, 0.25, 1
alphak = 7;           % Clear, 12.5, 25mm/hr atten = 0.15, 3, 7

SNx = 13.6+10.4;      % S/N for Pd=0.95, Pfa=1.0e-06
SNk = 13.6+10.4;

% Calculate the antenna gain
gxdb = 10*log10(4*pi*0.7*pi*dant*dant/(4*lamx*lamx));
gkdb = 10*log10(4*pi*0.7*pi*dant*dant/(4*lamk*lamk));

% Propagation factor
propxdb = 10*log10(lamx*lamx/(4*pi).^3);
propkdb = 10*log10(lamk*lamk/(4*pi).^3);

% Transmit power
ptdb = 10*log10(pt);

% Target RCS
sigmatdb = 10*log10(sigmat);     % Aircraft
sigmasdb = 10*log10(sigmas);     % Sea skimmer
```

```
prxdb = ptdb+2*gxdb+propxdb+sigmasdb-ltxx-lscanx-lmiscx-40*log10(r)-
2*alphax*r/1000;
prkdb = ptdb+2*gkdb+propkdb+sigmasdb-ltxk-lscank-lmisck-40*log10(r)-
2*alphak*r/1000;
% Threshold for detection
thermxdb = 10*log10(k0.672/tau);
thermkdb = 10*log10(k0.672/tau);

threshxdb = (SNx + thermxdb + nfxdb --nintxdb)*ones(size(r);
threshkdb = (SNk + thermkdb + nfkdb --nintkdb)*ones(size(r);

semilogx(r,prxdb-threshxdb,r,prkdb-threshkdb);
grid
%title('FIRE CONTROL RADAR: MAGNETRON: SKIMMER: Clear Air')
xlabel('Range (m)')
ylabel('Signal Level Above Smin (dB)')
axis([0,100000,-50,200]);

lx=find((prxdb-threshxdb)<0);
lk=find((prkdb-threshkdb)<0);
min(r(lx))
min(r(lk))
```

The detection ranges for the various options are listed in Table 14.15, which shows that only the X-band radar meets the specified range criteria of 15, 10, and 5 km for aircraft detection. For the sea skimmer, both the X-band and the Ka-band radars meet the detection requirements of 6, 4, and 2 km.

14.9.18 Effects of Multipath on Aircraft Detection

Figure 14.43 to Figure 14.46, generated using VCCALC (Fielding and Reynolds 1987), show that the constructive and destructive interference caused by multipath can reduce the signal level to below 0 dB under some circumstances. The implication of this is the possibility of missed detections even though the free space signal exceeds S_{min}.

Table 14.15 System performance

Target	Condition	X-band range (km)	Ka-band range (km)
Aircraft RCS = 1 m^2	Clear	67.3	42.2
	12.5 mm/hr	31.6	7.2
	25 mm/hr	14.6	3.9
Sea skimmer RCS = 0.1 m^2	Clear	40.3	29.5
	12.5 mm/hr	22.9	6.1
	25 mm/hr	11.6	3.3

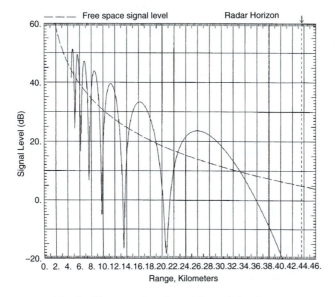

Figure 14.43 Multipath effects on aircraft tracking at X-band.

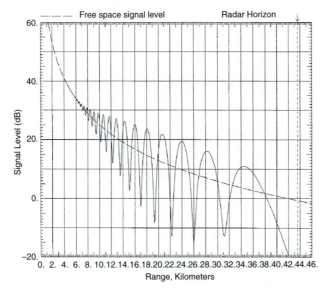

Figure 14.44 Multipath effects on aircraft tracking at Ka-band.

The depth of the fades is more pronounced at X-band, where detection may not occur between 13 and 14 km, and again at just below 10 km for aircraft targets. None of the fades at Ka-band dip so low at the range of interest for this radar.

Because the sea skimmer is flying so low, the fading structure is completely different, and at X-band works in favor of the radar, where constructive interference increases the signal amplitude beyond the free space level at ranges below 7 km. At Ka band, however, a deep fade appears between 4 and 6 km which could affect the detection probability at that range.

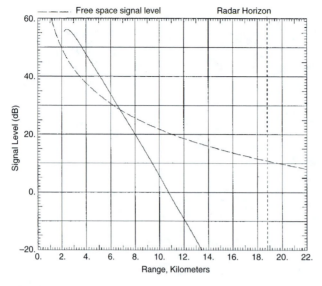

Figure 14.45 Multipath effects on sea skimmer tracking at X-band.

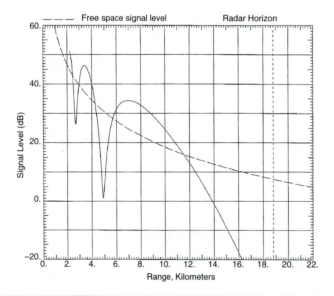

Figure 14.46 Multipath effects on sea skimmer tracking at Ka-band.

14.9.19 Detection Threshold and CFAR

Changes in radar characteristics with time (aging) and changes in the target background characteristics mean that a fixed detection threshold is not practical and a CFAR processor is required. Because no velocity information is available, the radar must transmit a block of pulses and examine all the Doppler gates in all of the range gates over the designated range of the target. Therefore

Table 14.16 Range/Doppler gate matrix showing the cell containing the target and the shaded region which highlights the bins used in the CFAR process

		Range gates									
	Bin	1	2	3	4	5					N
	1										
	2			X							
	3										
Doppler bins	4										
	125										
	126										
	127										
	128										

a cell-averaging CFAR processor that averages the returns from a particular Doppler bin across all of the range gates that span the designated range, as shown in Table 14.16, will have to be used. This averaging process excludes the cell under test, which proceeds sequentially from gate 1 to gate N (it is shown in gate 3 in the table).

14.9.20 Transition to Track

The mechanically operated FCR antenna moving at 60°/s (X-band) or 200°/s (Ka-band) during the search phase is unable to decelerate in time to keep the antenna pointing at the target. Therefore the angles, range, and Doppler bin at which detection takes place is recorded as the antenna sweeps past it and then returns to that designation more slowly.

If the target is not detected, a slow search is conducted in angles while the range gates broaden their search until the target is found or the system times out. The angular rates of this new search are such that the antenna can stop while still illuminating the target, and a transition to track mode is made. This transition process involves closing the Doppler loop, followed by the range loop and finally the angle loops.

14.9.21 Target Tracking

A split tracking gate straddles the sum channel range gate in which the detection took place, and early and late gate measurements are made only in the Doppler bin in which detection took place. As the Doppler is generally ambiguous, it cannot be used to prime the range filter, so a long (typically 50 ms or so) period is used to obtain a good range rate estimate to prime the range tracking filter.

The range tracking loop then closes using early and late gate signals from the appropriate Doppler bin, and automatic tracking in range occurs using the difference between these two signals to keep the gates centered on the target. This range tracking process is illustrated in Figure 14.47.

The angle tracking loops are then closed. The azimuth and elevation error signals from the same Doppler bin as is used for range tracking is sampled simultaneously with the sum channel signal. These error signals drive the angle servos to keep the antenna pointed at the target being tracked. Target dynamics that dictate the real and apparent acceleration and tracking loop bandwidth determine the tracking accuracy for both range and angles.

Assuming that the target is traveling at 280 m/s and flies past the radar at a height of 60 m and a ground range of 60 m, the peak velocities and accelerations experienced in polar coordinates can easily be determined using the MATLAB code discussed earlier in this chapter, as shown in Figure 14.48. The azimuth acceleration peaks at about 14 rad/s^2.

Assuming that thermal noise and dynamic lag are the primary sources of error, the optimum bandwidth, β_o (Hz), is

$$\beta_o = \left[\frac{1.28 \dot{\omega}_t^2 k_m^2 f_r \beta \tau SNR}{\theta_{3dB}^2} \right]^{1/5},$$

where

$\dot{\omega}_t$ = geometric acceleration at crossing (rad/s^2) [14],
k_m = monopulse gain constant (typically 1.6) [1.6],
f_r = pulse repetition frequency (Hz) [10 kHz],
$\beta \tau$ = IF bandwidth and pulse width [0.672],
SNR = pulse signal-to-noise ratio [10],
θ_{3dB} = antenna beamwidth (rad) [0.024 or 0.007].

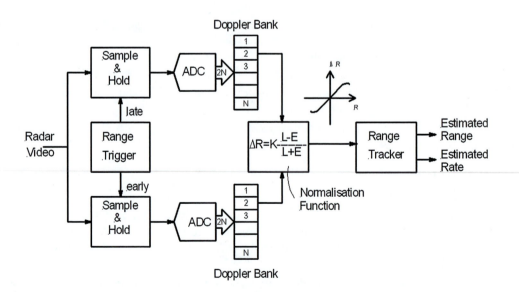

Figure 14.47 Schematic diagram of range tracking configuration.

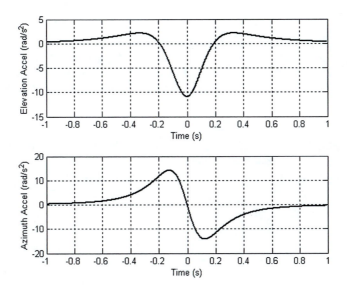

Figure 14.48 Angular acceleration for a crossing target.

The optimum angle servo bandwidth is 150 Hz for the X-band tracker and 245 Hz for the Ka-band unit. This is very high, as the specified crossing range is extremely short. If the bandwidth is expressed in terms of the equivalent noise bandwidth, β_n, the acceleration coefficient, K_a, can be calculated to be

$$K_a = 2.5\beta_n^2 = 0.63/t_o^2,$$

where t_o is the equivalent averaging time for the tracking loop.

The RMS angular tracking error caused by lag is

$$\varepsilon_a = \frac{\dot{\omega}_a}{K_a} = \frac{\dot{\omega}_a}{2.5\beta_n^2} = \frac{14}{2.5 \times 150^2} = 0.25 \text{ mrad},$$

and the RMS error caused by thermal noise can also be calculated to be

$$\sigma_\theta = \frac{\theta_{3dB}}{k_m\sqrt{2SNRf_r/\beta_n}} = \frac{0.024}{1.6\sqrt{2 \times 10 \times 10^4/150}} = 0.41 \text{ mrad}.$$

The alternative is to use a large angle servo bandwidth (≈ 350 Hz) that will cope with the geometric accelerations with minimal lag and to perform most of the filtering in Cartesian space. The bandwidth of such a filter will typically be between 1 and 4 Hz, which is sufficiently wide to account for target accelerations of up to 6g for the aircraft.

14.10 References

Barton, D. (1976). *Radar Systems Analysis*. Norwood, MA: Artech House.

Barton, D. (1988). *Modern Radar Systems Analysis*. Norwood, MA: Artech House.

Blake, L. (1986). *Radar Range-Performance Analysis*. Norwood, MA: Artech House.

Currie, N. and Brown, C. (1987). *Principles and Applications of Millimeter-Wave Radar*. Norwood, MA: Artech House.

Fielding, J. and Reynolds, G. (1987). *VCCALC: Vertical Coverage Calculation Software and Users Manual*. Norwood, MA: Artech House.

Morris, G. (1988). *Airborne Pulsed Doppler Radar*. Norwood, MA: Artech House.

Skolnik, M. (1970). *Radar Handbook*. New York: McGraw-Hill.

Skolnik, M. (1980). *Introduction to Radar Systems*. Tokyo: McGraw-Hill Kogakusha.

15

Radio Frequency Identification Tags and Transponders

15.1 PRINCIPLES OF OPERATION

Transponders were originally electronic circuits that were attached to some item whose position or presence was to be determined. The transponder operated by responding to a request received from an interrogator, either by returning some data from the transponder, such as an identity code, or returning the original properties of the signal received from the interrogator with a minimum time delay.

Because the interrogator signal is much stronger than the returned signal, the former would swamp the latter unless some characteristic of the response were different. This difference can usually be achieved by separating the interrogation and response temporally or by changing the frequency. It is also possible to encode the response with some form of spread-spectrum modulation that can be decoded by the receiver.

15.2 HISTORY

Transponders were developed for aircraft in World War II for identification friend or foe (IFF) application. Both commercial and military aircraft still use this technology and most air traffic control centers rely on the range and altitude information returned by commercial aircraft rather than raw radar information.

Since the first Gulf War, during which unacceptably large numbers of personnel were killed by their own side, transponder technology called the Battlefield Identification System (BIS) has been developed for ground vehicles. It relies on interrogator-transponder technology operating at 38 GHz.

Another important application of the transponder has been in the measurement of distance. Here, the interrogator sends a signal to the transponder which immediately responds on another frequency. By using the time of flight, the range can be determined with great accuracy. This technique is more accurate than using the skin echo, as the ultimate signal-to-noise ratio (SNR) remains very high. A tellurometer (see Chapter 5) uses this technique to measure ranges of hundreds of kilometers to an accuracy of a few centimeters.

The Global System for Mobile communications (GSM) network is being used increasingly as a form of transponder system, where the position of each phone can be determined with reasonable accuracy from field strength readings at various base stations. An application of this technology includes monitoring traffic flow conditions on major roads by tracking phone positions over long periods.

Transponder systems have become major players in the field of electronic identification. Within this application it is necessary to make the transponder as cheap as possible and to build the sophistication into the interrogator (reader). However, because it is not really feasible to perform low-cost frequency translation, these transponders (or tags) have given up the ability to supply range, and time slice the interrogation and response cycle to avoid swamping the response.

15.3 SECONDARY SURVEILLANCE RADAR

Secondary surveillance radar (SSR) is the modern term used in place of the old term IFF. Its purpose is to improve the ability of ground controllers to detect and identify aircraft, while also providing other information like flight level (barometric altitude). An SSR continuously transmits interrogation pulses from an antenna mounted above the normal radar antenna, as shown in Figure 15.1.

The transponder on board an aircraft continuously listens for the interrogation and sends back a reply containing information which depends on the mode of the interrogation. The aircraft is then shown on the air traffic control display as a "tagged icon" at the correct bearing and range. If the aircraft does not have a transponder, it is still displayed, but without the additional information.

Figure 15.1 Air traffic control radar with SSR antenna mounted above it.

Most of the modes used by SSR are reserved for the military, but some are dual purpose and some are for civilian aircraft, as follows:

- Mode 1: military two-digit mission code.
- Mode 2: military four-digit code.
- Mode 3/A: military and civilian four-digit identification code known as a squawk code. It is assigned by the air traffic controller.
- Mode 4: military reply code, depends on a 32-bit encrypted interrogation.
- Mode 5: military code, provides an encrypted message similar to that provided by mode S.
- Mode C: military and civilian, provides a 10-bit reply containing the aircraft barometric altitude.
- Mode S: military and civilian, contains a data packet standard for both interrogation and response. Each aircraft can be assigned a unique 24-bit identification for selective interrogation.

15.3.1 Interrogation Equipment

The air traffic control ground station consists of an active radar, known as a primary surveillance radar (PSR), and the interrogator system, known as a secondary surveillance radar (SSR), which is generally mounted above the PSR and pointing in the same direction. The SSR transmits interrogations continuously, between 400 and 500 times per second, at a frequency of 1030 MHz as it rotates. The interrogator includes mode information that can be interpreted by the transponder. Modes 1, 2, 3/A, and C are in common use. Mode 1 is used by military aircraft during different phases of a mission, while mode 2 is used to identify a specific mission. Mode 3/A identifies all the aircraft uniquely within the SSR coverage area, while mode C requests the aircraft altitude. Mode 4 is used by military aircraft for IFF purposes. Mode S is not a general broadcast, but addresses specific aircraft using their unique identification code and is used to reduce channel congestion.

The SSR transponder answers the interrogation at a frequency of 1090 MHz. This information and the echo returned by the PSR are processed and the aircraft information is displayed along with the synthesized radar return, as shown in Figure 15.2.

15.3.2 Transponder Equipment

The transponder consists of an L-band antenna, typically mounted on the bottom of the fuselage, and a transponder mounted in an avionics rack or on the instrument panel. Information from the aircraft static-pressure transducer is provided to the transponder through an altitude encoder. The transponder includes a facility to enter a four-digit beacon code and a mode selection switch. When the aircraft receives an interrogation, the transponder sends its reply at 1090 MHz.

15.3.3 Operation

The interrogator transmits three pulses, each 0.8 μs long, known as P1, P2, and P3. The time between P1 and P3 determines the mode, while P2 is used for sidelobe suppression. Mode 3/A uses a P1–P3 spacing of 8.0 μs and requests the beacon code that was assigned to the aircraft by the air traffic controller. Mode C uses a 21 μs spacing and requests the aircraft barometric

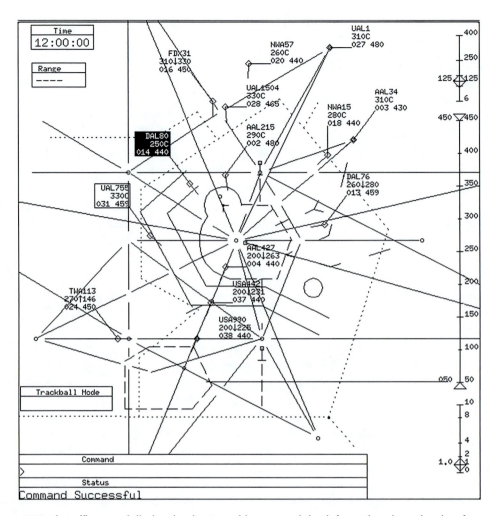

Figure 15.2 Air traffic control display showing tagged icons containing information about the aircraft.

altitude. Mode 2 uses a 5 μs spacing which is a request to military aircraft to respond with their military identification.

The transponder will only reply to the modes that it understands. After receiving the P3 pulse, it waits 3.0 μs before responding. Each response consists of 15 time slots, each 1.45 μs long, using on/off modulation of a 0.45 μs pulse in each slot. The pulses at either end, separated by 20.3 μs, are framing pulses and are always transmitted, while the remaining pulses contain information dependent on the mode. The range to the aircraft is determined by the round-trip time from the broadcast of the interrogation message to the receipt of the reply, that is why it is important that the delay in response be exactly 3.0 μs.

15.3.4 SSR Issues

15.3.4.1 Sidelobe Problems   At short range, the transponder signal in the antenna sidelobes can be sufficiently high to trigger a transponder response. This can cause ghosting, where the aircraft position is reported at two different bearings. If the leakage is severe, then a complete ring of returns may occur.

To combat this problem, the P2 pulse is transmitted 2 μs after the P1 pulse, not through the directional SSR antenna, but through an omnidirectional antenna. The power transmitted by this antenna is adjusted so that the power density is higher than any of the sidelobes, but lower than that of the main lobe. The transponder compares the signal levels of the P1 and P2 pulses, and if P1 is stronger, it knows that it has received a signal from the SSR antenna and will reply, otherwise it remains silent.

15.3.4.2 Congestion As the price of transponder equipment has declined, and with more stringent requirements that aircraft carry these devices, all commercial and most private aircraft now have them. In conjunction with almost continuous radar coverage, it is now common that one SSR will receive responses from aircraft that are triggered by another station. These are known as false replies, unsynchronous in time (FRUIT), and can result in severe cluttering of the air traffic control display. To get an idea of the number of aircraft involved, Figure 15.3 shows a snapshot of all of the transponder identified aircraft in the air around the United States prior to the 9/11 terrorist attack.

The use of mode S, in which each aircraft is addressed by its own completely unique 24-bit registration number, is one method that has been applied to reduce the problem. This is known as deFRUITing.

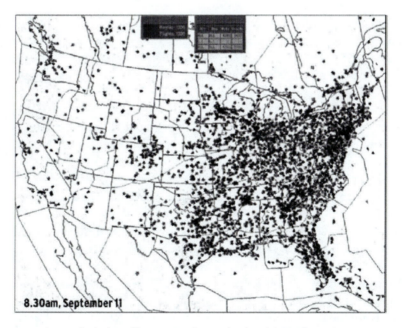

Figure 15.3 Transponder tracked air traffic over North America (Daniel 2002).

15.4 RADIO FREQUENCY IDENTIFICATION SYSTEMS

As a catchall name for all commercial interrogator-transponder systems, radio frequency identification (RFID) is now ubiquitous throughout the world. The following list considers a small number of these:

- Electronic article surveillance (EAS), in which the presence of an object is detected using an attached tag, is probably the fastest growing sector, with production approaching 10 billion units per year. Because of the size of the market, there is little incentive for standardization among manufacturers, and this can be confusing for end users.
- Animal tracking started with the use of magnetic tags to identify slaughter animals, but this has now extended to the labeling of pets. Tiny transponders, about the size of a grain of rice, are injected under the skin of a pet where they can be read by handheld scanners from a distance of a few inches.
- Shipping container and railcar tracking was one of the early successes of transponder technology and is fast becoming a global system. Electromagnetic coupled transponders are mounted on containers and can be read at speed using trackside interrogators.
- Vehicle access to garages has been available for many years and has become even more common with improved communications and encryption.
- Personnel access has also long been facilitated by radio frequency (RF) transponder technology, but until recently this has not had much market penetration because of its short operational range and the low cost and good reliability of the alternative swipe card technology. One exception to this trend is keyless personnel access to vehicles, which has grown significantly in recent years because it provides a good antitheft capability.
- Production control led by the automotive industry uses tags to identify and track components during the manufacturing process.
- Sports timing applications, particularly for long-distance running, has been very successful in combating cheating and improving the accuracy of race timing. This technology typically uses a small 125 kHz transponder attached to the laces of a runners shoe and is read by a mat as he runs over it. Trials in South Africa using 900 MHz units that can be read using an array of overhead antennas have shown that it is possible to generate timing accuracies of better than 0.2 s.
- Flat transponders about the size of postage stamps attached to documents provide authentication and access control for confidential material. Novel applications include enabling or preventing photocopying.
- Dairy tagging is now widely practiced throughout the world. Ear tags worn by all animals control access to feed stalls and identify individuals during milking so that a complete record of milk production efficiency can be obtained automatically.
- Petrol and chemical dispensing using RFID tags ensures that tankers only fill from the correct bowser. Because of the low cost of the hardware, it is now possible to extend this service to the corner petrol station. A tag is mounted next to the filler cap of the vehicle fuel tank where it ensures that the correct fuel is used, as well as providing input to an accounting and invoicing system.
- Toll roads are the latest industry to embrace the use of RFID technology in an attempt to combat congestion. This application is discussed in more detail later in this chapter.

- Transport systems can be monitored for efficiency by including tags in selected items that pass through. This can be used by postal and courier companies to ensure optimum efficiency. Tags are becoming so cheap as to be viable for use as replacements for postage stamps.
- Any technology that can improve the reliability and accuracy of automatic routing of airline baggage would save the industry millions of dollars per year. RFID systems are starting to replace paper tags for this process, though there are complications caused by other tags within the container.
- Criminal monitoring using bracelets or anklets that communicate with a nearby interrogator have been used to ensure that detainees remain under house arrest. This application is also discussed later in this chapter.
- Smart appliances, including refrigerators and grocery cupboards, that read the tags on products and warn the user of approaching use-by dates or even order fresh supplies via Internet access are now available. This technology can be extended to almost any appliance, including microwave ovens and even washing machines (Transponder News 1999).

15.4.1 Electronic Article Surveillance

Electronic article surveillance (EAS) systems are used to combat shoplifting by detecting goods that have not been authorized when they are removed from the retailer. They comprise a tag attached to the article and a sensor mechanism. Tags can only be neutralized by the retailer when he wishes to authorize the removal of the goods. In effect, these are single-bit RFID systems, as they convey presence but not identity.

There are four major technologies used for EAS systems:

- Microwave,
- Magnetic,
- Acousto-magnetic, and
- Radio frequency.

Market penetration is estimated at 6 billion tags per year, at about $0.10 each for the lowest cost tags. Magnetic and RF tags are very cheap and are generally attached permanently to the goods or packaging, while microwave tags are expensive and are removed from the merchandise using a special tool.

15.4.1.1 Radio Frequency Tags   These comprise a tuned LC circuit that resonates at an RF that absorbs energy from a field at that frequency. They can be deactivated by applying a high-power RF pulse that burns out a fuse in the circuitry.

15.4.1.2 Acousto-Magnetic Tags   This tag consists of a strip of magnetostrictive material and a strip of magnetic material of high coercivity. It resonates mechanically in the presence of a magnetic field of a particular frequency. This resonance can be detected by a receiver sensitive to the changing field created by the flexing magnetostrictive material. The tag is deactivated by modifying the magnetic bias on the strip of magnetic material. This technology is widespread for book protection in libraries, as the tags can be very thin, as seen in Figure 15.4, and therefore can easily be inserted behind the spines of books.

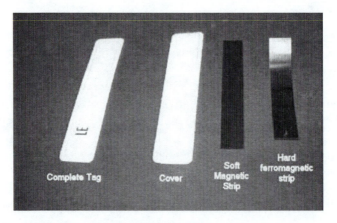

Figure 15.4 Photograph of a dismantled acousto-magnetic tag.

Figure 15.5 Microwave tag used in an electronic toll system [adapted from Bonsor (2001)].

15.4.1.3 Microwave Tags   Modern microwave tags (E-tags) operate at 2.45 GHz using the interrogator-transponder technique. The transponder is active, but low power, and offers a battery life of 10 years. Operational range is typically 4 m through dirt, glass, wood, etc. These are the devices that operate on many toll roads.

The Electronic Toll Collection (ETC) system, shown in Figure 15.5, works as follows (Bonsor 2001):

• As a car approaches a toll plaza, the RF field emitted from the antenna activates the transponder.
• The transponder broadcasts a signal back to the lane antenna with some basic information.

- That information is transferred from the lane antenna to the central database.
- If the account is in good standing, a toll is deducted from the driver's prepaid account.
- If the toll lane has a gate, the gate opens.
- A green light indicates that the driver can proceed.

Some ETC systems also display text messages that inform drivers of the toll just paid and their account balance.

A study has shown that a dedicated cash lane had an average transaction time of 10.5 s for a passenger car and 29.5 s for a commercial vehicle. A dedicated ETC lane processed 1000 vehicles/hr, with an average transaction time of 3.6 s/vehicle. The typical savings is 6.9 s per passenger car. The greater time savings, however, is realized with the elimination of queues because of the reduced transaction times. In the study, queues of more than 20 vehicles, which took up to 3 min to process had been observed at some toll plazas prior to ETC. With the introduction of ETC, there are virtually no queues at toll plazas where there once was often heavy congestion.

15.4.2 Multibit EAS Tags

These are generally of the LC tuned circuit design and have a series of capacitors that can be selectively shorted (mechanically or electrically) to alter the resonant frequency.

15.4.3 Magnetic Coupled RFID Transponder Systems

These low-cost items are the most common tags available today. They operate at frequencies of 125 kHz using antenna systems comprising a multiturn coil designed to collect energy from the reader's magnetic field as shown in Figure 15.6. Their one disadvantage is that because of the magnetic coupling, their range is limited to a couple of centimeters.

15.4.3.1 Operational Principles The transmitter radiates a reasonably powerful magnetic field at a frequency of 125 kHz. Using this energizing field, the transponder sends data back to the reader using one of the following techniques:

- Transmits a signal back at half the frequency while the transmitter operates in continuous wave mode.

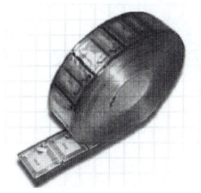

Figure 15.6 Printed magnetic coupled tags.

- Transmits back at the transmitter frequency after the transmitter has switched off during the decay period of the resonant LC receiver circuit (called flyback).
- The tag loads the transmitter field with a fluctuating load, and this is sensed by the transmitter circuitry by monitoring the amount of energy extracted from the field.

A new generation of tags that operate at high frequency (up to 29 MHz) with read/write capabilities have been available since 1998. These tags apparently have some anticollision properties that allow many tags to be in the reader field simultaneously. With these tags, reading and writing distance is generally limited to 20 cm, but some manufacturers claim operational distances up to 1 m. A summary of the operational principles of low-frequency magnetically coupled tags is shown in Figure 15.7.

15.4.4 Electromagnetic Coupled RFID Transponder Systems

Rather than using magnetic lines of force to couple the interrogator and transponder, this technology relies on the propagation of an electromagnetic signal. Because of the differences in propagation mechanism, it can operate over a much longer range.

Antennas are typically half-wave dipoles, so to reduce the transponder's size, the operational frequency must be high. However, higher frequency components cost more and the energy transfer is proportional to $1/f^2$, so high-frequency devices cannot transfer as much energy as their lower frequency counterparts. In addition, because the power density decreases with $1/R^2$, the energy transfer coefficient becomes extremely low rather quickly.

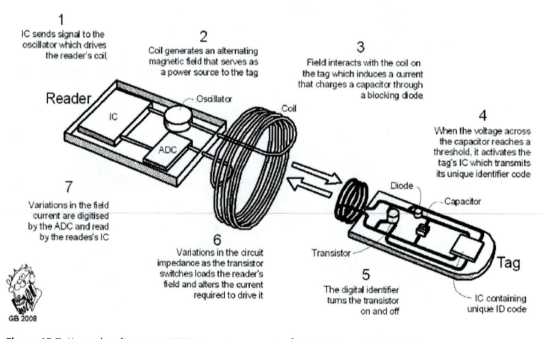

Figure 15.7 How a low-frequency RFID tag system operates [adapted from Want (2004)].

The power density, E (W/m^2), at a range, R (m), from the transmitter is

$$E = P_t G_t / 4\pi R^2,$$ (15.1)

where P_t is the transmitted power (W) and G_t is the transmit antenna gain (with respect to an isotropic radiator). The received power is the product of the power density, and the receiver antenna effective aperture, A_e (m^2):

$$P_r = E A_e.$$ (15.2)

The effective aperture of a half-wave dipole can be determined in terms of the receiver gain, G_r, and the frequency, f (Hz):

$$A_e = G_r \lambda^2 / 4\pi = G_r c^2 / 4\pi f^2.$$ (15.3)

Substituting equation (15.1) and equation (15.3) into equation (15.2) results in a closed relationship for the power received by the transponder:

$$P_r = \frac{P_t G_t G_r \lambda^2}{(4\pi R)^2}.$$ (15.4)

Assuming that antenna gains and matching remain constant for all frequencies of operation, then the minimum received power, P_{min}, determines the maximum operational range, R_{max}:

$$P_{min} = \frac{P_t G_t G_r \lambda^2}{(4\pi R_{max})^2},$$ (15.5)

$$R_{max} = \frac{(P_t G_t G_r)^{1/2} \lambda}{4\pi P_{min}^{1/2}}.$$ (15.6)

The maximum allowable power of the interrogator is determined by health and safety regulations for the general public, as shown in Figure 15.8.

As RFID use permeates our society, the various bodies that manage frequency and power safety issues worldwide have specified available frequency and power levels that can be used. The values for the United States are typical and are listed in Table 15.1.

A typical transponder requires a minimum received power of 0.1 mW, and in most practical applications the length of the antenna is much shorter than $\lambda/2$, making the gain $G = 1.5$. Substituting these values and the effective radiated power (ERP) into equation (15.6) determines the maximum range achievable in the various bands.

In the high frequency (HF) band, the maximum range is 830 m, and this reduces to 3.9 m in the ultra-high-frequency (UHF) band for a 1 W ERP, or 7.9 m if a directional transmit antenna is used. In the microwave band, this reduces still further to 3 m at 2.4 GHz and only 1.2 m at 5.8 GHz.

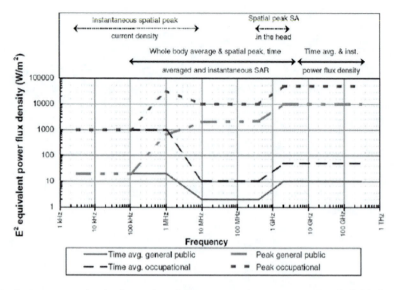

Figure 15.8 Equivalent power density for peak and time-averaged exposure to electric fields (ARPANSA 2002).

Table 15.1 U.S. RFID frequency regulations (Lahiri 2005)

LF	HF	UHF	Microwave
125–134 kHz	13.56 MHz 10 W ERP	902–928 MHz 1 W ERP or 4 W ERP with a directional antenna	2400–2483.5 MHz 4 W ERP 5725–5850 MHz 4 W ERP

The operational principles of backscatter modulation developed at Lawrence Livermore Laboratories are shown in Figure 15.9. Widespread implementation of this technique has allowed these tags to be manufactured for US$0.10 each.

15.5 OTHER APPLICATIONS

As the cost of RFID systems is reduced and their performance and security improves, more and more applications become feasible, from keyless entry for pets to inventory control; new uses appear daily. Law enforcement has been quick to embrace the technology with proposals to tag all expensive items so that they can easily be identified if stolen. A more novel application is the house arrest tag discussed below.

15.5.1 House Arrest Tag

In this application, offenders charged with serious crimes are electronically tagged while on bail awaiting trial in an effort to stop them from reoffending. The system is designed "to stop thugs

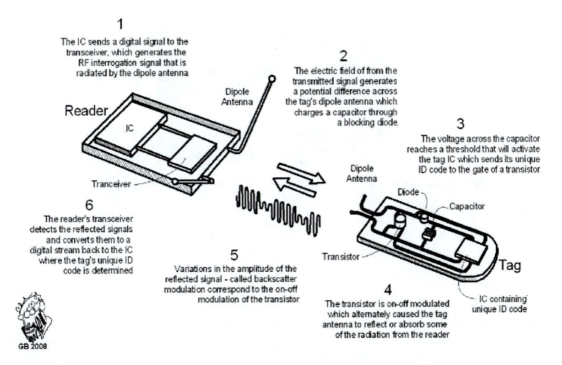

Figure 15.9 How a high-frequency RFID tag system operates [adapted from Want (2004)].

cocking a snoot" at police and the legal system, which is often forced to free them to reoffend. The move is in response to soaring levels of street crime and robberies in the United Kingdom and other western countries.

The offender wears an unobtrusive transmitter, shown in Figure 15.10, on his or her ankle or wrist that emits a signal that is picked up by a monitoring unit when the offender is within range. The monitoring unit is linked by phone to a central computer system where the information about the offender's presence or absence is stored. If the offender leaves the restricted area, the computer will be alerted and the appropriate steps taken. It is not possible to remove the tag and leave home unmonitored, and any tampering with the electronic monitoring unit will alert the central computer and be acted upon immediately.

15.6 SOCIAL ISSUES

Privacy advocates suggest that the identification of specific items linked by their RFID to a credit card would give marketing companies unprecedented access to an individuals purchasing profiles which would allow consumers to be targeted with specifically tailored sales pitches (with a resulting increase in credit card debt that society can ill afford).

Audit trails of commercial transactions would allow the police to trace the movement of individuals through logs maintained by the RFID tracking infrastructure. This would be in breach

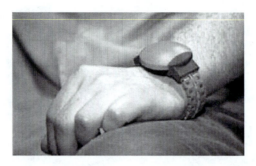

Figure 15.10 Electronic tag bracelet for bail condition monitoring (Courtesy BBC News, 26 February 2002).

of an individual's right to privacy, and laws are being formulated to control access to these logs.

A number of large-scale RFID trials, particularly those by Benetton, Wal-Mart, and Gillette, have been curtailed after pressure from privacy groups who were against the insertion of tags into individual items. Even placing the tags into packaging has been criticized, as it would allow criminals to determine (from your rubbish) that you have just purchased an expensive item.

It has been proposed that children be tagged at birth, which would allow parents, police, and even school administrators unprecedented and invasive knowledge to an individual's whereabouts—from the cradle to the grave.

Issues with regard to the ethics of electronic tagging of offenders is also under consideration. However, it is generally believed that keeping the offenders under what is effectively "house arrest" is less damaging than a stint in an overcrowded jail.

15.7 TECHNICAL CHALLENGES

RFID tags and readers are orientation dependent. Tags must be positioned properly relative to the readers so that antennas can exchange signals effectively. This can be overcome by using arrays of readers orientated appropriately.

RFID signals are easily blocked. Over short range they can be attenuated by packaging, particularly that made from metal foils or metallized plastic. Over longer ranges, because the signal levels are so low, they can be blocked by many common objects, even the human body. Again this can be solved by installing arrays of readers.

Tags are still too costly for inclusion into individual items. For this reason, mass market consumer retail businesses, which operate on low margin and high turnover, have yet to embrace the technology. In addition, competing technical standards and protocols prevent their universal adoption, although the International Organization for Standardization (ISO) is working to overcome this problem. The proposed 96-bit standard ID would allow each of the six billion human beings alive to acquire 1.3×10^{19} individually tagged items during their lifetime. This should be sufficient!

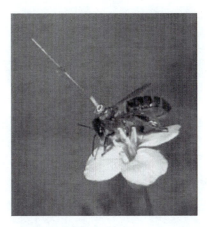

Figure 15.11 Harmonic radar transponder attached to a bee (Courtesy Andrew Martin).

15.8 HARMONIC RADAR

Harmonic radar is an interesting interrogator-transponder technique that is being used to track insects with good accuracy out to ranges of up to 1 km. In this application, a small transponder consisting of a half-wave dipole antenna and a Schottky diode is mounted on the insect. If the frequency is reasonably high, this can be extremely light. Systems have been manufactured at 10 GHz, where the antenna size is only 15 mm and the total mass is only a fraction of 1 g, as can be seen in Figure 15.11.

A dipole antenna has a reasonably significant radar cross section, but it is still much smaller than the clutter return amplitude from the ground, and so would be undetectable if operated in the conventional radar sense. However, the presence of the diode generates a component at the second harmonic which is then radiated in all directions. This can be detected by an appropriate receiver, and because reflections from the natural clutter do not occur at this frequency, there in no competing signal (Riley et al. 1996).

In a typical application, a moderately high-powered, pulsed X-band source is coupled to a rotating fan-beam antenna; a second antenna collocated with the first detects the harmonic. Filters on the antenna ports ensure that any second harmonic component is removed from the transmitter and only the harmonic is received. The output of the receiver is generally used to provide an input to a plan position indicator (PPI) for display and interpretation.

15.9 BATTLEFIELD COMBAT IDENTIFICATION SYSTEM

To reduce "friendly fire" casualty levels, a millimeter wave (38 GHz) interrogator transponder system for ground vehicles capable of identifying friendly combat ground vehicles at ranges of 150 to 5500 m ground-to-ground and 150 to 8000 m air-to-ground in under 1 s has been developed. The components of this system include a directional interrogator antenna coupled to a transceiver, an interface unit mounted on the shooter, and an omnidirectional antenna connected to the transponder on all friendly vehicles, as shown in Figure 15.12.

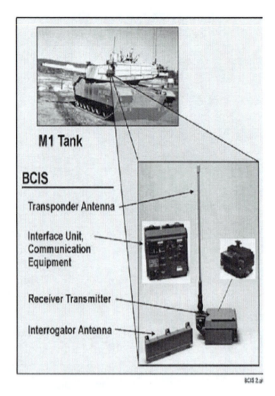

Figure 15.12 A 38 GHz BCIS battlefield IFF system.

The Battlefield Combat Identification System (BCIS) interrogation is started automatically on activation of the shooter platform's laser range finder or interrogation button. It sends an encrypted directional query to the targeted vehicle, which, if fitted with BCIS, will reply with an encrypted omnidirectional friend message. A light and voice confirmation informs the gunner that the target is "friendly" or, if no response is received, that the target is "unknown," which allows the gunner the discretion of choosing to fire or not (Military Analysis Network 1999).

15.9.1 Combat Identification: The Future

The long-range vision for combat identification is to enable military forces to identify all targets in the battle space of all combat mission areas. The fighting forces must be able to identify enemies, friends, and neutrals rapidly and positively, with the primary purpose of lowering combat attrition, increasing enemy losses. and minimizing the risk of friendly fire. Figure 15.13 illustrates the aspects of this vision that are purely transponder based. These will use the following networks:

- BCIS: Battlefield Combat Identification System.
- BCIS(–): lightweight handheld BCIS interrogator for use by soldiers.
- (A-G) BCIS: air-to-ground BCIS.

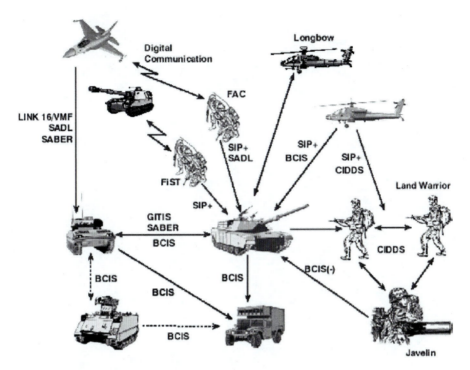

Figure 15.13 Battlefield IFF in the future (Military Analysis Network 1999).

- CIDDS: combat ID dismounted soldier (laser/RF-based system for soldiers).
- SINCGARS: GPS coordinate based "don't shoot me" net.
- (A-G) SBCI: the air-to-ground portion of the SINCGARS net.

15.10 References

ARPANSA (2002). *Radiation Protection Standard. Maximum Exposure Levels to Radiofrequency Fields—3 kHz to 300 GHz*. Canberra: ARPANSA. Available at http://www.arpansa.gov.au/pubs/rps/rps3.pdf.

Bonsor, K. (2001). How E-ZPass works. Viewed January 2008. Available at http://www.howstuffworks.com/e-zpass1.htm.

Daniel, M. (2002). Air traffic teams relive their 11 minutes of helpless horror. *Sydney Morning Herald*, Aug. 14. Available at http://www.smh.com.au/articles/2002/08/13/1029113929293.html.

Lahiri, S. (2005). RFID: a technology overview. Viewed January 2008. Available at http://www.informit.com/articles/article.aspx?p=413662.

Military Analysis Network. (1999). Battlefield combat identification system (BCIS). Viewed January 2008. Available at http://www.fas.org/man/dod-101/sys/land/bcis.htm.

Riley, J., Smith, A., Reynolds, D., Edwards, A., Williams, I., Carreck, N., and Poppy, G. (1996). Tracking bees with harmonic radar. *Nature* 379:29–30.

Transponder News. (1999). Current (future) trends in transponder systems. Viewed January 2008. Available at http://www.rapidttp.com/transponder/trends.html.

Want, R. (2004). RFID: a key to automating everything. *Scientific American* 290(1):56–65.

16

Tomography and 3D Imaging

16.1 PRINCIPLE OF OPERATION

The word tomography derives from the Greek word *tomos* meaning "section," so the process of tomography involves the generation of narrow sections through an object each made up of individual volume elements (voxels) with a cross section $\Delta x \times \Delta y$ and thickness, s, as shown in Figure 16.1b. This is at best a noninvasive or minimally invasive process that is performed using sensors outside the object of interest. In many applications, sequences of two-dimensional (2D) slices are combined to produce a pseudo-three-dimensional (3D) image. The process, when applied to X-rays, is referred to as computed axial tomography (CAT or CT).

Computed tomography only became feasible with the development of computer signal processing capabilities in the 1960s, but many of the basic principles were developed many years before that. In 1917, a mathematician, J. Radon, showed that the distribution of material or the material properties of an object can be determined if the integral values along any number of lines passing through a particular layer are known (Deans and Roderick 1983).

Such processing is common in medical applications where doctors require a 3D representation of the interior of the human body. Examples of imaging systems that use these techniques include

- 3D ultrasound
- 3D magnetic resonance imaging (MRI)
- Positron emission tomography (PET)
- X-ray CT scans.

Ultrasound technology was discussed briefly in Chapter 10. It uses the different elasticity characteristics of the target to produce an image, while MRI relies on the varying amounts of hydrogen (in general contained in water), and PET relies on the release of positrons from radioactive decay. However, the best understood, and possibly the most common, is X-ray CT imaging, which relies on density variations that affect the attenuation of high-frequency electromagnetic waves as they propagate through the target.

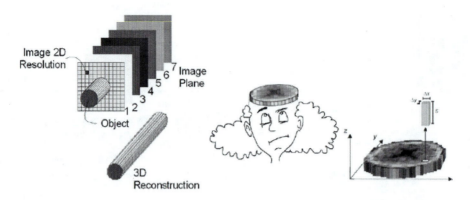

Figure 16.1 Computerized axial tomography: (a) general principle and (b) as applied to a head.

16.2 CT IMAGING

Between 1957 and 1963, a South African physicist, A. Cormick, independently developed a method of calculating radiation absorption distributions in the human body based on transmission measurements. He postulated that it would be possible to display very small differences in the absorption through the body, but never applied his research.

In the early 1970s, the principles of CAT scanning (also called computed tomography, or CT, scanning) were conceived, for a second time, by William Oldendorf and developed by Godfrey Hounsfield (Hounsfield 1973). Although often employing contrast media to enhance the quality of the images obtained, this test is barely invasive.

The most basic CT measurement developed in the 1970s involves the displacement and rotation of a collimated X-ray source (pencil beam) around the patient, as shown in Figure 16.2. The pencil beam source and detector are scanned together linearly across the patient to obtain a projection profile. The source and detector are then rotated by about 1° and the linear scan process is repeated. This translate-rotate motion is repeated until the source and detector have been rotated 180°. The whole mechanism (or the patient) is then stepped one step through the beam plane before the whole processes is repeated. This complex motion results in long scan times—approximately 5 min (Bronzino 2006).

To speed up the process, modern CT scanners consist of an X-ray source that produces a fan-beam which penetrates the patient and impinges on a bank of detectors, as illustrated in Figure 16.3. This complete assembly rotates around a central core to produce a sequence of intensity measurements over 360°.

In either case, a complex image reconstruction algorithm solves the equations derived from this massive set of measurements to produce an image that represents the density of the material in cross section through the patient.

The generation of clear images is compromised by movements of the patient during the scanning process, so faster scan methods have been developed. For example, continuously rotating CT systems that can image a complete slice in less than 1 s were first introduced in 1987. However, these systems still construct 3D images from individual "slices" through the patient.

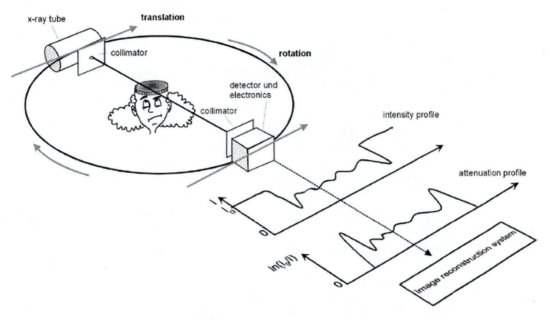

Figure 16.2 The simplest CT scan involves measuring the intensity of a pencil beam of X-rays from many different angular positions [adapted from Kalender (2005)]. Reproduced courtesy Publicis.

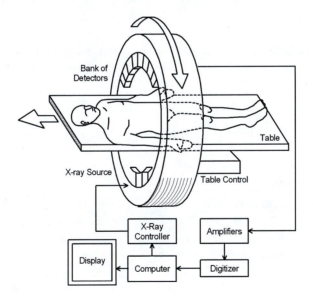

Figure 16.3 CT scanner schematic.

In the 1990s spiral scanners with multislice capabilities were introduced, producing even
faster 3D high-resolution images. This process is illustrated in Figure 16.4. The state of the
art is now a scan time of less than 0.5 s with a slice thickness of between 0.5 and 1 mm
(Kalender 2005). Such fine slices allow visualization of extremely thin sections of the brain, skull,
spinal cord, and spine, as well as other parts of the body, in two dimensions and with enough

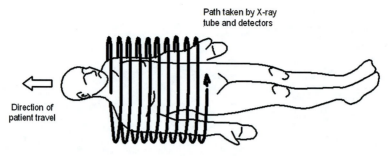

Figure 16.4 Spiral scan tomography.

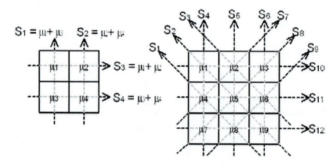

Figure 16.5 The solution of N^2 unknowns for an $N \times N$ image matrix can be determined by solving a system of linear equations as illustrated for (a) a 2×2 array and (b) a 3×3 array.

clear distinction between black, grey, and white areas of the image to allow pathologic diagnosis in many cases.

16.2.1 Image Reconstruction

As shown in Figure 16.2, measurements are made of the intensity of the X-ray beam and converted to a set of attenuation measurements. These are known as the "Radon transform" of the image. An inverse transformation must then be carried out to determine the distribution of attenuations for each pixel element $\mu(x,y)$ through the target.

The easiest process to understand is one in which there are N^2 unknowns in an $N \times N$ matrix of pixels, as shown in Figure 16.5. If sufficient independent measurements are made, it is possible to solve for all of the unknowns. In the simplest case of a target with only four elements, two measurements from two projections will yield a system of four equations and four unknowns that can easily be solved. The extension to a 3×3 matrix with nine unknowns can also be solved easily using the 12 equations, as defined schematically in Figure 16.5.

This process was used by early CT scanners in which the number of elements was limited. However, with the requirement for finer resolutions and more pixels, the computational overheads became unacceptably high. In modern CT scanners, the convolution-backprojection procedure is usually applied. It starts with a matrix loaded with zeros, and as each measurement is made, the projection value is added to all of the elements in the array along the direction of measurement. However, because each component contributes not only to the value at the desired point, but to the whole image, it is clear that a blurred image will result, as can be seen from the illustration in Figure 16.6.

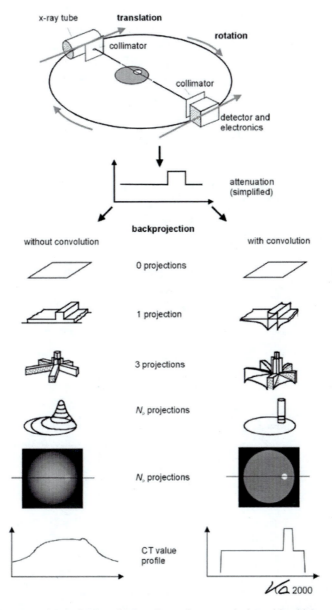

Figure 16.6 Image reconstruction by backprojection shows that convolution with a high-pass kernel is required to ensure a sharp image (Kalender 2005). Reproduced courtesy of Publicis.

To minimize the magnitude of this blurring, each projection is convolved with a kernel prior to backprojection. This convolution kernel represents a high-pass filter. Because convolution in the spatial domain is equivalent to multiplication in the frequency domain, it is possible to perform this convolution-backprojection process in that domain far more quickly and efficiently (Kak and Slaney 2001).

16.2.2 What Is Displayed In CT Images

As explained earlier, the CT computes the spatial distribution of the linear attenuation coefficient $\mu(x,y)$. However, because μ is strongly dependent on the energy of the X-ray photons, it makes a direct comparison between images from different systems impossible. Therefore the image is displayed in terms of its attenuation relative to that of water. These are referred to as CT units, or sometimes Hounsfield units (HU) in honor of the inventor (Sir Godfrey Newbold Hounsfield). For an element with attenuation, μ_T, its value in CT units or HU is

$$CT_{units} = \frac{\mu_T - \mu_{water}}{\mu_{water}} \cdot 1000. \tag{16.1}$$

On this scale, water, and consequently any water equivalent tissue with $\mu_T = \mu_{water}$ has a value of 0 HU, by definition. Air corresponds to a value of about −1000 HU because μ_{air} is close to zero. Bone and other calcifications with high atomic numbers and high density offer increased attenuation and therefore have higher CT values, typically up to 2000, as illustrated in Figure 16.7.

Most medical scanners use a range of CT values from −1024 to +3071, a total of 4096 values, as these can be represented by a 12-bit number. Because neither the imaging software nor the

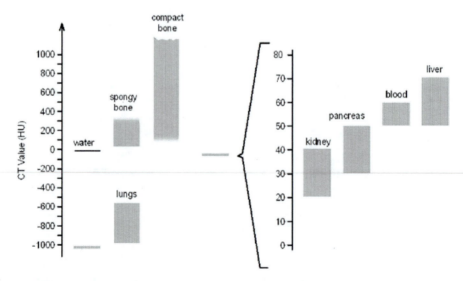

Figure 16.7 CT values for normalized attenuation of different materials [adapted from Kalender (2005)]. Reproduced courtesy Publicis.

eye can discern 4096 different shades of grey, CT ranges of interest for a specific imaging task are windowed and expanded to fill all shades from black to white. This is a process similar to that discussed for radiometric images in Chapter 4.

The contrast of a CT image is far superior to that of a conventional X-ray image, not because of the higher powers used, but because of the differences in the way the images are generated. In Figure 16.8, a comparison is made between a CT image and an X-ray image. The contrast between two adjacent pixels in the CT image is

$$\Delta CT = \frac{I_1 - I_2}{(I_1 + I_2)/2} \cdot 100$$
$$= \frac{63 - 35}{(63 + 35)/2} \cdot 100 = 57\%.$$

The X-ray intensity is determined by the sum of the attenuations along the whole path through which the beam travels, and therefore the contrast is determined by the difference between these two sums:

$$\Delta CT = \frac{\sum I_1 - \sum I_2}{(\sum I_1 + \sum I_2)/2} \cdot 100$$
$$= \frac{1738 - 1734}{(1738 + 1734)/2} \cdot 100 = 0.23\%.$$

16.2.3 2D Displays

Computed tomography scanning is primarily limited to the transverse plane because of the mechanics of the scanning system. Other image planes can be synthesized from the volume image constructed from these transverse slices. In Figure 16.9, the raw data for a scan through the patient's neck is shown, first as a longitudinal section marking the cuts, and then as the individual slices.

The best resolution that can be obtained for a CT image is determined by the diffraction-limited aperture of the X-ray source for the pencil beam system. In the fan-beam case, the resolution in the plane of the fan is determined by the spacing of the detectors in the array.

16.2.4 3D Displays

Three-dimensional displays represent a scan volume in a single image which has generally been manipulated to enhance a specific characteristic. This can only be done successfully when one high-contrast structure (like the skeleton) is to be displayed. Shaded surface displays (SSDs), maximum intensity projections (MIPs), and volume renderings (VRs) or perspective volume renderings (pVRs) are some of the common manipulations shown in Figure 16.10.

16.3 MAGNETIC RESONANCE IMAGING

Felix Bloch and Edward Purcell discovered the magnetic resonance phenomenon independently in 1946. In the period between 1950 and 1970, nuclear magnetic resonance (NMR) was developed

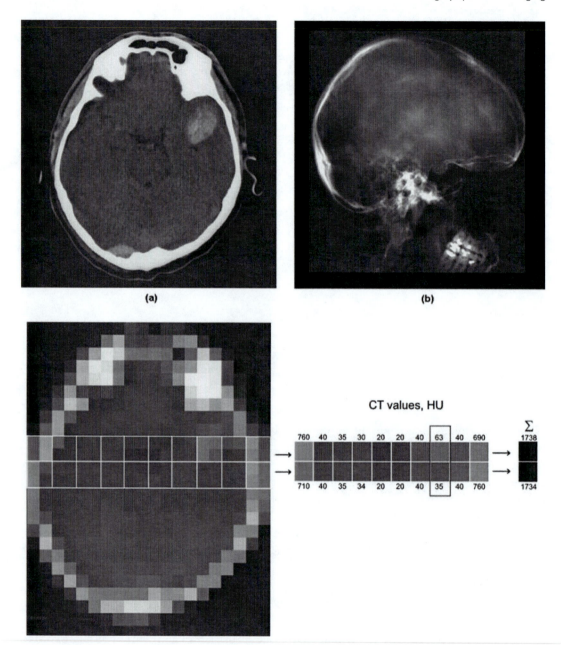

Figure 16.8 Contrast of CT and X-ray images (Kalender 2005). Reproduced courtesy Publicis.

and used for chemical and physical molecular analysis. In 1971 Raymond Damadian showed that the nuclear magnetic relaxation times of tissues and tumors differed, thus motivating scientists to consider magnetic resonance for the detection of disease.

Magnetic resonance imaging was first demonstrated on small test tube samples in 1973 by Paul Lauterbur. He used a backprojection technique similar to that used in CT scans. In 1975,

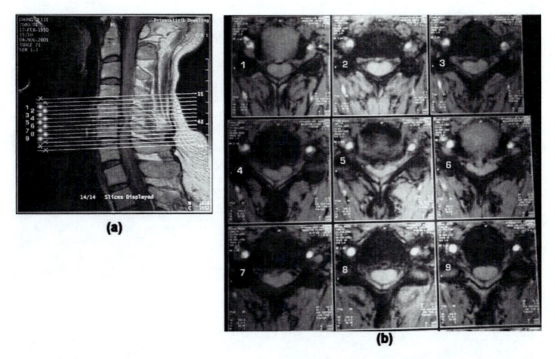

Figure 16.9 CT image of the spine showing (a) a longitudinal section indicating the slices and (b) individual images of each of the slices.

Richard Ernst proposed MRI using phase and frequency encoding, and the Fourier transform. This technique is the basis of current MRI techniques. A few years later, in 1977, Raymond Damadian demonstrated an MRI called field-focusing nuclear magnetic resonance. In that same year, Peter Mansfield developed the echo-planar imaging (EPI) technique (Hornak 2007).

Magnetic resonance imaging uses the principle of NMR and an assembly of powerful magnets to produce images generated by various elements within the patient. It is most commonly used to image hydrogen, as the human body consists predominantly of that element.

16.3.1 Nuclear Magnetic Resonance

Atomic nuclei have an angular momentum arising from their inherent property of rotation or spin. Since the nuclei are electrically charged, the spin corresponds to a current flowing around the spin axis which, in turn, generates a small magnetic field, as illustrated in Figure 16.11.

Each nucleus with an odd number of nucleons, and hence a nonzero net spin, has a magnetic moment, or dipole, associated with it. In general, the orientation of the dipoles is random, but if they are placed in a magnetic field they will become aligned to it.

Protons, or hydrogen nuclei, with a spin of ½ can align parallel to the field or antiparallel, as shown in Figure 16.12. The two orientations have slightly different energy (Zeeman splitting), with the spin-up state (parallel) having the lower energy. Their effects almost cancel, but the lower state has a slight excess that is exploited by MRI.

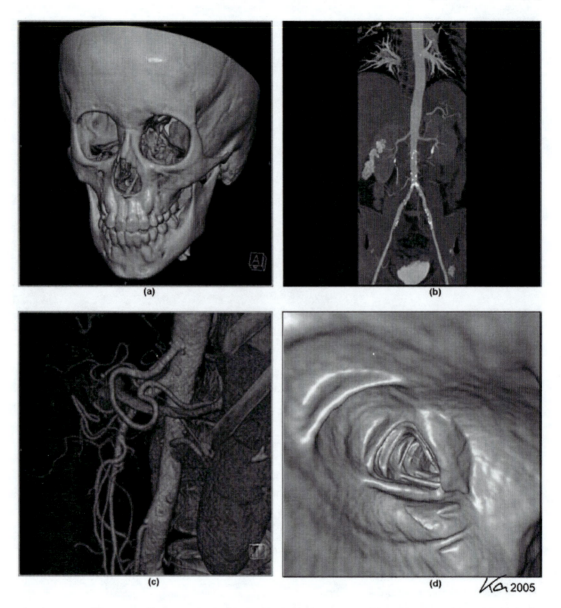

Figure 16.10 Different 3D display methods: (a) SSD for skeletal structures, (b) MIP for CT angiography, (c) VR for abdominal structures, and (d) pVR for a virtual colonoscopy (Kalender 2005). Reproduced courtesy Publicis.

When a magnetic field is applied, the whole population of nuclei has a bulk magnetization vector, M, aligned to it. This is defined as the z-direction. By applying a small rotating magnetic field in the x-y plane, the nuclei can be tipped away from the z-direction and made to rotate. The field rotation rate is tuned to the natural precession frequency of the nuclei (hence magnetic resonance), which, for hydrogen nuclei in a magnetic field of 1 Tesla (10^4 Gauss) is 42.57 MHz.

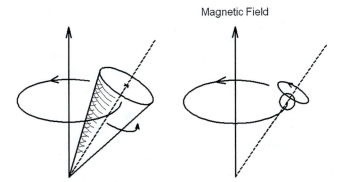

Figure 16.11 Comparison between (a) spinning top precession and (b) hydrogen atom precession.

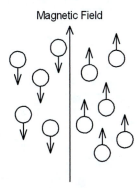

Figure 16.12 Hydrogen atoms align to an external magnetic field.

The relationship between the applied magnetic field, B (Tesla), and the resonant frequency, f (Hz), is known as the Larmor relationship:

$$f = \gamma B, \tag{16.2}$$

where γ is the gyromagnetic ratio, which for hydrogen is 42.57 MHz/T.

Other materials have different resonant frequencies in the same field—(^{31}P) has a resonance at 17.24 MHz and (^{23}Na) at 11.26 MHz—so it should be possible to "tune in" to any specific species to observe their response in isolation. However, for imaging purposes, only hydrogen is used in practice because its concentration is high and therefore the image process is more sensitive (Hornak 2007).

As the power in the rotating field pulse is increased, the bulk magnetization vector, M, will continue to rotate away from the direction of the static magnetic vector until it is orthogonal to it, rotating in the x-y plane only. When the pulse ends, the magnetization vector continues to rotate for a time and in so doing generates a small voltage in the field coils. Initially all the rotations remain synchronized, but after a short period, interactions between adjacent nuclei that begin to spin at slightly different rates result in a loss of energy, and this causes the total spin to decay until it is once again aligned with the static field, as shown in Figure 16.13.

16.3.2 Imaging Process

The imaging process exploits the time taken for the spin to decay, as this gives an indication of the proton, and hence water, density in the body. As can be seen in Figure 16.14, in addition to the fixed magnetic field generated by the main coil, three gradient coils can apply varying fields in the x, y, and z directions. Manipulation of these fields is used to identify the spatial position of the hydrogen concentration.

As shown in Figure 16.15, if no gradient field is applied along the magnetic axis, the Fourier transform of the decaying Larmor signal shows a single broadened spectrum. If a magnetic field gradient, G_x, is applied in the x direction, the Larmor spectrum will have characteristics of the variations in proton density profile in the direction of the gradient:

$$f(x) = \gamma(B + xG_x). \tag{16.3}$$

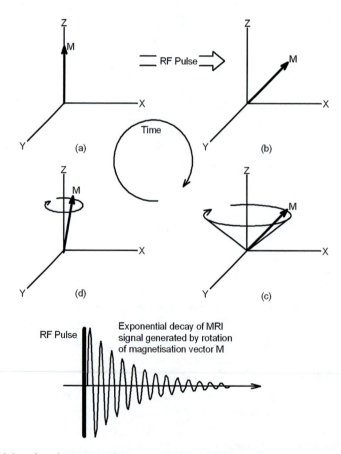

Figure 16.13 Principles of nuclear magnetic resonance showing (a) the unexcited magnetization vector aligned with the external field, (b) the result of the application of a radio frequency (RF) pulse to rotate the magnetization vector, (c) the rotating magnetic field vector, and (d) the magnetic field vector rotation decaying [adapted from Paykett (1982)].

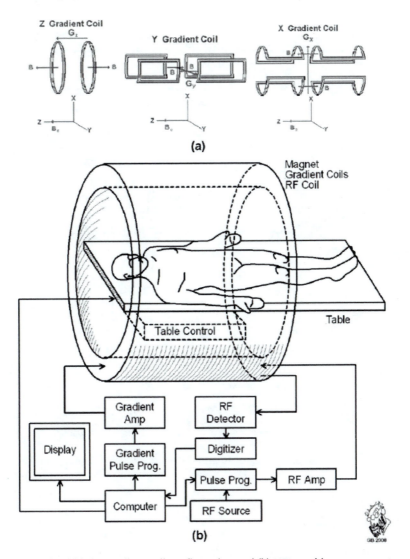

Figure 16.14 Schematic of (a) the gradient coil configuration and (b) MRI machine components.

By rotating the gradient field and producing proton density profiles from various angles, a computer can perform the inversion function to re-create in two dimensions the internal structure of the body. This process (using the backprojection method) is illustrated in Figure 16.16.

To obtain high-resolution images, many measurements are made over 360° by applying the appropriate excitation current to two or sometimes all three gradient magnets. For example, if a cut is to be made through the *xy* plane, then the following fields are generated:

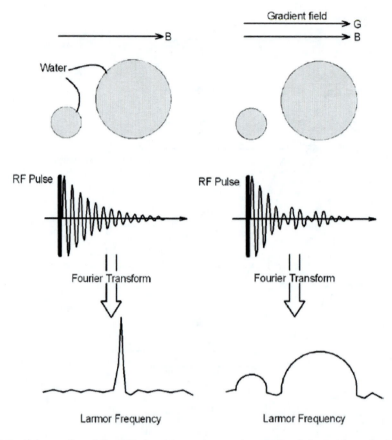

Figure 16.15 Spatial encoding of the MRI signal to produce an image slice [adapted from Paykett (1982)].

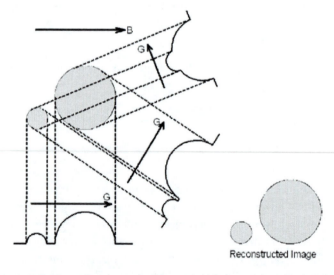

Figure 16.16 Image reconstruction using the projection method [adapted from Paykett (1982)].

$$G_y = G_f \sin \theta$$
$$G_x = G_f \cos \theta. \tag{16.4}$$

Unlike conventional CT scans, because the direction of imaging is determined by the magnetic gradient, it is possible to produce images in any plane—axial, coronal, or sagittal—or at an angle by adjusting the current flowing through the x, y, and z coils.

16.3.3 Imaging Resolution

In most imaging systems, diffraction limits the resolution; that is, the resolution is determined by the wavelength divided by the angle subtended by the receiver aperture. MRI is the only imaging system in which the resolution is not a function of the wavelength (Bronzino 2006).

If Figure 16.15 is examined, it is obvious that the gradient field, G, serves to translate the temporal response of the receiver output into a spatial one, therefore the resolution will be a function of the gradient. In addition, as discussed in Chapter 11, the resolution of a spectral analysis process is directly proportional to the observation time, therefore the resolution is proportional to the observation time and the gradient.

A typical 1.5 T MRI machine has a gradient of $G = 0.1$ Gauss/mm which is observed for $T_{obs} = 8$ ms. The gyromagnetic moment for hydrogen is $\gamma = 42.57$ MHz/T, which equates to $\gamma' = 26{,}747$ rad/s/Gauss.[1] Therefore the spatial gradient is

$$\gamma' G T_{obs} = 26{,}747 \times 0.1 \times 8 \times 10^{-3}$$
$$= 21.4 \text{ rad/mm}.$$

Finally, the smallest resolvable feature, δx (mm), is equal to one spatial cycle, therefore

$$\delta x = 2\pi / \gamma' G T_{obs}. \tag{16.5}$$

Substituting for the values determined previously $\delta x = 0.29$ mm.

16.4 MAGNETIC RESONANCE IMAGES

This use of proton density makes it possible to produce images of tissues that are comparable, and in some cases superior in resolution and contrast, to those obtained with CT scanning. Examples of this capability for 2D imaging are shown in Figure 16.17 and for 3D reconstruction in Figure 16.18. Moreover, since macroscopic movement affects NMR signals, the method can be adapted to measure blood flow and other dynamic processes.

16.5 FUNCTIONAL MRI INVESTIGATIONS OF BRAIN FUNCTION

Functional MRI (fMRI) investigations of brain function rely on the fact that oxygenated and deoxygenated hemoglobin molecules behave slightly differently in a magnetic field. Magnetic

---

1 1 Gauss = 0.0001 Tesla.

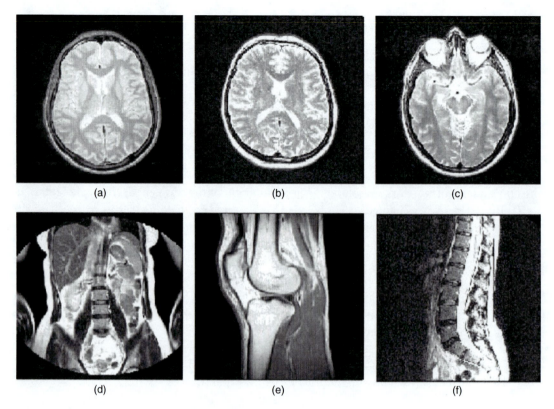

Figure 16.17 Magnetic resonance images of various parts of the body showing its capability to produce images of soft tissue.

resonance images can then be made to show the oxygen use in the brain using the blood-oxygen-level-dependent (BOLD) signal. Images made using this technique, while the subject is observing a specific scene or thinking specific thoughts, are compared to those made by a control to determine where a specific brain function occurs. The image in Figure 16.19 shows the difference between the subject looking at a face and looking at a blank screen.

16.6 POSITRON EMISSION TOMOGRAPHY

Positron emission tomography (PET) involves the injection of radioactive molecules made from radioactive isotopes with short half-lives, such as ^{11}C, ^{15}O, or ^{13}N, into the body where radioactive decay takes place, releasing a positron, as illustrated in Figure 16.20. Images of the processes occurring within the body are then made by detecting the gamma radiation that is created by the annihilation of a positron and an electron. In this case the radiation is in the form of two photons, each with an energy of 511 keV, traveling in opposite directions (Freudenrich 2000).

The gamma ray detector used in a PET scanner consists of an annulus of scintillation crystals each connected to a photomultiplier tube, as shown in Figure 16.21. If a gamma ray strikes one

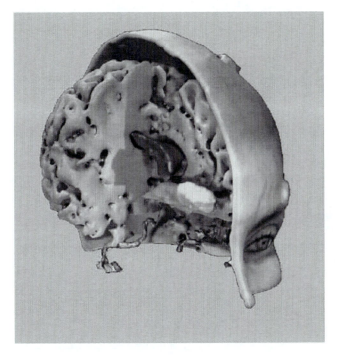

Figure 16.18 3D MRI image re-creation of a human brain.

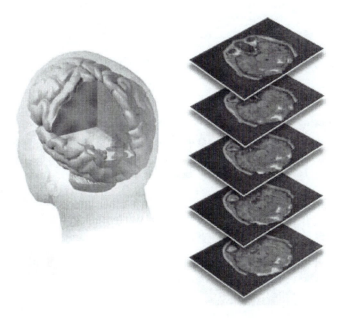

Figure 16.19 Functional MRI of the areas of the brain used to recognize a face.

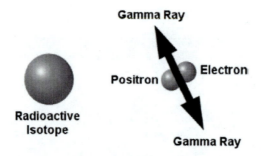

Figure 16.20 Emission of a gamma ray following annihilation of a positron and electron.

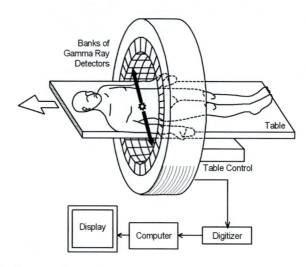

Figure 16.21 Schematic diagram of a PET scanner showing the trajectory of a gamma ray pair.

of the crystals, it produces a photon of light that is converted to an electron and amplified by the tube to produce a measurable signal. A coincidence detector ensures that only pairs of hits that occur within about 25 ns are recorded.

A coarse 2D image is made of an axial slice through the body by combining the directions and relative time delays in all of the measured events. The scanner is then moved and the process repeated until a complete 3D image has been built up.

Depending on the type of molecule injected, PET can provide information on different biochemical functions. For example, if the molecule that is radioactively tagged is glucose, the scan will show an image of glucose metabolism, or how much energy the body is using in a specific area. This has been useful in studies of brain function.

PET scans do not have the same high resolution as that of fMRI, shown in the previous section, but they are more sensitive in identifying physiological processes that use chemicals which do not have a good MRI signature. An example of this capability is the use of radioactive phosphorus to look for abnormal bone growth indicating the possible presence of a tumor.

16.6.1 Examples of the Use of PET Scans

Because the brain is a major consumer of energy, radioactive glucose can be used to highlight regions of activity. In Figure 16.22, two images of the brain are shown in which the hotter regions indicate areas of activity for a normal brain and one of a person suffering from Alzheimer's disease. Figure 16.23 shows another PET scan of the brain in which the different regions are highlighted as the patient speaks, hears, sees, and thinks about words.

16.7 3D ULTRASOUND IMAGING

Progress in ultrasound imaging has been phenomenal in the last decade as signal processing techniques and speed have increased. The fuzzy monochrome 2D slices are a thing of the past with the introduction of 3D and even four-dimensional (4D) imaging capabilities.

16.7.1 2D Medical Ultrasound

Ultrasonic scanning for medical diagnoses uses the same principle as sonar. Pulses of high-frequency ultrasound, generally between 1 and 5 MHz, are created by a piezoelectric transducer and directed into the body. As the ultrasound traverses various internal organs, it encounters changes in acoustic impedance which cause reflections. A typical medical ultrasound device consists of a transducer with characteristics that can be altered, a central processing unit (CPU) to process the information, and various peripherals such as keyboard, printer, and display, as illustrated in Figure 16.24.

In the B-scan mode, a linear array of transducers is used to scan a plane in the body. Using the angle information and the range obtained from the round-trip time, a polar image can be

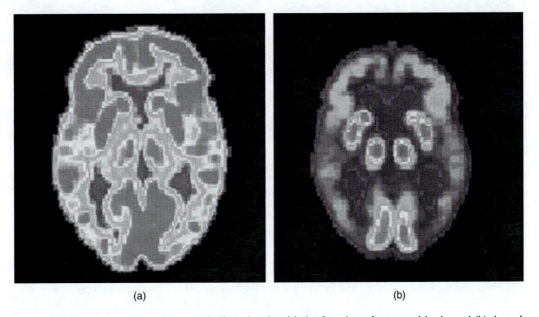

(a)   (b)

Figure 16.22 PET images of glucose metabolism showing (a) the function of a normal brain and (b) that of a person with advanced Alzheimer's disease (NIA 2007).

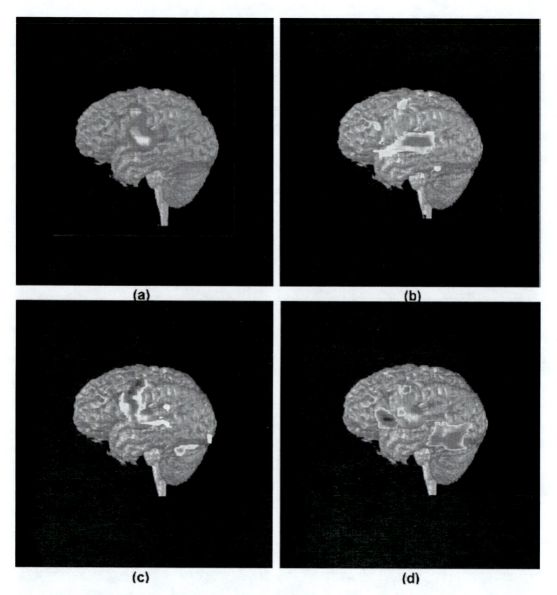

Figure 16.23 PET scan showing the different portions of the brain used in processing words. (a) speaking words, (b) hearing words, (c) seeing words written and (d) thinking about words (National Institute on Aging 2007).

formed. This is processed and displayed as a 2D plot with intensity encoding for reflected signal strength.

There are many different probe types, three of which are shown in Figure 16.25. The shape of the probe determines the field of view and its application. Because the penetration depth of high-frequency (high-resolution) ultrasound is limited, probes are often designed for insertion into the body via its various orifices.

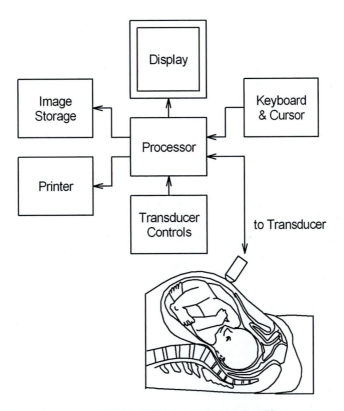

Figure 16.24 2D ultrasound components [adapted from Freudenrich (2001)].

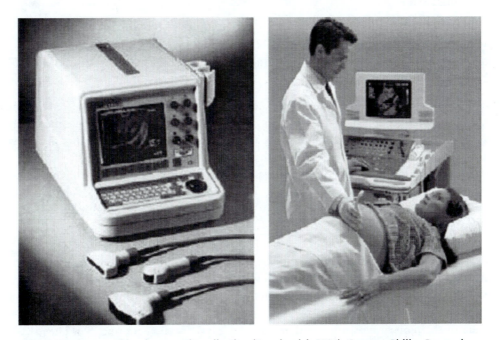

Figure 16.25 2D ultrasound hardware and application (Freudenrich 2001). Courtesy Philips Research.

In addition to the standard B-scan imaging mode, a number of other modes can be selected by the sonographer. For example, the A-scan mode uses a single transducer to measure along a line through the body and the echoes are plotted as a function of time. This technique is used for measuring the distances or sizes of internal organs. The M-scan mode is used to record the motion of internal organs, as in the study of heart dysfunction.

Greater resolution is obtained in ultrasonic imaging by using higher frequencies. A limitation of this property of waves is that higher frequencies tend to be much more strongly absorbed.

16.7.1.1 Medical Applications   Most medical applications of ultrasound were developed specifically to reduce the risks to fetuses from ionizing X-rays. They include the following:

- Measuring fetus size to establish a due date.
- Determining fetus position to see whether it is breech or head-down for birth.
- Checking the placenta to see that it is properly formed and not obstructing the cervix.
- Counting the number and the sex of fetuses.
- Detecting whether a fertilized egg has implanted in the fallopian tubes (ectopic pregnancy).
- Determining the volume of amniotic fluid.
- Monitoring the fetus during specialized procedures such as amniocentesis.

Nonobstetric uses for ultrasound are also common:

- Looking for tumors on ovaries and breasts.
- Imaging the heart to identify abnormal structures or functions.
- Measuring blood flow using Doppler (see Chapter 10).
- Seeing kidney stones.
- Early detection of prostate cancer.

An example of a 2D ultrasound image of a 12-week-old fetus is shown in Figure 16.26. This should be observed in conjunction with a visible image of a similar-age fetus, shown in Figure 16.27.

16.7.1.2 Dangers of Ultrasound Use   Two potential dangers exist. They are localized heating due to the absorption of energy, and the formation of bubbles due to cavitation, where the ultrasound induces dissolved gases to leave solution. Research has shown that birth weights of babies that have been scanned regularly are lower than those of babies that have not been scanned. They are also more likely to be left handed and their speech development is delayed (Volkin and Dargan 2006).

16.8 3D EXTENSION

If the relative positions of the images made by a 2D ultrasound scanner are known, it is possible to build a 3D image. In general, the transducer is often still moved across the body by hand because skilled sonographers perform the function better and are less intimidating than mechanical positioners.

Some of the pioneering work was done at the University of California, San Diego (UCSD) on a suitably modified Acuson 128XP/10 with a C3 transducer. However, in the last few years a

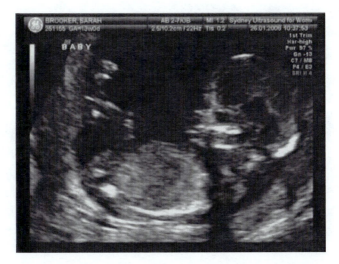

Figure 16.26 2D Ultrasound scan of Charlotte Rose Tapp when she was a 13 week fetus. (Courtesy Sarah Brooker and Mike Tapp.)

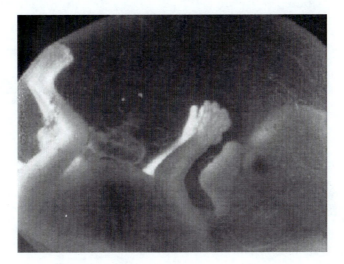

Figure 16.27 Photograph of a fetus of similar age (13 weeks).

number of manufacturers have produced commercial versions of the 3D imager, such as the one shown in Figure 16.28.

The process of 3D image generation starts with the sonographer obtaining a 2D scan and highlighting the region of interest. Further images are made at an angle to the scanned slice by moving the transducer. Rendering software then builds up an image over the strongly reflecting interface between the amniotic fluid and the fetus skin.

This process has now reached maturity, and exceptionally good images are generated as a matter of course, as can be seen in Figure 16.29 and Figure 16.30. Confirmation that this

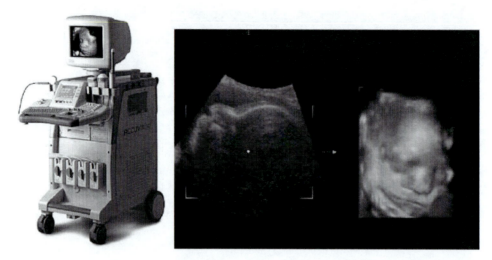

Figure 16.28 Modern 3D ultrasound system with a sample image (Courtesy Madison Accuvix).

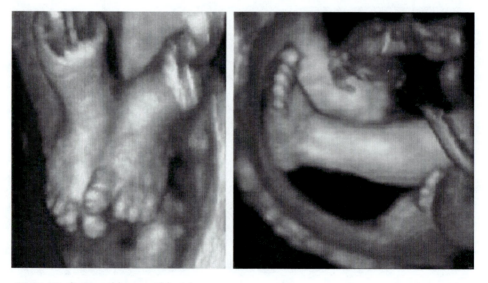

Figure 16.29 3D ultrasound images of feet.

rendering process can be accurate is demonstrated by the before and after pictures shown in the newspaper cutting reproduced in Figure 16.31.

The scanning and processing speed of modern ultrasound machines is now so fast that manufacturers are offering movies. This is referred to as 4D ultrasound because it includes a time axis.

16.8.1 Ultrasonic Computed Tomography

Because the pulsed echo ultrasound process can only see tissue interfaces, an alternative imaging technique called ultrasonic computed tomography has been developed. It transmits through the

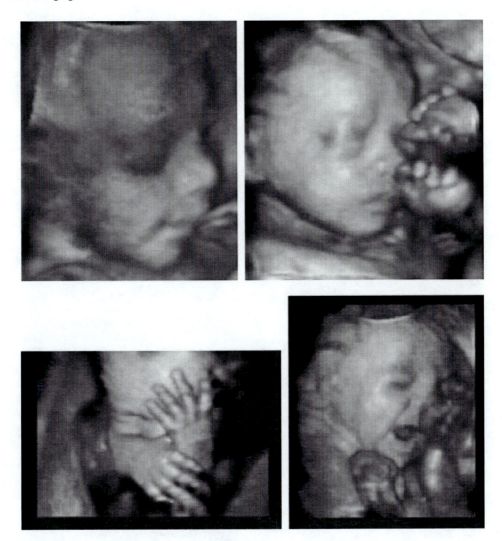

Figure 16.30 3D ultrasound images of hands and faces.

target and uses the attenuation and the propagation time to estimate both the attenuation coefficient and the acoustic refractive index of the object.

16.9 3D SONAR IMAGING

Sophisticated 2D sonar arrays such as the one developed by Thomson Marconi Sonar (TMS) in Sydney, NSW, Australia, can produce short-range 3D images with voxol resolutions down to $1 \text{ mm} \times 1 \text{ mm} \times 1 \text{ mm}$. This sensor was developed to produce high-resolution images of objects such as sea mines in turbid water where the visibility is extremely poor.

As shown in the photograph in Figure 16.32, this sensor consists of a group of three uniformly spaced transmitters which insonify the target with high-frequency (>1 MHz) chirp sound pulses.

Watching your baby smile in the womb

By TIM HILFERTY

NO longer do parents have to wait until their baby is born before they decide whom it takes after.

Three-dimensional ultrasound images, first tested in Australia, have improved at such a rate that parents can see their baby smile months before birth.

These amazing pictures show a baby just before and after birth.

The before picture, taken by the latest in 3-D ultrasound machines — the Siemens Omni — is as detailed and as clear as any baby snap in a family photo album.

But it may be some time before NSW parents have the option of viewing the face of their unborn child.

While the technology is slowly being introduced into hospitals in the UK, the US and Europe, so far only one 3-D machine is in operation in NSW and only on a trial basis.

Dr Henry Murray, an obstetrician at Nepean Hospital, said a machine was being tested there to study its potential uses.

The machine costs between $200,000 and $400,000.

They work by taking two-dimensional images from two different angles, and then creating one image using a sophisticated computer program.

Dr Murray admitted he felt a

sense of wonder when he looked at the face of an unborn baby.

"You still feel that way, it is one of the joys of obstetrics," he said.

"You can get some pretty amazing images with the 2-D machines but this goes one better than that."

Where the machine does excel is in picking up facial abnormalities, and deformities in the arms and legs.

"It can show the surgeons what they are up against," Dr Murray said.

He said hospitals in some countries were using 3-D ultrasounds as a "novelty" for excited parents-to-be.

"That's not something we would encourage. Although there is no evidence that ultrasounds are harmful, we don't like doing them unless they are absolutely necessary."

Andrew Hartmann, managing director of Acuson Australia, which distributes ultrasound machines, said crystal-clear images of babies' faces were rare.

"Three things have to happen; the baby has to be reasonably still, it has to be facing the right way, and it has to be smiling," Mr Hartmann said.

He said the Omni machine, which took the US images, would probably not be exported to Australia.

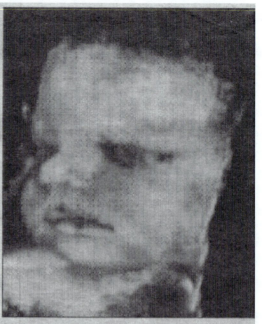

Before . . . a scan of a baby in the womb before birth.

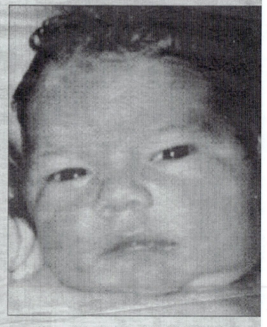

After . . . the same baby shortly after being born.

Figure 16.31 Image comparison: 3D ultrasound of a baby in the uterus and shortly after birth (Hilferty 2001).

Figure 16.32 Complete (a) nonscanned imaging sonar and (b) 32-transducer receiver tile. Courtesy TMS.

A sparse phased array made up of 84 tiles, shown in the figure, each made up of a random pattern of 32 hydrophone receivers, receives the echo.

After a single chirp, the received amplitude and relative phase information from each of the 2688 receivers is processed to produce high-resolution 3D images such as those shown in Figure 16.33 and Figure 16.34. To achieve a range resolution of $\delta R = 1$ mm in sea water, where $c \approx 1522$ m/s, requires a chirp bandwidth of

$$\Delta f = c/2\delta R = 1522/(2\times10^{-3}) = 761\,\text{kHz}.$$

The received echo data are first compressed using the equivalent of a 2D matched filter or autocorrelation for each of the hydrophone outputs. The algorithm then needs to re-create the acoustic focal plane for each 1 mm of range, or every 1.3 μs. This is a process akin to that performed by the lens of an optical system and can be accomplished using a 2D fast Fourier transform (FFT) over the complete array. Finally, to produce moving images, an update rate of at least 10 frames/s is required.

It is obvious that there were a number of issues that needed to be addressed to realize this sensor. In terms of the hardware to generate and receive such a wide-band chirp, nonresonant transducers are required, or at least those with a very low Q. By their nature, their sensitivity will be low. This factor, in conjunction with the high attenuation in water for frequencies greater than 1 MHz, will result in a short operational range. However, the main issue is one of processing power. It is obvious from the description of the algorithms that an immense computational effort is required to produce moving images in real time.

Figure 16.33 Photograph and 2D projection of a 3D image of a G clamp (Mountford 2000).

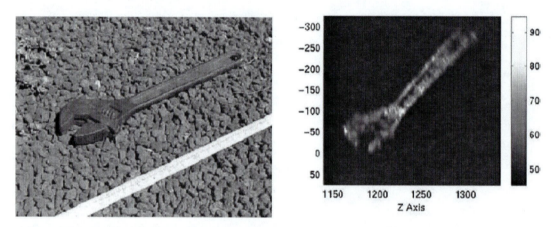

Figure 16.34 Photograph and 2D projection of a 3D image of a monkey wrench (Mountford 2000).

16.10 GROUND PENETRATING RADAR

Ground penetrating radar (GPR) operates by transmitting a wideband low-frequency electro-magnetic signal into the earth and then listening for reflections in the normal manner as shown in the illustration in Figure 16.35. A typical Global Positioning System (GPS) signal may span the frequency range from 100 MHz to 1 GHz or higher. This can be generated using stepped frequency methods generated by direct digital synthesis (DDS) or using a fast impulse or a fast rising/falling edge.

As with the previous example, the main problem with GPR is to couple this wideband energy into an antenna because most antennas are resonant, and so have bandwidths of less than 10%.

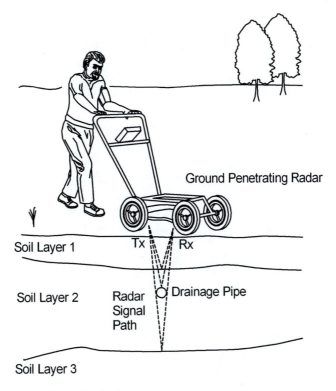

Figure 16.35 Ground penetrating radar deployment.

One method of broadening the bandwidth of an antenna is to load it resistively. This also has the effect of reducing its efficiency. GPR antennas often have efficiencies of less than 1% (the rest of the power is dissipated as heat). To compensate for this low efficiency, high transmission powers and good receiver sensitivity are required.

A low frequency (<2 GHz) is selected because the absorption of electromagnetic (EM) radiation by rock is proportional to frequency. However, the first issue is to ensure that as much energy as possible is transmitted into the solid.

The reflection coefficient, ρ, as the signal passes into the ground from the air is

$$\rho = \left| \frac{Z_1 - Z_o}{Z_1 + Z_o} \right| = \frac{\sqrt{\varepsilon_{r1}} - 1}{\sqrt{\varepsilon_{r1}} + 1}, \tag{12.35}$$

where $Z_o = \sqrt{\mu_o/\varepsilon_o}$ is the characteristic impedance of air and $Z_1 = \sqrt{\mu_1/\varepsilon_1 \varepsilon_{r1}}$ is the characteristic impedance of the solid surface of the ground.

The reflection coefficient as the electromagnetic wave propagates, at normal incidence, from one nonmagnetic material to another within the solid is given by

$$\rho = \left| \frac{Z_2 - Z_1}{Z_2 + Z_1} \right|$$

$$\rho = \left| \frac{\sqrt{\frac{\mu_o}{\varepsilon_o}} \left(\frac{1}{\sqrt{\varepsilon_{r2}}} - \frac{1}{\sqrt{\varepsilon_{r1}}} \right)}{\sqrt{\frac{\mu_o}{\varepsilon_o}} \left(\frac{1}{\sqrt{\varepsilon_{r2}}} + \frac{1}{\sqrt{\varepsilon_{r1}}} \right)} \right|$$

$$\rho = \left| \frac{\frac{1}{\sqrt{\varepsilon_{r2}}} - \frac{1}{\sqrt{\varepsilon_{r1}}}}{\frac{1}{\sqrt{\varepsilon_{r2}}} + \frac{1}{\sqrt{\varepsilon_{r1}}}} \times \frac{\sqrt{\varepsilon_{r2}} \sqrt{\varepsilon_{r1}}}{\sqrt{\varepsilon_{r2}} \sqrt{\varepsilon_{r1}}} \right|$$

$$\rho = \left| \frac{\sqrt{\varepsilon_{r1}} - \sqrt{\varepsilon_{r2}}}{\sqrt{\varepsilon_{r1}} + \sqrt{\varepsilon_{r2}}} \right|, \tag{12.36}$$

where

   Z_1 = characteristic impedance of material 1, $\sqrt{\mu_o / \varepsilon_o \varepsilon_{r1}}$,
   Z_2 = impedance of the material 2, $\sqrt{\mu_o / \varepsilon_o \varepsilon_{r2}}$,
   Z_o = characteristic impedance of free space, $\sqrt{\mu_o / \varepsilon_o}$.

Absorption of EM radiation by solids is determined by their relative dielectric constant, ε_r, and the loss tangent, $\tan\delta$, of the material. For most materials this is a function of frequency.

The attenuation in decibels for propagation through an unbounded dielectric material is

$$\alpha_d = 27.3 \sqrt{\varepsilon_r} \tan\delta \frac{d}{\lambda_o}, \tag{12.37}$$

where

   α_d = one-way attenuation (dB),
   ε_r = relative dielectric constant,
   $\tan\delta$ = loss tangent,
   d = distance (m),
   λ_o = wavelength (m).

As the GPR is wheeled over the ground, reflections from the surface and from underground objects produce patterns, shown in Figure 16.36a and Figure 16.36b. Large flat objects form reasonably linear constructs, while small objects produce hyperbolic returns because of the wide beamwidth of the antenna in the direction of travel. As is the case in synthetic aperture radar (SAR), an effectively narrow beam can be generated by processing the along-range data, a process called focusing.

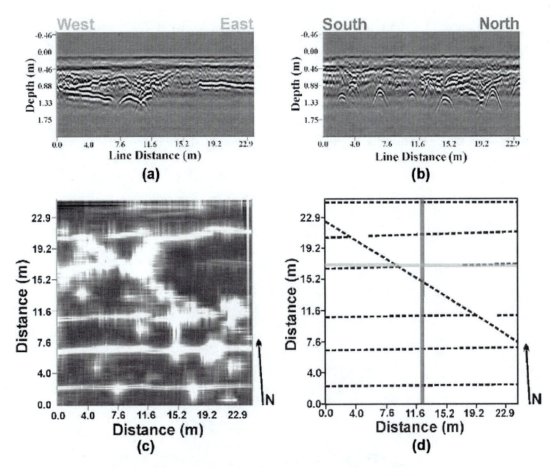

Figure 16.36 Ground penetrating radar results for agricultural drainage pipe location: (a,b) GPR images along orthogonal axes, (c) a reflectivity map for depths between 0.9 and 1.4 m, and (d) an interpreted map of the area showing the positions of the two cuts.

Table 16.1 Some applications of GPR

Engineering	Geotechnical	Mining	Environment
Bedrock profile	Karst topography	Reef delineation	Buried tanks and drums
Sinkholes	Low-density zones	Fault delineation	Contamination plumes
Leaks from services	Sedimentary layers	Depth of weathering	Geological structures
Void detection	Old excavations	Water fissures	Dam situation
Service detection	Depth of fill	Dykes	Bridge scour
		Fracture mapping	

If a number of parallel passes are made across an area of ground, as illustrated in Figure 16.36d, and the range and phase data are stored along with accurate registration information, then it is possible to use what is effectively a phased array to focus the cross-range axis as well. This results in a 2D image of the form shown in Figure 16.36c.

Applications for GPR are myriad, as can be seen from the list in Table 16.1, these include engineering, geotechnical surveys, mining exploration and, of course, environmental and forensic investigations.

16.10.1 3D Imaging Using GPR

The 2D image presented can be displayed with time (depth) along the third axis to produce a complete 3D image of the underground volume. Such displays have become one of the indispensable tools of the modern archaeologist, as they eliminate the damage caused by the exploratory trenches that were employed in the past.

Figure 16.37, made by Dean Goodman in Japan, shows GPR data taken in an alfalfa field. The top layers are devoid of useful information, but the lower levels show a 1100-year-old circular tomb and contents. The linear structure adjacent is a more modern fence line, only 500 to 600 years old.

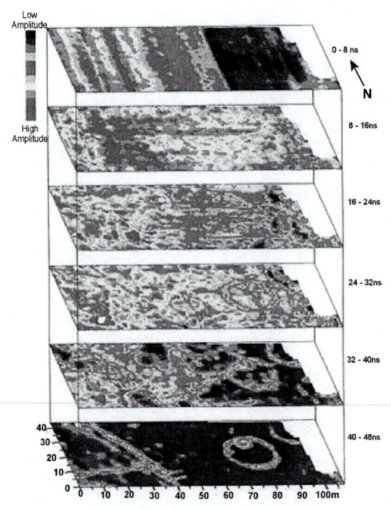

Figure 16.37 A 3D GPR image showing a 1100-year-old circular burial moat adjacent to a more modern (500- to 600-year-old) fence line (Goodman and Conyers 1997). Courtesy AltaMira Press.

16.11 WORKED EXAMPLE: DETECTING A RUBY NODULE IN A ROCK MATRIX

A GPR designed to find nodules of ruby embedded in rock generates a transmitter pulse with an amplitude of 500 V and a duration of 0.5 ns. What is the received signal level from the ruby nodule described in the following section?

The ruby nodule is 100 mm in diameter and embedded 2.5 m into the rock mass, as illustrated in Figure 16.38. The characteristics of both materials are listed in Table 16.2.

If the radiated signal amplitude (E-field) is unity and the reflection coefficient as the electromagnetic wave enters the rock is

$$\rho = \frac{\sqrt{\varepsilon_{r1}} - 1}{\sqrt{\varepsilon_{r1}} + 1} = 0.2,$$

then the amplitude of the reflected signal is 0.2 and the transmission coefficient is $1 - \rho$, so the amplitude of the transmitted signal is $1 - 0.2 = 0.8$. The reflection coefficient as the electromagnetic wave strikes the nodule will be

$$\rho = \left| \frac{\sqrt{\varepsilon_{r1}} - \sqrt{\varepsilon_{r2}}}{\sqrt{\varepsilon_{r1}} + \sqrt{\varepsilon_{r2}}} \right| = 0.26,$$

so the amplitude of the reflected signal will be $0.8 \times 0.26 = 0.208$. Back at the surface, the transmission coefficient is still 0.8, so the amplitude of the signal that enters the receiver antenna is $0.208 \times 0.8 = 0.1664$.

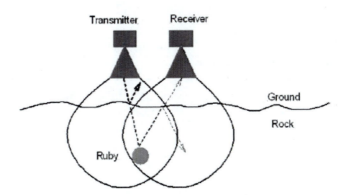

Figure 16.38 Operational scenario for ruby detection using GPR.

Table 16.2 Properties of the rock and the ruby

Characteristic	Rock	Ruby
ε_r	2.25	6.6
$\tan\delta$	0.005	0.001

The power is proportional to the square of the amplitude, so the received echo power compared to the transmitted power in decibels is

$$P_{rec}/P_{tx} = 20\log_{10}(0.1664) = -15.5\,\text{dB}.$$

The propagation velocity, c^* (m/s), is reduced by the square root of the dielectric constant of the rock:

$$c^* = \frac{c}{\sqrt{\varepsilon_r}}$$
$$= \frac{3\times10^8}{\sqrt{2.25}} = 2\times10^8\,\text{m/s}.$$

The range resolution is a function of the pulse width of the signal transmitted and the propagation velocity in the rock:

$$\delta R = c^*\tau/2 = 0.05\,\text{m}.$$

The transmitter power, P_{tx} (W), is proportional to the square of the voltage divided by the circuit impedance. For $Z = 50\,\Omega$ and $V = 500$ V, the transmitter power is

$$P_{tx} = 10\log_{10}(V^2/Z) = 37\,\text{dBW}.$$

Assuming that the antenna transmits uniformly over the lower hemisphere, it will have a gain of 3 dB. This will be reduced by 20 dB for an efficiency of 1% to −17 dB for both receiver and transmitter.

For a rectangular pulse with a duration, τ (s), the spectrum will have the form shown in Figure 16.39. For $\tau = 0.5$ ns, the maximum frequency at the first zero is 2 GHz and the average frequency over the band 0 Hz to 2 GHz will be 374 MHz.

For $f_{ave} = 374$ MHz, the wavelength $\lambda_{ave} = 0.8$ m will be used in the range equation.

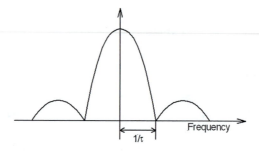

Figure 16.39 Idealized spectrum of a short-pulsed GPR.

Because the diameter of the nodule is small compared to the wavelength, the Rayleigh formula is used to calculate the scattering cross section. This is modified because the effects of the dielectric have already been considered:

$$\sigma_{dB} = 10\log_{10}\frac{128\pi^5}{3}\frac{a^6}{\lambda^4} = -33 \text{ dBm}^2.$$

The attenuation per meter (one-way) is

$$\frac{\alpha_d}{d} = 27.3\sqrt{\varepsilon_r}\tan\delta\frac{1}{\lambda_o}$$

$$= 27.3\times1.5\times0.005\times\frac{1}{0.8} = 0.256 \text{ dB/m}.$$

For a total distance traveled of $2.5 \times 2 = 5$ m, the attenuation will be $0.256 \times 5 = 1.28$ dB, which is hardly significant.

Applying the radar range equation, including the losses due to attenuation and transmission coefficients, etc., the received power is

$$P_{rec} = P_{tx} + 2G_{ant} + 10\log_{10}\frac{\lambda^2}{(4\pi)^3} + \sigma_{dB} - L$$

$$= 37 - 2\times17 - 34.9 - 33 - 15.5 - 1.28 = -81.7 \text{ dBW}.$$

Assuming a 50 Ω input impedance, the received echo will have a root mean square (RMS) amplitude of 0.58 mV.

The matched filter bandwidth needs to be about 2 GHz, so the thermal noise level will be

$$N = 10\log_{10}kTB$$

$$= 10\log_{10}(1.38\times10^{-23}\times290\times2\times10^9) = -111 \text{ dBW}.$$

A wideband amplifier will have a noise figure of about 4 dB, so the final noise level will be −107 dBW. The received signal-to-noise ratio (SNR) will therefore be

$$SNR = -81.7 + 107 = 25.3 \text{ dB}.$$

This should be sufficient to see the target quite easily.

Because the antenna beamwidth is very wide, target angular resolution is poor. As the radar unit is dragged over the ground, the apparent range to the nodule will change and a hyperbolic echo will result, as discussed in the previous section.

16.12 References

Bronzino, J., ed. (2006). *Medical Devices and Systems*. Boca Raton, FL: CRC Press.
Deans, S. and Roderick, S. (1983). *The Radon Transform and Some of Its Applications*. New York: Wiley.

Freudenrich, C. (2000). How nuclear medicine works. Viewed January 2008. Available at http://health. howstuffworks.com/nuclear-medicine1.htm.

Freudenrich, C. (2001). How ultrasound works. Viewed January 2008. Available at http://health. howstuffworks.com/ultrasound1.htm.

Goodman, D. and Conyers, L. (1997). *Ground-Penetrating Radar: An Introduction for Archaeologists*. Walnut Creek, CA: AltaMira Press.

Hilferty, T. (2001). Watching your baby smile in the womb. *Daily Telegraph*, February 1.

Hornak, J. (2007). The basics of MRI. Viewed January 2008. Available at http://www.cis.rit.edu/ htbooks/mri/.

Hounsfield, G. (1973). Computerized transverse axial scanning (tomography). *British Journal of Radiology* 46:1016–1022.

Kak, A. and Slaney, M. (2001). *Principles of Computerized Tomographic Imaging*. Philadelphia: Society of Industrial and Applied Mathematics.

Kalender, W. (2005). *Computed Tomography*, 2nd ed. Erlangen: Publicis.

Mountford, G. (2000). 3D imaging sonar. Private correspondence. Sydney, NSW, Australia.

National Institute on Aging. (2007). Alzheimer's disease: unraveling the mystery. Viewed March 2008. Available at http://www.nia.nih.gov/NR/rdonlyres/A294D332-71A2-4866-BDD7-A0DF216DAAA4/0/ Alzheimers_Disease_Unraveling_the_Mystery.pdf.

Paykett, I. (1982). NMR imaging in medicine. *Scientific American* 246(5):78–88.

Volkin, L. and Dargan, R. (2006). Study shows potential dangers of ultrasound in fetal development. Viewed January 2008. Available at https://www.asrt.org/content/News/IndustryNewsBriefs/Sono/ studyshows062408.aspx.

Index